DEVELOPMENTAL
BIOLOGY

DEVELOPMENTAL BIOLOGY

N. J. BERRILL

McGRAW-HILL BOOK COMPANY

NEW YORK

ST. LOUIS

SAN FRANCISCO

DÜSSELDORF

JOHANNESBURG

KUALA LUMPUR

LONDON

MEXICO

MONTREAL

NEW DELHI

PANAMA

RIO DE JANEIRO

SINGAPORE

SYDNEY

TORONTO

This book was set in Elegante by Westcott
&Thomson, Inc., printed on permanent paper
by Halliday Lithograph Corporation, and
bound by The Book Press, Inc. The designer
was Barbara Ellwood; the drawings were done
by BMA Associates, Inc. The editors were
Jeremy Robinson and Laura Warner. John F.
Harte supervised production.

Library of Congress Catalog Card Number
71-142966

07-005020-1

4567890HDBP79876543

**DEVELOPMENTAL
BIOLOGY**

CONTENTS

Preface   *vii*

Part I   The Assembly of Cell and Organism   2
  1   Synthesis and Assembly   *13*
  2   The Structural Basis of Form and Action   *35*
  3   The Cell Cycle   *57*
  4   Cortical Patterns and Inheritance   *77*
  5   Multicellularity   *101*
  6   Polarity, Form, and Differentiation   *125*
  7   The Development of Multicellular Organisms   *149*
  8   Morphogenetic Fields and the Plant Meristem   *165*

Part II   The Nature of Animal Development   182
  9   Gametes and Gametogenesis   *195*
 10   Fertilization   *217*
 11   Cleavage, Polarity, Determination, and Gradients   *237*
 12   Cleavage and Nuclear Activity   *261*
 13   The Egg Cortex and Development   *279*
 14   Developmental Mechanics or Formative Movements   *303*
 15   Embryonic Induction   *341*

Part III   Organization, Reconstitution, and Differentiation   362
 16   Cell Assembly and Interaction   *365*
 17   The Limb as a Developing System   *385*
 18   Morphogenesis of the Vertebrate Eye   *407*
 19   Metamorphosis, Hormones, and Genes   *423*
 20   Genes, Prepatterns, and Determination   *453*
 21   Cytodifferentiation   *473*
 22   Malignancy, Differentiation, and Development   *491*

Index   **517**

PREFACE

Developmental Biology is a newly capitalized term denoting a long-standing area of biological interest and investigation that has gone under various and more restrictive names during the past century. From the very beginning the subject has embraced far more than the development of eggs and, through the years, has had its ups and downs. The present upsurge of interest has been stimulated mainly by the recent remarkable advances in molecular and cell biology, which are obviously of basic importance to an understanding of both developmental and organismal biology. Moreover, there are fashions in science, as elsewhere, and interest in a particular field waxes and wanes, particularly as one generation of investigators succeeds another. Accordingly, at the present time, developmental biology has acquired a new look and a new promise.

The account offered here is intended primarily for advanced undergraduate or beginning graduate students, on the assumption that they will already have some acquaintance with a diversity of animals and plants and will have had an introduction to cell biology, and perhaps a little formal embryology. Effort has been made to present text and illustrations that are self-sufficient, but many inexpensive aids are available for students who lack the suggested background. The dominant theme is that of self-assembly and directed assembly of organized substance, with molecular and cell biology brought in only to the extent they are needed for the development point of view. Another theme is that the egg is a cell with a general capacity for development, but reinforced by special properties that make developmental expression feasible.

Inquiry and concepts are stressed throughout the book, rather than accomplishments and answers, the treatment reflecting the fact that at all levels the inquiry is open-ended. Names and dates are included to some extent, both in the text and in lists of readings at the end of chapters, partly to emphasize the continuity of investigation and conceptual thinking throughout the history of the subject and more importantly to serve as leads into current literature, at least to the time of publication. Quotations have been freely used to indicate the always personal character of thought and work. Concepts are listed at the end of each chapter.

I am much indebted to various members of the biology department of Swarthmore College, particularly Dr. Norman Meinkoth, Dr. Robert Savage, and Dr. L. Livingstone, for their interest and support during the writing of this book. I also wish to express appreciation to Dr. L. G. Barth, Dr. James W. Lash, and Dr. Fred Wilt for their most helpful readings of the manuscript.

N. J. BERRILL

DEVELOPMENTAL
BIOLOGY

PART ONE

THE ASSEMBLY OF CELL AND ORGANISM

What is development? No one has completely defined it, any more than the organism itself has been fully defined. In all cases a cell or a group of cells becomes separate from the organism as a whole, either physically or physiologically, and progressively becomes a new complete organism or a new part thereof. A fern spore settles and develops into a fern gametophyte. An insect egg may become a caterpillar which transforms into a pupa and emerges as a butterfly. The stump of a salamander leg regenerates a new limb. A microscopic cell, the human egg, proliferates and develops into a giant animated organism able to contemplate its own nature and origin.

All cells arise from preexisting cells. There is no other way for a cell to be produced, unless man eventually succeeds in synthesizing cells from simple elements or molecules in the laboratory. At this late stage in the history of the planet, however, a single cell may be a bacterial or a protistan organism which through division gives rise to successive new generations. Or it may be a constituent cell of a multicellular organism—animal or plant—and by multiplication contribute to the growth of the whole. In such a system it may become a relatively very large cell indeed, an egg, and by a series of divisions give rise to an integrated cell assembly, a multicellular organism of the next generation.

Both in the prolonged evolution of organisms and in the development of all kinds of individual organisms at the present time we see matter in action in its most complex form, as processes of self-assembly, directed assembly, functional activity, disassembly, reconstitution, and replication. The main problem in studying such phenomena is to perceive what is actually going on behind the facade of the changing visible form. The philosopher A. N. Whitehead has said that process itself is the reality; in a recent text, development is stated to be an expression of the irreversible flow of biological events along the axis of time. Although abstract, these are undoubtedly valid concepts that emphasize the dynamic and sequential character of reality and development. For development is a clearly ordered process whereby structural and functional organization of the system, whether of a single cell or of a multicellular organism, becomes progressively expressed. Apart from energy supply and utilization, the fundamental phenomena appear to be the production of differentiated structural substances and the organization of such materials into patterns distinctive of cell, tissue, or organism. The end point is highly structured matter in action. The aim of developmental biology is to discover what processes are involved in producing this culmination, how they are related, and how they are controlled.

The study of development is accordingly a continuing effort to interpret the end results in terms of preceding events and underlying processes. These have meaning only with respect to what is finally, or transiently, produced. There is in fact a teleological aspect to development which has been expressed in the phrase that the end is fore-shadowed in the beginning, as though development has a goal which the developing system may attain by devious means. This has been unacceptable to many biologists; yet the semblance is apparent and is not explained by being ignored. To ignore it is to underestimate the organism, especially the history of organisms in general and of each organism in particular.

Development and evolution may both be regarded as a process of attaining successively higher levels of organization of matter by means of self-assembly and directed assembly. During evolution matter progressively evolved from prebiotic assemblies to procaryotic cells to eucaryotic cells to multicellular organisms, collectively constituting ecosystems of increasing magnitude and complexity, successive generations being linked in time by replication. In development, as in evolution, the organism passes from one level of organization to a higher level with the emergence of new properties at each such translation. But whereas evolution has occupied several billion years of planetary time, development expresses a comparable phenomenon in hours, days, or weeks.

THE CELL

In the beginning is the cell; yet the study of the assembly and constitution of subcellular components and extracellular material, in addition to having inherent interest relating to the general concept of self-assembly of matter, throws light on the assembly and construction of the cell as a whole. The cell itself exhibits extreme differentiation within the limits of its own boundary; this differentiation calls for intensive study of the special roles of nucleus and cortical cytoplasm, and their interactions. The development of multicellular organisms, small or large, usually begins with a single cell, as egg or spore, or from a group of cells in some way set aside from the parental organism, and such development calls more for study of the interactions among cells and of the pervading wholeness of organismal assemblies. The understanding of the cell, in other words, is both a goal and a point of departure. At the same time, study of the multicellular state at its own level gives perspective to all that underlies it.

The concept now widely accepted is that there is organizational structure in cells and that this structure has a profound influence on the chemical activities that occur within a cell. A cell is truly structure in action, in the same indivisible sense that the human organism has been defined as a "minding body," rather than an association of mind and body.

DEVELOPMENTAL BIOLOGY

In studying this subject, we are dealing with possibly the most fascinating phenomenon to be encountered on this planet, a continuing joy to those who have watched it for long; and we are also involved in a continuing inquiry into the true nature and significance of what we see, a perpetual process of defining problems in crucial terms and of questioning living systems through experimentation—in fact, a dialogue between man and nature, including his own, for questions raise new questions, and there are no final answers.

Most of the earlier and much of the recent attention concerns the development of the animal egg, i.e., the transformation of an exceptionally large and usually otherwise specialized single cell into a relatively complex and multicellular organism. Processes of growth and differentiation are typically involved. These terms, as well as the term *development,* are used sometimes broadly and sometimes in a more restricted sense, so that precise meanings need to be kept in mind. This emphasis is made here because some ambiguity is found as soon as one starts to study this subject. Developmental biology has come to mean the study of two distinct levels of phenomena: **(1)** the growth, differentiation, and, in general, development of a simple cell into a highly structured cell or cell-organism; and **(2)** the growth, differentiation, and development of a cell into a multicellular, integrated organism. The former process is more precisely known as *cytodifferentiation,* the latter more generally as *development* itself. Differentiation thus has a double reference.

According to J. T. Bonner, in his essay on the evolution of development, "It so happens that all genetic systems, perhaps partly because they evolved at an early stage in evolution, involve single cells and single nuclei. Therefore, the variative transmission stage is one of having small units. Because there has also been a selective pressure for large organisms, the inevitable result is that a gap between small reproductive cells and the large adult must be filled by development. But each step in this development is in itself adaptive to its particular environment. . . . To put it another way, one cannot produce a large complex organism from a single zygote cell in one magic explosion. The very fact that it is so large and intricate means that it must be built slowly with care and system. . . . Therefore we must remember one of the oldest lessons in embryology: it is simply that one step, one set of conditions leads to another, and that developmental progress is a causal sequence of events. This is the epigenesis of Aristotle and William Harvey. In modern times, the only significant addition that has been made to this basic scheme of sequential events is that they are controlled and guided by the activities of the genes."[1]

Nearly a century and a half earlier Von Baer demonstrated development to be always from the homogeneous to the heterogeneous, from the general to the special. He based his conclusions on persistent and accurate microscopic observations on the eggs and embryos of a wide variety of animals, and thereby established the modern developmental biological approach, in which facts and phenomena precede the formulation of concepts, rather than the reverse procedure. Nevertheless, following the publication of Darwin's great generalization concerning evolution, the urge to generalize became

[1] J. T. Bonner, The Unsolved Problems of Development: An Appraisal of Where We Stand, *Amer. Sci.,* **48:**514–527 (1960).

too strong, and Haeckel's theory of 1891, his biogenetic law that ontogeny is a recapitulation of phylogeny, became widely acclaimed. Developmental biology became, in the hands of some, a study of life histories inspired by the idea that they represent condensed versions of the evolutionary history of the species. Yet it was one of Haeckel's students, Wilhelm Roux, who introduced the experimental method that typifies modern developmental biology.

The present account, therefore, is concerned principally with the cell from three points of view: the cell as a cell, the cell as an egg capable of undergoing development into a complex multicellular organism, and the cell as a constituent of cell populations comprising multicellular systems.

QUESTIONS AND ANSWERS

When asking questions of nature, at whatever level, the framing of the question is of the utmost importance, because the answer, if any, will be in the same terms as the question. If the question relates to the role of the genes in development, the answer will further our understanding of that role. If the question relates only to cytoplasmic events during development, something may be discovered concerning those events but little will be learned about related gene activity. In view of the multilevel organization of organisms and the developmental transformations, such limitations need to be kept in mind. Questions and answers related to one level are not likely to illuminate another level, except to raise questions of another kind. The relationship between levels must be in the question if it is to appear in the answer. We must also recognize that a fuzzy question gets a dusty answer. In fact, putting a question to nature, either observational or experimental, that is simple, clear, and novel is perhaps the most difficult task the mind can undertake. When we consider the slow progress in conceptual understanding of developmental phenomena since studies were first made, we may be tempted to conclude that the whole developmental process surpasses human comprehension, as did Hans Driesch, one of the great pioneering developmental biologists, who evoked an entelechy, or guiding principle, for every cell in a developing system, as a means of escape from his intellectual frustration; or we may decide that molecular biology holds the key and that enlightenment will soon come. In any case the challenge persists, in spite of the great technological advances of recent years.

It is also important to put our questions to those organisms most likely to respond, while expedience dictates that such organisms be widely obtainable and readily maintained or cultured in the laboratory. Classical genetics, for instance, has been based mainly on *Drosophila* because fruit flies are easy to breed and have a life span so short that geneticists can follow many generations within the course of a year. On the other hand, the developmental approach to *Drosophila* has been difficult compared to the study of the larger eggs and embryos of many other insects. In contrast, the developmental stages of amphibians have been ideal for analytical and experimental embryology, but amphibians are poorly suited for breeding experiments because of their life span of several years and also because of the survival hazards innate in their metamorphic life cycle. Consequently classical genetics mainly related germ-cell chromosomes with adult characters of the fruit fly, while experimental embryology yielded much information concerning cytoplasmic developmental events of amphibian embryos (comparable study of developing sea urchin eggs has been equally productive). Now a new look at genes by way of microbes and viruses has brought insight of another order. Choosing the best organism to work with, apart from availability and ease of maintenance, is almost as important to progress in understanding as is framing the biological question in crucial terms.

LEVELS OF ORGANIZATION

All biologists recognize that there are various levels of organization within living systems. They range from simple molecular to macromolecular, to the organelles of cells, such as membranes, mitochondria, plastids, and nuclei, etc., to the cells as such, to cell assemblies, to specialized organs and systems, to multicellular organisms as a whole, in some instances to organismal assemblies constituting colonial or superorganisms, and in a somewhat different sense to interacting communities of discrete organisms comprising ecosystems and the ecosphere as a whole. It is also recognized that each level may be studied as such and that patterns and relationships may be worked out at each level without significant reference to the other levels. Accordingly we have molecular biology, cellular biology, organismal biology, population biology, and environmental biology, each primarily concerned with its own particular field. At the same time there is a continuing debate between reductionist and organismal biologists concerning the role of molecular biology to biology as a whole. The reductionists adopt the precept that a fuller understanding at the molecular level will eventually yield explanations of all that happens at higher organizational levels; organismal biologists believe that the organism as a whole needs to be studied at its own level.

The properties of each biological level of organization, from the whole ecosystem down through its constituent organisms, are intimately interrelated with the properties of each subordinate level, down to the smallest. The nature of an organism is clearly related to the nature and number of cells comprising it. The general and particular nature of each cell is as closely related to the nature of its component parts, particularly the organelles. Organelles in turn consist of membranes, granules, and amorphous materials, and these structural components can be separated into large molecular components, which can be dissociated into smaller ones, and so on down to atoms and even to subatomic particles. There is no real break in the descending or ascending series of organizational levels.

EMERGENCE OF NOVELTY

The emergence of new properties when a higher level of organization is attained is evident at all stages in the evolution of matter, both living and nonliving. It is evident at every step in the ascending series of the elements, from the simple hydrogen atom to the unstable, disintegrating radioactive elements at the upper end of the scale. Passing from the atomic to the molecular state, the properties of molecules relate to the atomic constituents but far transcend them.

The phenomenon of emergence of novelty at every new level of organization pervades the whole hierarchy of matter, from subatomic particles to societies of organisms. At every new level not only does a host of new properties appear but the new units of organization interact with one another and with their environment according to rules of their own. Even to formulate the problems they present, the activities and interactions must be studied at their own level if the laws of pattern and order at that level are to be even partially understood. Only then is there a prospect that they can be related to or interpreted in terms of their constituent assemblies. Atoms of hydrogen and oxygen combined as molecules of water vapor, for instance, exhibit certain properties and forms of behavior of their own; but as cohering units constituting liquid water, they have general and unique properties which are vital to life. So in a comparable way, when large numbers of individual cells cohere as multicellular assemblies, properties emerge which are inexpressible by single cells; interrelationships are established, new organizations become possible, and novelty again appears. Novelty is unpredictable except in principle, and accordingly each level of organization in both the living and nonliving exists in its own right and demands attention as such.

THE LIMITATIONS OF MATTER

Throughout the living kingdom, animal and plant alike, remarkable resemblances are seen at all levels from the molecular to the grossly anatomical. Much of this is homologous, such as the skeletal limb patterns of a mouse and a bat, resulting from common ancestry and consisting of the same identifiable components adapted to different functions. On the other hand, structures that are remarkably alike are frequently formed in clearly unrelated organisms, such as the eye of the octopus and that of the vertebrate; while at the molecular level hemoglobins repeatedly and independently appear in unrelated animals and even in plants, although not necessarily serving the same function. As in any architectural venture, whether it is to be a building or an organism, the properties and limitations of the available materials are reflected in the product.

To begin with, there are fewer than 100 stable elements, representing great yet limited diversity. The heavier elements are mostly toxic, and the unstable, radioactive elements are generally lethal. At the molecular level the limited range of usable elements is reflected in limitation of the sorts of molecules available to organisms for the construction of adaptive systems. Iron, copper, and manganese have peculiar powers of catalysis, and consequently compounds of these elements with porphyrin protein compounds are widely used as such in the living kingdom.

The special and very varied properties of the nonmetals, however, are involved in all the most complex molecules known to exist. There are only some 13 nonmetals, and among these the properties of carbon are unique. Organisms apparently make the fullest possible use of the special properties of a very limited supply of suitable elements. Organisms may be said to be built up of standard parts with unique properties. Such parts are units of matter and energy which can exist only in certain possible arrangements. Accordingly, in biological engineering, as in other engineering, any functional problem can have only one or another of a very few possible kinds of solution. Organisms can form long chains of carbon atoms, such as long molecules of fibrous proteins, though there are only a few ways in which they do this. The fibers and membranes formed by such proteins may be greatly strengthened by connecting the threads by cross-linkages of sulfur (as in vulcanization), a method widely employed in vertebrates to make horn, nails, and hair; an alternative method is to link the long molecules together by means of polyphenolic substances, as in tanning, a method employed to harden the exoskeleton of insects, the silk of silkworms, and the anchoring threads of mussels. More important, many fibers possess the potential property of contractility, and because of this the essential character of muscular contraction appears to be the same in every kind of organism; in fact, fewer possible ways seem to exist for building up movement mechanisms from available components than there are for hardening cuticles.

All proteins are constructed from various combinations among only 20 kinds of amino acids, and although the number of possible combinations and permutations is almost beyond reckoning, the actual number of combinations is not infinite. The possibilities of development and organization are accordingly both dictated and limited by the nature of the available building materials. Because of this condition, restriction appears in the kinds of cells, of tissues, of organs, and of organisms. At all levels of organization some patterns can exist while others cannot. At the same time, for instance, in the construction of an organ such as the eye, the functional requirements are such that only a few basic types of design are workable; therefore whether the building material is the living substance of vertebrate or mollusk or the nonliving material of a camera, the basic construction must be much the same for a significant image to be formed. In all such cases natural selection determines which of several possibilities actually survive.

EMERGENCE AND EPIGENESIS

Development typically begins with a cell and ends with a mature organism. At one extreme, developmental biology is concerned with the differentiation of the cell and merges with cell biology as such. At the other, it is concerned with the cellular, structural, and functional characteristics of the organism as a whole and so becomes an integral part of organismal biology, with particular emphasis on the maintenance and restoration of structural integrity and cellular constitution. In between lies the elaborate process whereby a single reproductive cell, or sometimes a small mass of somatic cells, develops and grows into a multicellular organism, often of great size and complexity. Thus developmental biology embraces the cell and the organism as well as the conversion of the one into the other, although the process of conversion is primary. During this process the developing organism passes from level to level of organization, and much of our problem is to understand the process of conversion of properties at one level to those of the next. At every stage, one set of conditions leads to another, so that we are dealing with a causal sequence of events. This sequence, always proceeding from the general to the particular, is characteristic of virtually all developmental phenomena and is generally known as *epigenesis.*

As the developing organism proceeds from level to level in the organizational scale, something new emerges at each step. This is the principle underlying the phenomenon of epigenesis. A simple example is that when an egg cell or a spore undergoes successive divisions, the property of cohesiveness becomes apparent, together with all the complexities of the multicellular situation. The nature and origination of intercellular cohesion on the one hand and the consequences of such cohesion and its variability thus become vitally important to an understanding of the developmental process. At a later stage in the development of most animals some tissue shifts to a location internal to the remainder, so that environmental conditions become different for the two components and in addition there is opportunity for interactions between the two. Each new circumstance leads to new possibilities, and development is primarily the orderly and ordered sequence of transformations.

EVOLUTION OF THE CELL

Matter has an obvious capacity to assemble itself into compounds of aggregates of greater size and complexity, particularly when subject to some influx of energy. The synthesis of organic compounds from simple gas mixtures in the laboratory has given strong support to Oparin's hypothesis of the abiogenic synthesis of such compounds under primitive earth conditions. Apparently, any one of the simple carbon compounds could have given rise to the whole range of organic substances as long as water and some form of nitrogen were present under reducing conditions, and some form of energy of cosmic or earthly origin was present in sufficient quantity. In the laboratory amino acids polymerize to polypeptides similar to natural proteins; activated sugars polymerize to polysaccharides; and even low-molecular-weight nucleic acids have been produced in the absence of enzymes. The list of compounds assembling under primitive earth conditions is long and includes aldehydes, ketones, alcohols, amides, amines, purines, pyrimidines, nucleosides, nucleotides, some nucleotide polymers, porphyrins, glycosidic polymers, and amino acid polymers (proteinoids).

Of all such substances, proteins have played a fundamental and exclusive role in the further evolution of organic compounds and the origin of life. Part of this fundamental role results from the presence of both positively and negatively charged amino acid groups, so that both positively and negatively charged sites exist on the protein molecule. These together with residual valencies allow protein molecules to come together and form complexes of a colloidal, hydrophilic nature, i.e., protein particles with a diffuse shell of water molecules around them. Such particles may coalesce to form *coacervates*, in which the outer layers of bound water keep the oppositely charged molecules apart, while the opposite charges on the different molecules draw them together. Such systems are strongly affected by changes in the environment. From chemical and physical systems such as these the first self-sustaining cells are thought to have evolved by various processes of self-assembly. However this may have occurred, procaryote cells, lacking a nuclear membrane and now represented by bacteria, blue-green algae, and certain fungi, evolved long before the larger, more complex eucaryote cells with their well-developed internal membrane systems (including the nuclear membrane), which constitute all plant, animal, and protistan organisms.

Leaving aside any question of "directiveness" or "purposiveness" in evolution, the cell as a free-living, self-sustaining, self-replicating entity represents the culmination of possibly 3 billion years of evolutionary activity. Progressive understanding of the nature of the cell has been accompanied by progress in comprehending the prebiotic and early biotic events leading to the evolution of the eucaryote cell. One of the highly speculative concepts now emerging is that the cell in its fully perfected eucaryote form evolved from the procaryote cell as the result of several symbiotic unions with other independent self-replicating units during an early evolutionary period when the external medium retained some of the organic properties of the primeval planetary soup postulated by Oparin. In brief, it seems likely that many kinds of self-replicating organic complexes existed as long as the medium itself produced adequate supplies of organic compounds which self-assembled under the impact of solar or chemical energy. But as proficiency in utilizing such substances increased and consumption exceeded production, the medium became impoverished, and natural selection favored those that joined forces in symbiotic union. Chloroplasts and mitochondria, in this view, were essentially independent subcellular organisms that have been incorporated as integral parts of eucaryote cells, in much the same way that many coelenterate animals now incorporate unicellular photosynthetic algal cells within their tissues. It has also been suggested that basal granules and centrioles, associated with cilia, flagella, and aster formation, are also evolutionary incorporations. Virus particles, too, seem to belong to the same ancient evolutionary phase, although their present relationship to cells, whether eucaryote or procaryote, is invasive and transient.

The cell, in evolution, is the crucial accomplishment. It was the climax of assembly processes occurring during the primary phase of evolution, and in fact it became the assembly unit for all that followed. It remains the assembly unit for all living multicellular organisms, and it is the general though not exclusive starting point for the development of new multicellular individual forms of life.

CONCEPTS

Matter exhibits *levels of organization,* from subatomic particles to atoms, molecules, cells, organisms, and ecosystems. Developmental biology is mainly concerned with the *relationships among molecules, cells, and organisms.*

At each higher level of organization, there is an *emergence* of new properties; in other words, at each level of organization the *whole is more than the sum of its parts.*

Matter undergoes processes of *self-assembly.*

Living substance is built up from comparatively few elements, each with unique properties, thus combining *diversity* with *limitations.*

Limitation in kinds of structural and metabolic materials results in *restriction in possible solutions* to any problems of biological engineering, whether at the molecular, cellular, or organismal level.

Development is a multilevel phenomenon, and during development the organism passes from one level to another.

As an organism passes through the succession of developmental stages, the progress is from the general to the particular, with emergent properties appearing at each step. This is the phenomenon of *epigenesis.*

READINGS

ALLISON, A. C., 1969. Silicon Compounds in Biological Systems, *Proc. Roy. Soc. London, Ser. B,* **171**:19–30.

BONNER, J. T., 1960. The Unsolved Problems of Development: An Appraisal of Where We Stand, *Amer. Sci.,* **48**:514–527.

GROBSTEIN, C., 1967. "The Strategy of Life," Freeman.

HENDERSON, L. J., 1913 (reprinted 1964). "The Fitness of the Environment," Beacon Press.

KEOSIAN, J., 1964. "The Origin of Life," Reinhold.

OPARIN, A. I., 1964. "The Chemical Origin of Life," Charles C Thomas.

STERN, H., and D. L. NANNEY, 1965. The Origins of Molecular Order, chap. 15 in "The Biology of Cells," Wiley.

CHAPTER ONE

SYNTHESIS
AND ASSEMBLY

The simplest organisms are single, free-living cells, and the cell is the basic unit of organization of all multicellular organisms. The procaryote cells—namely, bacteria, blue-green algae, etc.—are essentially one-envelope systems organized in depth. They consist of central nuclear components surrounded by cytoplasmic ground substance, with the whole enveloped by a plasma membrane. Neither the nuclear apparatus nor the respiratory enzyme systems are separately enclosed by membranes, although the inner surface of the plasma membrane itself may serve for enzyme attachment. The procaryote cell, in fact, is so small [a typical bacteria has a diameter of about 0.5μ (micron)] that simple organic molecules could traverse the cell 100 or 200 times a second, and diffusion alone suffices to take care of metabolic transport needs.

Eucaryote cells are essentially two-envelope systems and are very much larger than procaryote cells. Secondary membranes envelop the nucleus and other internal organelles, and to various extents they pervade the cytoplasm as the endoplasmic reticulum. The two cell types represent two major steps in the evolution and organization of living matter. The basic processes involved in the assembly of procaryote and eucaryote cells, however, appear to be fundamentally the same and primarily concern the synthesis and assembly of proteins. Although protein assemblies alone cannot constitute even the simplest cell, there is much truth in the statement that proteins, ions, and water are the major components of all living matter.

Energy is acquired from environmental sources and is expended to create order and action. The study of the acquisition and expenditure of energy constitutes a part of the field of biochemistry. Stern and Nanney, in a profound introduction to this subject, "The Biology of Cells," summarize the situation as follows: "Despite the complexity of enzyme systems essential to the acquisition of energy, the fundamental principles are very few. In the formation of living substance cells must counter two thermodynamic trends—oxidation and hydration. The one is due to the abundance of oxygen, the other to the abundance of water. To overcome these trends living systems have evolved two chemical configurations—the reduced nicotinamide group and the pyrophosphate bond. With these, cells can build all the organic molecules of life; the exceptions to this are incidental to the general principle. All cells can form these organic compounds given a source of energy, and for most cells this source is carbohydrate. Carbohydrate, however, is not given to the living world, but is made by it. And the capacity to synthesize this fuel of life resides in a limited number of cells, those with photosynthetic pigments. Different though the processes of synthesis and degradation may appear to be, their ultimate energetic transactions rest on one and the same mechanism—the capacity of porphyrin molecules to catalyze the transfer of electrons. Out of the catalytic capacity of enzymes grow the diverse

CONTENTS

Protein synthesis

Genetic control: the operon

Protein molecules

Construction and reconstruction of collagen, a structural protein

Aging of collagen, and quaternary structure

Virus assembly

Concepts

Readings

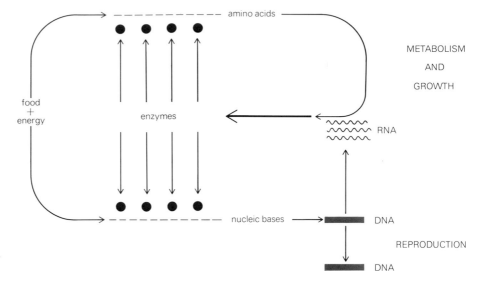

FIGURE 1.1　General scheme of metabolism, growth, and reproduction. [*After Lwoff, 1959.*]

organic molecules of life, out of the special catalytic capacity of porphyrin enzymes grow the two forms of chemical energy which make the diversity possible." [1]

　　The importance of protein has been emphasized by the realization that all enzymes are proteins and also by the analysis of cell ultrastructure made possible by the electron microscope. At the same time the vital role of water and electrolytes, quite apart from intense metabolic cell activity, must never be overlooked. In other words, though structural proteins are the basic and predominant molecular units for the building of living structure—which is the most obvious event of differentiation or development—we need to keep in mind that the organism, cell, and cell components are in a dynamic state, and that electron micrographs, in which ultrastructure inevitably appears static, are pictures of a moment. The synthesis and assembly of proteins, however, are responsible for the more stable structure of cells and their components, and also for much of the action.

PROTEIN SYNTHESIS

Proteins are manufactured only within living cells from some 20 kinds of amino acids (apart from the increasingly successful human efforts to emulate the process in the laboratory). In the primeval phase of evolution it is probable that amino acids in the environmental organic medium underwent spontaneous polymerization to form peptide and possibly polypeptide chains, in a process of self-assembly unaided by more than an influx of energy. Yet it has been calculated that even if the whole earth had been made of nothing but amino acids which had arranged themselves

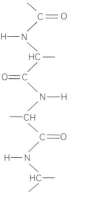

FIGURE 1.2　Polypeptide backbone in linear form. Open dashes indicate amino acid side chains.

[1]H. Stern and D. L. Nanney, "The Biology of Cells," p. 300, Wiley, 1965.

randomly and completely 10 times a second, in the whole period of the earth's history there would have been little chance of producing even once a single molecule of insulin. Nature had to evolve the exquisite machinery of the system now generally known as the "central dogma," in order to direct within the confines of a cell the special assemblies of amino acids to form long-chain proteins with specific and useful properties. This system, whereby biological information flows from DNA to RNA and then to protein, but not in the reverse direction, is the ideological core of present molecular biology.

Proteins are the primary basis both for ordered structure and for activity in living cells and organisms. All known enzymes are proteins. As a class proteins are immensely versatile, although all of them form some relatively stable three-dimensional structure. Since such structures cannot be copied without being unfolded or otherwise taken apart, they must be replicated by other means. Nucleic acid, in addition to serving as a template for more nucleic acid, is able to direct protein synthesis in the guise of messenger RNA (mRNA); thus nucleic acid (DNA) indirectly produces enzymes. Enzymes not only are necessary for various steps in the process, but, as DNA polymerase, they catalyze the formation of more DNA when some is already present as a primer. Similarly, RNA polymerase catalyzes the formation of RNA. How such a mutually dependent system could have evolved is hard to conceive.

The nature of the DNA code is now generally familiar. In brief, the information required for assembling amino acids into the specific sequences characteristic of proteins is contained in the nucleotide sequence of the chromosomal DNA double helix. Each triplet of DNA nucleotide bases, or codon, directs the formation of a particular amino acid but by way of a corresponding anticodon, or triplet of RNA nucleotide bases.

Transcription is the synthesis of the so-called messenger ribonucleic acid (mRNA), which is formed by using one of the two strands of the DNA as a template. Messenger RNA has nucleotides arranged in a sequence that can assemble amino acids in a defined order.

Translation, which follows transcription, involves utilization of the nucleotide sequence in messenger RNA to form polypeptide chains of proteins. This assembly process takes place on ribosomes.

Ribosomes are roughly spherical particles, with a diameter close to 200 Å (angstroms), composed of protein and RNA, with the latter predominating. The assembly of amino acids occurs not directly but through the mediation of the much shorter molecules of soluble RNA (sRNA), also known as transfer RNA (tRNA), consisting of from 70 to 80 nucleotides.

The first step is the activation of an amino acid by an activating enzyme specific for that particular amino acid and also for a specific tRNA molecule. The tRNA molecule apparently acts as an acceptor for an activated amino acid and serves as an adaptor for carrying the amino acid to the site of protein synthesis on a ribosome. There is at least one specific type of tRNA for each of the 20 kinds of amino acids.

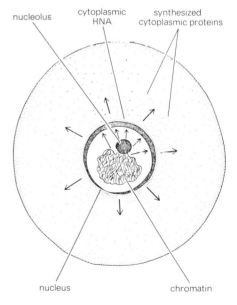

FIGURE 1.3 Diagram of passage of ribonucleic acid from the nucleolus through the nuclear membrane into the cytoplasm to synthesize cytoplasmic protein.

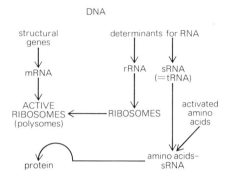

DNA

structural genes   determinants for RNA

mRNA   rRNA   sRNA (=tRNA)

ACTIVE RIBOSOMES (polysomes) ← RIBOSOMES   activated amino acids

protein   amino acids– sRNA

FIGURE 1.4   Origin and role of the three classes of cellular RNA.

(Ribosomal protein is very heterogeneous. For example, there are at least 20 different proteins in the 30s, or lighter, ribosomal subunits. These proteins, together with RNA, will combine with intact 50s subunits and function in protein-synthesizing systems.) The basic event occurs on the ribosome, where the information encoded in the linear polymer structure of the messenger RNA is read out. Since amino acids are added one at a time by the respective tRNA adaptors, it is evident that the ribosome moves along the messenger strand as they are added.

As a rule single ribosomes are inactive, although one ribosome alone acting on a messenger strand, as a polysome, is known to be capable of producing a single polypeptide chain. Active ribosomes usually occur in clusters of five or six, known as polyribosomes. It is known that several ribosomes working together on the same mRNA strand enable several polypeptide chains to be assembled at the same time. Ribosomes combine with mRNA at chain ends only, while after combination of ribosome and mRNA, ribosomes are found at various positions along the length of the message molecule. As synthesis of a peptide chain is finished, ribosomes leave the polysome, i.e., ribosomes read the mRNA chain sequentially.

GENETIC CONTROL: THE OPERON

Bacteria have yielded the most information and the greatest insight in recent years concerning the nature and control of synthetic processes, and present understanding of the operative machinery of the cell relates mainly to these procaryote cells. To what extent this may apply to the far more complex eucaryote cells of multicellular and true protistan organisms is not yet clear. The selection of bacteria, *Escherichia coli* in particular, for such studies, however, was logical since they are the simplest of all organisms and their DNA is merely a ring that contains about 5,000 genes. This DNA directs the synthesis of structural proteins that build the edifice, and of the enzymatic proteins that control the many chemical activities of the cell; it also

FIGURE 1.5   Ribosomal movement along messenger RNA, culminating in release of completed polypeptide molecule and of ribosome. Ribosomes attach to one end of the message, corresponding to the amino-terminal end of the protein, and, in conjunction with other components of the protein-synthesizing machinery, begin to move along the message as the peptide bonds are formed. New ribosomes continually attach to this end of the message and proceed in this way, so that at any one time the mRNA will carry many ribosomes along its length. This complex of ribosomes and mRNA is called a *polysome*. At the end of the genocopy in the mRNA, the translation machinery meets a codon that signals termination and release of the completed polypeptide chain. [*From* MOLECULAR ORGANIZATION AND BIOLOGICAL FUNCTION, *edited by John M. Allen: Fig. 3 (p. 25) by Alexander Rich. Harper & Row, 1967.*]

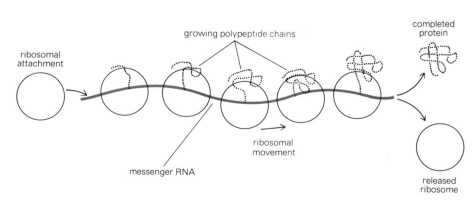

growing polypeptide chains

completed protein

ribosomal attachment

messenger RNA

ribosomal movement

released ribosome

directs the synthesis of agents, probably proteins, that control its own functioning, agents called repressors. The extension of this concept to include eucaryote cells is complicated by the fact that although the bacterial DNA has relatively few genes and lies naked in the cytoplasm, that of the more evolved cells is covered with proteins and contains tens of thousands, even millions, of genes. In any case we have a useful and powerful working hypothesis concerning control mechanisms at the genetic level.

It is assumed that there are two types of genes, structural genes and regulator genes. In the words of Jacob and Monod, who formulated the hypothesis: "The synthesis of enzymes in bacteria follows a double genetic control. The so-called

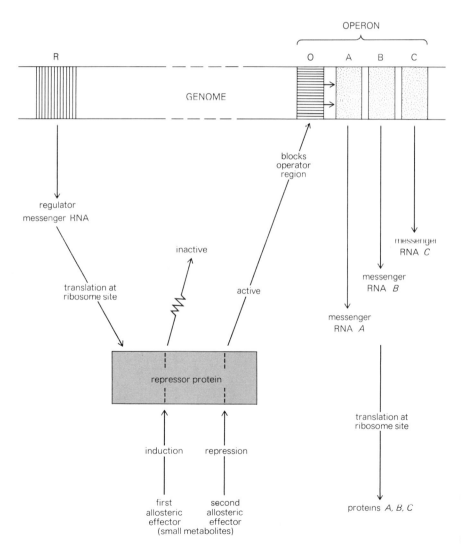

FIGURE 1.6 Control of enzyme synthesis by gene activation and repression. Binding the effector to the repressor allows it to activate or inactivate the operator according to the type of system. The repressor acts on an operator gene which controls all the linked structural genes involved in the synthesis of serially acting enzymes. The genetic loci shown are the regulator gene (R), the operator gene (O), and three structural genes (A, B, C). [From CELLULAR DIFFERENTIATION by J. Richard Whittaker. © 1968 by Dickenson Publishing Company, Inc., Belmont, California. Used by permission of the publisher and author.]

structural genes determine the molecular organization of the proteins. Other, functionally specialized, genetic determinants, called regulator and operator genes, control the rate of protein synthesis through the intermediacy of cytoplasmic components or repressors. The repressors can be either inactivated (induction) or activated (repression) by certain specific metabolites. This system of regulation appears to operate directly at the level of synthesis by the gene of a short-lived intermediate, or messenger, which becomes associated with the ribosome where protein synthesis takes place."[1]

For example, the gene of E. coli which controls the production of the enzyme β-galactosidase has been intensively studied. This enzyme is not synthesized by E. coli in significant amount until a β-galactoside such as lactose is added to the medium. A thousand times more of the enzyme is then produced, and this depends on the synthesis of the appropriate messenger RNA in the usual way. This is an example of an *inducible* enzyme. On the other hand, E. coli cells grown in the absence of amino acids contain all the enzymes necessary for their synthesis; yet if the particular amino acids are added to the medium, these enzymes disappear. They are *repressible* enzymes, inhibited by the presence or feedback effect of their own end products.

According to the hypothesis, when the DNA sequence of the structural gene is transcribed into the corresponding messenger RNA, its expression is governed by an adjoining specialized DNA segment called the *operator.* Moreover, a single operator segment may control the activity of several adjacent structures simultaneously. Thus in an inducible system the regulator gene normally produces something which inhibits the structural gene, but when a certain metabolite appears the regulator gene reacts with it to make it inactive and so unable to repress. In contrast, the regulator gene in a repressible system produces something which does not inhibit the structural gene until the gene interacts with the metabolite produced by that structural gene. In addition to regulator and structural genes an operator gene is supposed to be present which is a sort of designator of a group of closely associated genes. The operator gene together with its adjacent structural genes is known as an *operon.* The operon, in turn, is controlled by a macromolecular compound called the *repressor,* whose structure is controlled by an appropriate regulator gene. Cells clearly do have the capacity to turn specific synthesis on and off, and some such regulatory system must prevail.

It is likely that the regulating mechanisms operating in bacteria have their equivalent in, for instance, mammalian cells, but in these structurally more complex cells of higher organisms they are probably differently organized and there must be agencies superimposed on them which play a major role in processes peculiar to these cells, such as differentiation. For the regulatory problems posed by differentiated organisms not only are of an order of complexity immeasurably greater than that in microorganisms, they are also of a different nature. Even in a relatively simple

[1] F. Jacob and J. Monod, Genetic Regulatory Mechanisms in the Synthesis of Protein, *J. Mol. Biol.,* **3:**318–356 (1961).

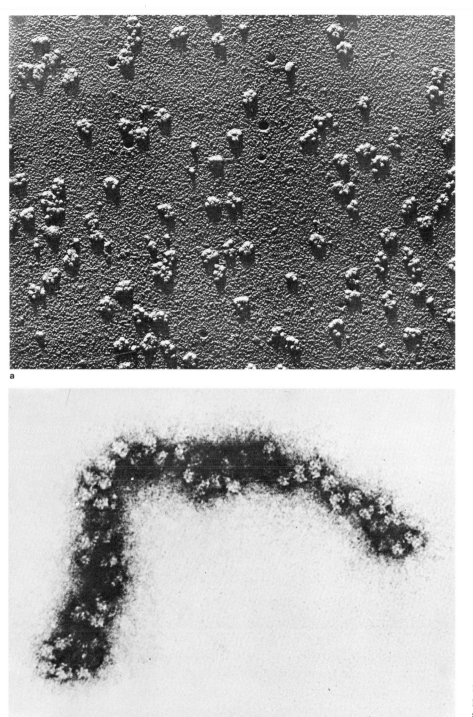

FIGURE 1.7   Electron micrographs of (**a**) reticulate polysomes and (**b**) polysomes synthesizing myosin, from chick embryos. [*Courtesy of A. Rich.*]

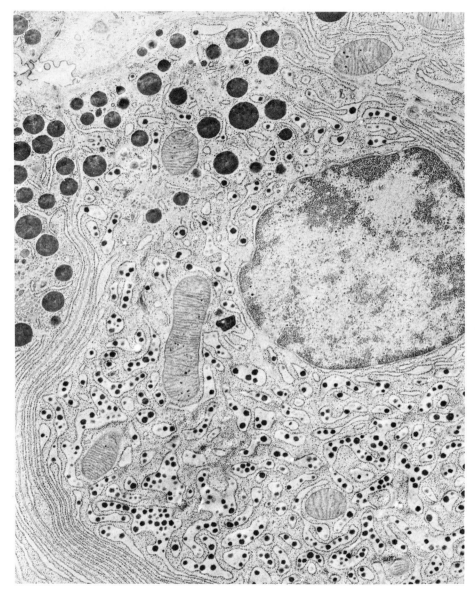

FIGURE 1.8 Part of bat pancreas secretory cell, showing nucleus, gaps in nuclear membrane, mitochondria, endoplasmic reticulum, and secretory granules. [*Courtesy of K. Porter.*]

complex organism, the mold *Aspergillus*, it is estimated that there are two regulator genes for every structural gene; in more complex forms the number will be far greater.

Although this topic will need to be discussed later since it relates in various ways to development phenomena and controls at all levels, some features of the nucleus of eucaryote cells should be noted here. To begin with, as already stated, the nuclear material is segregated from the cytoplasm by a double-layered nuclear membrane. Nuclear products may reach the cytoplasm through small pores in the membrane, clearly visible in electron micrographs; and they inevitably mix with the cytoplasm when the nuclear membrane and endoplasmic reticulum disintegrate during the early stages of mitosis.

The nucleus may be regarded as an organelle of particular constitution, for it consists of more than dispersed or condensed chromatin within a well-defined membrane. The chromatin itself is a composite of four components, namely, DNA, a small amount of RNA, the characteristic basal chromosomal proteins or histones, and nonhistone proteins. In addition to these components there is the nucleolus, densely staining during the cell interphase but disappearing during mitosis. Both the histones and the nucleolus have been assigned a vital role in the regulatory system. RNA accumulates within the nucleolus. While newly synthesized RNA is found in the chromatin itself, longer-synthesized RNA is found mainly in the nucleolus; RNA synthesized even earlier becomes lost to the nucleolus and appears in the ribosomes. In addition to its other possible functions, the nucleolus appears to be the factory for ribosomal synthesis, specifically for ribosomal protein synthesis and for assembly of RNA with ribosomal protein. The embryos of a mutant form of the African toad *Xenopus*, which lacks the nucleolus, cannot make ribosomes. Histones on the other hand are closely bound to the chromosomal structure and more or less envelop it; considerable evidence suggests that they may act most importantly as the nonselective repressors of the genetic system. Trying to prove this, together with searching for other types of repressors, is a continuing activity in cell and molecular biology.

PROTEIN MOLECULES

Protein molecules have long been studied as enzymes. These were essentially the soluble proteins extracted by biochemists; the insoluble, structural material was discarded. Both kinds of proteins, however, are vital to the cell. Whether assemblies of proteins are structural or contractile fibers, or whether they are globular proteins which catalyze chemical processes, all proteins are composed of polypeptide chains.

A fairly typical globular protein, serum albumin, has a molecular weight of 70,000, with dimensions of about 150×40 Å. If extended it would appear as a helical polypeptide about 1000 Å long. Therefore it must be coiled. In fact, every globular protein is a coiled structure. The precise and constant nature of the coiling depends on various forces, including hydrogen bonding, electrostatic attraction, and disulfide linkages between adjacent cysteine units, particularly the last. There is virtually no end to the specific configurations possible for chains consisting of hundreds of amino acid residues, and each configuration will have special properties of its own. Such properties determine how the particular molecule can serve as a specific enzyme or as a special unit of structure.

This has been well described by Anfinsen: "The restatement of linear genetic information in the form of three-dimensional protein structure results from a rapid

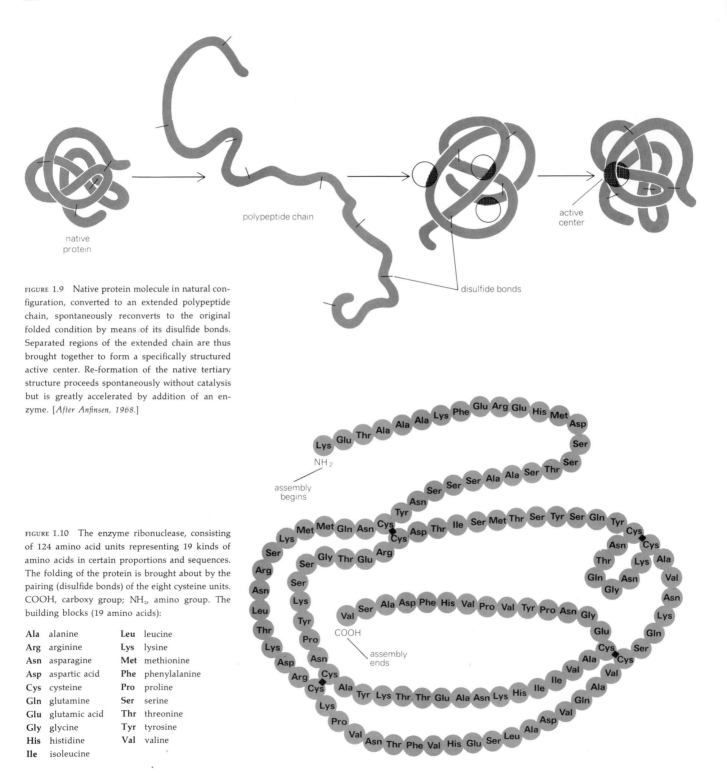

FIGURE 1.9 Native protein molecule in natural configuration, converted to an extended polypeptide chain, spontaneously reconverts to the original folded condition by means of its disulfide bonds. Separated regions of the extended chain are thus brought together to form a specifically structured active center. Re-formation of the native tertiary structure proceeds spontaneously without catalysis but is greatly accelerated by addition of an enzyme. [*After Anfinsen, 1968.*]

FIGURE 1.10 The enzyme ribonuclease, consisting of 124 amino acid units representing 19 kinds of amino acids in certain proportions and sequences. The folding of the protein is brought about by the pairing (disulfide bonds) of the eight cysteine units. COOH, carboxy group; NH₂, amino group. The building blocks (19 amino acids):

Ala	alanine	**Leu**	leucine
Arg	arginine	**Lys**	lysine
Asn	asparagine	**Met**	methionine
Asp	aspartic acid	**Phe**	phenylalanine
Cys	cysteine	**Pro**	proline
Gln	glutamine	**Ser**	serine
Glu	glutamic acid	**Thr**	threonine
Gly	glycine	**Tyr**	tyrosine
His	histidine	**Val**	valine
Ile	isoleucine		

and spontaneous interaction of amino acid side chains with each other, with the completed polypeptide backbone, and with the environment, without necessity for additional genetic information. The achievement of this unique geometry might be visualized as a rather helter skelter process. An almost infinite number of sets of interactions are possible as an extended polypeptide chain coils upon itself. . . . The only obvious driving force during this approach to native conformation is the selection of progressively more stable conformations with ultimate fixation of geometry in the form possessing the most favorable free energy of conformation, the native protein. Thus, unlike the complex predetermined path of successive changes occurring during differentiation, the cell must rely, in its first steps of development, on a relatively random process involving explicit information—the amino acid sequence of a polypeptide chain. . . . It has been suggested as an alternate mechanism that a polypeptide chain may progressively assume a three-dimensional conformation similar or identical to that which it occupies in the completed protein molecule, as a synthesis from the NH_2-terminus toward the COOH-terminal end of the chain. However the weight of evidence . . . appears to be consistent with a process in which tertiary structure appears only upon completion of translation of the genetic quantum of information."[1]

Such evidence includes the renaturation of scrambled ribonuclease molecules. After complete separation of the four disulfide bonds in the native ribonuclease protein, the reduced random chains were allowed to reoxidize under conditions leading to a random mixture of disulfide bonds. Exposure to conditions favoring disulfide interchange then induced rapid rearrangement of the disulfide bonds as in native protein. This resynthesis of ribonuclease in vitro took about the same time as in nature, i.e., 2 minutes.

Accordingly, the amino acid sequence coded for by a gene in turn codes for a specific three-dimensional structure. This conversion from linearity to spatial organization (secondary to tertiary structure) appears to be spontaneous. The internal positioning of hydrophobic and hydrophylic side chains, respectively, seem to be particularly important in determining tertiary structure, whether or not covalent disulfide bonds are present. This seems to be true for the globular and fibrous proteins alike.

Once formed, with secondary and tertiary spatial organization established, the protein molecule persists in a dynamic state. There is little doubt that a protein molecule represents an electronic continuum capable of both energy and information transfer from one part to another in a reversible manner, by way of the peptide linkages which separate one side chain from another and by way of the saturated carbon atoms that separate the side chain functional groups from the polypeptide chain. Any change, in any chemical group or bonds, propagates throughout the protein chain, as **(1)** a change in electron densities, **(2)** a change of partners within or among side groups, or **(3)** a change of bondage with and induced structuring of adjacent water molecules. All of these affect conformation and activity. When we consider

[1] C. B. Anfinsen, Spontaneous Assembly of the Three-dimensional Structure of Proteins, *Develop. Biol.*, suppl. 2.

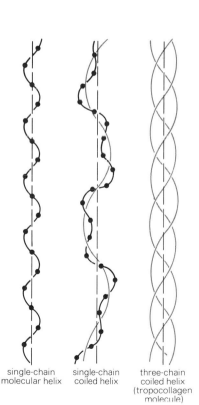

single-chain molecular helix single-chain coiled helix three-chain coiled helix (tropocollagen molecule)

FIGURE 1.11 Single-chain polypeptide helix, single-chain coiled helix, and three-chain coiled helix.

that protoplasm exists mainly in states akin to a gel, we see that the continuum represented by protein molecules extends to the cell as a whole and underlies its general sensitivity and responsiveness.

Although the concept of one gene–one enzyme seems to be roughly true, it is well known that single enzymes commonly exist in multiple molecular forms, called *isozymes*, within the cells or tissues of a single organism. Moreover, the tissue patterns are tissue-specific, each kind of adult tissue having a characteristically different isozyme pattern from the others. A good example is the enzyme lactate dehydrogenase, which exists as five isozymes. In this case two genes make two different polypeptides, which are the building blocks of the enzyme. Each enzyme molecule is a tetramer, consisting of four polypeptides. The five may be all of one kind of polypeptide or all of the other kind, or of any one of the three possible combinations of the two kinds of polypeptides in groups of four (4 : 0, 3 : 1, 2 : 2, 1 : 3, or 0 : 4). Since polypeptides are genetic products, the different proportions of isozymes found in different kinds of cells appear to be due to differences in activities of the two genes involved.

CONSTRUCTION AND RECONSTRUCTION OF COLLAGEN, A STRUCTURAL PROTEIN

The patterning (crystallinity), rigidity (strength), and plasticity (adaptability) of a structural protein can be seen in collagen. A fibrous protein found external to the cells which secrete it, collagen is widespread as a component of connective tissue throughout the animal kingdom. It is exploited for all manner of purposes by different organisms and in various ways and places in one and the same organism. It is also readily extractable and has been intensively studied outside the living system.

Information derived from x-ray diffraction studies, electron microscopy, chemical analysis, and other means has shown how collagen fibrils are built up from long-chain protein molecules. The properties of this seemingly indispensable structural protein depend both on its chemical composition and on the physical arrangement of the individual molecular units. These units are joined together to form the long fibrils constituting the fibrous collagen. Under the electron microscope the fibrils appear cross-striated, showing periodic bands about 640 Å apart.

The discovery that collagen fibrils can be disassembled in acid into their component molecules and then reassembled into their original form has been enlightening. It is also thought-provoking that this fact has been known since 1872 but was not followed up until 70 years later. The more recent studies show that such reconstitution is complete in its finest detail, for bands with the same periodicity reappear. Also, in native collagen, the polarized molecules, i.e., those with chemically and electrically distinct "heads" and "tails," are now known to be lined up facing in the same direction but overlapping by about one-quarter of their length. This overlapping is responsible for the 640- Å periodicity.

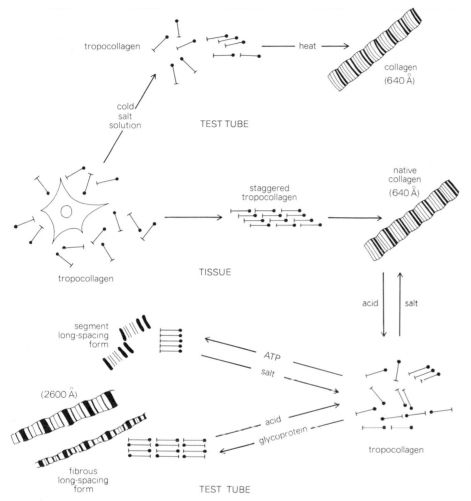

tropocollagen

cold salt solution

TEST TUBE

heat

collagen (640 Å)

tropocollagen

staggered tropocollagen

native collagen (640 Å)

TISSUE

acid | salt

segment long-spacing form

ATP

salt

(2600 Å)

acid

glycoprotein

fibrous long-spacing form

TEST TUBE

tropocollagen

FIGURE 1.12 Reconstitution of collagen. Collagen fibrils reconstitute spontaneously when an acid solution of native collagen molecules is neutralized. The fibrous long-spacing form of collagen is produced by adding glycoprotein to an acid solution of native collagen. The segment long-spacing form of collagen is produced by adding ATP to an acid solution of collagen. The central block suggests a possible way in which collagen fibrils are formed extracellularly, in vivo, from secreted tropocollagen. [*After Gross, 1956.*]

Accordingly, the ease with which collagen can be disassembled merely by subjecting it to a weak acid medium, such as acetic acid, without apparent degradation of the molecules, offers opportunities to investigate experimentally the agencies that operate in the process of reassembly, or polymerization. The molecules, or monomers, may be precipitated from solution simply by neutralization or by adding monovalent salts, which indicates that the physical state of the collagen depends on ionic and pH conditions of the surrounding medium in vitro and probably also in vivo.

The assembly process, however, is by no means a simple polymerization of precursor molecules of a particular length. The composition of the medium is all-important. On the one hand, it may be varied considerably without affecting the fine structure of the assembled collagen fibrils; on the other, certain additives result in the assembly of fibrils with a fine structure pattern that is markedly different. If glycoprotein is added to an acid solution of native collagen, for instance, fibrils assemble which have a band spacing with a period of about 2800 Å, forming the so-called fibrous long-spacing form of collagen. The significance of the glycoprotein is not its particular chemistry but the negative charge of its large molecules.

A somewhat similar but nonfibrous banded material forms when adenosine triphosphate (ATP) is added to the acid solution of monomers; this material is called *segment long-spacing collagen.* Moreover, any of the three structurally different forms of collagen can be disassembled and reassembled into either of the other two forms; the basic unit, named *tropocollagen,* from which any one of the three can be constructed is threadlike, about 2900 Å long and 50 Å wide. This macromolecule consists of three polypeptide chains, each of which is composed of about 1,000 amino acids linked together. Each such chain is twisted into a left-handed helix, and the three helices are in turn twisted around one another to form a right-handed superhelix, held together by hydrogen bonds. The three forms of collagen vary in structure according to the alignment of these macromolecular units.

Native collagen fibrils consist of units that overlap about one-quarter of their length, the overlap representing the short-band periodicity, and they face the same way. The fibrous long-spacing form consists of similar units that do not overlap, thus producing the periodicity of the full length of the unit, and they do not all face the same way. Though the nonfibrous long-spacing form also has units that do not overlap, they all face the same way. As structural proteins within organisms, the native form of collagen has obvious advantages of strength, from both overlap and molecular orientation. Of the several possible configurations, natural selection has undoubtedly operated in favor of the most useful, although by subtle and indirect means of medium control. The properties of collagen depend, therefore, on the amino acid constitution of the protein molecule, and on the specific organization at the supramolecular level of environmental control.

AGING OF COLLAGEN, AND QUATERNARY STRUCTURE

The collagen that has been studied most has been extracted from mammals. Its condition appears to change according to the age of the animal. A certain fraction of the total amount of collagen taken from young animals can be dissolved in cold neutral solutions. Such dissolved collagen molecules polymerize spontaneously to form typical cross-striated fibrils with native periodicity simply upon being warmed to the mammalian body temperature. The more rapid the growth of the animal, the greater is the amount of collagen which can be extracted, but after starvation of the animal for a few days this soluble collagen disappears from the tissues. The collagen normally extractable in cold solution appears to be newly synthesized by the cells

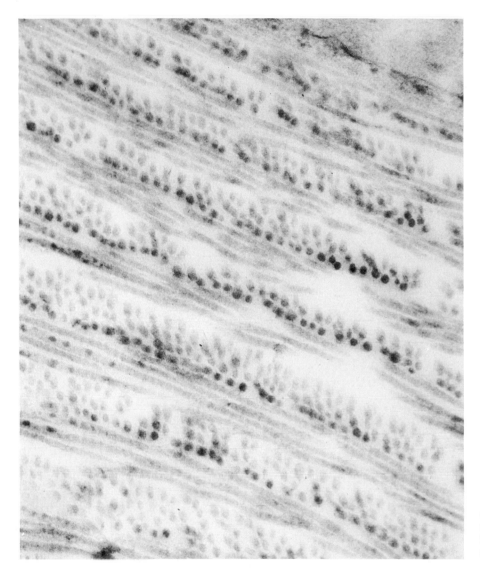

FIGURE 1.13 Ply layers of collagen fibrils (from tadpole cornea), characteristic of basement membrane of epidermis. Alternate layers appear at right angles to one another. [*Courtesy of P. Weiss.*]

and is accordingly still soluble because it is not yet tightly assembled. As it becomes older, dilute acids are required to extract it, i.e., to disassemble it; aged collagen is too tightly bonded to be dissolved at all. This is a progressive change that may be characteristic of many other structural proteins, and much of the aging process in general may be related to it.

Collagen fibrils commonly organize further to form multilayer sheets of considerable strength, e.g., in cornea, bone, and the basement layer of amphibian skin.

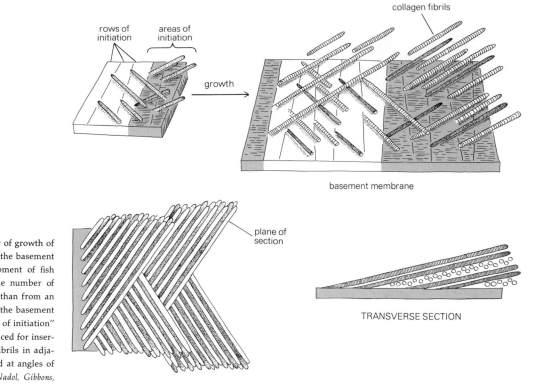

FIGURE 1.14 The "scindulene" theory of growth of the basement lamella. Thickening of the basement lamella with growth during development of fish larva results from an increase in the number of collagen fibril rows per layer, rather than from an increase in the number of layers. As the basement membrane expands, additional "rows of initiation" in the basement membrane are produced for insertion of more fibril rows per layer. Fibrils in adjacent "areas of initiation" are oriented at angles of 105 to 110° to one another. [*After Nadol, Gibbons, and Porter, 1969.*]

In cornea the successive layers of fibrils are laid down at right angles to one another in a sort of crystalline array. The fibrils of one layer are aligned along the dark cross-striations of the fibrils below, which are oriented at right angles. A similar arrangement is seen in the organization of fibrils in the epidermal basement membrane of the frog, although here each successive ply consists of several layers of fibrils. The stacking up of layers of different orientation appears to be a building principle carried over even in vitro, from the property of fibers with regular periodicity, to aggregate in such a way that homologous segments become aligned with one another. It is seen in collagen, fibrin, and myofibrils. The sudden change in direction by 90° in passing from one layer (cornea) or one ply (basement membrane) to the next is harder to explain.

We have here the concept of biological crystallizations as at least a contributing mechanism in forming or developing cell and organismal structure. The collagen system may represent a simple example of the spontaneous association of high-polymer building blocks whereby cell membranes, endoplasmic reticulum, mitotic spindles, and other structures are formed, dissolved, and re-formed under the influence of varying physical-chemical forces. The main business of synthetic enzymatic mechanisms would then be the production of the building blocks and the control of low-molecular-weight metabolites and electrolytes of the internal medium. How a number of proteins, differing in chemical composition, may collaborate in an apparently spontaneous construction of an organism of great complexity, with distinct parts and with a built-in adaptive motor mechanism, may be seen in the processes of assembly of viruses.

VIRUS ASSEMBLY

It is a tremendous step from the construction of either extracellular or intracellular protein aggregates to the organization of a whole cell. A cell must have at its disposal some 2,000 enzyme species in order to secure energy and to synthesize all its building blocks; it must possess the genetic information for the synthesis of all these enzymes, as well as sufficient space in which to organize them in appropriate structural systems. Even in a small procaryote cell such as the colon bacillus, the energy requirements, as shown by the ATP turnover rate, are impressive. At least 2 million ATP molecules are broken down per second in order to achieve the biosynthesis of all the cell components. Fortunately virus particles offer opportunity for analysis at a more comprehensible, subcellular level, for although they cannot metabolize, grow, or undergo division, they do reproduce their kind, and in a manner now fairly well understood. Even though small virus particles may be 8,000 times smaller than the smallest true microorganism, they may still be regarded as a suborganism of considerable complexity. Whether simple or complex, they all have a central core of nucleic acid enclosed by a protein envelope.

The simpler types may be represented by the poliomyelitis and tobacco mosaic viruses. The polio virus consists of a protein shell built up out of 60 structurally equivalent, asymmetric protein subunits of approximately 60 Å in diameter, packed together in such a way that they form a spherical shell of about 300 Å in diameter, enclosing the nucleic acid. Analysis shows that the shell contains about 49,600 amino acids and the nucleic acid about 5,200 nucleotides. Assuming that three nucleotides are necessary to code for one amino acid, the maximum number of amino acids that can be organized by the nucleic acid is 1,700. The viral nucleic acid cannot, therefore, build the shell directly but only through the assembly of smaller units. These identical subunits actually consist of 620 amino acids each, building blocks of a particular shape which fit together to form a hollow sphere. Once the subunits have been made, they can do nothing but assemble to form the shell or coat.

The tobacco mosaic virus, TMV, is somewhat different. It has the form of a rod, and is a cylindrical assembly of protein subunits with a single helical long molecule of RNA inside the cylinder. Treatment with acetic acid separates the particle into

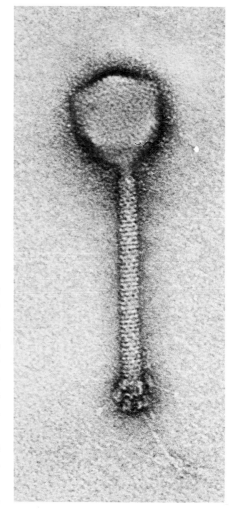

FIGURE 1.15 Virus particle (*Bacillus subtilis* bacteriophage). [*Courtesy of A. K. Kleinschmidt.*]

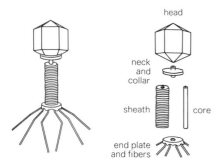

FIGURE 1.16 The parts and structure of the T bacteriophage. [*From* CELLULAR DIFFERENTIA-TION *by J. Richard Whittaker.* © *1968 by Dickenson Publishing Company, Inc., Belmont, California. Used by permission of the publisher and author.*]

an RNA molecule and some 2,000 protein subunits which are all alike. When these components are gently disassociated and then brought together under appropriate conditions, they reassemble to form complete and infectious virus particles. We see here not only a spontaneous reassociation between a nucleic acid and a specific protein but also the principle of polymeric limitation, namely, that there are limits to the combination of protein molecules to form larger construction units of a particular size and shape, and also limits to the aggregation or combination of such units to form the tubular sheath around the nucleic acid thread. More explicitly, we see the possibility that the size, as well as the shape, of supramolecular aggregates may be determined by intrinsic properties of the subunits, and by progressive change in the combining powers of the aggregate as its size increases.

Bacterial viruses, or "phage," have evolved special structural means of entering the bacterial cell. The most studied is the T_4 virus that infects the colon bacillus *Escherichia coli*. Although still far below the level of a cell in organization and size, it is considerably further up the organizational scale than the tobacco mosaic virus. Whereas the RNA core of the tobacco mosaic virus possesses only five or six genes, which are all that is necessary to direct the manufacture of its constituent materials, the T_4 virus contains a DNA core comprising more than 75 genes, the thread being highly coiled within a protein membrane to form a polyhedral head. A short neck connects the head with a tail consisting of a springlike contractile sheath, or tube, which surrounds a central core and is attached to an end plate. Six slender fibers and six short spikes protrude from the end plate. Altogether the virus is a highly organized structure, not unlike a landing module for the moon.

Apart from the nucleic acid thread, each kind of morphological unit consists of a protein specifically distinct from the others. Thus the membrane of the head consists of subunits of a molecular weight of about 80,000. The subunits of the sheath have a molecular weight of about 50,000, each sheath consisting of about 200 subunits; the tail spikes and fibers each seem to be a bundle of polypeptide chains of a molecular weight of about 100,000. Altogether the virus is a functioning, adaptive structure with functionally differentiated parts. It becomes anchored to the bacterial wall by its tail fibers so that the tail and head are perpendicular to the surface, as if guyed by the tail fibers. Within a few minutes after attachment, the sheath shortens to less than half its original length and the core pierces the wall of the bacterium, allowing the DNA molecule coiled in the head to pass into the bacterium.

The course of events within the cell is well known but will be summarized here. In brief, the bacterial DNA is disturbed, manufacture of bacterial protein stops, and the viral genes take over. Among the first proteins to be made are the tools of the trade, so to speak, the enzymes needed for viral DNA replication. This happens within the first 5 minutes after breakthrough. Three minutes later another set of genes directs the synthesis of the structural proteins that will form the head components and tail components. The first completed virus particle appears 13 minutes

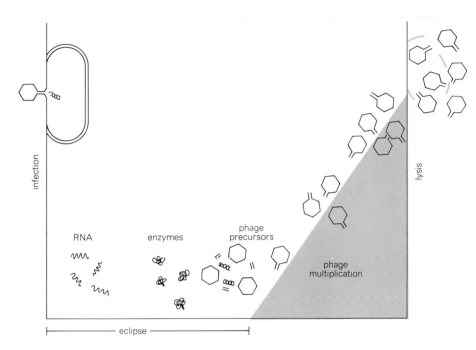

FIGURE 1.17 Schematic representation of the steps in virus (phage) infection, involving sequential synthesis of RNA, new enzymes, and virus particle precursors. Multiplication of virus is followed by lysis of the infected cell. [*After Staehelin, in R. Weber (ed.), "Biochemical Animal Development," vol. 1, p. 460, Academic, 1965.*]

after infection, and the process continues for another 12 minutes until about 200 such particles have been completed and the raw materials of the bacterial cell have been virtually exhausted. A viral enzyme, a lysozyme, then appears and attacks the cell wall, liberating the new viral particles.

Here, in a relatively simple yet nonetheless complex form, is a picture of assembly of a suborganism. The primary question is whether the assembly is entirely one of self-assembly of the various components produced under genic direction, or whether the assembly process itself is also gene-controlled.

The experimental procedures leading to an analysis of this problem are highly sophisticated and need not be detailed here; it is enough to say that they involve the study of many conditionally lethal mutations in the T_4 virus that lead to the accumulation of recognizable virus components under certain conditions. These experiments provide information regarding the function of many genes that affect the construction of the virus particle.

1   They indicate that a remarkably large number of gene products is required for the whole process.

2   They indicate that construction follows three major pathways, which lead independently to the formation of head, tail, and fibers, and are followed by steps in which the completed components, or subassemblies, are assembled into infectious viral particles.

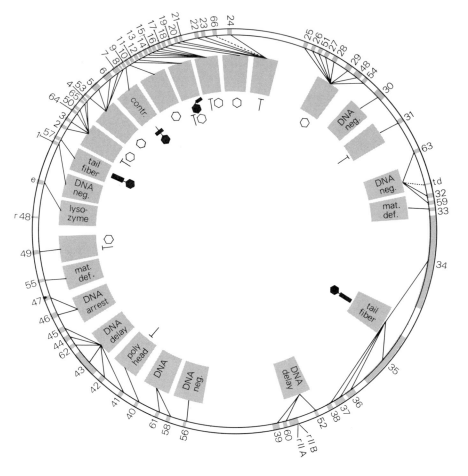

FIGURE 1.18 Genetic map of the T₄ virus showing relative positions of more than 75 genes already identified on the basis of mutations. Diagrams show viral components of cells infected by mutants defective in the particular gene. The labels and symbols indicate defective phenotypes as follows: *DNA neg.,* no DNA synthesis; *DNA arrest,* DNA synthesis arrested after a short time; *DNA delay,* DNA synthesis commences after some delay; *mat. def.,* maturation defective (DNA synthesis is normal, but late functions are not expressed). A hexagon indicates that free heads are produced; an inverted T, that free tails are produced; *tail fiber,* that fiberless particles are produced. Gene 9 mutants produce inactive particles with contracted sheaths; gene 11 and 12 mutants produce fragile particles which dissociate to free heads and free tails. [*After Edgar and Wood, 1966.*]

3 The sequential order in later pathways appears to be imposed entirely by the structural features of the various intermediates. It is controlled at the level of gene product interaction rather than directly by gene action. Altogether, three factors—the kinetics of intracellular phage production, the nearly identical times of appearance of the T₄ virus late proteins, and the finding of several extract complementative groups corresponding to single gene products—argue against the concept of assembly sequence control by sequential gene induction. These factors support the view that all the morphogenetic precursor components are synthesized simultaneously and independently throughout the latter half of the latent period within the infected cell. Many genes are involved in the production of primary products, but these interact, or self-assemble, to form the various subassembly units, and these in turn self-assemble to form the final invasive and self-reproductive product.

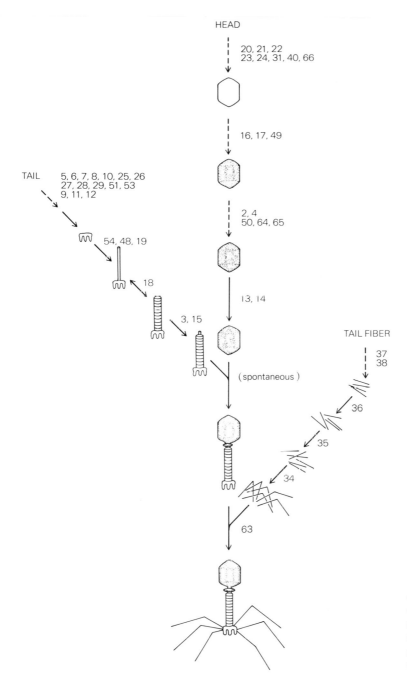

FIGURE 1.19 Pathway of morphogenesis for T_4 virus has three main branches leading independently to the formation of heads, tails, and tail fibers, which subsequently combine to form complete virus particles. The numbers refer to products of identified genes. [*After Wood, Edgar, et al., 1968.*]

CONCEPTS

Given a source of energy, amino acids spontaneously polymerize and form peptide chains, but the ordered sequence of amino acid residues into long, specific polypeptides and proteins requires the special assembly machinery of nucleic acids and ribosomes.

Polypeptide chains cross-link to form proteins. These are coiled or helical configurations of specific tertiary structure which function as specific enzymes or as units for further assembly.

Proteins form quaternary structures through association or assembly of tertiary subunits, such structures usually exhibiting a regular periodicity.

The physical and chemical state of the solvent stablilizes the quaternary structure to a great extent, with electrolytes and the structurization of water playing an important role.

The particular configuration of both tertiary and quaternary assembly of protein structure depends on the specific types, numbers, and sequences of the constituent amino acid residues.

All genetic control is primarily in the determination of kinds and quantities of amino acids produced. Information contained in the specific amino acids, together with that expressed by their specific sequential ordering, yields the information operating in subsequent self-assembly processes at higher levels.

READINGS

ANFINSEN, C. B., 1968. Spontaneous Assembly of the Three-dimensional Structure of Proteins, *Develop. Biol.,* suppl. **2.**

_____ , 1967. Molecular Structure and Function of Proteins, in J. M. Allen (ed.), "Molecular Organization and Biological Function," Harper & Row.

ANDERSON, T. F., 1967. The Molecular Organization of Virus Particles, in J. M. Allen (ed.), "Molecular Organization and Biological Function," Harper & Row.

BONNER, J., 1965. "The Molecular Biology of Development," Oxford.

EDGAR, R. S., and W. B. WOOD, 1966. Morphogenesis of Bacteriophage T_4 in Extracts of Mutant-infected Cells, *Proc. Nat. Acad. Sci.,* **55**:498–505.

GROSS, J., 1956. The Behavior of Collagen Units as a Model in Morphogenesis, *J. Biophys. Biochem. Cytol.,* suppl. **2.**

KELLENBERGER, E., 1966. The Genetic Control of the Shape of a Virus, *Sci. Amer.,* December.

LWOFF, A., 1959. "Biological Order," Wiley.

NADOL, J. B., J. R. GIBBONS, and K. PORTER, 1969. A Reinterpretation of the Structure and Development of the Basement Lamella, *Develop. Biol.,* **20**:304–331.

RICH, A., 1967. The Structural Basis of Protein Synthesis, in J. M. Allen (ed.), "Molecular Organization and Biological Function," Harper & Row.

WHITTAKER, J. R., 1968. "Cellular Differentiation," Dickenson.

WOOD, W. B., and R. S. EDGAR, 1967. Building a Bacterial Virus, *Sci. Amer.,* July.

_____ , _____ , et al., 1968. Bacteriophage Assembly, *Fed. Proc. Amer. Soc. Exp. Biol.,* **27**:1160–1166.

CHAPTER TWO
THE STRUCTURAL BASIS
OF FORM
AND ACTION

Cells and their more distinctive components are membrane-bound. Metabolic exchange between the cell and its external environment and also between organelles (such as mitochondria, lysosomes, and the nucleus itself) and the surrounding cytoplasm is governed by the properties of the membranes. Cell membranes, which are closer to being solids than they are to being any other form of matter, are typically several molecular layers thick. As such they restrict the passage of substances into and out of cell and organelle alike, preventing random diffusion. In addition to being barriers to diffusion, they also regulate the *rate* of movement of both water and solutes across cell boundaries, as well as the *direction* of molecular movement. Apart from the double-helix basis and the conformation of chromosomes with the nucleus, the overall structure of a cell depends on the presence and properties of membranes, microfibrils, and microtubules. Cell structure and cell function go hand in hand.

CONTENTS

The eucaryote cell

Membrane

Molecular organization of membrane

Mitochondria

Chloroplasts

Contractile filaments: actomyosin

Microtubules

Nucleation centers

Reassembly of a cell

Concepts

Readings

THE EUCARYOTE CELL

Eucaryote cells are generally categorized as two-envelope systems, in contrast to the one-envelope system of bacteria and other procaryote cells. Primarily the distinction is that in eucaryotes the nucleus as well as the whole cell is bounded by a membrane, whereas in procaryotes the nuclear material lies naked to the surrounding matrix. This distinction, however, is probably symbolic of the extensive development of internal membrane system in eucaryote cells.

To begin with, the whole cell is bounded by plasma membrane of the same general character as that of bacteria. An additional surface layer, or wall, consisting of polysaccharide or mucoprotein may or may not be present, depending on the type of cell. The nucleus, which consists of the chromosomal DNA assemblies together with associated RNA and proteins, the nucleolus, and the nuclear sap, or matrix, is bounded by its nuclear membrane. This effectively segregates the contents significantly but not completely from the surrounding cytoplasm. The membrane is seen to be double in electron micrographs and also to be studded by pores. The pores together amount to about 10 percent of its surface and lead directly from the nuclear interior into the cytoplasmic matrix.

The cytoplasm lies between nuclear membrane and plasma membrane, both of which should properly be considered part of the cytoplasm itself. It consists of matrix, membrane, membrane-enclosed inclusions such as mitochondria, lysosomes, and plastids, and other inclusions such as fibrils and granules which are not membrane-enclosed. Altogether the cell is a highly organized assembly of nucleic acids,

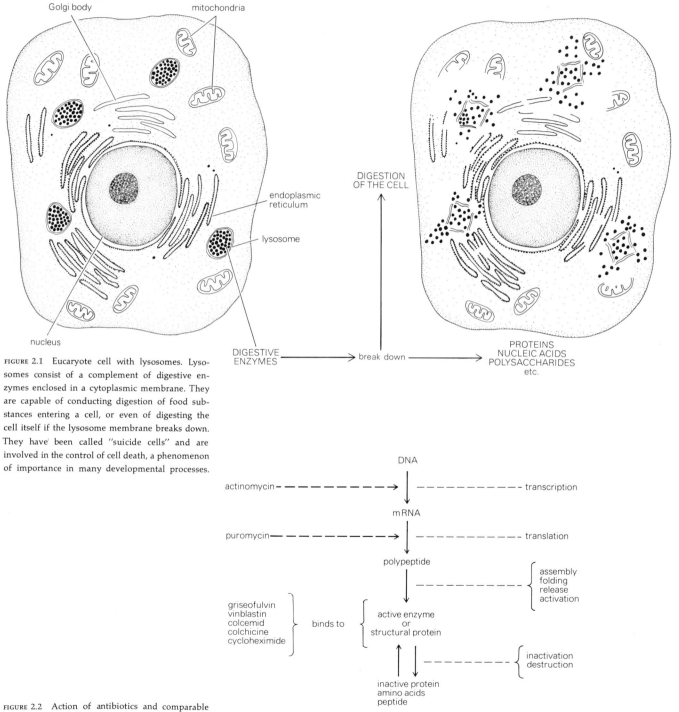

Golgi body   mitochondria

endoplasmic
reticulum

lysosome

nucleus

FIGURE 2.1   Eucaryote cell with lysosomes. Lysosomes consist of a complement of digestive enzymes enclosed in a cytoplasmic membrane. They
are capable of conducting digestion of food substances entering a cell, or even of digesting the
cell itself if the lysosome membrane breaks down.
They have been called "suicide cells" and are
involved in the control of cell death, a phenomenon
of importance in many developmental processes.

DIGESTION
OF THE CELL

DIGESTIVE
ENZYMES   ⟶   break down   ⟶   PROTEINS
                                        NUCLEIC ACIDS
                                        POLYSACCHARIDES
                                        etc.

DNA

actinomycin ─ ─ ─ ─ ─ ─ ─ ─ → ─ ─ ─ ─ ─ ─ ─ ─ ─ transcription

mRNA

puromycin ─ ─ ─ ─ ─ ─ ─ ─ → ─ ─ ─ ─ ─ ─ ─ ─ translation

polypeptide

                                          ⎰ assembly
                          ─ ─ ─ ─ ─ ─ ─ ─ ⎱ folding
                                          ⎰ release
                                          ⎱ activation

griseofulvin
vinblastin
colcemid   binds to   active enzyme
colchicine   or
cycloheximide   structural protein

                                          ⎰ inactivation
                          ─ ─ ─ ─ ─ ─ ─ ─ ⎱ destruction

                                inactive protein
                                amino acids
                                peptide

FIGURE 2.2   Action of antibiotics and comparable
compounds on protein synthesis.

STAGES IN THE CONTROL OF METABOLIC ACTIVITY

proteins, carbohydrates, lipids and phospholipids, electrolytes, and water. This statement of course applies to eucaryotes and procaryotes alike. The distinction relates to size and organization. Both types are also alike in their essential processes—in coding for proteins by the DNA-RNA-ribosome route, in controlling diffusion both ways between cell interior and external environment, and in conducting multifarious metabolic activities within the cell.

In other words, all cells are concerned with the three basic mechanisms of sequence coding, catalysis, and diffusion control. Wherever chemical linkages are formed or broken, enzymes are present. Where they are localized, they appear to be functionally related to the biochemical activities of whatever the local structure may be, whether it is connected with transient changes in chemical bonding during passage through membranes, the transformation of small molecules, or the more ambitious processes of macromolecular syntheses. This is clearly the case for the membranous organelles—mitochondria, plastids, and lysosomes.

In bacteria both enzymes and ribosomes are scattered through the matrix, although there is some evidence that respiratory enzyme complexes may be more or less associated with the plasma membrane. In eucaryotes the enzymes are typically localized in some manner:

1   They may be tightly bound to membranes, e.g., the enzymes of electron transportation systems in mitochondria and chloroplasts.
2   They may be enclosed by but not bound to membrane, as in lysosomes (which are typically sacs about 0.6 μ in diameter densely filled with hydrolytic enzymes that spill out in injured cells and accelerate decomposition and removal) and also as in certain circumstances such as metamorphosis and other selective destructions.
3   They may be bonded to discrete but nonmembranous structures, e.g., DNA and DNA polymerase and also the sliding-actin and myosin components of the actomyosin of muscle cells.

Apart from the presence in eucaryote cells of such vital organelles as mitochondria and chloroplasts, the most striking features distinguishing them from procaryote cells are size and the great development of the internal membrane system known as the endoplasmic reticulum, two features that may be related both functionally and evolutionarily. The question of size, in fact, applies to all stages and levels of both development and evolution, from subcellular components to cells as such and so on to multicellular organisms and their social aggregations.

Bacteria are typically about 0.5 μ in diameter, too small, for instance, to contain a single lysosome. A typical eucaryote cell is about 20 μ in diameter, or about 64,000 times greater in volume than bacteria. However, since the volume of a sphere increases as the cube of its radius, whereas the surface area increases only as the square, any increase in volume is accompanied by a decrease in the surface area/volume ratio. Thus a fortyfold increase in diameter (or 64,000× increase in volume) is accompanied by a fortyfold decrease in the surface/volume ratio. Inasmuch as the plasma membrane of a cell serves for passage or transport between cell interior and exterior, such an

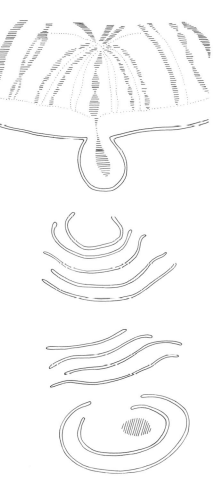

FIGURE 2.3 Nuclear blebbing occurs when a chromosomal fragment approaches the nuclear membrane. A bleb grows and breaks free, flattening into a cytoplasmic membrane. Such membranes probably synthesize the protein of a secretory granule.

increase in the surface/volume ratio would have dire consequences for a cell unless compensatory changes in cell organization were made, since the rates of diffusion, as of the passage of oxygen and carbon dioxide through the plasma membrane, limit the rate of all other cell activities. Obviously compensatory changes have been made or the eucaryote cell as we know it could not exist. To quote Stern and Nanney, "The effective adaptation of cells to the challenge of diffusion appears to reside in the organization of the endoplasmic reticulum. The deep and extensive invaginations of the plasma membrane increase the surface area available for exchange of solutes between cell interior and environment. The spaces between the membranes of the reticulum limit the diffusion of solutes and channel them to different regions of the cell. The nuclear membrane may be far more important as a mechanism for facilitating exchange of solutes between the contents of the membrane-bounded canals and the nuclear apparatus than for separating that apparatus from the cytoplasmic ground substance."[1]

Even so, eucaryote cells metabolize and multiply at a very much slower rate than do bacteria under similar conditions. The endoplasmic reticulum varies greatly, however, in the extent of its development among different cell types, being best developed in cells most actively involved in protein synthesis, as in enzyme-secreting cells of the pancreas. In such cases, particularly, continuity is apparent between nuclear membrane and endoplasmic reticulum and (via the smooth-membrane Golgi apparatus at least) between the reticulum and the plasma membrane. The combined structure, in other words, appears to be a single continuous entity, to some degree regionally specialized.

MEMBRANE

The cell is a polyphasic system consisting of membranes, filaments, particles, and various matrices, all directly or indirectly in dynamic relationship with the external environment of the cell. Membrane bounds the cell as a whole and extends in varying form through the cell interior. Robertson supports a unitary theory of the cell which "suggests that a cell is a three-phase system consisting of the following: first, a nucleocytoplasmic matrix phase that is continuous via the nuclear pores with the whole of the 'unstructured' part of the cell; second, a membrane phase, continuous from the nuclear membrane to the outside surface membrane through the endoplasmic reticulum; and third, an outside phase, brought into the cell by invagination. . . . The membrane always maintains its asymmetric polarity, with the same surface always directed toward the cytoplasmic matrix. Membranes bounded on both sides by cytoplasmic matrix do not exist."[2]

[1] H. Stern and D. L. Nanney, "The Biology of Cells," p. 451, Wiley, 1965.

[2] J. D. Robertson, The Organization of Cellular Membranes, in J. M. Allen (ed.), "Molecular Organization and Biological Function," p. 80, Harper & Row, 1967.

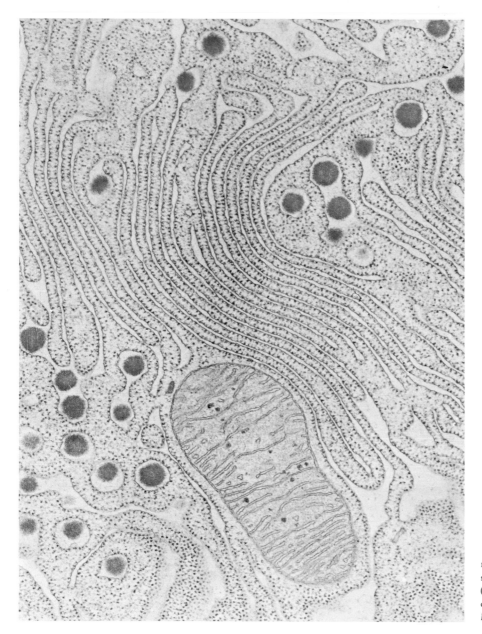

FIGURE 2.4 Electron micrograph of part of a cell, showing endoplasmic reticulum with ribosomes (*top*), a mitochondrion with cristae (*center*), and secretory granules (*above and below*). [*Courtesy of K. Porter.*]

There is no doubt that membrane structures are vital to the life of the cell; in fact, without membrane there is no cell. It not only separates compartments where special activities proceed with an efficiency otherwise not possible, but it is an active participant in much that goes on. Although there is circumstantial evidence for local variation in membrane constitution, there is little or no doubt that it is basically similar wherever it is encountered. Until very recently, more was known about what a membrane does than about what a membrane is, for membrane is so thin that it lies near the limits of resolution of even the electron microscope and has never been seen through the light microscope. Cell membrane has several functions:

1   It forms cell compartments; it separates the whole cell interior from the external environment; it isolates the nucleus from cytoplasm; and it isolates the interior of organelles, particularly mitochondria, lysosomes, and plastids, from surrounding cytoplasm.

2   It has properties of selective permeability which enable it to keep out many molecules and ions, and in general to determine the direction of flow of compounds, even against a concentration gradient.

3   It serves as an organizational basis for sequential or group associations of enzymes, particularly within mitochondria and on the cytoplasmic surface of the endoplasmic reticulum.

4   It can form repeating lamellar structures of diverse form and function, such as chloroplast lamellae, the stacked discs of outer segments of retinal rods, the spirally rolled myelin sheath formed around nerve axons by Schwann cells, the internal structure of yolk platelets, and, not least, the stacking of the cisternae of the Golgi apparatus and the endoplasmic reticulum itself, in cells where these are exceptionally well developed.

5   It is itself an active structure with some capacity for flow, particularly from the plasma membrane, as evaginations and invaginations, giving rise to extruded or secreted vesicles and to intruded or pinocytotic vesicles, respectively.

In electron micrographs, membrane appears as a pair of dense lines, each about 20 Å thick, separated by a lighter zone about 35 Å across. The concept of a "unit membrane" refers to this feature as a basic structure common to membrane systems generally. Both protein and phospholipids have long been known to be components of membrane. The question has been: How are they organized to form membrane structure with the dimensions and properties observed? Both proteins and lipids, especially phospholipids, are capable of forming films, and the cell membrane is evidently a complex which calls attention to possible states of matter in living substance.

MOLECULAR ORGANIZATION OF MEMBRANE

Both protein and phospholipid molecules are polar compounds which orient in an electrically charged field and in relation to one another. Both have polar and apolar groups within a single molecule and tend to form an adsorbed film along nonaqueous surface. Phospholipids, however, are much smaller than proteins and accordingly are capable of easily orienting themselves within a relatively short distance. When protein, phospholipid, and water are mixed, films are produced in which the polar groups of the phospholipids are attached to protein, and in electron micrographs such films appear as double membrane, much like the natural membranes of cells. The small phospholipid molecule may well be the key component in the structural organization of cell membranes.

How protein, phospholipid, and water are organized to give the structure of the unit membrane is still debatable. Some investigators consider the two dense layers to be primarily protein and the light intermediate layer to be organized phospholipids. Alternatively, and probably more likely, the light layer may be highly hydrated and each dense layer may comprise the complete protein-lipid complex.

According to Robertson, the "molecular pattern for the unit membrane structure . . . is a core of lipid covered on the inside and outside by a monomolecular film of non-lipid. . . . The unit membrane is an asymmetric structure. The diagram demonstrates, further, that in the cell membrane structure the lipid is in the smectic state; that is, the lipid molecules are in a close-packed bilayer arrangement with the polar groups directed outward, and covered by monomolecular films of non-lipid. It shows further, that there are only two layers of lipid molecules in the thickness of the membrane, with two other molecular layers on the outside. Thus a limit is placed on the thickness of the membrane, which depends on the molecular species involved."[1]

It is possible that different types of component lipids and phospholipids may be to membrane structure what amino acids are to protein structure, namely, that the secondary, tertiary, and quaternary structure of membranes is determined by the structure and space-filling properties of the building blocks, their ratio, and their two-dimensional patterns. This may turn out to be a useful concept, as a working hypothesis, although it has been opposed.

Although membrane grows and can rapidly be replaced, little is known concerning membrane assembly. One suggestion is that small complexes of membrane are organized in the cytoplasm and are inserted as blocks into the preexisting membrane. With regard to endoplasmic reticulum, smooth membrane apparently derives from the so-called rough endoplasmic reticulum with its attached ribosomes. In any case, since the different enzymes associated with the mature membrane appear at different times during maturation, a one-step, self-assembly process seems ruled out. Moreover the different turnover rates of lipids and proteins in the membrane make it unlikely that there is a continuously forming basic membrane into which specialized functional components could be inserted. Here the question which arose concerning the sequential assembly of components during virus construction is seen to apply again: Is the sequential order in the assembly process determined by gene action, or is it determined exclusively at the level of interaction of gene products? The conclusion reached concerning virus assembly was that the sequential order in the assembly pathway is indeed imposed entirely by structural features of the intermediate gene products, and that products used in the early and late stages of assembly are produced at one time and cannot do otherwise than assemble in certain sequences. This same principle probably holds also for the assembly of structures such as membrane and organelles.

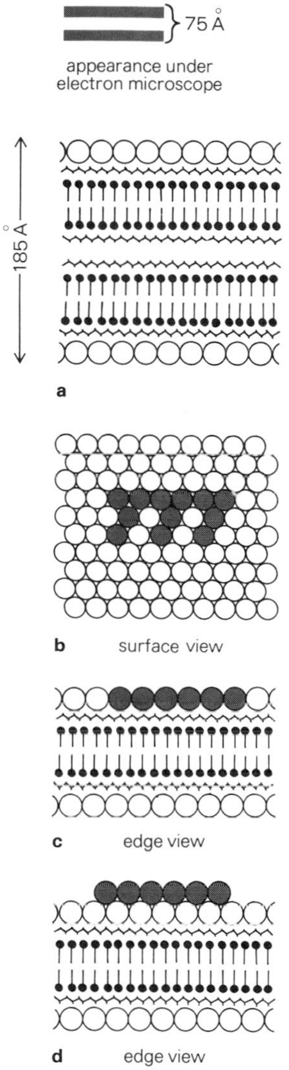

[1] *Ibid.*, p. 80.

FIGURE 2.5 Diagrammatic representation of (**a**) membrane structure and of (**b–d**) respiratory enzyme assemblies in terms of "unit membrane" hypothesis. In (**c**) the assembly is shown as a part of the membrane, and in (**d**) it is attached laterally to the membrane. [*From* MOLECULAR ORGANIZATION AND BIOLOGICAL FUNCTION, *edited by John M. Allen: Fig. 14 (p. 129). Harper & Row, 1967.*]

FIGURE 2.6   Model of a mitochondrion showing outer membrane, inner membrane extending inward as folds, or cristae, and possible arrangement of circular DNA within the mitochondrion. The DNA molecules may be attached to portions of the membrane. [*After Nass, copyright 1969 by the American Association for the Advancement of Science.*]

MITOCHONDRIA

Much of our present understanding of the general nature of membrane comes from study of the mitochondrial membrane in particular. Since mitochondria are found in all aerobic cells in the animal and plant kingdoms and are the site of the organized enzyme systems responsible for the energy-transforming reactions involved in oxidative phosphorylation (in other words, since they are the "power plants" of the cell), they have received great attention. They are absent only in bacteria, in which the plasmalemma serves the same biochemical functions as the mitochondrial membrane. Aerobic bacteria contain a cytochrome system attached to the plasma membrane. As cells increase in size and complexity (i.e., eucaryote cells as compared with procaryote cells), the oxygen requirements also increase and intracellular membrane systems become necessary to compensate for the decreasing ratio of surface area to volume.

Each kind of cell apparently has its own characteristic number of mitochondria (e.g., a rat liver cell has about 1,000, a renal tubule cell about 300), and often mitochondria of specific shape, which are passed on by division when the host cell divides. Typically a mitochondrion is a sausage-shaped organelle about 1 μ long and one-third as wide. There are an outer and an inner membrane; the outer membrane is smooth, the inner membrane is folded to greater or lesser degrees to form internal membranes, or cristae, which partially divide the interior into matrix-filled compartments. In the living cell, mitochondria may be seen, by phase microscopy, to swell, coil, branch, fragment, coalesce, etc., and typically to be in constant movement.

Although the inner and outer membranes both appear to have unit-membrane structure, there is evidence that the electron-transport enzymes and phosphorylating enzymes, or respiratory enzyme assemblies necessary to sustain the life of a cell, are associated with the inner membrane. Altogether the mitochondria constitute a unique kind of membrane system, containing large amounts of phospholipid and protein; and at least 25 percent of the protein associated with the membrane is enzymatic.

An unsettled question, basic to the nature of membrane, is this: In what structural way are the catalytic enzyme proteins incorporated with the structural enzymes? When cells are homogenized, the cytochromes and flavoproteins of the respiratory enzyme chain are found exclusively in the insoluble particles presumably derived from the membrane. Such particles or fragments usually contain the complete complement of enzymes necessary for electron transport, and often the capacity to couple phosphorylation of ADP to electron transport. On the other hand most of the enzymes associated with the Krebs cycle are found in the soluble fraction, i.e., from the matrix or the intermembrane space. A single mitochondrion from a liver cell has approximately 17,000 such respiratory assemblies, or about 650 per μ^2.

We are faced here with an ever-recurring question: To what extent does analysis of cell components studied outside the cell or organelle of which they are parts reflect this actual organization and activity when the components are integrated with the living system? A mitochondrion may be pictured as a multienzyme complex; but this only emphasizes the difficulty in separating any single mitochondrial function from the rest of the mitochondrion in order to understand its function, for protein-protein interactions are such that enzymes in isolation may act differently, both qualitatively and quantitatively, from the same enzymes bound to or incorporated in their normal locational sites. Even so, it may be that all membrane systems should be looked upon as specialized types of multienzyme systems. In any case we are left with the question: How is it that a membrane can come to differ chemically, and therefore functionally, along its length?

Mitochondria are now known to be semiautonomous and to carry out the biosynthesis necessary for their own replication, for they contain most of the components needed for the synthesis of DNA, RNA, ribosomes, and activating enzymes. They are able to grow and to divide. Yet, in spite of the strong implication that they may represent the modified descendants of once-symbiotic bacteria, they are by no means fully autonomous organelles. For whereas the mitochondrial DNA appears to encode for a limited number of the mitochondrial structural proteins and possibly for some of the ribosomal components, the large majority of mitochondrial proteins are encoded by nuclear genes. Electron micrographs show that fission of mitochondrial DNA begins before division of the mitochondrion into two daughter organelles. Whatever the evolutionary origin of mitochondria may have been, they appear to constitute a satellite genetic system, at least in part independent of nuclear DNA, and capable of high rates of division and replication.

CHLOROPLASTS

Chloroplasts, of primary importance for all photosynthetic eucaryote cells, may be regarded as multienzyme membrane systems, comparable to mitochondria although of different and greater structural complexity. A typical chloroplast of a higher plant cell is about 10 μ in diameter and 5 μ thick—as large, in fact, as certain kinds of whole animal cells. It is essentially a lamellar system which includes grana, the photosynthetic structures proper. The grana are in turn stacks of vesicles and are connected in a complex manner.

In section a granum is seen as a stack, usually in cylindrical array, of membrane elements which are structurally continuous with simpler intergrana, or stroma, elements that traverse and connect with other grana. At magnifications less than × 100,000 the grana elements are seen as alternating light and dark layers; at higher magnification each dense layer is seen to be double. A granum is accordingly thought to comprise repeating, basic membrane units, each unit being a triple molecular sandwich of protein-lipid-protein containing chlorophyll. Although a unit membrane appears to be the basic structural element in a chloroplast granule, the functional unit is a pair

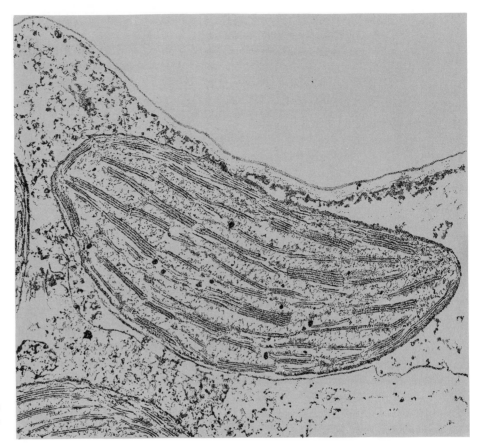

FIGURE 2.7   A proplastid in a dark-grown (etiolated) maize leaf, showing typical lamellate organization, together with many ribosomes. [*After Bogorod.*]

of these unit membranes united at their ends to form a flattened closed system, or lamellar unit.

Much is known, of course, concerning the photosynthetic processes as such. The problem is to associate the various phases and aspects of photosynthesis with specific components and general organization of the chloroplast. The structural and molecular complexity is such that only speculation is possible. In our present context, however, we are more concerned with the developmental aspect of the chloroplast.

Chloroplasts first appear as small, spherical proplastids about 1 μ in diameter, to be seen in the tips of plants grown in the dark. These bodies have a crystal-lattice-like structure containing the chlorophyll precursor, protochlorophyllide, attached to protein; they also contain DNA, RNA, and ribosomes, as in mitochondria. Upon exposure to light this prolamellar body dissociates into a group of loosely packed vesicles. These disperse through the plastid and align in rows, each row fusing to form several long, flat vesicles. At points along these vesicles, additional saclike structures are formed. During these events, protochlorophyllide reduces to chlorophyllide, and rapid synthesis of chlorophyll follows.

By such means the grana are progressively built up. Obvious questions are: Why does the plastid grow only so far in the dark, and how does light act to stimulate production of mature chlorophyll? One speculation, involving the operon hypothesis, is that light somehow regulates the production of specific informational RNA through derepression of particular genes or segments of the plastid DNA. The high degree of organization existing in the chloroplast membrane is such that it seems improbable that it could have resulted merely through random assembly from a mixture of its components. An alternative concept is that when molecules make an extremely good fit in an organized structure, one kind may serve as a template on which an adjoining, or bound, molecule is synthesized. *Each structural-synthetic event, in this view, determines the next such event; and the entire sequence of events determines the functional capability as well as the structure of the whole.* This would be epigenetic development at the molecular level of assembly.

The coding for structural proteins in mitochondria and plastids by their respective genomes may be seen as a carry-over from their independent evolution and incorporation of long ago, if this has been the case. Regardless of the question of origin of the mitochondrial and plastid DNA systems, however, the localization of such systems within isolated membranous organelles in the cell (particularly since both mitochondrial and plastid DNA differ significantly from nuclear DNA) effectively relieves the nucleus of at least some organizational and substantial burden. Yet since both organelle and nuclear DNA are involved in the full development of mitochondria and plastids, the possibility arises that there is a sequential genetic control of the assembly process as a whole. This is purely speculative.

The question has been raised regarding the assembly of membrane as such, not only as it concerns the exceptionally highly organized membrane structures of mitochondria and plastids. Thus, if it is assumed that membrane is formed by a stepwise process leading from proteins to polyprotein complexes, from complexes to functional subunits, and finally to the membrane continuum, how is such a precision assembly achieved?

A spontaneous self-assembly process implies that proteins to be assembled, once liberated from the sites of polyribosome synthesis, spontaneously interact in an aqueous environment of the cytoplasm to produce progressively more complex macromolecular structures. This may well be the case, as we have already seen to some extent in the self-assembly phenomena of collagen. Otherwise, if built-in specifications of the components to be assembled are insufficient to provide for precise assembly, external devices for assembly must be present and operative, and if so, we need to know what they are and by what mechanism gene action controls them. In any case, since different enzymes appear at different time points in the development or maturation of the endoplasmic reticulum membrane, a one-step, self-assembly process can be ruled out.

CONTRACTILE FILAMENTS: ACTOMYOSIN

Paracrystalline organization, so evident in collagen, is seen to perfection in the highly organized, yet motile structure of muscle fibrils and other contractile cell structures. In muscle cells the chemical energy of ATP is converted to mechanical energy, involving the interaction of two protein macromolecules, actin and myosin. X-ray diffraction studies and electron micrographs show that the contractile structure of striated muscle is built up from overlapping arrays of actin and myosin filaments arranged in cross section in double hexagonal lattice patterns. The thick myosin-containing filaments are situated at the lattice points of the hexagonal pattern, and the actin filaments are symmetrically placed at points between them. In addition, a helical array of cross-bridges projects from the surface of the thick filaments with a regular spacing, or pitch. The fine structure of flight muscle in insects is similar except that the cross-bridges have a different pitch. Each such projection ends in a globular region associated with ATP activity. In the two dimensions of a cross section, therefore, an organized crystalline structure is evident, with precise spacing of the components. In the remaining dimension, i.e., the longitudinal axis of the fiber, there is freedom to move; and the actin and myosin filaments slide relative to one another, the energy necessary to the process being supplied by ATP. Hydrolysis of ATP to ADP by actomyosin is accompanied by dissociation of this protein complex into actin and myosin. These components then recombine when the ATP is exhausted. We are concerned here not with the energetics, however, but with the phenomena of assembly and disassembly associated with a complex of two species of protein acting together.

Actin exists in two forms, globular or G-actin and fibrous or F-actin, which are interconvertible. The globular form is a monomer which is stable as such in the absence of electrolytes, but which polymerizes to form the double-helix F-actin when neutral salts are present in the surrounding medium. The globular form has a molecular weight of about 60,000 and measures about 300×30 Å, but in the presence of salts and ATP these units aggregate to form a filament about 1μ long with a molecular weight of 40 million. Myosin is a long, thin protein consisting of two parts—a more globular, heavier subunit and a lighter, helical subunit—which mainly determine the general shape and solubility properties of myosin. Together the two subunits assemble to form spindle-shaped aggregates in media of low ionic strength. Myosin threads by themselves will not contract in the presence of ATP, but they will do so if actin is present as well.

The exquisite paracrystalline organization of the whole assembly is clearly as necessary to the contractile function as the energy basis associated with ATP-ADP interchange. It is also clear that we are dealing with a rapidly reversing equilibrium between a monomer and a high-molecular-weight polymer, and that the state of equilibrium is closely dependent on the ionic strength and pH of both the natural and the experimental media, as the case may be. The solvent plays the main role

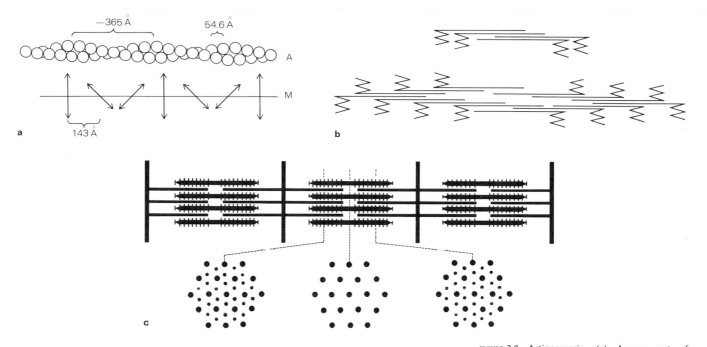

in stabilizing the protein structure. The main driving force of polymerization most likely is the formation of ionic and hydrophobic groups which become transferred from contact with the active phase into the interior of the polymer. Consequently whenever large structures are built up from smaller protein subunits, the process should be accompanied by increase in volume and entropy because of reduced contact of protein surfaces with the aqueous medium. As foreseen many years ago, the structurization of water itself plays a vital role in the organization and operation of living fibrillar systems.

We have seen that the information carried in the amino acid sequence of a polypeptide chain dictates the spatial arrangement of the chain within the tertiary structure. More indirectly the same information must provide for the assembly of the folded protein molecules into higher-order aggregates, for the high-weight polymers are not simply random assemblies of the monomers. The organization of subunits to form a specific superstructure is systematic, commonly leading to the formation of helical structures whose gross organization is generally directed by the ionic conditions of the medium during the self-assembly process. Remarkably, however, only three stable polypeptide conformations—the alpha helix, the collagen conformation, and the beta pattern—are used to construct the highly organized structures of a large range of fibrous tissues. From these basic molecular conformations, three-dimensional networks of widely varying properties and functions are built up.

FIGURE 2.8 Actinomyosin. (a) Arrangement of G-actin monomers in actin filament (A). The pitch of the helix and the subunit repeat differ from those of myosin (M). (b) Aggregation of myosin molecules to form filaments whose structural polarity reverses at midpoint; the straight components form the backbone of the filaments, the folded components the cross-bridges. Orientation in opposite senses in the two halves could generate sliding forces toward the center. (c) General structure of striated muscle, showing overlapping array of actin- and myosin-containing filaments. [*After Huxley, copyright 1969 by the American Association for the Advancement of Science.*]

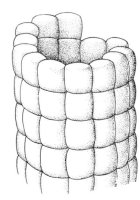

FIGURE 2.9 Model for the assembly of a microtubule. The globular subunits assemble in a helical array, yet stack up one on top of another so that the wall of the microtubules is made up of linear strands. [*After Tilney, 1968.*]

MICROTUBULES

Microtubules may well be the most widely distributed of all cellular filaments. They occur adjacent to the cell wall of plant and many animal cells, in various elongate cells such as neurons, in projections from cell surfaces (as in axopodia, cilia, and flagella), and as the fibers of mitotic spindles. As many as 3,000 such fibers have been counted in a cross section. They are undoubtedly important in the overall strategy of the cell.

Apart from the many functions microtubules appear to serve, we are concerned here mainly with their nature and assembly. Typically they are straight, hollow cylinders approximately 240 Å in diameter. In some cases they have been resolved into 12 or 13 subfilaments. However, the basic components are globular units of about 40 Å in diameter, which seem to be stacked and arranged at a slight inclination, to form a hollow helical array.

These subunits, or monomers, are a protein, provisionally named *tubulin,* a word which states where it is found but not what it is. All of these monomers so far analyzed have a molecular weight of 60,000 and essentially the same amino acid composition; in these respects and in some other features they closely resemble the actin subunits of muscle. If tubulin and actin monomers are identical or at least very closely related proteins, it is notable that virtually the same subunit can be employed in the assembly of such differently organized and functioning structures as microtubules and muscle fibrils. As is true of collagen and myosin, the assembled microtubule can be disassembled into its subunits and reassembled.

Microtubules show assembly at two levels: **(1)** the assembly of a microtubule from monomer or dimer subunits, and **(2)** the association of a number of microtubules as an organized complex comparable to the oriented spacing of actin and myosin filaments in a muscle fibril. They have been studied mainly as components of bacterial flagella, as the cilia of ciliates and flagellum of sperm, and as axopodal elements of heliozoans. The microstructure and chemistry of cilia and flagella are fundamentally the same throughout the animal and plant kingdoms. Only the flagella of bacteria are basically different. They are also much simpler.

In flagellated bacteria the flagella consist of microtubules about 140 Å in diameter which resolve into 8 to 10 longitudinal strands of subunits named *flagellin.* They are accordingly smaller than the microtubules constituting flagella and other structures in eucaryote cells. Bacterial flagella can be disassembled to produce a solution of flagellin monomers, from which new flagella may reassemble spontaneously under certain conditions. Reassembly occurs more readily if small bits of native flagella are present in the solution to serve as seeds, or nuclei, recalling the initiation of crystals by comparable seeding with small crystals in concentrated solutions of particular inorganic substances. If the flagellin solution and the fragments of native flagella are both from bacteria with straight flagella, the reassembled flagella are straight. If, however, the flagellin solution is from straight flagella but the seeding pieces are from bacteria with curly flagella, the reconstituted flagella are curled. Thus the nature of the assembly depends in part on the innate properties of the subunits but also partly on the conformation of the base on which they assemble, when such a base in present.

Among the great diversity of single-celled protozoan organisms, some offer exceptional opportunity for the study of microstructure, for they often combine relatively great size with high complexity and can be readily cultured in the laboratory. Heliozoans (with their stiff-rayed axopods) and suctorians (with their feeding tentacles) have been exploited with regard to microtubule assembly and array. In the heliozoan *Actinosphaerium*, numerous long pseudopodia, or axopodia, 5 to 10 μ in diameter at the base, extend as much as 400 μ into the surrounding water. If prey adheres to an axopod, it is carried to the cell-body surface by a process of melting or withdrawal of the axopod material, after which a new axopod forms. In other words, during the process of obtaining food, slender extensions from the cell surface are continually being assembled, disassembled, and reassembled.

A cross section through an axopod, particularly near its base, reveals a remarkable structure, the axoneme, consisting of numerous microtubules arranged in two interlocking coils or spirals, the number of constituent tubules decreasing as sections are cut progressively toward the axopodial tip. Individually each tubule has the standard dimensions already described. Apart from the double spiral pattern of its arrangement of constituent tubules, the spacing of tubules is comparable in preciseness to that of the filaments of the actomyosin fibrils of muscle cells. Each tubule is separated from its immediate neighbors by 70 Å, and each coil from adjacent coils by 300 Å. Single tubules appear to traverse the entire length of the axopod.

Experimentally it is found that subjecting the live cell to hydrostatic pressure, low temperature, or the antimitotic drug colchicine causes the axonemes to disappear, with corresponding resorption of the whole axopod into the cell body. During the resorption process electron micrographs show no sign of the microtubule structure, either in array or individually. The protein subunits clearly disassemble. When normal conditions are restored, new microtubules rapidly reassemble from their subunits and assume the characteristic regular spiral spacing. The reassembly and arraignment precede the actual outgrowth of axopodia and therefore may cause such outgrowth to take place at their tips. Evidence indicates that the same monomers are used over and over during the recurrent process of normal axopodial formation by the cell. The actual assembly of monomers to form a tubule has been compared to the assembly of subunits to form the hollow stem of the TMV virus. The spacing to form the double coil of tubules is probably attained through the presence of short and long links between tubules, although these have not been clearly seen in the electron micrographs.

Microtubules may or may not form crystalline arrays of the sort just described. They have been found in most types of cell so far examined where cell shape departs radically from the spherical or cuboidal, and the microtubules may themselves be

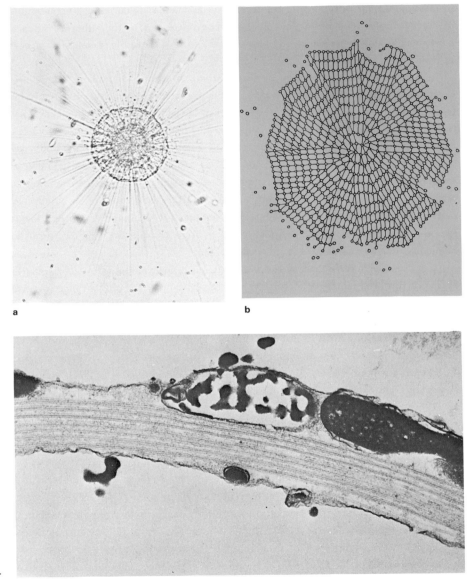

FIGURE 2.10 (a) The heliozoan *Actinosphaerium*. Slender axopodial processes radiate from the cell body. Within each axopodium is a birefringent core, or axoneme. [*Courtesy of L. G. Tilney.*] (b) Microtubule array of an axopodium. Cross section of normal axopodia shows double spiral or helical arrangement of microtubules. [*From Tilney, 1968, after MacDonald and Kitching.*] (c) Longitudinal section shows a re-forming axopodium after 10 minutes of recovery following dispersion of subunits by ultrasonic treatment. [*Courtesy of L. G. Tilney.*].

mainly responsible for the development and maintenance of the form. The natural shape of a free body with liquid properties is spherical, as seen in a suspended drop of water or oil or in a soap bubble. When free bodies are clustered together, again as seen most readily in a mass of soap bubbles, the exposed surfaces remain curved but adjoining surfaces become more or less flat. Surface tension and mutually adhesive forces are responsible for these configurations.

When other shapes appear, as in most cells, then other agencies must exist which either produce or maintain the distortions. Microtubules seem to be the most likely agent. They are commonly seen to be lined up parallel with and close to cell walls and in general to conform to whatever particular shape a cell may exhibit, as already noted for *Actinosphaerium* axopodia and suctorian tentacles. They are also seen in the disclike, nonnucleate blood platelet. The discoidal form of platelets is produced by a band, the marginal band, which encircles the cell and is composed of microtubules. When platelets are subjected to low temperature, the microtubules depolymerize and the cell assumes a spherical form. When the higher temperature is restored, microtubules begin to reappear, the marginal band gradually forms as the tubules associate with one another, and the discoidal shape returns. That microtubules are alone responsible for developing and maintaining particular cell shapes is at present a likely possibility, the evidence being mainly circumstantial. In any case, what determines the location, quantity, and particular orientations of the microtubules, relating to cell shape, remains unknown.

Microtubules, however, are not merely structural. They have long been known as the basic active elements in all cilia and flagella except those of bacteria, all of which exhibit a rhythmic contractile beat. All have a diameter approximating 0.2 μ and accordingly represent an assembly much smaller than the axopod of *Actinosphaerium*. All exhibit in cross section the familiar 9 + 2 array of tubules with remarkable uniformity. The ciliary membrane is continuous with the plasma membrane of the cell and shows the same unit structure. The complex of tubules, or axoneme, consists of two central tubules and a ring of nine outer double tubules that extend continuously throughout the length of the organelle.

The two central tubules are enclosed by a sheath. Each of the nine outer units, however, consists of two closely adjoined microtubules, with a pair of short arms extending from each pair in a clockwise direction as seen looking toward the base of a cilium. Further, the outer and central units are held together at frequent intervals along the length of the axoneme by radially oriented links which are joined to one another longitudinally by secondary fibrils.

Altogether we see an array very different from that of the *Actinosphaerium* axopodium, yet comparable to it. Here again each microtubule, at each of the outer rings of doublets, appears to be composed of between 10 and 13 chains of globular subunits, forming protofilaments. Experiments in the isolation and fractionation of cilia from ciliate protozoans such as *Tetrahymena* and *Paramecium* show that to a considerable extent they can be disassembled and subsequently caused to reassemble, at least in part. Evidence suggests that the contractile components are the pair of projecting arms of each doublet tubule, and that they consist of or contain an ATP-sensitive, myosinlike protein similar to the so-called myxomyosin associated with the streaming movements in noncellular slime molds, but distinct from contractile proteins related to the actomyosin.

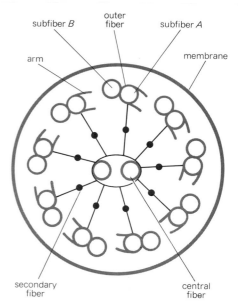

FIGURE 2.11 Diagram of structure visible in transverse section of a cilium, showing typical 9 + 2 organization of microtubules with arms and secondary fibers. [*After Gibbons, 1967.*]

NUCLEATION CENTERS

Cilia and flagella are associated with basal bodies. In animal cells, at least, the microtubular mitotic spindle is associated with centrioles similar to the kinetosomes, or ciliary basal bodies, although larger. In flagellate protistans the centriole does double duty as flagellar basal body and mitotic center. Apart from any capacity the microtubule components may have for self-assembly, disassembly, and reassembly, the basal bodies and centrioles serve as nucleation centers for microtubule production.

Basal bodies and centrioles have essentially the same structure and function. Each consists of a short microtubule array of nine triplets, without the two central tubules characteristic of cilia and flagella as such. The basal body undoubtedly controls the production of the associated cilium or flagellum, as the case may be; the tubules of each doublet in the outer ring are continuous with two of the three tubules forming a triplet in the basal body. The development, or assembly, of the basal body has been followed to some extent. The earliest stage consists of nine singlet microtubules surrounding a central cartwheel-like system of fibrils; it is regarded as a distinct developmental stage, since it occurs only at stages when new basal bodies are being formed. The second and third microtubules are added subsequently. The first and second later extend as a doublet into the cilium or flagellum. Evidence indicates that growth involves sequential addition of material to the tip of the axoneme, beginning at the basal body.

Basal bodies may contain both DNA and RNA. The evidence is questionable. Because of their almost identical structure and their capacity for replication the centrioles may do so as well. There is no evidence that the basal bodies contain ribosomes. On the other hand, studies of mutants based mainly on the protozoan *Tetrahymena* show that chromosomal genes are involved in cilia formation. Accordingly there is probably much the same control system as in mitochondria and plastids, with the basal body DNA-RNA coding for at least some aspect of the ciliary assembly (possibly the organization of the array of microtubules) and with chromosomal DNA perhaps coding for the structural protein subunits.

It is significant that when outer doublets of the ciliary flagellar axoneme are isolated, depolymerized, and allowed to redevelop in vitro without the basal body, only single microtubules are formed. Axoneme growth is coincident with plasma-membrane protrusion, and current opinion is that a cilium or flagellum grows from the tip, perhaps requiring an enzyme associated with the tip membrane. In any case the basal granule, whether of cilium or flagellum, may be regarded as a precisely organized semiautonomous organelle that is still imperfectly understood.

Centrioles are also centers associated with the microtubule formation, in connection with either microtubule assembly or orientation, specifically with the asters and spindle of the mitotic-division apparatus of the cell. Because of their small number as well as small size they are less amenable to analysis than the basal bodies studding

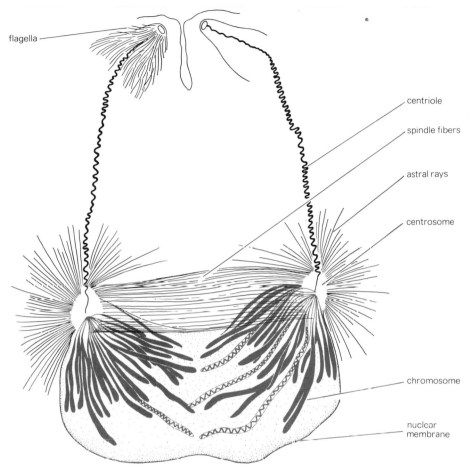

flagella

centriole

spindle fibers

astral rays

centrosome

chromosome

nuclear membrane

FIGURE 2.12 Late division stage of the termite flagellate *Pseudotrichonympha*, showing fiber systems associated with centrioles. The anterior end of the cell has already divided, and from each part a long, coiled centriole extends posteriorly to the end of a large central spindle. Distal ends of the spindle centrioles are surrounded by a centrosome, through which astral rays, chromosomal fibers, and central spindle fibers pass. [*After Cleveland, 1957.*]

the surface of ciliate protozoans. Yet they are essential to the division process, and they also are unmistakably self-replicating bodies. Somehow, by means entirely unknown and not even conceived, a mature centriole gives rise to a new, smaller centriole which appears at right angles to the parental structure, usually in keeping with the division cycle of the cell as a whole.

Although centrioles operate as organizational centers for aster and spindle formation during mitosis, their role is less clear and also less precise than the role of basal bodies relative to cilia formation. During the assembling and arranging of microtubules around a centriole, for example, some tubules approach the centriole at an oblique angle and some may miss it altogether. The whole phenomenon of cell division is in fact both very complex and, when protistan organisms are included, very variable indeed. Suffice it to say here that cell division typically involves a process of mitosis. It is concerned primarily with nuclear division. It is under the influence and possibly under the direct control of centrosomal centers that may or may not contain a discernible centriole. It also is a process of cleavage of the cell as a whole, involving growth of and constriction by the plasma membrane and peripheral protein filaments, which divide the cytoplasmic body. Two fundamental processes in fact appear to be involved in the complete division of a cell: **(1)** replication of all organelles, primarily of the nucleus itself, together with mitochondria, plastids, and basal body, i.e., the (9 + 2) assemblies (all these possess nucleic acids to some extent); and **(2)** an incisive extension of the plasma membrane. These two processes normally accompany each other temporally but are nevertheless essentially independent.

REASSEMBLY OF A CELL

The assembly process demonstrated in the experiments with virus particles suggests that comparable assembly processes might be possible with cells. Such assembly has been performed with the common amoeba, *Amoeba proteus,* by Jeon, Lorch, and Danielli (1970), who have shown that new, viable amoebae may be produced with the membrane of one cell, cytoplasm from one or more other cells, and the nucleus from a third cell. None of these components is viable by itself, and when any one of them is missing the reassembled cell does not live.

The basic procedures of reassembly are:

1 Removal of the nucleus from an amoeba
2 Removal of the cytoplasm to the extent that the remaining part cannot survive even when a nucleus has been inserted
3 Injection of the desired cytoplasm to refill the above
4 Insertion of a nucleus

The procedures used for the first two steps consist of enucleation and removal of cytoplasm either by microsurgical methods or by centrifuging cells at high speed. By these means the original membrane can be emptied of 90 percent of its original cytoplasm, together with its nucleus. The refilling with new cytoplasm and a new nucleus are performed microsurgically. When all components come from the same strain of cells, reassembly is said to be easy; 80 percent of the reassembled amoebae behave and reproduce normally and are indistinguishable from normal amoebae of the same strain. Accordingly we now have the technical ability to assemble amoebae which contain any desired combination of components. This system may be used, for instance, to test the viability of nuclei or mitochondria that have been exposed to various treatments, to investigate the compatibility of components reassembled from different known genetic strains, or to examine the effect of cell aging on cell components.

CONCEPTS

Eucaryote cells, in contrast to one-envelope-system procaryote cells, are essentially double-envelope systems in which internal membrane is extensively developed for the dual purpose of segregating special activities and controlling diffusion.

Increase in cell size is accompanied by decrease in the surface/volume ratio. The development of internal membrane systems may be a compensatory response to the consequences of such decrease in ratio of surface area to volume.

The membrane of mitochondria contains ordered sequences of respiratory enzymes. It is possible that all membrane contains or is associated with enzyme assemblies.

Highly organized cell components when depolymerized are capable of reconstitution through self-assembly processes. Such assembly is accompanied by volume and entropy increases because of reduced contact of protein surfaces with aqueous solvent.

Discrete self-replicating organelles such as mitochondria, chloroplasts, and basal bodies contain DNA and RNA. They are semiautonomous inasmuch as their genes direct the synthesis of some but not all of their structural components.

Protein filaments and tubules, both static and motile, may be assembled, disassembled, and reassembled during cell activities.

Microtubule orientation determines or is associated with the shape of cells.

There is a constant turnover of the molecular and macromolecular constituents of cell structures.

READINGS

ADELMAN, M. R., et al., 1968. Cytoplasmic Filaments and Tubules, *Fed. Proc. Amer. Soc. Exp. Biol.,* **27.**

BOGORAD, L., 1967. The Organization and Development of Chloroplasts, in J. M. Allen (ed.), "Molecular Organization and Biological Function," Harper & Row.

CLEVELAND, L. R., 1957. Types and Life Cycles of Centrioles of Flagellates, *J. Protozool.,* 4:230–241.

DIPPELL, R., 1962. Ultrastructure of Cells in Relation to Function, in W. H. Johnson and W. C. Steere (eds.), "This Is Life," Holt, Rinehart and Winston.

EHRET, C. F., 1960. Organelle Systems and Biological Organization, *Science,* **132** (July 1).

GIBBONS, I. R., 1967. The Organization of Cilia and Flagella, in J. M. Allen (ed.), "Molecular Organization and Biological Function," Harper & Row.

HARRINGTON, W. F., and R. JOSEPHS, 1968. Self-association Reactions among Fibrous Proteins: The Myosin ⇌ Polymer System, *Develop. Biol.,* suppl. **2.**

HUXLEY, H. E., 1969. The Mechanism of Muscular Contraction, *Science,* **164:**1356–1366.

JEON, K. W., I. J. LORCH, and J. F. DANIELLI, 1970. Reassembly of Living Cells from Dissociated Components, *Science*, **167**:1626–1627.

LEHNINGER, A. L., 1967. Molecular Basis of Mitochondrial Structure and Function, in J. M. Allen (ed.), "Molecular Organization and Biological Function," Harper & Row.

NASS, M. M. K., 1969. Mitochondrial DNA: Advances, Problems and Goals, *Science*, **165**:25–35.

PICKEN, L., 1960. "The Organization of Cells," Oxford.

PITELKA, D. R., 1969. Centriole Replication, in "Handbook of Molecular Cytology," North-Holland Publishing Company, Amsterdam.

RACKER, E., 1968. The Membrane of the Mitochondrion, *Sci. Amer.*, February.

RANDALL, J., et al., 1967. Development and Control Processes in the Basal Bodies and Flagella of *Chlamydomonas reinhardi, Develop. Biol.*, suppl. **1**.

ROBBINS, P. R., 1968. The Biochemical Organization of Cytoplasmic Membranes, *Develop. Biol.*, suppl. **2**.

ROBERTSON, J. D., 1967. The Organization of Cellular Membranes, in J. M. Allen (ed.), "Molecular Organization and Biological Function," Harper & Row.

TILNEY, L. G., 1968. The Assembly of Microtubules and Their Role in the Development of Cell Form, *Develop. Biol.*, suppl. **2**.

WOODWARD, D. O., 1968. Functional and Organizational Properties of Neurospora Mitochondrial Structural Protein, *Fed. Proc. Amer. Soc. Exp. Biol.*, **27**.

CHAPTER THREE

THE CELL CYCLE

The living cell is never static. In its immature stage it usually is in the process of either growth, leading to another division, or cytodifferentiation, leading to specialized structure and function. At the least it is in continuous process of maintaining itself, which also is a form of growth since the cell, as an open-ended steady-state system, is a flux of material moving into and out of the organization.

CONTENTS
Energy
Size
Growth
Division
Cycle
Cloning and synchronization
Concepts
Readings

ENERGY

The cell at its own level of organization is a unity, although our mental and experimental equipment is better able to cope with it as a plurality of distinctive organelles. The reciprocal relations between the parts determine the character of the whole, just as the essence of a clock lies in the organization of its components, or the nature of an atom results from the relations between orbital electrons and its nucleus. All are dynamic entities in which both order and energy are essentially one. Energy for the chemical and mechanical work of cells is obtained mostly from the partial or complete oxidation of carbohydrates, which are also the source of carbon skeletons in macromolecular synthesis. All cells must convert part of their carbohydrate supply into compounds such as ATP or NADH, for such energy donors are essential to the joining of carbon skeletons and to their modification. Yet different cell types vary in their use of carbohydrate, and they also vary the pattern of its use during their development; growing cells need a large supply because they are synthesizing new constituents, while, in contrast, muscle cells need little carbon for synthetic purposes but require great amounts of ATP as fuel for mechanical work.

All exchange with the immediate cell environment is necessarily effected or controlled by the plasma membrane in various ways: diffusion of simple molecules such as oxygen, carbon dioxide, and water; electrolyte pumping, as in the sodium pump; membrane-enzyme-controlled transfer; and mechanical intake by pinocytosis or output by evaginative vesiculation. The ratio of total exchange surface to enclosed mass of metabolically active material clearly is vital.

SIZE

Cells are characteristically of a certain size, each according to its specific nature and to the class of organism to which it belongs. As a rule the cells of plants are larger than those of animals. The cells of amphibians are larger than those of mammals, tissue for tissue. In all vertebrates liver cells are much larger than blood cells. Among protistans the range in size is enormous. Yet each kind of cell, specific for tissue and for species, remains within certain narrow limits, since when a cell divides into two, the daughter cells have roughly half the size of the parent cell, while the daughter cells grow to approximately double that size before they in turn divide. Accordingly in the case of cells that continue from one division to another, size fluctuates between a certain minimum and a maximum twice that value. The size characteristic of a particular kind of cell is determined both by its species specificity and its tissue type. Little is known concerning the nature of this determination.

Cell size is related to nuclear content, irrespective of type. Thus, larval kidney cells, lens epithelial cells, and red blood cells, etc., of haploid, diploid, and triploid salamanders, respectively, vary in size directly with degree of ploidy, depending on whether the nucleus has one, two, or three sets of chromosomes. In certain giant cells, large size is correlated with the presence of numerous nuclei, e.g., striated muscle, and giant species of *Amoeba* and *Paramecium.* In other giant cells, such as eggs and certain large unicellular algae such as *Acetabularia,* the single nucleus is both extremely large and exceptional in other ways.

Whatever the cell may be, a certain characteristic size prevails. When a cell attains its particular maximum, it may divide or it may persist at this maximum for a period of time and finally die. Either event, however, can be postponed almost indefinitely by preventing a cell from attaining its maximum size. Injured cells usually repair themselves quickly. If a large part of an amoeba is cut off, the part retaining the nucleus rapidly heals over and continues to grow. As long as this procedure is repeated, healing and growth continue but cell division fails to take place. Only after growing to the critical maximum size does division occur. Evidently a nucleus with a particular chromosomal content can support a certain maximum mass of surrounding cytoplasm. When this mass is exceeded, cell division is triggered. A factor in this triggering may be the changing ratio of cell surface area to cell volume.

The significant relationship appears to be the nucleocytoplasmic balance, each component of the system influencing the other, the nucleus as the controlling center and agency for primary production, and the cytoplasm as the region of terminal productions and feedback to the nucleus.

GROWTH

Because of the spectacular linear growth of neurites, particularly the axon of neurons, nerve cells have long served for studies of cell growth, beginning with Ross Harrison's study of the outgrowing processes of isolated cells taken from the spinal cord of salamander embryos. Using his hanging-drop preparation, a technique that laid the foundation for the present vast field of tissue-culture experimentation, he was able to see the early stages of such outgrowths free of the influence of adjoining tissue: "These observations show beyond question that the nerve fiber develops by the outflowing of protoplasm from the central cells. This protoplasm retains its amoeboid activity at its distal end, the result being that it is drawn out into a long thread which becomes the axis cylinder. No other cells or living structures take part in this process. The development of the nerve fiber is thus brought about by means of one of the very primitive properties of living protoplasm, amoeboid movement, which, though probably common to some extent to all the cells of the embryo, is especially accentuated in the nerve cells at this period of development."[1]

The growth of the nerve-cell axon is probably but a special case of cell growth, exceptionally amenable to experimentation, and the territorial sequence of protein synthesis, from chromosomal DNA to polysome activity in the cytoplasm, is from the nucleus outward through the nuclear membrane, with membrane and other material displaced progressively outward as new membrane and material continue to be produced near the nuclear surface.

The mature neuron generates neuroplasm continuously in its cell body and then conveys it into the nerve fiber as a cohesive column advancing at a daily rate of about 1 mm (axonal flow). This has been established by experiments of Weiss resulting in "damming" of flow in constricted nerve, supplemented by electron-microscopic and cinematographic data. Neurotubules and neurofilaments extend in linear, latticelike arrangement down the length of an axon. Employing tritiated hydrogen as a radioactive marker, both Leblond and Weiss have followed the course of labeled amino acids as a crest of protein traveling down the length of an axon. This is not a stream of substance *inside* a stationary axon but a movement of the *whole axon itself* as a semisolid column.

These and many other experiments have led to the concept, stated by Weiss, that *the mature neuron is in perpetual growth and grows forth as axon, consuming its substance on the way.* In fact from 6 to 11 percent of the cell protein moves out daily from the cell body, amounting to an average renewal rate of the whole solid content of each central nerve-cell body almost once a day. Weiss had earlier made the following inferences from the phenomenon of axonal flow:

1 The macromolecules of the neuron, particularly enzymes and other proteins, do not persist indefinitely, but on the contrary, their population undergoes continuous degradation and dissipation.

2 Their renewal through the manufacture of new macromolecules, especially protein, does not occur ubiquitously, but is strictly a preserve of the eccentric cell body around the nucleus.

3 Accordingly, the enzyme and structural protein requirements of the axon, depleted by steady dwindling along the line, must be steadily replenished from that localized central source if the whole unit is to be kept alive.

4 Since the mass of a long axon can be more than 100 times as great as that of its cell body, the rate of that steady supply stream must bear a direct relation to the rate of consumption in the axon.

[1] R. G. Harrison, Observations on the Living Developing Nerve Fiber, *Proc. Soc. Exp. Biol. Med.,* 4:142 (1907).

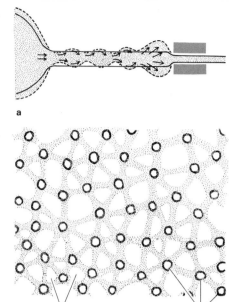

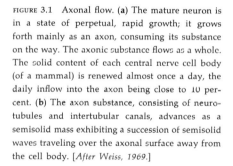

FIGURE 3.1   Axonal flow. **(a)** The mature neuron is in a state of perpetual, rapid growth; it grows forth mainly as an axon, consuming its substance on the way. The axonic substance flows as a whole. The solid content of each central nerve cell body (of a mammal) is renewed almost once a day, the daily inflow into the axon being close to 10 percent. **(b)** The axon substance, consisting of neurotubules and intertubular canals, advances as a semisolid mass exhibiting a succession of semisolid waves traveling over the axonal surface away from the cell body. [*After Weiss, 1969.*]

Besides the 1-mm flow rate, a second faster rate of centrifugal traffic in nerve fibers, from 10 to 100 times as fast, courses through the intraaxonal channels as a flow within flow, carrying centrally synthesized transmitter substances to their terminal action stations at neuromuscular and neuroneural junctions.

Stimulation of growth in nerve cells, expressed either as growth and cell divisions in embryonic cells or as enhanced neurite outgrowth in nondividing nerve cells, can be accomplished by the so-called nerve-growth factor. This is a protein that has been isolated in pure form from snake venom and from submaxillary salivary glands of mice; it has a constitution similar to that of serum albumin and a molecular weight of at least 20,000. It exerts a marked stimulation of mitosis (i.e., cell growth and multiplication) of sympathetic nerve cells of embryos of mice and chicks, resulting in a manifold increase in the volume of nerve tissue. Its presence in an embryo is necessary for normal development of the nervous system. Increased RNA synthesis becomes evident within 2 hours, followed by increase in protein and DNA synthesis. The nerve-growth factor is consequently said to derepress (activate) genes that regulate macromolecular synthesis.

A more recently discovered nerve-growth factor, which may be different from the first, affects the outgrowth of neurites rather than the proliferation of cells, and is some million times more active. In fact the amount of this nerve-growth factor which evokes the appearance of 100 neurites from a single ganglion, or nerve-cell cluster, seems to be about 10 molecules, so that since each neurite appears to arise from a different neuron, each molecule of nerve-growth factor must affect several cells. It is likely, therefore, that proteins, either as such or as carriers, can act as regulators of metabolic functions in host cells and may be important in the control of growth and differentiation.

Macromolecules are now known to penetrate the cell membrane readily by means of membrane movements associated with vesicle formation, a process generally described as endocytosis, micropinocytosis, pinocytosis, or phagocytosis, depending on its dimensions. Endocytotic vesicles or vacuoles containing foreign macromolecules are believed to receive intracellular digestive enzymes by fusing with lysosomes. The ingested macromolecules mostly undergo intracellular digestion, but small fractions escape destruction and reach specific sites of action. Macromolecules tend to be adsorbed heavily to the surface of living cells, complete within seconds, much of which is taken up by a cell within the hour. Surprisingly, larger macromolecules are taken up more readily than smaller ones, in the case of both basic proteins and other compounds. Thus ferritin in the form of large aggregates is taken up faster than small aggregates, while supramolecular complexes of DNA penetrate cells more readily than molecular DNA solutions. Similarly antigens in particulate form or as complexes with RNA are more active than their soluble counterparts. Moreover, cell membranes possess specific sites that interact with particular proteins and other compounds prior to uptake.

RELATIVE RATE OF UPTAKE
OF DIFFERENT PROTEINS

Albumin	1
Ferritin	65
Lysine-rich histone	150
Poly-L-lysine	400
Crude histone	800
Arginine-rich histone	1,000
Poly-D-lysine	1,500
Poly-L-ornithine	2,000

DIVISION

When cells increase in size sufficiently they generally divide. Division is itself a necessary form of automatic regulation. If it did not occur, the ratio of the surface area to the internal volume would steadily drop and the supply of nutrient from the medium would become more and more inadequate to provide for the metabolic needs of the interior. A cell economy which is viable at one surface/volume ratio would become completely nonviable at another—an example of the "scale effect," a hazard well known in planning industrial chemical plants on the basis of laboratory experiments. As the cell becomes larger, some concentrations decrease while others increase, because diffusion both inwards and outwards becomes more difficult. The whole chemical balance is changed, and as a result something happens to trigger off the division of the cell. The details are not known, but it is significant that the mass of DNA per cell tends to remain roughly constant.

All cells have their individual period of existence, the cell cycle. They begin in the division of a parental cell, and they cease at the end of a period of maturation and special functioning, or else lose their individuality in becoming a pair of daughter cells. Single-cell organisms, the protists, maintain various degrees of differentiated structure throughout the process of division, and only partial reconstruction of new individuals is evident. In multicellular organisms, cells fall roughly into three categories:

1 Cells that retain full capacity for division and exhibit little or no special differentiation, serving principally as limiting membranes and/or as reserves for tissue replacement
2 Cells that exhibit considerable general differentiation, such as liver cells, the gastrodermal cells of hydras, and the photosynthetic cells of plants, but are still able to undergo division
3 Cells that when mature have lost their capacity to divide, notably cells with extreme structural or chemical specialization

Eucaryote cells divide by mitosis, a complex process involving chromosomal assortment typically associated with cytoplasmic cleavage. There is little doubt that the two events are closely and causally connected, although the nature of the connection is not known.

Mitosis, an obviously dynamic process that begins after DNA replication is complete, involves:

1 In *prophase,* the contraction or condensation of the chromosomes. Time-lapse photography shows a sudden contraction of the nucleus, followed by the disappearance of the nucleolus and nuclear membrane.
2 In *metaphase,* the final synthesis and arrangement of the spindle apparatus linking the chromosomes to the two centrioles.
3 In *anaphase,* the movement of the chromosomes toward opposite poles, with the kinetochores, which are the sites of spindle-fiber attachment, leading the way.

4   In *telophase,* the return of the chromosomes to their dispersive state, the re-formation of the nuclear membrane, and the reappearance of the nucleolus. Cytokinesis begins in anaphase and overlaps the mitotic anaphase and telophase.

In phase-contrast films the chromosomes seem to slide along the spindle fibers on their course toward the poles. A possible explanation is that the movement is caused by microtubules sliding past one another. According to this hypothesis, as the mitotic apparatus forms, microtubules of opposite polarity grow from opposite sides of the spindle to lie alongside one another. By anaphase, these tubules have slid sufficiently so that their head ends no longer overlap near the poles, while their

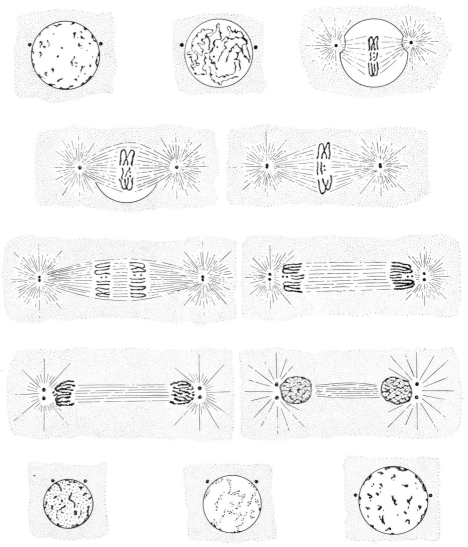

FIGURE 3.2   Mitosis in the early cleavage of *Drosophila,* illustrating the history of the centriole. [*After Huettner, 1933.*]

tails are pulled to the center cleavage furrow or cell plate, and then remain slightly overlapped. Accordingly there should be twice as many continuous microtubules near the equatorial cell plate as near the poles, while in metaphase the number should be equal in the two regions. Independent analyses confirm that such is essentially the case.

Since the total DNA thread in eucaryote cells is enormously long (that of a single human cell being estimated to exceed 1 m if fully extended), the DNA must necessarily be folded and stabilized in some carefully ordered manner, possibly determined by the chromosomal histones. Once DNA synthesis begins, however, it seems that the whole mitotic sequence must pass to completion, although it may be temporarily blocked before prophase. On the other hand, cells (sea urchin eggs) will continue cleavage if the mitotic apparatus is sucked out during metaphase.

Cell division underlies all the phenomena of multicellular development, inasmuch as cell differentiation usually begins as mitosis ends, and cell multiplication is a condition, though not a cause, of morphogenesis in the development of multicellular organisms. In any case division initiates a cell cycle; and in both tissue growth and egg development, as in the growth of protistan populations, one division follows another after an interphase period of shorter or longer duration. Two organizational states alternate—the mitotic and the interphase—and the question has arisen as to what extent new macromolecular materials must be formed and to what extent old material may be disassembled and reassembled. Obviously new material is necessary to account for the increase in size which occurs in cells during interphase, with the exception of eggs which are essentially giant cells that successively subdivide following fertilization. On the other hand the disappearance of internal membrane structures, notably the nuclear membrane and endoplasmic reticulum, during the onset of mitosis and their reappearance at the close, together with the regular formation and disappearance of the extensive fiber systems comprising the mitotic apparatus, suggest a use and reuse of protein building blocks for the sake of economy of substance and energy.

The relation between nuclear membrane and the endoplasmic reticulum is already fairly well established, the reticular membranes arising as stacks from the nuclear membrane and successively moving outward toward the cell periphery. Whether or not the plasma membrane derives from the same system is still debatable.

Nuclear membrane probably transforms continuously into endoplasmic reticulum during interphase, but the rate of transformation depends on the physiological state of the cell. The massive dissolution of the membrane which occurs at the onset of mitosis is an acceleration of this process. After the disappearance of the nuclear membrane (it converts entirely into reticular material), the endoplasmic reticulum as a whole becomes dispersed and stored in the cytoplasm for possible subsequent use. Following mitosis, daughter-cell chromosomes are at first devoid of a surrounding membrane. Elements of the endoplasmic reticulum gradually surround the chromosomes and eventually fuse to form a new continuous double nuclear membrane.

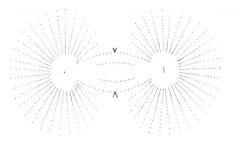

FIGURE 3.3 Diagram of giant resting first-cleavage metaphase spindle of the flatworm *Polychoerus*, to show centriolar orientation. [*After Costello, 1961.*]

Mitosis and interphase are in sharp contrast in virtually every way. Autoradiographic analyses show that during mitosis there are no DNA synthesis, a cessation of RNA synthesis, and a great decrease in protein synthesis. Lack of RNA and DNA synthesis probably results from the condensed state of the chromatin at mitosis. Reduced protein synthesis is another matter. Polyribosomes, necessary for protein synthesis, disaggregate during metaphase, and this disruption is clearly responsible. After mitosis the polyribosomes reassemble, while at the same time decondensation of the chromosomes and reestablishment of the nuclear membrane takes place. As the polyribosomes reappear, the rate of protein synthesis increases. The polyribosome assembly following mitosis occurs independently of any new RNA synthesis and evidently depends on the messenger RNA passed from mother cells to daughter cells.

Many efforts have been made to extract the specific molecules which are functional in mitosis, beginning with the isolation of the mitotic apparatus as a whole from dividing sea urchin eggs, employing controlled proteolytic digestion of the cell, by Mazia and Dan (1952). The problem has been to identify the protein, among the many extracted, which is the basis of the active structure. Over 3,000 microtubules have been counted in the spindle, each comparable to a single microtubule in the outer doublets of a cilium or flagellum. Treatment of isolated mitotic apparatus with mild hydrochloric acid dissolves less than 10 percent of the protein in the apparatus but results in the selective morphological disappearance of the microtubules, the same treatment being equally effective in dissolving the outer doublet tubules of sperm flagella. The extracted protein appears essentially the same in the two cases. This perhaps is not unexpected, since in certain protozoans, particularly the symbiotic flagellates living in the intestine of termites, e.g., *Barbulanympha,* centrioles serve as organization centers for both flagella and mitotic spindle fibers simultaneously.

However long the interphase of a cell may be, the mitotic cycle itself is passed through rapidly and has not been easy to study in the living state. The use of agents that interfere with the organization of gelated structures, such as the mitotic spindle, is now changing the situation, and experimentally controlled mitosis can now be studied combining polarization microscopy with phase and electron microscopy. Suitable material, as always, is crucial to what may be seen, and the newly spawned, unfertilized eggs of the marine polychaete worm *Pectinaria,* which remain in metaphase of the first maturation division for several hours, have proved satisfactory. Various chemical agents are known which cause reversible dissolution of spindle structure. One of these, the antibiotic griseofulvin, causes spindle disappearance within a few minutes, followed by rapid recovery when the cell is returned to normal seawater. Accordingly there is clear evidence that the spindle microtubules can be disassembled into their subunits and can quickly reassemble when normal conditions are restored.

Central bodies, or centrioles, are clearly involved in the organization of both spindle and asters in animal cells, especially in cleaving eggs. They are, however, notoriously difficult to study, and their fine structure has only relatively recently come to light, as described in the preceding chapter. Virtually nothing is known

GRISEOFULVIN, 10^{-5} M   SEAWATER

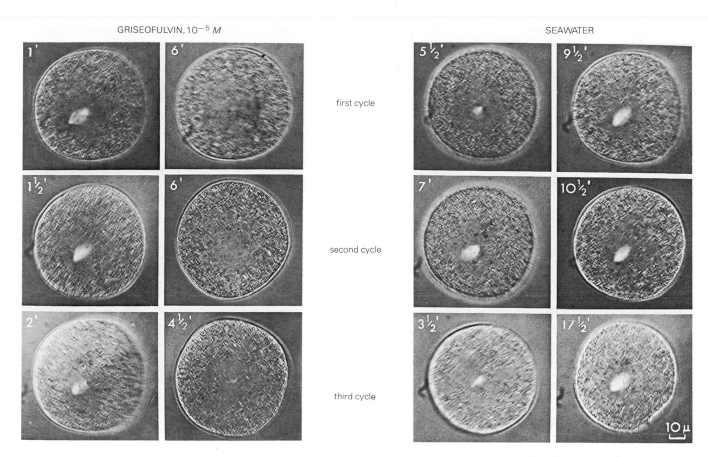

first cycle

second cycle

third cycle

FIGURE 3.4 Spindle dissolution in oocyte of polychaete *Pectinaria* when exposed to solution of the antibiotic griseofulvin in seawater, followed by reconstitution of spindle shortly after return to pure seawater. The experiment was repeated with the same egg three times. In each case spindle dissolution occupied about 5 minutes and recovery about 8 minutes. [*Courtesy of S. E. Malawista.*]

concerning the nature of their operation or of their chemistry. The difficulty arises from their extremely minute size and the impossibility of isolating them; nor are they visible in the living cell.

In nearly all cells mitosis is followed by cleavage, i.e., the cytoplasmic bulk divides into two parts (cytokinesis), each containing a daughter nucleus, together with a single or paired centriole in animal cells. This cytoplasmic division, or separation, is performed differently in the animal and plant kingdoms, however, although in both cases mitosis and cleavage appear to be closely and causally connected. The nature of the connection is not known. Most of the work on the process of cleavage in the animal cell has been done with sea urchin and salamander eggs. Earlier theories that visualized the mitotic apparatus itself as responsible for cleavage (through extension of the spindle or through mutual repulsion of the two expanding asters) are probably incorrect, since eggs are frequently able to cleave even when most of the mitotic apparatus has been destroyed.

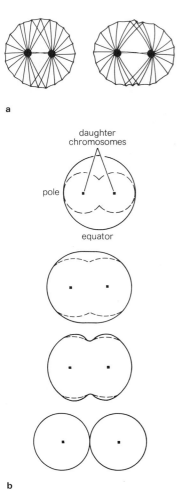

a

daughter
chromosomes

pole

equator

b

FIGURE 3.5 Models of cleavage. (a) Dan's theory: model shows two characteristic features of eggs in early part of cleavage, a bending of the polar astral rays to give a "fountain figure," and a flattening of the equatorial region, which is about to develop into a furrow. A "suction force" is evoked, in this theory, as a result of the absorption of equatorial cytoplasm by the two asters. (b) The expanding-membrane theory of Mitchison, whereby an initial expansion at the poles allows the equatorial region to contract and produce first an elongation and then a constriction.

Two current theories assign the active role in cytoplasmic division to the cell cortex: **(1)** A contracting ring of cortical material forms in the equatorial furrow marking the plane of separation of the daughter cells; and **(2)** the active parts of the cell surface are the polar regions, where an expansion in area is taking place. According to Swann and Mitchison (1958), daughter chromosome groups initiate molecular changes in the cell surface, leading to cortical expansion or growth. Robert Chambers, who introduced, developed, and exploited the micromanipulation of living cells during the first half of this century, showed in 1942 that a fragment of a membrane (in this case, the fertilization membrane) if strewed with carbon particles continues to expand in area; i.e., such expansion is intrinsic, not an inflation. Similarly, kaolin particles attached to the plasma membrane of a sea urchin egg show that a localized expansion occurs during cleavage, starting as a wave at each pole of the egg and spreading toward the equator. The furrow contracts initially and then greatly expands.

Preceding the actual appearance of the cleavage furrow, streaming movements arise in the cytoplasm at the two opposite poles of the mitotic apparatus, resulting in an elongation of the cell in this axis. These subcortical currents sweep around the two asters and add gelating material to the inwardly growing cortex, mainly at the floor of the cleavage furrow. Remarkably, the cleavage furrow will pass unchanged through a needle placed in its path.

Cleavage of a cytoplasmic surface following mitosis is seen in a wide diversity of cells, and there is evidence that the cleavage process differs considerably in different types. Little is known concerning the process in the relatively very small cells of animal tissues, although surface membrane expansion has been noted in the division of neuroblast cells. In sea urchin eggs, the expansion of the cell membrane and cortex during division without active equatorial constriction seems to be well established. In the very much larger eggs of amphibians, the furrow itself appears to be the active region, although as a result of growth, but again without contraction. In any case, simply converting one sphere into two spheres without any total volume change involves a 26 percent increase in surface area. Growth and expansion, however, may well be related phenomena.

Another type of egg in which the formation of the cleavage furrow has been closely studied is that of the squid. This is a molluscan egg of about the same volume as amphibian eggs but more elongated and much more yolky. The cleavage furrow in fact cuts into the egg only a comparatively short distance, beginning in the center of the superficial layer of cytoplasm, or blastodisc, and proceeding toward the edges. The opportunity here is that it is possible to obtain a temporal sequence by taking a series of sections of the furrow from the blastodisc edge toward the center. Electron micrographs show the earliest indications of a furrow appear as a flattened region with longitudinal surface folds running toward the furrow, and a dense layer of fibrils, each of about 70 Å in diameter, immediately below the plasma membrane running parallel to and below the furrow. Contraction of the fibrils, which are anchored at their ends, cuts through the cytoplasm. As the furrow divides the cytoplasm, most of the longitudinal surface folds unfold and add their plasma membrane to the newly forming surface of the dividing cells. Contracting fibrils can cut through the cytoplasm, however, only if the cell surface is curved, i.e., raised above the level of the cutting device. Experimental flattening of the egg surface either prevents furrow formation or stops it immediately if it has already begun. Release of pressure permits resumption of furrow formation almost at once.

Although the hypothesis of spindle elongation as a division force has been abandoned, together with that of mutual repulsion by the asters, the asters clearly play some role in establishing the cleavage furrow of a dividing cell. Recent experiments of Rappaport support the view that furrow establishment requires joint action of a pair of asters upon equatorial cell surface. "According to this mechanism, two dimensions are important in determining whether furrowing occurs. One is the distance from the spindle to the surface and the other is the distance between astral centers. The latter dimension is important because the size of the zone that can be simultaneously influenced by both asters decreases as the distance between asters increases. . . . When a block is placed between one aster and the equator, furrowing fails although all other geometrical relations are within normal limits. If the equatorial region must be affected by both asters before a furrow can develop, it should be possible to prevent furrowing by increasing the distance between asters. But asters that are so far apart that they cannot induce a furrow when the distance to the surface is normal should be able to establish a furrow in surface that is closer than normal to them. In this investigation the distance between asters and the distance from the asters of the surface were simultaneously altered before the position of the furrow was established. Very short spindle-to-surface distances were produced by perforating flattened cells; they allowed study of combinations involving large interastral distances and short spindle-to-surface distances."[1] It was found that asters spaced so far apart that they could not produce a furrow in surface located at approximately the normal distance could induce furrowing if the surface were abnormally close. Thus the deficiency occasioned by increase in one dimension was remedied by decrease in the other.

Both cleavage (cytokinesis) and chromosomal segregation (mitosis), and their interrelationship, remain topics of great importance and much obscurity. For instance, what makes sister chromatids almost invariably go to opposite poles during mitosis, as shown by tritiated thymidine—autoradiographic experiments with cultures of embryonic mouse cells? One interpretation is that when a conserved unit of a chromosome (that unit of a two-unit chromosome that has been synthesized at the immediately preceding replication period, its partner having been synthesized during an earlier replication) is used as a template for the first time, it is permanently attached to a structure distinct from that to which its parent chromosome was attached. Grobstein suggests that the centriole produced in preparation for the next mitosis is the structure

[1] R. Rappaport, Aster-equatorial Surface Relations and Furrow Establishment, *J. Exp. Zool.*, **171**:65 (1969)

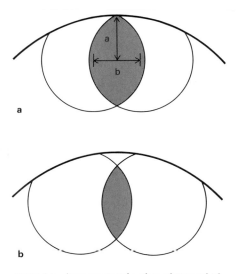

FIGURE 3.6 Aster-equatorial surface relations which would obtain if cleavage-furrow establishment were a consequence of joint action of the asters. (a) The area affected by both asters reaches the surface. (b) The effect of moving the asters apart is shown. Astral diameter and distance from the astral center to the surface are unchanged. [*After Rappaport, 1969.*]

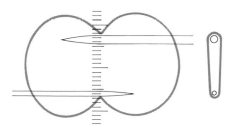

FIGURE 3.7 Direct measurement of maximum tension exerted by furrow of dividing sea urchin egg. The tension required does not exceed 1.5×10^{-3} dyne, which is no more than that of an actomyosin thread. Experiments show that the mechanism of division must lie in or very near the cell surface. [*After Rappaport, 1969.*]

to which all "conserved units" that are being used as templates for the first time are attached. Thus one set of chromatids with a similar history or time of synthesis would go to one pole, and the other set with a different history would go to the other pole. This is a plausible working hypothesis but is as yet only speculative.

Centrosomes, with centrally located centriole, appear to be vitally associated with the formation of both aster and spindle fibers in animal cells and in many protists, particularly flagellates. Asters clearly play a prominent role in cytokinesis in animal cells. It is puzzling, therefore, that cell division in multicellular plants is effected without centrosomal or astral systems. Spindles form, chromosomes segregate toward the spindle poles, and a new fibrillar wall, the phragmoplast, appears at right angles to the polar axis, between the two groups of chromosomes. Polysaccharide vesicles appear in the midline of the phragmoplast and coalesce to form the first structure of the new cell wall.

CYCLE

The general nature of the cell cycle is best seen in free-living cells, e.g., in *Gonyaulax,* one of the marine dinoflagellates notorious for causing "red tide." *Gonyaulax polyedra* is one of many microorganisms responsible for the luminescence of the ocean, the luminescence usually being observed only upon stimulation, such as agitation of the water, which causes the cells to emit bright flashes of light. Cultures have been maintained in the laboratory for many years.

In this organism, the several phases of the cell cycle follow an invariable sequence attuned to the light and dark of each 24-hour day. In brief, photosynthesis and growth occur during the daylight hours, luminescent glow is greatest about midway between dusk and dawn, i.e., with a peak from 11 to 12 P.M., while cell division occurs during the 5-hour period spanning the end of the dark period and the beginning of the light. The cycle as a whole is circadian and is maintained for 2 or more weeks in continuous dim light, controlled by a sort of internal rhythm or clock mechanism. A similar endogenous biological clock which "gates" the specific event of cell division in the cell developmental cycle is evident in populations of *Euglena, Paramecium, Tetrahymena,* and other protists.

The molecular events occurring during a cell cycle are difficult to analyze for a single cell or for a cell population if the cells are at various stages of the cycle. A synchronous cell population, i.e., one in which all cells are doing the same thing at the same time, is practically essential for many such studies. The noncellular slime mold, the myxomycete *Physarum polycephalum,* presents remarkable opportunity in this respect. It differs from most other organisms, including the cellular slime molds, inasmuch as the growth phase of the life cycle consists of a mass of protoplasm of indefinite form containing up to several million nuclei. This mass, which streams

FIGURE 3.8 A scanning electron micrograph of *Gonyaulax.* Magnification: ×2,000. [*Courtesy of Beatrice Sweeney.*]

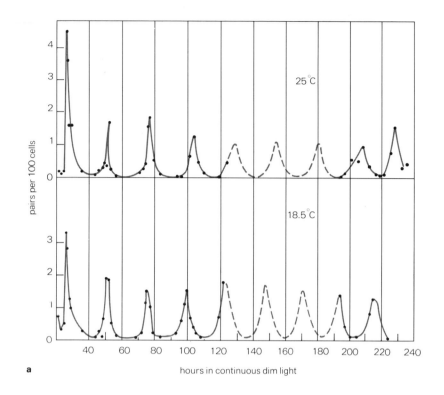

a

hours in continuous dim light

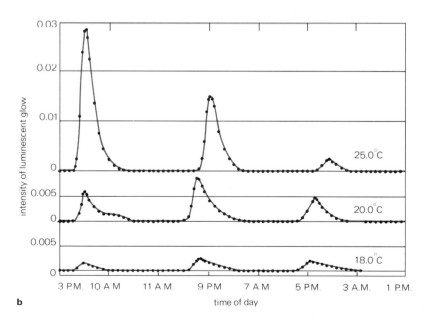

b

time of day

FIGURE 3.9 (a) Rhythm of cell division in *Gonyaulax* maintained in continuous dim light, as measured by percentage of paired cells present. (b) Rhythm of luminescent period maintained in darkness, for three different temperatures. [*After Sweeney and Hastings, 1958.*]

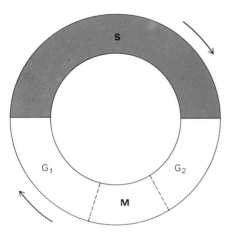

FIGURE 3.10 The cell cycle, or mitotic cycle, showing relative duration of phases in a growing cell. *S*, synthesis of DNA; G_1, the presynthetic phase; G_2, the postsynthetic phase; *M*, the period occupied by mitosis.

to and fro, is called a plasmodium, and in many of its properties it is like an enormous multinucleate amoeba, although biochemically it may be regarded as a single cell. As long as it continues to grow, the mass of cytoplasm remains continuous, bounded by a single continuous plasma membrane, although the nuclei divide every few hours.

Synchrony is attained if numerous small plasmodia, each containing fewer than 100 nuclei, are placed together on a flat surface in a culture medium lacking constituents necessary for growth. Such pieces coalesce; following the addition of growth media to the culture, the first synchronous division of the nuclei of the newly formed macroplasmodium occurs after about 5 hours. Subsequent mitoses occur at about 8-hour intervals. The questions which arise are: What molecular events lead to nuclear division, and what makes the divisions so highly synchronous? Possible explanations are that some regulatory biological clock within each nucleus or some cytoplasmic factor is responsible. Experiments combining two plasmodia which are out of phase with each other, and other experiments in which large and small pieces of two plasmodia are combined, respectively, show that the critical factor for mitosis is not formed in the nucleus and does not trigger mitosis at a predetermined time. The stimulator is accordingly believed to accumulate in the cytoplasm and to be transferred to the nucleus shortly before the onset of mitosis.

Experiments utilizing antibiotics, such as puromycin (which inhibits protein synthesis at the ribosomal level), actinomycin-D (which inhibits DNA-primed RNA synthesis), and others with similar action, show that all the essential structural proteins for both mitosis and nuclear reconstruction following telophase are completely synthesized by 15 minutes before metaphase. Further, all proteins determining the duration of mitosis are synthesized up until about 5 minutes before metaphase. The striking structural changes that accompany mitosis and nuclear reconstruction do not involve synthesis of new proteins after late prophase and are presumably brought about by conformative changes in proteins already present.

DNA replication in the plasmodium begins immediately after mitosis, continues for nearly 3 hours, and is synchronized at the molecular level. The rate of synthesis of RNA decreases during mitosis and during mid-interphase, with two peaks of precursor incorporation occurring during interphase. The entire sequence of DNA synthesis, DNA-primed RNA synthesis, mitotic protein synthesis, and the mitotic event is an invariable order and is evidently closely controlled. For easy reference the entire cycle is commonly subdivided into several phases: the mitotic phase, and an interphase consisting of three periods, the G_1, S, and G_2 phases. The S phase is the period of DNA synthesis. It is preceded by the G_1 and G_2 phases (short for gap 1 and gap 2).

Since DNA synthesis takes place almost entirely within the chromosomes, its production can be closely followed by means of radioactive tracers, notably tritiated thymidine. This is chosen because most or all of the cell's thymine is used in DNA synthesis and the available pool of this precursor is small during early interphase. The DNA synthesis thus takes place during the middle of the interphase, the S phase, preceded and succeeded by G_1 and G_2, respectively, when no DNA synthesis is detectable. These pre- and post-DNA synthesis stages have no definable positive qualities except that they seem to be partially independent of the S phase in duration. G_1, however, may be regarded as a period of synthesis and mobilization of the substrate and enzymes necessary for DNA production. It is the most variable part of the mitotic cycle; i.e., changes in cycle duration are reflected in the time of increase or decrease of the G_1 phase. G_2, the period between the end of DNA synthesis and "prekinesis," probably represents the period of production of spindle and aster proteins; it has a relatively high energy requirement, possibly for the mitotic event.

The phases of mitosis are closely similar in all types of tissue cells, and so are the phases of cellular aging. Assuming that both mitosis and aging are under gene control, the implication is that in all cells one set of genes directs mitosis and another similar set of genes directs aging, although each particular type of cell differentiation will be directed by an equally specific set of differentiation genes.

When a cell divides, it does not thereupon necessarily have to make a choice between a path leading to another division and one leading to differentiation. Much depends on the particular nature of the differentiation. Extreme structural differentiation (as in nerve cells, retinal and other sensory cells, the sting cells of coelenterates, and mature sperm cells) and also extreme chemical differentiation (as in keratinizing epidermal cells, scleratinizing plant cells, mature red blood cells, intestinal goblet cells, and others) are seemingly incompatible with mitotic division. The cells that give rise to such types are mainly unspecialized; after a series of mitotic divisions, they follow a course of differentiation and have a life expectancy of a few weeks for an epidermal cell, a few months for an erythrocyte, or from one to many years for neural cells, depending on the species. In such cases the flow to death may be rapid, moderate in rate, or barely perceptible, but is nevertheless continuous.

A different situation prevails in a category of tissue cell types that are differentiated but perhaps not excessively so, particularly generally differentiated kinds such as the chloroplast-containing plant cells, gastrodermal cells of coelenterates, and cells of liver, retinal pigment, cartilage, cardiac tissue, etc., of vertebrates, not to mention the variously and strikingly differentiated unicellular protistan organisms. In all such cases, successive cell divisions occur, at least in certain circumstances, in the presence of a notable degree of differentiation and often specialization, with only temporary disruption and disturbance of structure and function. Normally, cells in this group divide rarely and the special character of the cells persists. When such cells are cultured outside the body, however, they tend to lose their visible characteristic features as the rate of cell division increases. In fact, when cell proliferation in such a culture attains its maximal rate, the specific character of the individual cells is no longer recognizable. They come to resemble embryonic cells. The question is: To what extent is the original type of differentiation retained through cell division, and for how long?

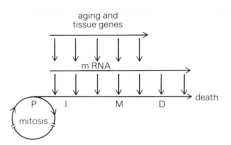

FIGURE 3.11 Diagram of phases in life cycle of tissue cells. *P*, proliferative phase of mitosis; *I*, immature cells preparing for tissue function; *M*, mature cells; *D*, dying cells. [*After Bullough, 1967.*]

CLONING AND SYNCHRONIZATION

Cell cultures, particularly of mammalian cells, are studied as synchronized populations, i.e., populations with all cells in the same phase of the cell cycle, this situation having been brought about by exposure to thymidine followed by blockading of the cell population with colcemid, which arrests all cells in metaphase, after which the culture is grown in fresh medium and all cells continue their regular cycle but starting at the same point.

A further degree of standardization of cell populations is attained by cloning, which eliminates the possible condition that a cell culture originating from a number of tissue cells consists of cells of subtly diverse kinds. Cloning consists of isolating a single cell from a culture and establishing a new culture which consists only of the descendants of that cell. The genetic and other constitutional characteristics are therefore, at least initially, the same for all the cells of that particular clone. Cloning and synchronization procedures are accordingly generally employed in the study of cell populations, whether of tissue cells of multicellular organisms or of protistan organisms.

Cartilage cells, muscle cells, and retinal-pigment cells from chick embryos, cloned and subcloned many times, retain their essential specific nature in spite of loss of visible expression of the differentiated cell character. The so-called dedifferentiation is more apparent than real and has been termed *modulation* to distinguish it from real loss of specific identity. It is possible, for example, for cartilage cells to multiply for at least 20 generations in mass culture conditions—i.e., two clonal passages grown to 1,000 cells each, without detectable function as cartilage cells—and yet express their original differentiated phenotype when restored to a permissive medium. It appears that as long as cells pass from mitosis to mitosis at maximal rates, too little time is available for full synthesis and operation of the structural or functional proteins associated with differentiation. Pigment cells and young skeletal and cardiac-muscle cells even retain their function during clonal conditions.

Individual cells of synchronous mitotic mammalian cultures proceed through the cell or mitotic cycle with different rates. Cells more than 90 percent in metaphase at the start may enter the S phase anywhere from 3 to 12 hours after division. The period of DNA synthesis through mitosis ($S + G_2 + M$) remains relatively constant, but the G_1 phase varies over more than half the cycle as a whole. Variation in G_1 seems to account for most of the variation in interphase time and might be due to genetic, nutritional, or volume changes within a population. Neither cell size nor genetic constitution, however, appears to be responsible for the variation. One suggestion is that scarce cellular components connected with growth-rate determination become unequally segregated during cell division. Another is that cells normally enter a G_0 state for variable periods between mitosis and G_1. The existence of such a G_0 state is strikingly evident in the cell cycle of mammalian ova, where the cell (last-generation oogonium) is virtually arrested in this state, i.e., between mitosis and

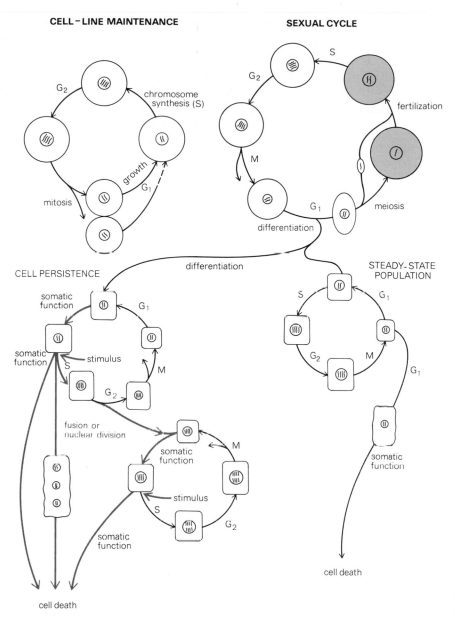

CELL–LINE MAINTENANCE

SEXUAL CYCLE

FIGURE 3.12 Types of cell cycles. [*After Stern and Nanney, 1965.*]

the onset of oocytal growth. It remains arrested for days or even years according to whether the cells are those of mice or men. To proceed it needs to be stimulated by the hormone estrogen. The G_0 state, however, also designates the condition of nondividing cells generally.

Tissue-specific differentiation, compatible with continuing cell divisions, appears late in the G_1 phase. Thus cultures of synchronized human lymphoid cells synthesize immunoglobulin proteins as part of the essential elements of the antibody response. Little immunoglobulin appears immediately before, during, and immediately following mitosis; i.e., it appears in neither the G_2 nor the M phase, but production is greatest during the late G_1 and the S phases of the cell cycle.

Once a cell enters a path to differentiation, whether this path leads only to eventual death or to possible resurrection of the mitotic event, the cycle expressed as G_1, S, G_2 seems no longer appropriate, although DNA synthesis may still take place. Discussion of the course of cytodifferentiation in somatic cells is taken up in Chapter 21, in another setting.

In summary, to quote Stern and Nanney: "Since no cell is eternal, all cells must have a finite life history, and that history ends in one of three ways—division, fusion, or death. Not only is there a beginning and end to the history of a cell, its history bespeaks a high degree of temporal organization. Not only is there a beginning and end to the history of a cell, but its course is marked by characteristic patterns. Some features of the patterns are basic to all cell types; others are elaborations of a basic succession of intracellular events, and are prominent in the cells of higher forms. Whether simple or elaborate, such patterns are expressions of an inherent capacity of cells to regulate their activities along the axis of time. In a strict sense, no cell has the same structural and metabolic configuration for two intervals in its history, however close and however small these intervals may be. A fixed characterization of a mature cell is an approximation; all cells are continuously undergoing directional change. The challenge of explaining how cells progressively alter with time is far from being met."[1]

[1]H. Stern and D. L. Nanney, "The Biology of Cells," p. 479, Wiley, 1965.

CONCEPTS

Cell size depends on genomic specificity, on degree of chromosomal ploidy, and on histogenetic type: cell size is species-specific and tissue-specific for cells with normal nuclear constitution.

Cell growth is a continuous process of synthesis of material which moves outward from the vicinity of the nucleus toward the cell periphery.

Cell division maintains constancy of cell surface/volume ratio within narrow limits, and restores standard nucleocytoplasmic balance.

Macromolecules are capable of entering cells and stimulating growth.

A variable degree of incompatibility exists between continuing cell division and cell differentiation, depending on the degree of specialized differentiation.

Cytoplasmic protein aggregations such as polyribosomes and endoplasmic reticulum disassemble and reassemble during and following cell division, respectively. Conversely, spindle microtubules and aster filaments assemble and disassemble during and following cell division.

Cleavage is primarily a cortical contractile-filament activity associated with the formation of a new plasma membrane. It is localized through the formation and positions of the asters.

Cell cleavage and mitosis are independent processes normally linked to each other.

The stimulator triggering mitosis accumulates in the cytoplasm and is transferred to the nucleus shortly before the onset of mitosis.

In continuously dividing cells, the cell cycle divides into a mitotic phase and an interphase, with the period of DNA synthesis restricted to a more or less middle part of the interphase.

During the cell cycle of protistan organisms, newly formed cells pass through a growth and differentiation phase and enter a new mitotic phase. In most kinds of somatic cells of a metazoan organism, newly produced cells take a course leading either to another mitosis or to differentiation and eventual death.

Cells that are moderately specialized but still capable of returning to a course of mitotic divisions lose visible expression of their phenotype but retain their basic character—a phenomenon termed *modulation*.

READINGS

ARNOLD, J. M., 1968. An Analysis of the Cleavage Furrow in the Egg of *Loligo pealii, Biol. Bull.,* **135**:413–419.

BULLOUGH, W. S., 1967. "The Evolution of Differentiation," Academic.

―――, 1962. The Control of Mitotic Activity in Adult Mammalian Tissue, *Biol. Rev.,* **37**:307–342.

―――, 1952. The Energy Relations of Mitotic Activity, *Biol. Rev.,* **27**:133–168.

CHALKLEY, H. W., 1951. Control of Fission in *Amoeba proteus* as Related to the Mechanism of Cell Division, *Ann. N.Y. Acad. Sci.,* **51**:1303–1310.

CHAMBERS, R., 1938. Structural and Kinetic Aspects of Cell Division, *J. Cell. Comp. Physiol.,* **12**:149–165.

COON, H. G., 1966. Clonal Stability and Phenotypic Expression of Chick Cartilage Cells in Vitro, *Proc. Nat. Acad. Sci.,* **55**:66–73.

COSTELLO, D. P., 1961. On the Orientation of Centrioles in Dividing Cells, and Its Significance: A New Contribution to Spindle Mechanics, *Biol. Bull.,* **120**:285–312.

CUMMINS, J. E., and H. P. RUSCH, 1968. Natural Synchrony in the Slime Mould *Physarum polycephalum, Endeavour,* **27**:126–129.

GRANT, P., 1968. Informational Molecules and Embryonic Development, in R. Weber (ed.), "Biochemistry of Animal Development," Academic.

GROBSTEIN, C., 1959. Differentiation of Vertebrate Cells, in J. Brachet and A. Mirsky (eds.), "The Cell," Academic.

HARRIS, H., 1968. "Nucleus and Cytoplasm," Clarendon Press, Oxford.

HIRAMOTO, Y., 1956. Cell Division without Mitotic Apparatus in Sea Urchin Eggs, *Exp. Cell Res.,* **11**:630–636.

HUETTNER, A. F., 1933. Continuity of the Centrioles in *Drosophila melanogaster, Z. Zellforsch.,* **19**:119–134.

INOUÉ, S., 1964. Organization and Function of the Mitotic Spindle, in R. D. Allen and N. Kamiya (eds.), "Primitive Motile Systems in Biology," Academic.

MALAWISTA, S. E., H. SATO, and K. G. BENSCH, 1968. Vinblastine and Griseofulvin Reversibly Disrupt the Living Mitotic Spindle, *Science,* **160**:770–771.

MAZIA, D., 1961. Mitosis and the Physiology of Cell Division, in J. Brachet and A. Mirsky (eds.), "The Cell," vol. III, Academic.

———— and K. DAN, 1952. The Isolation and Biochemical Characteristics of the Mitotic Apparatus of Dividing Cells, *Proc. Nat. Acad. Sci.,* **38**:826–838.

PICKEN, L., 1960. "The Organization of Cells," Clarendon Press, Oxford.

RAPPAPORT, R., 1969. Aster-equatorial Surface Relations and Furrow Establishment, *J. Exp. Zool.,* **171**:59–68.

RUSCH, H. P., 1970. Some Biochemical Events in the Life Cycle of *Physarum polycephalum,* in "Advances in Cell Biology," Appleton-Century-Crofts.

STERN, H., and D. L. NANNEY, 1965. "The Biology of Cells," Wiley.

STREHLER, B. L., 1962. "Time, Cells, and Aging," Academic.

SWANN, M. M., and J. M. MITCHISON, 1958. The Mechanism of Cleavage in Animal Cells, *Biol. Rev.,* **33**:103–135.

SWEENEY, B. M., and J. W. HASTINGS, 1958. Rhythmic Cell Division in Populations of *Gonyaulax polyedra, J. Protozool.,* **5**:217–224.

WEISS, P., 1969. "Panta' Rhea"—And So Flow Our Nerves, *Proc. Amer. Phil. Soc.,* **113**:140–148.

————, 1967. Neural Dynamics, *Neurosci. Res. Program Bull.,* **5**:371–400.

WENT, H., 1966. The Behavior of Centrioles and the Structure and Formation of the Achromatic Figure, *Protoplasmatologia,* **6**(Gl):1–109.

WESSELLS, N. K., and W. J. RUTTER, 1969. Phases in Cell Differentiation, *Sci. Amer.,* March.

CHAPTER FOUR

CORTICAL PATTERNS
AND INHERITANCE

Two points of view tend to dominate the study of organisms—the reductionist and the holistic. A current reductionist viewpoint is that a better and better understanding of the interactional properties of the molecular components is all that is necessary to comprehend differentiation and development. If a single fundamental process does exist which leads to cell differentiation, and by implication to development and maintenance of the multicellular organism, variable genic activity together with its regulation is the obvious candidate for such a universal and exclusive role. All depends on the genic control of synthesis of innumerable diverse structural and enzymatic proteins, and all else follows from their subsequent interactions, except for possible genic control of the timing of synthesis initiations.

CONTENTS

The whole and its parts
   Cortical differentiation
   The basis of pattern
   Cortical reassembly
   Determination of pattern
   Cortical inheritance
   Nuclear and cytoplasmic interaction: I
Cortex and nucleus interaction
   The plant-cell wall
   Determination of form in *Acetabularia*
   Nuclear and cytoplasmic interaction: II
Concepts
Readings

THE WHOLE AND ITS PARTS

Those who support the holistic view, which in varying degree emphasizes the "organism as a whole," no longer deny the general validity of the molecular approach but doubt whether it represents the whole story. As Nanney has said, "Genic regulation is a beautiful truth, but it is not all we know or all we need to know."[1] This outlook offers more promise of understanding some of the differentiation and developmental phenomena than does a more doctrinaire molecular biological standpoint, for cells are made from preexisting cells, which always contain many components besides the nucleic acid sequences. To what extent, if any, essential biological information is encoded and transmitted by materials and mechanisms other than the nucleic acid templates needs to be answered and, if possible, at the unicellular level, before an attempt is made to analyze how a single cell, typically an egg, is able to give rise to a complex multicellular organism. To what extent does organized cell structure itself influence the course of differentiation?

As a prelude to this question, namely, whether preformed cell structure plays an essential role in cell heredity, Sonnenborn has stated the premise of the molecular approach as follows: "Higher levels of structure within the cell, above that of the polypeptide product of cistron or gene, are accounted for by three factors: the physico-chemical properties of the reactants, their random collisions, and the ionic and molecular constitution of the cell 'soup' in which the collisions occur. These factors are held to determine how gene products are built up into multipolypeptide enzymes and structural proteins, how multienzyme systems come together in proper sequential

[1] D. L. Nanney, Cortical Patterns in Morphogenesis, *Science*, **160**:496 (1968).

a

b

FIGURE 4.1 (a) Ciliate showing terminal cytostome and rows of kineties, each kinety consisting of a row of kinetosomes, or basal bodies, with filaments constituting a kinetodesma. (b) Photo of dorsal side of *Tetrahymena*. [*Courtesy of D. L. Nanney.*]

arrangment, how enzymes and substrates come together and yield further products, how smaller ribosomes combine to form larger ribosomes, and so on. The 'self-assembly' hypothesis in its most extreme form thus ultimately traces the building of *all* cellular structure to molecular contributions from milieu to genes and to random collisions of previously unarranged reactants. While some molecular biologists . . . seem to have great confidence in the full adequacy of this hypothesis, others adopt it tentatively with the express purpose of seeing how far it can be carried, how much structure can be accounted for without invoking additional factors. That some degree of structure can already be explained in this way seems evident. The question then is: Do we know, or can we discover whether the hypothesis is sufficient to account for *all* cell structure?"[1]

Cortical Differentiation

Ciliate protozoans have long been studied with some such thought as this in mind. They are highly appropriate organisms for several reasons: they are unicellular and readily multiply in laboratory cultures, they are usually comparatively large (some are even very large indeed), and they exhibit cortical structural patterns that are highly differentiated. The ciliate cortex, which may be as deep as 2 μ, contains rows of cilia, each cilium with its own basal body, or kinetosome; the rows of cilia typically orient with regard to body shape and locomotory axis, and locally form specific assemblies, particularly feeding organelles. Ciliates are also characteristically equipped with two kinds of nuclei per individual, the micronucleus, which appears to be inactive most of the time but is necessary for mitotic division, and the macronucleus, which is essential to the general functioning of the cell-organism and for processes such as growth and regeneration.

Very small fragments of the whole organism are capable of reconstitution, at least in relatively large ciliates such as *Stentor, Blepharisma,* and *Spirostomum.* The importance of the cortex in this connection is shown by certain experiments. When all or virtually all of the endoplasm of a stentor is withdrawn through a small incision in the surface, leaving only the cortex and the nucleus, endoplasm is promptly restored and normal growth and reproduction follow. On the other hand, if all the cortex is stripped off, leaving only endoplasm and nucleus, the cell becomes spherical, survives for a while, but finally dies. However, if a piece of cortex is left on the endoplasm, it gradually spreads around it, reconstituting the visible markers of its gradients in the form of gradation in stripe width, and eventually regenerates and reproduces normally. Inasmuch as the kinetosomes are apparently self-replicating bodies and form precise, complex patterns in the cortex, namely, characteristic infraciliatures, the question arises whether such cortical patterns perpetuate themselves more or less independently of nuclear genes.

[1] T. M. Sonnenborn, Does Performed Cell Structure Play an Essential Role in Cell Heredity?, p. 166 in J. M. Allen (ed.), "The Nature of Biological Diversity," McGraw-Hill, 1963.

The complexity of the cortical structure of ciliates is such that a certain terminology has arisen in describing it. The infraciliature is the whole cortical complex of cilia, basal granules, and fibrils. A row of granules, or kinetosomes, is normally paralleled on the right by a long filament called a *kinetodesma*, which consists of a bundle of parallel fibrils, each fibril connecting to a kinetosome at one end and ending freely within the kinetodesma at the other. Each fibril leaving a kinetosome does so on the right side and passes anteriorly to merge with the adjacent bundle.

The Basis of Pattern

Together, a row of kinetosomes and its associated kinetodesmas is known as a *kinety*. Most of the differentiations of the cortex—cilia, flagella, trichocysts, membranelles, cirri, etc.—seem to owe their origin and maintenance to the kinetosomes. Further, certain kinetosomes appear to hold dominant positions within the infraciliature (these positions vary according to species and patterns), and they seem to be responsible for initiating, directing, and maintaining the normal fully developed cortical pattern. In most, if not all, ciliates studied, a particular kinety appears to hold a key position, for instance, relative to mouth or oral structure. This is called *kinety 1*, and other kineties are numbered or labeled in relation to it.

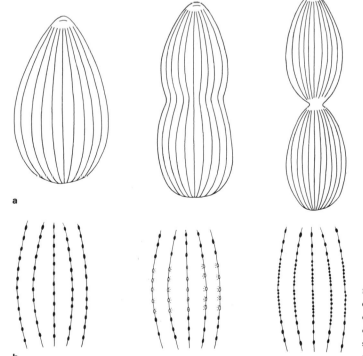

FIGURE 4.2 **(a)** Fission of a simple ciliate, involving extension of kineties, constriction, and formation of a new cytostome. **(b)** Extension of middle region of kineties, involving multiplication of kinetosomes, followed by growth (typical of dorsal side of most ciliates).

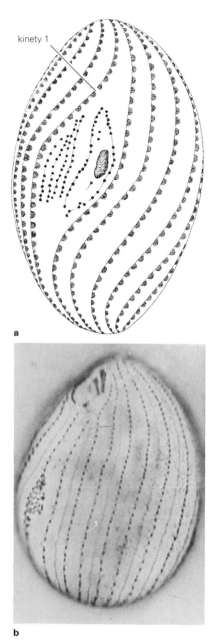

a

b

FIGURE 4.3   (a) Formation of successive rows of new kineties from kinety 1 to give rise to a "field," or oral primordium, associated with an expanding area of the cortex as a whole (typical of the ventral side of many ciliates). (b) Photo of *Tetrahymena*, showing kinety 1 with oral field on left, and cytostome at upper left. [*Courtesy of D. L. Nanney.*]

It is clear that the unit of cortical pattern—the kinetosome–basal body plus cilium plus filament—is an asymmetric structure, as shown by the right side and anterior direction of outgrowth of the basal filament. These units naturally organize into longitudinal rows exhibiting a corresponding right-left and anteroposterior pattern. Two-dimensional sheets of cortex containing rows of kineties therefore possess a related right-left asymmetry and an anteroposterior axis, which together characterize the organism as a whole. How do all these structures reproduce themselves?

The various ciliates lend themselves in different ways to experimental or analytical investigations. Thus medium-sized ciliates, such as *Paramecium* and *Tetrahymena*, readily cultured in pure clones through innumerable generations, have been mainly studied genetically in relation to structure and other properties. *Stentor*, on the other hand, particularly the large species *Stentor coeruleus* and *Stentor polymorphus*, are less readily bred but serve magnificently for microsurgical studies.

Stentors have the shape of a cone. About 100 pigmented stripes of graded width alternating with rows of cilia associated with cortical contractile fibers extend from the narrow base of the cone to the wide saucerlike broad end. The broad end exhibits a whorl of concentric stripes and ciliary rows. Cell shape apparently is largely a function of this cortical pattern, since disarrangement of the pattern results in a corresponding abnormality of cell shape. The broad end contains the feeding organelles, which consist of a nearly complete circle of ciliary membranelles terminating in a gullet and cytostome, or cell mouth. A contractile vacuole lies to the left of the gullet. *Stentor* is also notable in having an unusually large macronucleus in the form of a string of beads, together with micronuclei which become active during the process of conjugation. Growth or replacement of the cortical pattern occurs whenever a single individual divides into two and whenever an individual is experimentally cut up or otherwise surgically manipulated.

Small fragments which are only $\frac{1}{123}$ the size of the largest individual can re-form feeding organelles and narrow tail ends in proportion to their size. A small part can become a whole if it contains one macronuclear node, or bead, and a proportionate extent of cortex. However, in such cases, after cortical healing, the feeding organelles have about the same length and breadth as those of the largest animals, and the pigmented stripes are also disproportionately wide. These appear to be units of structure, macroscopic rather than macromolecular, which are necessarily of a certain size, the same in large individuals as in small ones, just as the cilium–kinetosome–basal filament complex is a structural unit that can vary little in size. This principle of utilization of standard parts pervades the whole of the organic world. A large ciliate has the same type and size of ciliary-complex unit as a small ciliate but many more of them. And what holds true for cell components holds for the cell as a whole; an elephant has cells of essentially the same kinds and sizes as a mouse, but correspondingly more of them. The limits and possibilities are set for each level of construction.

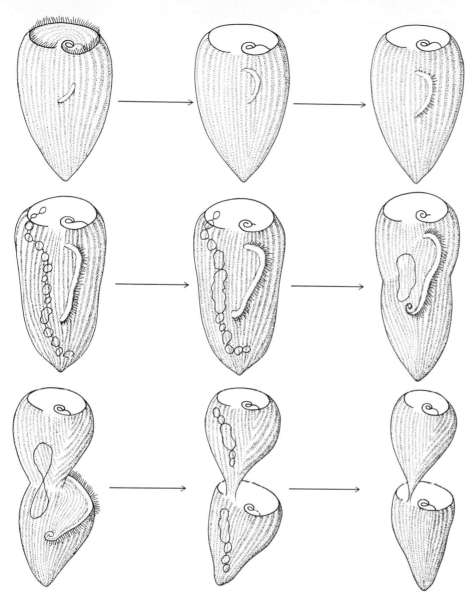

FIGURE 4.4 Stages in division of *Stentor*, showing contraction and division of the macronucleus, and formation and growth of new primordium leading to development of a complete system of kineties and cytostome in posterior half of the individual, accompanied by constrictive separation of anterior and posterior daughter individuals. [*After Tartar, 1962.*]

Cortical Reassembly

Stentors can be cut without actual separation into pieces, with remarkable results. According to Vance Tartar, who has carried out most of the operations performed on *Stentor:* "Stentors can be cut repeatedly with a glass needle until the ectoplasmic structure consists of forty or more stripe patches too small for further cutting and lying in random arrangement. When the head and tail are first excised, the *Stentor* is left in the maximum state of inostropy attainable. The original cell axis is completely obliterated and the normal pattern of the striping is entirely disrupted. Yet minced

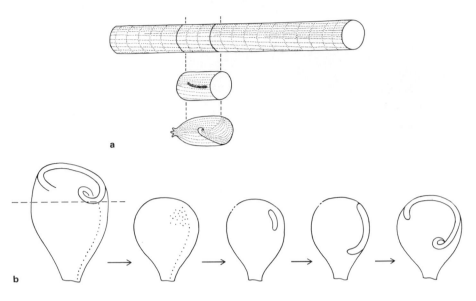

FIGURE 4.5 (a) Diagrammatic representation of basic cortical pattern in *Stentor*. The long tube represents the essentials of a line of stentors, unseparated and without cell elaborations, with polarized longitudinal stripes at graded distances from one another, and with hypothetical transverse connections. The short tube represents a section of the long tube, with oral primordium forming where wide and narrow stripes are adjacent. The individualized unit shows posterior striping gathered together to form the holdfast, and the migrating primordium carrying lateral stripes to cover the anterior end. [*After Tartar, 1962.*] (b) Decapitation followed by formation of a new primordium and development of a new anterior end. [*After Weisz, 1954.*]

stentors regenerate and can reconstitute the normal form within a day, i.e., the minced *Stentor* when relaxed has a faceted appearance which is one of many observations which leads us to believe that the cell shape is determined by the pattern of striping. Although this pattern is quite chaotic, each patch retains its portion of the ectoplasmic striping which carries an intrinsic polarity. . . . The patches then gradually reorient themselves with their stripes running parallel and homopolar, and as they realign the stripes join together in continuous runs, provided that the spacing or pigment bands are of about the same width. . . . Further reconstruction consists in the development of the oral primordium and the progressive realignment and rejoining of stripe patches. Any parts which fail to fit into the normal pattern are eventually resorbed. . . . There is no evidence of an imposed axis which reorientates all the patches like a magnet, and the action of an external field is ruled out by the continuous rotation of the specimen. Rather, the impression is of a gradual resolution of the problem by the patches themselves which in their reorientation gradually come to settle upon a single polar axis."[1]

Determination of Pattern

Whether in normal division, experimental fragmentation, or cortical mincing, a new complex of feeding organelles arises from an oral primordium whenever such a complex needs to be formed. A new oral primordium develops in stentors in a certain region of the cortex determined or indicated by the mutual relationship of longitudinal stripes. Stripes vary in width, and at the broad end of an individual especially fine

[1] V. Tartar, Morphogenesis in *Stentor, Advance. Morphogenesis,* 2:17 (1962).

stripes are always seen to the right of the widest stripes. A new oral primordium always begins in the fine-stripe area next to the wide stripes. The pigmented stripes themselves, whatever their function, are in this connection probably no more than indicators of cortical geography and regional cortical differentiation. They indicate that a particular locality has properties of its own, and that an oral primordium develops in a precise situation in the system as a whole, specifically located in a normal individual but found elsewhere in modified individuals.

Many experiments have been performed to explore this situation, and always wherever wide-stripe areas are placed next to fine-stripe areas an oral primordium, foreshadowing the development of oral membranellar structure, appears in the adjacent fine stripes or at the junction of the two contrasting areas. For example, if a stripe of fine striping is implanted among the wide stripes at the back of the cell and regeneration is induced by cutting off the original broad end with its oral structure, a new primordium forms not only at the front end at the junction of broad and fine stripes, as expected, but also on the back side, where fine and broad stripes now lie close together, with the result that a doublet stentor is produced in place of a singlet.

Moreover, if wide striping is experimentally placed to the right of a normal fine-stripe zone, the new oral primordium is correspondingly reversed and gives rise to membranes which coil to the left instead of to the right. The adjacent stripe pattern, therefore, seems to be involved in the manner in which the fine structures in the primordium are put together; reversed primordium sites produce reversed polarity in the band of oral membranes.

The positioning of structures according to the right-left or anteroposterior axes of the cell-organism is also shown by the nucleus. The long beaded-chain macronucleus of a stentor normally has a specific position within the cell, to which it returns if displaced. If however the left-right stripe pattern is reversed, the location of the macronucleus is also reversed. Tartar suggests that the problems of morphogenesis in a form like *Stentor* may well be expressed in terms of a basic, cortical, cytoplasmic pattern, particularly since it has been demonstrated that in tiny fragments any portion of this pattern retains its intrinsic polarities and is capable of developing a complete new individual. A strong case clearly exists for this approach, for even the nucleus is subject to cortical controls. At the same time there is evidence that the proteins utilized in the synthesis of the oral apparatus of *Stentor* are being synthesized all

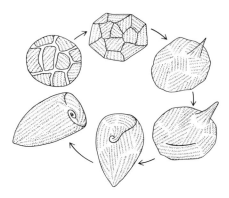

FIGURE 4.6 Reconstruction and regeneration in a minced stentor. The head and tail are excised first, then the cortex is cut with a glass needle until it is reduced to an aggregate of random patches. The sequence of recovery and reorientation of patches to reconstitute a new individual, with a new primordium regenerating a typical head structure, is shown clockwise from operation diagram. [*After Tartar*, 1962.]

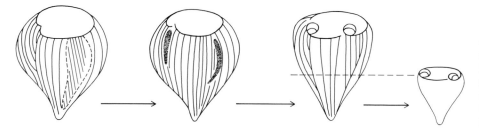

FIGURE 4.7 Grafting an extra fine-line zone into the wide-stripe region of a decapitated stentor leads to double oral regeneration, one from the graft and one from the host primordium. A second decapitation is followed by double regeneration from the posterior piece to form a doublet. [*After Tartar*, 1962.]

the time and not merely as needed, apparently through the agency of nuclear RNA. Cortex and nucleus evidently constitute a mutually interacting system, and neither can be said to be fully autonomous. The important regulatory event must involve agents or conditions which promote extensive multiplication of kinetosomes at the locale where stripes contrast and consequently the outgrowth of fibers and cilia from them.

Ciliate protozoans constitute an enormous group of widely diversified types, each of which lends itself to morphogenetic analysis in one way or another. Only the largest are suitable for microsurgical experiments. Others are more appropriate for breeding analyses, for studies of normal processes associated with division and conjugation, and for the culture and examination of naturally occurring abnormal forms. Normal division, or fission, has been closely observed, for example, in *Euplotes*, where it occurs at 24-hour intervals, i.e., it exhibits a regular diurnal, probably circadian rhythm characteristic of many other protozoan organisms.

Division typically has two meanings: (1) the actual act of cleavage or separation of the cell mass into two parts, and (2) the whole process, of which the final cleavage is merely the culmination. Thus in *Euplotes*, and others, division in its broader sense is a very complicated process involving the establishment of two small cortical "morphogenetic fields" about 8 hours before final separation of the cell into two daughter cells. A pair of small "daughter patterns" appears and spreads respectively over the anterior and posterior halves; the eventual cleavage of the cell body as a whole is in effect a separation of two already-existing individuals.

The most complex process, the most significant, and the one that is still imperfectly understood, is the origin of a new oral primordium and its development into the complex feeding apparatus of gullet, vestibule, and membranelles, whether in a broad stripe–narrow stripe contrast area of a normal or manipulated stentor or during the general division process of a paramecium. Details vary somewhat, according to the specific nature of the structures to be assembled, but the basic events are similar among the various forms studied. New short rows of kinetosomes appear, which give rise to a so-called anarchic field, from which the whole oral assembly develops. The kinetosomes of kinety 1 in normal individuals appear to give rise to the original kinetosomes of the new primordium.

Cortical Inheritance

Although *Paramecium* is unsuitable for performing the sort of operations that have been so successful on *Stentor*, it has been possible by more indirect means to alter the precisely regular normal pattern of structure, to produce individuals with additional mouths, gullets, anuses, etc., without loss of viability. The altered structural pattern persists indefinitely, even through the sexual process of conjugation. A brief description of the normal pattern accordingly becomes necessary here. In *Paramecium* there are six major groupings of kineties, located as one dorsal and five ventral fields.

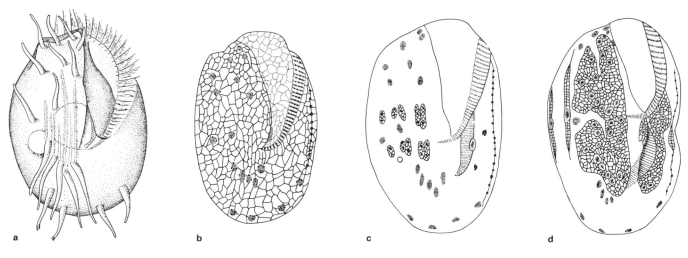

a   b   c   d

FIGURE 4.8   Ventral surface of *Euplotes*. (a) Special-ized (fused) cilia forming localized cirri and oral membrane. (b) Silver line–stained basis of ciliary pattern. (c) Onset of division: degeneration of old cirri bases and formation of a double set of new expanding areas of network pattern. (d) Greatly expanded new areas representing organization of two daughter individuals. [*After Chatton and Seguela*, Bull. Biol. France et Belge, **74**:*349 (1940)*.]

The whole dorsal surface, in fact, is one great field of longitudinal and parallel kineties, except where they converge near the two ends of the body. The five ventral fields are differently oriented with regard to the cell axis and to one another. One of them, the circumoral field, occupies the middle region of the ventral surface, and its kineties curve inward and define the vestibule leading to the gullet and mouth. The sites of other locally specialized features, such as the cell anus and the contractile vacuoles, are also precisely situated relative to the ventral field complex. In all cases, the anteroposterior axis and the right-left orientation of kineties are clearly defined.

One of the simplest departures from the normal which has been induced in *Paramecium* is the inversion of one or more rows of the numerous longitudinal kineties, so that the inverted kineties show fibers emerging on the left and extending backwards, instead of emerging on the right and extending forwards. This partly deranged pattern has been perpetuated through several hundred generations, as the result of replication of basal units with the same orientation as those of the original inverted rows.

Doublet individuals are commonly seen in cultures of various ciliates, i.e., two apparent individuals virtually complete in total structure and organelles except that they are conjoined like Siamese twins and usually possess a single macronucleus between them. Doublets of this nature may be produced in *Stentor* by grafting a piece of fine-stripe cortex from one individual into a wide-stripe cortical area of another; when regeneration follows amputation of the anterior end of the host, a double-bodied individual forms. In cultures of *Paramecium* and where doublets turn up as a result of abnormal fission or of conjugation, the remarkable happening is that the doublets propagate as doublets by regular fission just as in singlets, and even undergo conjuga-tive reproduction. Moreover it is possible to mate doublets with singlets, so that a vast array of experimental possibilities is opened up.

Primarily the question is the nature of the inheritable basis which enables a

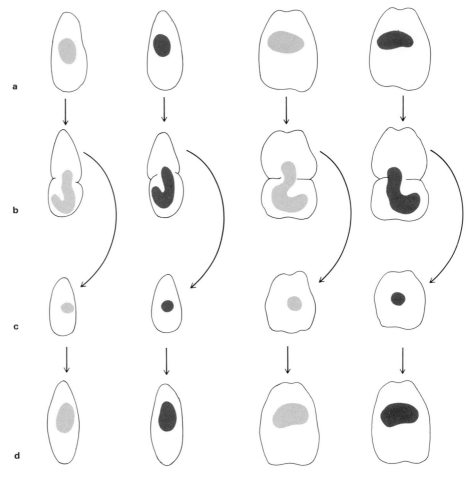

FIGURE 4.9   The size of the macronucleus of *Para-mecium* is determined by the cytoplasm, independently of nuclear origin. The light macronuclei are derived from singlet individuals, the black macronuclei are derived from doublet individuals. (a,d) Adult cells before and after fission, respectively. (b) Unequal division of macronucleus between daughter cells in the case of both singlet and doublet individuals. (c) Fission products that have received much less than half of the dividing macronucleus. (d) The small piece of macronucleus, regardless of its origin from singlets or doublets, grows into a normal-sized macronucleus in singlets and into an oversized macronucleus in doublets. [*After Sonnenborn, 1963.*]

doublet to reproduce itself as a doublet and to behave generally as a single individual although with duplicated cortical systems, in contrast to the self-replicating singlet individual. The difference between the two forms is the same kind of difference seen between normal individuals and those with one or more inverted kineties, but is of a vastly more complex order. The question is essentially the same in the two cases: Is the inheritance of the new organization in any way under the control of, or determined by, the nucleus; or is the cortex as a whole, whether doublet or singlet, a truly self-perpetuating pattern, even though dependent on nuclear genes for the synthesis of some of its raw materials? We are concerned, in other words, with the nature of the inheritance of pattern as distinct from substance, and whether the innate tertiary or quaternary configurative potentials of the gene-determined proteins is a sufficient or an insufficient basis to work from. This question is crucial.

Paramecium, over the years, has been subject to intensive genetic experimentation and analysis, with not only gene markers but also the so-called kappa and killer traits used as cytoplasmic markers. Neither nuclear genes nor nuclear size, nor any components free to move as part of the fluid endoplasm, are found to have any bearing on the control of the difference between singlets and doublets. Only the cortex itself seems to be or to contain whatever genetic basis there may be. Such a conclusion, however, is so potentially important that positive evidence is required for its acceptance. Positive evidence, for example, could be the perpetuated transmission of a piece of special cortical pattern grafted into a normal cortical system, were it possible to make grafts such as have been done in *Stentor,* which is not the case. Nature often performs a particular type of experiment which can be exploited.

In such an instance, following conjugation between a doublet and a singlet, the singlet carried away a piece of a doublet's cortex at the time of separation, and incorporated it as part of its own cortex. This particular freak was isolated and gave rise to a clone of like individuals, all of which were intermediates between singlets and doublets, with two sets of vestibules, two gullets, and the two ventral kinety fields, but with a single dorsal surface. That is, a piece of cortex pulled off from oral cortex of one cell and incorporated on the surface of another resulted in the inheritance and development of a complete additional oral region along the whole length of the animal on its ventral side. Accordingly, a small piece of cortex, as a natural graft, contains the genetic basis of a large but delimited part of the cell cortex.

Another ciliate, *Tetrahymena,* has joined the ranks of the exploited, for companion studies with those on *Paramecium.* Within a single culture strain at least 20 distinguishable types can be recognized solely on the basis of the number of ciliary rows, or kineties, even though they have the same nuclear constitution and exhibit the same basic pattern of cortical organization. These are known as *corticotypes,* and relatively stable clones of corticotypes 16 to 21 can be maintained. To quote Nanney, in a summary of experiments employing these different types: "Clearly the cortical patterns are maintained by mechanisms which require further exploration. And the most powerful device for elucidating hereditary mechanisms is the breeding test. Conjugation is the chief device for genetic analysis in ciliates. Because two cells come together, exchange nuclei, and become genetically alike in the process, each conjugating pair constitutes in effect a reciprocal cross. Although the nuclei are all alike with respect to all conventional genic markers, they are housed in cytoplasmic structures derived from different sources. The expected consequences of a reciprocal cross is that differences caused by genic differences will disappear, and that progeny, regardless of their cytoplasmic housing, will come to be alike. If, in contrast, the differences persist for an appropriate number of generations after conjugation, it is clear that they are not caused by differences in nuclear genes but must be related in some way to cytoplasmic properties. This in fact is the result obtained when crosses are made between different corticotypes of *Tetrahymena.* When a mating is arranged be-

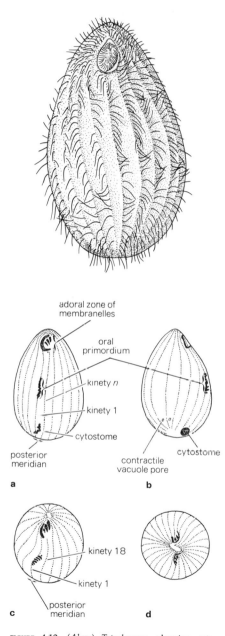

FIGURE 4.10 (*Above*) *Tetrahymena,* showing cytostome and rows of cilia (kineties). (*Below*) Semidiagrammatic view of silver-staining structures: (**a**) ventral view, (**b**) right lateral view, (**c**) apical view of singlet, and (**d**) apical view of doublet cell. [*After Nanney, copyright 1968 by the American Association for the Advancement of Science.*]

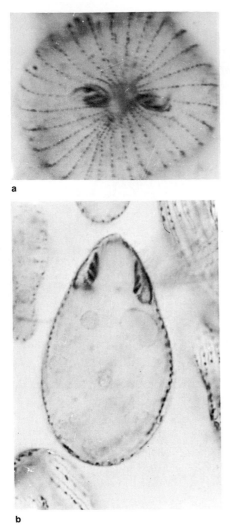

a

b

FIGURE 4.11   **(a)** Polar view of a "duplex" form with two sets of oral structures and 29 instead of the normal 18 ciliary rows. **(b)** Longitudinal optical section of a duplex form. [*Courtesy of D. L. Nanney.*]

tween a cell of corticotype 16 and another of corticotype 20, for example, the type-16 cell remains type-16 and produces a clone of predominantly type-16 progeny . . . meanwhile the original type-20 cell produces a clone of predominantly type-20 progeny. The fact that the cells have conjugated, and have exchanged nuclei, has no detectable effect on their subsequent behavior. These results establish unequivocally that cortical differences are not the consequences of conventional genic differences. They suggest cytoplasmic mechanisms for the maintenance of cortical states."[1]

The question, therefore, focuses strongly on the nature of cortical organization. In ciliates we see precise patterns of kinetosomes, kineties, and their organellar assemblies. Although kinetosomes have not actually been observed to give rise to daughter granules, any more than centrioles have been observed to give rise to daughter centrioles, in both cases the strong circumstantial evidence of propinquity seems to show that they do.

Various experiments, mainly on *Stentor,* and many observations on the developing cortical organization in ciliates undergoing fission or exhibiting different states during different phases of a complex life cycle (typical of parasitic forms) indicate that the visible cortical pattern is not primary but reflects a more fundamental and far more subtle cytoplasmic organization. For instance, when two oral primordia, one at an early and one at a late stage of differentiation, are transplanted into different areas of a host stentor, they do not go on developing as before but show a strong tendency to synchronize to the same phase of development. In some way they become subject to an agency outside themselves, even though all the elements necessary for growth and development are clearly available. The general conclusion from studies on these various ciliates and other kinds of protozoans is that preformed structures play an essential role in determining the organization of new structures. This is a concept that will be seen to be of great importance in connection with the phenomena of regeneration of parts in multicellular organisms.

Nuclear and Cytoplasmic Interaction: I

Typically a large cell has a large nucleus, at least within certain limits of cell size. The cells of haploid and triploid salamanders, for instance, are smaller and larger, respectively, than those of the normal diploid individuals, and their respective nuclei are smaller and larger also. Yet all such cells are small when compared with a stentor and most of the protozoans. One device whereby giant cells can be maintained is seen in protozoans such as the larger rhizopods (i.e., heliozoans, foraminiferans, and radiolarians), where the single cell body of the fully grown individual contains multiple nuclei, each nucleus in effect being in charge of a certain volume of cytoplasmic territory, although a continuous membrane-cortex represents the single integrity of

[1] D. L. Nanney, Cortical Patterns in Morphogenesis, *Science,* **160**:496 (1968). Copyright 1968 by the American Association for the Advancement of Science.

the whole. The giant amoeba, *Chaos chaos,* similarly has many nuclei scattered through its interior instead of one.

The nuclear duality of ciliates is another means by which ultralarge cells are maintained. The minute, diploid micronucleus, active only in mitosis, is vital to reproduction, whether during fission or conjugation. The enormously larger macronucleus divides amitotically during fission; during conjugation it disappears, and a new macronucleus develops from a descendant of a micronucleus. In either event the final size of the macronucleus is closely correlated with the size of the cell-organism of which it is a part. Whether it has the beaded appearance of a *Stentor* macronucleus or the more compact oval shape of *Paramecium* or *Tetrahymena,* it corresponds to a multinuclear condition, for the large size reflects the presence of a large number of genomes. Thus when a stentor or a paramecium undergoes fission and the macronucleus is seemingly divided into two in a grossly imprecise manner, each daughter cell receives a component already consisting of many complete sets of chromosomes. The macronucleus itself plays no active part in the fission process, other than to be passively divided transversely into two parts; yet when fission is over, it is clearly responsible for the normal nuclear processes involved in growth, maintenance, and regeneration of the cell. Multiplication of genomes within a single macronuclear envelope, whatever its configuration, makes possible the maintenance and restoration of the extremely large size of the cell.

These phenomena suggest that the total quantity of nuclear material, particularly DNA, whether packaged in one bundle or many, determines the size of the cell it maintains. This is literally only a half-truth. We have already mentioned that the positioning of the *Stentor* macronucleus changes according to changes made in cortical structure. In *Paramecium,* doublets have a cell size and a macronucleus each about twice the size of the cell and macronucleus of a singlet. Conjugation between doublets and singlets gives rise to clones of doublets and of singlets, but in either case the macronucleus may have descended from a singlet or from a doublet parent. However this may be, the macronucleus grows to the size characteristic of the cell, doublet or singlet, in which it finds itself. Cell size, in other words, determines the size of the nucleus, although nuclear size depends on the nucleic acid content and on the degree of dispersion, or hydration, of the constituents. It is increasingly clear, however, that nucleus and cytoplasm represent an interacting system which must be regarded as a whole. This is strikingly evident in studies of the giant, unicellular, green alga *Acetabularia,* discussed in the following section.

CORTEX AND NUCLEUS INTERACTION

The protozoans are but one group of unicellular organisms that exhibit remarkable and diverse form associated with and possibly determined by cortical filamentous substructure. Ciliates present a visible, self-perpetuating kinetosome-infraciliature organization which expresses and makes possible experimental and observational analysis of cortical differentiation and development. At the same time the very existence of such a visible, propagative, dynamic organizational basis tends to distract attention from the fact that in other groups complex form is attained by far less obvious means, and that while filaments, microtubules, and, at least in ciliates, kinetosomes clearly play a part, they may be subject to controlling agencies of a much more subtle nature.

Certain green marine algae, lumped together (probably improperly) as the Siphonales, exhibit the unicellular condition, inasmuch as they are not divided into cells; yet they grow to considerable size and attain remarkable and diverse forms. All such organisms, and many others, are grist for this particular mill, for each exhibits in its own specific way all the problems of differentiation and development that are found at the cell level throughout all forms of life. The leading contributor, however, is the mermaid's cap, i.e., the various species of *Acetabularia* which are common below the tide in the warmer seas. It is remarkable for its size, its shape, and its life span.

The Plant-cell Wall

Plant cells are notably different from animal cells, whether in unicellular or multicellular forms, in the nature of the cell wall. In fact, the cell wall in plants is generally considered to be outside the living membrane, the plasma membrane of the cell proper. In animal cells the plasma membrane is itself the boundary between the cell interior and the adjacent environment. In plants a secondary wall which is typically formed external to the plasma membrane has much to do with the shape of the cell. The nearest to a comparable situation in the animal organism is the secretion of macromolecular monomers by cells as a reinforcing material. Thus collagen assembles into quaternary, pseudocrystalline structure adjoining the basal surface of epithelia. In plants such reinforcement is the rule. There is some debate whether the supporting wall is to be regarded as a part of the living plant cell or is truly outside it in the sense that the collagenous basement membrane in animal tissues is external to the adjoining cells. The question is mainly semantic.

Plant cells do possess a plasmalemma, or plasma membrane, of apparently unit-membrane structure, through which cytoplasmic vesicles containing precursors of wall compounds discharge their contents. Wall proteins, both enzymatic and structural, are extracellular secretion products. The material which becomes assembled external to the plasma membrane is generally considered to be built up in two phases: a primary wall is formed during the early growth of the cell, and the "secondary" wall component is laid down on the inner face of the primary wall as the cell matures. Patterning is evident in the primary wall in three dimensions. It is seen in stratification and in the disposition of structural elements in the plane of the wall.

This process of deposition and structurization has been analyzed in the cells of higher plants, in which the structural elements are mainly cellulose microfibrils, although other compounds related to cellulose may also be present as matrix components. In young cells about to undergo growth in length, the microfibrils in the primary wall tend to be oriented either randomly or transversely to the axis of future growth. As a cell elongates, however, these microfibrils orientate more and more in line with the axis. Cell shape, resulting from specific directions of cell growth, appears to be determined by the changing pattern of distribution of the microfibrils.

The general situation is, therefore, not unlike that of the infraciliature of ciliates, although the latter complex is clearly a component part of the living cortex. At the same time it compares with the tertiary and quaternary self-patterning of collagen, although collagen is protein whereas cellulose is a polysaccharide assembled by polymerization of carbohydrate (glucose) monomers. Since the microfibrils are laid down external to the cell cytoplasm and their orientation progressively changes in conformity with the growth orientation of the cell and therefore with cell shape, the question arises: Is the orientation of these fibrils determined by agencies other than their own innate properties?

A partial answer comes from the discovery of microtubules. They are regularly present in plant cells, characteristically in the cortical region of the cytoplasm close to the cell wall, and they are now known to be concerned in some intimate way with the deposition of cell-wall material. They do not appear to be concerned with cellulose synthesis as such, however, but specifically with **(1)** determining where synthesis shall occur and **(2)** orienting the microfibrils produced. How these functions might be accomplished is not known. What, in turn, orients the microtubules, especially in the absence of nucleation centers such as kinetosomes, also remains obscure. This is perhaps one of the most fundamental problems of cell biology insofar as it concerns cell differentiation and organismal development. Not all plant cells have walls of cellulose, however; some, such as *Acetabularia,* have a cell wall composed primarily of mannan, which is a polysaccharide or polymer of the sugar mannose and which is present mainly as granules or very short rods instead of filaments. Yet the form of the cell is striking indeed.

Determination of Form in *Acetabularia*

The problems concerning the development and determination of form and of other developmental features are more readily approached in *Acetabularia* than in almost any other kind of single plant cell. *Acetabularia* may be classed as a single cell because the whole organism is contained with a single, continuous cell wall and moreover contains but a single nucleus. Yet it can grow to lengths exceeding 5 cm under certain conditions, and *Acetabularia* may live as individuals for several years. These are remarkable achievements for a one-celled organism.

The mature plant consists of three parts: **(1)** rhizoids, which are rootlike structures that attach the individual to the substratum; **(2)** a narrow stalk, several centimeters long, which contains a central vacuole; and **(3)** a cap, usually about 1 cm across,

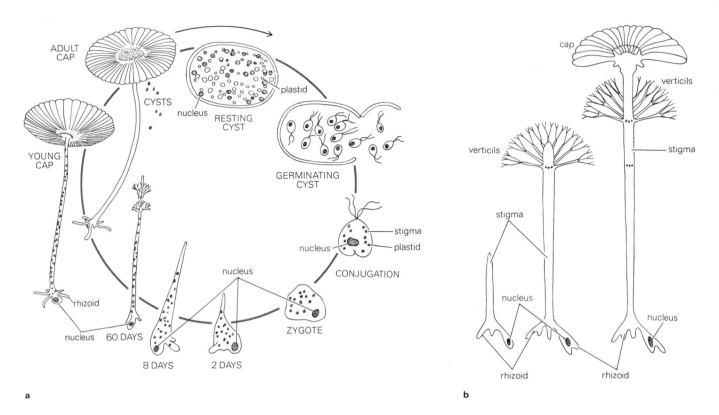

a   b

FIGURE 4.12 (a) Life cycle of *Acetabularia.* (b) Growth of *Acetabularia,* showing young capless stage with rhizoids containing nucleus in one branch; an intermediate stage with a whorl of sterile verticils; and an older stage with fully formed cap, a distal whorl of verticils, and a ring of stigma representing the attachment of an older whorl now lost. [*After Brachet, 1965.*]

at the distal end of the stalk. The proportions vary, however, according to external circumstances, particularly with regard to light. Throughout most of the life cycle there is a single large nucleus, about 150 μ in diameter, i.e., about as large as many mammalian and other kinds of eggs, which occupies one of the basal rhizoids. When the cap attains its final size, the nucleus undergoes successive divisions and thousands of daughter nuclei migrate up the stalk into the cap, where they become surrounded by small masses of cytoplasm and an enclosing membrane, and begin to form cysts. As the cap degenerates and cysts are set free, the cyst contents of nuclei and cytoplasm, after further nuclear division, give rise to motile, haploid gametes. These unite in pairs to form diploid zygotes, which settle and start to grow into new individuals, at first as a small stalk, later as a larger stalk equipped with one or more distal rings of small branches, and finally as a longer stalk with a typical cap.

Acetabularia, in its several species, represents an organism of extreme value for experimental investigation. It is readily obtained and is easily maintained in laboratory cultures. It has single-cell structure of an essentially filamentous nature but with a striking differentiation into the three parts just mentioned. The single nucleus, during the vegetative stage, is so large and accessible that it can be removed intact, washed repeatedly to remove any adhering cytoplasm, and reinserted into the same or another individual. The caps of different species are highly distinctive and can be used as indicators of genetic determinants. And the organism has a remarkable ability to regenerate amputated caps. The relatively enormous size of the well-grown *Acetabularia* makes possible many kinds of experiments that would be most difficult, if not impossible, with smaller organisms.

There is an extreme contrast between the mature acetabularia and the newly formed zygote, for the latter is merely a small diploid cell, formed by the fusion of a pair of flagellated gametes. It is only about 50 μ in diameter, or less than half the diameter of the nucleus alone of the mature form. The zygote undergoes a real development into an organism of comparatively giant proportions. How does a single nucleus sustain such an enormous assembly, and how does it determine the general and specific character of the cap, if such it does? The facts that the solitary nucleus resides at the extreme basal end of the elongate organism and that the specifically diagnostic features are manifested at the extreme distal end, with the long tubular stem as a linear channel of communication between them, make possible experimental approaches that are not feasible in other situations.

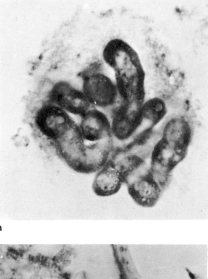

a

Cap formation is accordingly a complex morphogenetic event involving net synthesis of protein, together with synthesis of specific enzymes and synthesis of specific polysaccharides. The polysaccharide composition of the cap wall differs from that of the cell wall of the stalk; and when the cap is being formed, not only are the characteristic cap polysaccharides produced but also the enzymes necessary for their synthesis. Cap formation is thus a typical example of cellular differentiation, involving precise regulation and localization of specific protein synthesis.

The first sign of cap development is a swelling of the stalk apex, in which special proteins and other substances are initially distributed in a ring. Later the ring differentiates in defined regions, giving rise to a number of local swellings and ridges which correspond numerically to the future cap partitions. The formative process involves disassembly as well as assembly, however, for the differentiation of the initial ring into local segments is accomplished mainly by lysis, presumably by lytic enzymes. These enzymes have become localized, with precise spacing, and must accordingly be under spatial control themselves and in turn must be responsible for the spatial patterning fundamental to the developing cap.

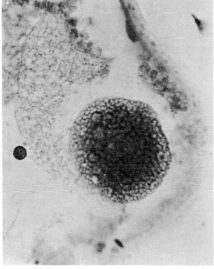

b

FIGURE 4.13 Nucleus of *Acetabularia*. (**a**) A normal nucleus, with sausage-shaped nucleoli. (**b**) A nucleus in which the energy production in the cytoplasm has been reduced; the nucleolus is spherical and contains a vacuole. [*Courtesy of J. Brachet.*]

Nuclear and Cytoplasmic Interaction: II

Early experiments on *Acetabularia*, mostly by Hämmerling, showed that the stalk quickly regenerates an amputated cap, and that it can do so even when the rhizoidal base with its contained nucleus is also amputated. In fact cap regeneration can take place many weeks after the nucleus has been removed from the cell. The capacity for regeneration decreases from the apex of the stalk toward the base, and the regulated synthesis of specific enzymes which accompany cap formation when the nucleus is present also takes place in its absence. Thus we might conclude that the nucleus is not directly or immediately concerned with the reconstitution phenomenon. Yet

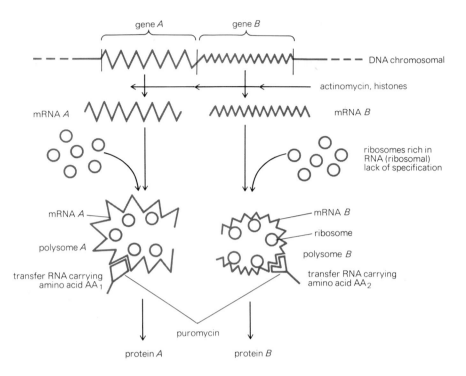

FIGURE 4.14 Schematic representation of genic control of protein synthesis, and action sites of the antibiotics actinomycin and puromycin. [*After Brachet, 1965.*]

two of the enzymes involved in the synthesis of cap polysaccharides are specified by the DNA in the cell nucleus. Moreover, all the information necessary for the production of the cap passes from the nucleus to the cytoplasm long before the cap is normally produced, for removal of the nucleus provokes extremely premature development of the cap in very young plants. This may be many weeks sooner than is normal, showing that the information is already present in the cytoplasm but that in some way its use is inhibited by the presence of the nucleus, possibly by nucleus-dependent continued growth of the stalk.

The width of the stalk is such that grafting experiments are possible between two species of *Acetabularia* which differ in details of cap structure. Two pieces of stalk, from different species, are simply pushed together end to end so that the naked cut cytoplasms come into contact and fuse. When a nonnucleated piece of stalk of one species is thus grafted to a nucleus-containing base stem of another, the first cap subsequently regenerated usually shows the character of the species represented by the stalk but a second regeneration exhibits the character of the nucleus-containing part, or at least is of a mixed or intermediate character, suggesting that some determinative influence spreads from the nucleus. Further experiments supported this interpretation: in binucleate grafts (i.e., grafts containing one nucleus from each of *Acetabularia mediterranea* and *Acetabularia crenulata*), intermediate-type caps regenerate which undergo no further change; but when trinucleate grafts are made, containing

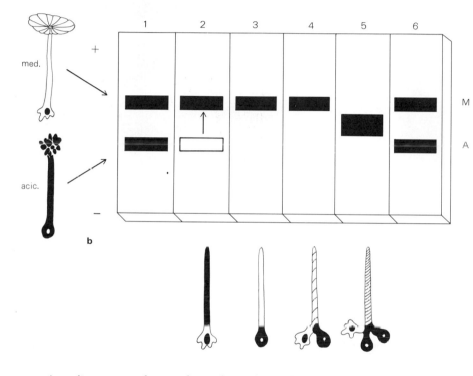

FIGURE 4.15 Interspecific grafting of nucleus and cytoplasm. (a) Cap of *Acetabularia mediterranea* (*left*) and *Acetabularia crenulata* (*right*), and intermediate-type cap formed from *A. mediterranea* base stalk containing *A. crenulata* nucleus. (b) Origin of specific proteins (as shown by zymogram patterns of acid phosphatase) in the two species *Acetabularia mediterranea* and *Acicularia schenkii*. Nuclei and cytoplasm of the two species were combined in various proportions. Each species contains a single acid phosphatase with a characteristic electrophilic mobility in starch gel. Strip 2 shows the conversion of the *acicularia* enzyme to the *mediterranea* type when a *med.* nucleus is substituted for an *acic.* nucleus. Strip 3 shows that the *med.* enzyme persists even after the *med.* nucleus is replaced by an *acic.* nucleus. Strip 4 shows that only *med.* enzyme is found in hybrid cytoplasm regenerated in the presence of one nucleus from each species. Strip 5 shows the intermediate phosphatase band frequently observed when two *acic.* nuclei and one *med.* nucleus collaborate in the production of new cytoplasm. This band apparently represents an intermediate stage in the conversion of *acic.* enzyme to *med.* enzyme. [*After Markert, 1963.*]

one *A. mediterranea* nucleus and two from *A. crenulata*, a morphological shift occurs in the direction of crenulate-type caps—presumably because of the imbalance between nuclear substances. With fused cytoplasmic systems containing nuclear material from *A. crenulata* and *A. mediterranea* in ratios ranging from 4 *cren.*/0 *med.* to 0 *cren.*/4 *med.*, the mixed nuclear types again produced mixed results.

The specific quality of the nucleus influences a regenerating cap's species character, although these experiments are merely indicative. Other experiments are:

1 Removal of a cap from a mature individual about to undergo nuclear subdivision and migration inhibits such activity until a new cap has been re-formed and grown to full size.

2 When a cap is removed, the nucleus shrinks to a fraction of its original size; when a mature cap is grafted on top of a young stem, the juvenile nucleus precociously undergoes divisions. Again we see a nucleocytoplasmic interacting system.

a

b

FIGURE 4.16 (a) Caps regenerated by enucleated stalks of normal individuals. (b) Caps regenerated by enucleated stalks in the presence of actinomycin. [*Courtesy of J. Brachet.*]

3   If a young capless cell is cut into three parts, the top segment continues to grow and develops a cap. The basal part containing the nucleus also grows in length and forms a new cap. The middle part of the stem remains alive but neither grows nor regenerates. Yet if the cutting off of the middle part is delayed for a time, following the cutting of the top part, the middle part, when later cut off, is found to have acquired the capacity to grow and to form a new cap.

4   If a nucleus is extracted, repeatedly washed to remove traces of basal cytoplasm, then inserted into the isolated middle piece of another cell, this otherwise inert piece grows and forms a cap at its distal end and rhizoids at its basal end.

These experiments support the conclusion that a growth and form-determining agent diffuses from the nucleus and travels toward the distal tip of the stem, where it becomes progressively concentrated. Alternatively, the presence of the nucleus establishes or reinforces the polarity of the middle stem piece, a gradient in so-called formative substances is established, and typical polarized cell differentiation develops.

Whatever the control mechanism for cap or rhizoid formation may be, the stalk itself is richly supplied with chloroplasts and mitochondria, which play a dominant energizing role. Both function largely independently of the nucleus, and energy production with regard to rate of oxygen consumption and rate of photosynthesis remains normal in enucleated fragments for several weeks. Such fragments are also capable of net synthesis of proteins for at least that period. Chloroplast DNA, different from nuclear DNA, continues to be synthesized in enucleated fragments, and so does transfer RNA, ribosomal RNA, and chloroplast RNA. In fact, very small droplets of *Acetabularia* cytoplasm containing mitochondria and chloroplasts form an enclosing membrane and exhibit internal cytoplasmic streaming for many weeks in isolation. Such is the cytoplasmic channel between basal nucleus and distal cap. How does the nucleus get its messages through to the site of morphogenesis, and how does a cap communicate with the nucleus?

The system open to analysis is almost ideally arranged: a basal nucleus which either produces or initiates the formation of "formative substances"; a long, tubular stalk through which formative substances or some other kinds of signals must travel; and a distal end where the more complex structural processes of assembly (morphogenesis) occur. It should not be forgotten that the stalk itself is a formed structure with characteristic length and diameter ratios.

In any analysis of such a differentiating or developing system it is important to recognize that controlling factors may be either permissive or determinative. Insufficient oxygen, for instance, will slow down or inhibit any process depending on oxygen, although oxidative processes as such may have little bearing on the morphogenetic event. In *Acetabularia* and other photosynthetic organisms the normal functioning of chloroplasts is a necessary condition for morphogenesis but does not account for its occurrence. Enucleate posterior parts have no morphogenetic capacity but contain chloroplasts. Anterior parts which have morphogenetic capacity depend on light; they do not grow in red light, which does not support photosynthesis. Yet the light relationship is not merely a switching on and off of photosynthesis, since the interruption of exposure to red light by only 1 minute of blue light per day is enough to induce cap regeneration. What the specific light receptors may be and how they are involved are not known.

The main problems, however, are the activity of the nucleus in relation to cap formation and the actual course of cap development. The nucleus, as we have seen, is remarkably large, although it diminishes in relative size as the cell increases in size. This perhaps suggests that cell size as a whole tends to outgrow nuclear growth and reaches a limit because of this. There is also the fact, already mentioned, that the cap removal induces a shrinkage of the nucleus, again indicating a precise mutual size relationship between nuclear and cytoplasmic volume. However this may be, the nucleus is also remarkable in that it contains many very large nucleoli rich in RNA, but little discernible DNA. A typical diploid nucleus, at least, must be present but is probably so small proportionately to the massive accumulation of RNA that it is virtually undetectable. Accordingly the build-up of the enormous *Acetabularia* cell must depend on the accumulation of nuclear RNA rather than on polyploidal growth of the chromosomal DNA.

Messenger RNA's from the nucleus pass into the cytoplasm and inevitably move toward the tip of the stalk, since, if for no other reason, that is the only direction open to travel. On the way they become involved in synthetic metabolism of the cytoplasm and the activity of chloroplasts. For instance, a particular photoperiodism,

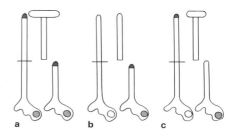

FIGURE 4.17 The distribution of messenger RNA. (a) Normal cell: mRNA manufactured in nucleus accumulates at tip of decapitated stalk and new cap regenerates. (b) Cell treated with ribonuclease: regeneration of enucleate pieces is blocked permanently, but pieces that still contain a nucleus are blocked only temporarily. (c) Cells treated with actinomycin: enucleated stalk regenerates new cap utilizing mRNA already present, but nucleated part does not; i.e., actinomycin combines with DNA and prevents synthesis of messenger RNA but has no effect on any RNA's already made. Puromycin, which blocks the actual assembly of amino acids into protein, prevents enucleate pieces from regenerating, but this effect is only temporary for nucleated pieces. Thus cap formation requires synthesis of proteins. [*After Brachet, 1965.*]

or photosynthetic rhythm, can be set up by exposing the alga to regular periods of light and dark, and this rhythm continues for a few weeks in enucleate fragments. If fragments containing a nucleus are combined with enucleate fragments having a different photoperiodism, the rhythm of the nucleated fragment prevails. In some way the nucleus is involved in the nature or operation of the elusive biological clock.

The main morphogenetic action occurs at the stalk apex, (1) as a continuing growth of the stalk during early stages, (2) as a production of rings of sterile branches at or near the tip during juvenile stages, and (3) as the development of the prospectively fertile cap. It is notable that the nucleus does not divide until the cap is fully developed and mature—i.e., that it does not divide at all during the growing period of the cell—and also that the immature cap can be redeveloped from the stalk apex repeatedly following its removal. Messenger RNA's from the nucleus accumulate at the stalk tip throughout the period of growth, and it is assumed that they carry the genetic information that will control the synthesis of the specific proteins which are essential for the formation of the cap.

In any case, two groups of so-called morphogenetic substances originating in the cell nucleus have been distinguished: (1) one by which species specificity, such as cap shape, becomes determined and (2) one that is necessary for the realization of such specificity but about which no more is known.

Indirect evidence of RNA activity at the tip is of three kinds:

1   Treatment of the enucleated upper part of the stalk with ultraviolet light, which is known to damage nucleic acids, destroys the capacity to develop a new cap.
2   Treatment of the same region with ribonuclease, the enzyme that breaks down RNA molecules, also inhibits cap regeneration.
3   Treatment with antibiotics such as D-actinomycin, which specifically inhibit the synthesis of messenger RNA, stops the accumulation of the substances essential for tip growth and cap regeneration even in cells that still retain their nucleus.

Assuming that mRNA is involved, if not mainly responsible, for the original development or the replacement of a cap at the distal end of the long stalk, the question arises: How does such a messenger emanating from the nucleus go through the mass of ribosomes of the stalk cytoplasm before it sees fit to combine with ribosomes concerned with morphogenesis.

Whatever the explanation, morphogenesis takes place at the stalk apex, and experiments with labeled substances show that the apex is the main center of cytoplasmic protein synthesis. Here polyribosomes are the most abundant. But protein synthesis occurs in the absence of morphogenesis and therefore cannot be solely responsible. Synthesis of cytoplasmic RNA also occurs mainly in the apical region, and, in fact, so does the laying down of new cell-wall material, so that altogether the situation becomes complex indeed. It is evident, however, that the apex has the composition and structural properties necessary for morphogenesis.

Altogether a tenuous chain of command is discernible between nuclear DNA and apical, cortical morphogenesis; yet it is but a beginning toward understanding the whole event. The linear coding for specific proteins is clearly an essential part of the morphogenetic process, but the construction process proper is essentially a cortical phenomenon, and there is every evidence that subtle cortical properties are as much in control of the outcome as is the nature of the materials brought to it or manufactured within it for incorporation. Genetic coding goes hand in hand with patterning properties of the cell cortex which are so little comprehended that it is difficult to speak of them. Nevertheless they exist and cannot be ignored.

CONCEPTS

Cortical organization is visibly represented by complex fiber systems supporting cilia and other organelles in ciliate protozoans.

The formative units of visible cortical pattern in ciliates are kinetosomes, or basal bodies, which may be scattered and inactive or which may assemble into pseudocrystalline patterns, mainly rows, of fiber-producing bodies.

A small fragment of cortex, in association with the nucleus, can give rise to an entire cortical complex.

Cortical fiber structures, if disassembled, are capable of reorganizing into normal patterns in the presence of the nucleus.

Regional, specialized cortical structures differentiate according to local nonvisible cortical properties.

Cortical pattern can be inherited, in part directly, in part indirectly, i.e., perpetuation and growth of cortical structure are at least semiautonomous.

The mechanism responsible for the inheritance and maintenance of cortical pattern, although expressed in the activity of kinetosomes, is not known.

Development of widely separated specialized cortical regions or organelles is synchronized by forces, of unknown nature and source, operating within the cell.

However independent the nucleus and certain cytoplasmic structures including the cortex appear to be, they mutually influence one another and form a truly interacting system.

Form may be established and maintained by extracellular self-assembling macromolecules and by intracellular self-assembling microfilaments, microtubules, and kinetic networks.

Morphogenetic control systems may be permissive or determinative.

Nuclear messenger RNA may determine specific structural protein synthesis at the cell cortex.

Cortical cytoplasm may determine nuclear activity.

Cytoplasmic structural differentiation is epigenetic.

READINGS

BRACHET, J. L. A., 1965. *Acetabularia, Endeavour,* **24,** September.

EHRET, C. F., 1960. Organelle Systems and Biological Organization, *Science,* **132:**115–123.

EPHRUSSI, B., 1953. "Nucleo-cytoplasmic Relations in Micro-organisms," Clarendon Press, Oxford.

GIBOR, A., 1966. *Acetabularia:* A Useful Giant Cell, *Sci. Amer.,* November.

HÄMMERLING, J., 1967. The Role of the Nucleus in Differentiation, Especially in *Acetabularia,* in E. Bell (ed.), "Molecular and Cellular Aspects of Development," Harper & Row.

HARRIS, H., 1968. "Nucleus and Cytoplasm," chap. 1, Oxford.

LWOFF, A., 1950. "Problems of Morphogenesis in Ciliates," Wiley.

MARKERT, C., 1963. The Origin of Specific Proteins, in J. M. Allen (ed.), "The Nature of Biological Diversity," McGraw-Hill.

NANNEY, D. L., 1968. Cortical Patterns in Morphogenesis, *Science,* **160:**496.

PRESTON, R. D., 1968. Plants without Cellulose, *Sci. Amer.,* May.

SONNENBORN, T. M., 1963. Does Preformed Cell Structure Play an Essential Role in Cell Heredity?, in J. M. Allen (ed.), "The Nature of Biological Diversity," McGraw-Hill.

TARTAR, V., 1962. Morphogenesis in *Stentor, Advance. Morphogenesis,* **2:**1–26.

———, 1961. "The Biology of *Stentor,*" Pergamon.

WEISZ, P., 1954. Morphogenesis in Protozoa, *Quart. Rev. Biol.,* **29:**207–227.

WERZ, G., 1969. Morphogenic Processes in *Acetabularia, Colloq. Ges. Biol. Chem.,* **20:**167–186.

———, 1965. Determination and Realization of Morphogenesis in *Acetabularia,* in "Genetic Control of Differentiation," Brookhaven National Laboratory, Symposium in Biology, no. 18.

CHAPTER FIVE
MULTICELLULARITY

Development, as commonly understood, is the progressive attainment of form (morphogenesis) and regional differentiation of structure and substance (histogenesis and cytodifferentiation) by multicellular organisms. The developmental processes as seen in protistan organisms and the individual maturing cells of multicellular organisms are specifically those of cytodifferentiation, which is only one aspect of development in the broader sense. Comprehension of the inclusive developmental process depends not only on knowledge of the manifold major and minor events concerned in the process but also on points of view. Which came first, for instance, the chicken or the egg? This was a long-debated question which polarized the approach, although a third option now seems more plausible, that neither chicken egg nor chicken came first but that the two evolved together. In other words, in the study of development we are dealing usually with eggs and adults, and at least one biologist regards development as that which is necessary to fill the gap between them. Molecular biologists regard the cell, and therefore the egg, as primary, and look upon the developed organism as a product of what is set in motion at or shortly after development begins. Evolutionary biologists are more aware that natural selection acts mainly upon developed organisms, although they recognize that all stages of development and life cycle, including the egg itself, are subject to selective forces. The perspective varies. The reality includes all these phases.

CONTENTS

The multicellular state
The role of extracellular material
Morphogenesis in cellular slime molds
  Aggregation
  Differentiation
  Culmination
Cell adhesion and communication
  Cell movement
  Contact inhibition
  Cell adhesion
  Cell communication
Concepts
Readings

THE MULTICELLULAR STATE

The multicellular state is a state which has been attained in evolution by unicellular organisms, or is attained in development, as a rule, by the multiplication of the isolated single reproductive cells of a multicellular organism. It may also be acquired through the coming together of numerous separate cells (aggregation) to form a unified mass. However it comes about, its primary feature is the adjoining or joining of cell to cell. The properties of cell surfaces and cell walls become paramount.

The multicellular condition is often regarded simply as a state that contrasts with the unicellular condition; this attitude is implied in the phrase "the commitment to multicellularity," as though the multicellular condition were a great but single advance. We tend to ignore the frequency with which, outside of the Metazoa, unicellular organisms become multicellular, for instance, among ciliates (*Epistylis*), flagellates (*Volvox*), blue-green and other algae, and bacteria. Each can contribute to an understanding of the means by which cells cohere as multicellular entities, but each does so in its own way in terms of its special cortical and extracellular materials. The same holds for the two overwhelmingly successful cases, the animals and plants. As multicellular organisms they should be considered separately as in-

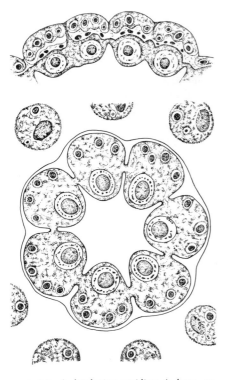

FIGURE 5.1 A developing gonidium (a large, somatic reproductive cell) of *Volvox globator*, showing cell bridges between dividing cells which persist in attenuated form in the fully developed multicellular organism. [*After Janet, 1912, Limoge.*]

dependent natural experiments in multicellularity, to see what is common to both and what is unique in each. Even the two great kingdoms probably should be regarded as having multiple unicellular origins; in plants, the various groups of algae, particularly, and in animals the sponges, coelenterates, and possibly other groups may well have evolved independently as multicellular organisms. Just as it is unsafe to generalize concerning the nature of the cell from studies made exclusively on either procaryote or eucaryote cells, so generalizations concerning the multicellular state based on either plants or animals exclusively should be made with caution. There is much that they have in common and much that is different between them; study of both is enlightening.

A comparable situation is seen in a broader study of life. How far can we generalize concerning the nature of life from studies of earthly life alone? The discovery of even the lowliest forms of life on another planet (it is still hoped that there may be life on Mars) would supply a basis for comparison and give some indication of which properties may be universal and which are special to life on the planet earth. General concepts concerning multicellularity will undoubtedly emerge from studies relating to a variety of organisms, but in the present state of the science they need to be formulated in connection with specific types. Concepts and specifics must be considered together. At the same time certain general features are obvious from the start.

THE ROLE OF EXTRACELLULAR MATERIAL

Cells may be held together by an enveloping mass of extracellular material, with or without cell contact or cell junction of some kind, as in blue-green algae, *Volvox*, cartilage, etc., or they may adhere to one another by their adjoining surfaces, as in a mass of soap bubbles. The basic questions concern (1) the nature of cell interfaces and junctions, (2) the nature of the apparent supracellular agency that confers an individuality or integrity upon a conjoined mass of cells, and, not least perhaps, (3) the fundamental role of the multicellular condition.

Desmosomes, the most obvious kind of intercellular junctions, appear to play an important role with regard to cell support. In the larval epidermis of the newt, for example, an acid mucopolysaccharide material within or lying upon the external cell-unit membrane layer is seen to be continuous with pillars, partitions, or strands which bridge the extracellular gap at the adhesion sites. Intracellular components of desmosomes include a dense plaque anchored to and parallel with the internal cell-membrane surface, and numerous tonofilaments which approach the plaque from the filamentous cytoskeleton of the cell. Although a few filaments actually enter the plaque, the majority loop at some distance from it and return toward main filament tracts of the cell skeleton. The orientation of filaments as they approach desmosomes appears to reflect the relationship of a given adhesion site to the overall cell architecture. No filaments have been seen to pass into or through cell membranes. In this view, the cell membrane may be looked on as separating intracellular supportive structure from an extracellular adhesive mechanism. In fact throughout the protistans individual size and longevity appear to go together, with nuclear replication involved in some manner (multinucleate, polyploidal, or nucleolar) in maintaining the system. Yet these are limited accomplishments. The surface/volume ratio sets a restriction on growth and activities. It is notable that in the spectacularly large *Acetabularia* and in the multinucleate but noncellular slime molds, e.g., *Physarum*, a relatively large surface/volume ratio is maintained through extreme departure from a spherical shape—in *Acetabularia* by the long, filamentous stalk and shallow cap, in *Physarum* by the very thin sheetlike and filamentous extension of the protoplasmic mass.

Increase in individual size and accompanying longevity have clearly been of great evolutionary value, even though an unseen microbial world remains as rampant as ever. We regard the evolution of the eucaryote cell as a great advance from the procaryote cell; yet it is plausible to see in it an elaborate adaptation to its own relatively large size, even in comparatively small and more or less spherical kinds. A sufficiently high surface/volume ratio is maintained through the agency of extensive internal membrane systems, i.e., the endoplasmic reticulum as a whole, together with membranous organelles. Compared with the procaryote cell, size is increased, complexity is increased, and longevity is increased, yet basal metabolic rate is greatly reduced. Apart from the outstanding but comparatively few cases of gigantism among protistan organisms (each with its own special devices for maintaining an overlarge cell), the size of cells in general apparently represents the comfortable upper limits attainable. Multicellularity has been and remains the primary and obviously most successful means of increasing organismal size while maintaining standard eucaryote cell surface/volume ratios.

Just as the adaptive response of the emerging eucaryote cell to the need for maintaining adequate surface/volume relationships has been by an internal extension of membrane systems, so multicellularity has made possible the maintenance of increasingly large organisms. These, however, are not simple three-dimensional protoplasmic masses, any more than are the eucaryote cells themselves. The membrane, which by definition has a relatively enormous surface and little thickness, remains the primary structure, only now it consists of individual cells adjoined side by side to form epithelial layers typically one cell thick but of possible almost unlimited extension. Epithelial structure is the basis of most simpler forms of multicellular animals. It is the predominant state at least in the early stages of animal development, and it is also typical of much of the structure of both lower and higher plants. Other cell conformations are possible, however, and in most animals and plants are present.

In all cases the multicellular state leads to the concept of "the organism as a whole," for all multicellular organisms, whether they consist of a few hundred cells or a hundred billion, are clearly integrated as unified systems. Is the organism as a whole essentially a social cell state, or are the individual cells significantly or mostly under the control of a supracellular agency which overrides cell individuality, although contributed to by the constituent cells? If the first is true, then, how do the cells communicate in order to maintain an organized cell society? If the latter is true, what is the nature of the supracellular guidelines? There are no easy answers.

The essential feature of becoming multicellular is that the products of cell division, whether of a subdividing egg, a spore, or the zygote of a colonial protist, cohere. Whether cells cohere or not depends primarily on whether the affinity for one another of the new cell surfaces formed during cell division is greater than their affinity for the aqueous environment. Such affinity varies according to the specific nature of the cells and the electrolyte content of the medium. Essentially the same situation holds for cells that are initially free but subsequently come together as aggregates, a phenomenon characteristic not only of the cellular slime molds but also of disaggregated tissues of metazoa such as sponges and vertebrates and even in some cases of the early development of eggs.

The development and maintenance of form generally call for supportive material or structure. Within the individual cell, cortically located microtubules and other filaments play an important if not exclusive role in this respect. In multicellular assemblies they may still retain a contributory role, but extracellular substances assume a major function. The question is whether these extracellular substances are of primary or secondary importance. The multilayered, pseudocrystalline collagenous basal membrane underlying the epidermis of vertebrates is an example; the conjoined cellulose walls of plant cells, similarly consisting of a multilayer assembly of oriented (in this case cellulose) fibrils is another. To what extent are these organized, oriented extracellular structures responsible for the form and integrity of the cellular systems they clearly support? According to Picken, "The importance of extracellular materials and structures in conferring rigidity, in maintaining shape, and in permitting the range of cell action to extend in space far beyond that of the single cell, cannot be overestimated. Indeed, although the existence of multicellular organisms is not uniquely determined by extracellular materials, it is hardly an exaggeration to say that these are chiefly responsible for the shape of multicellular organisms."[1] That they play such a role is unquestionable; yet this is not to say that they are the controlling agent of the growth and multiplicative processes primarily responsible for the pattern and integrity of a specific multicellular system, although this has been suggested.

However this may be, the supportive role of such materials is obvious, and, moreover, each particular substance has its own innate properties which make possible, limit, or exclude certain structural organizations. Much of the difference between animal and plant tissues, apart from the absence or presence of chlorophyll, may result from the basic difference in the nature of the extracellular substances, even though polysaccharides are present in each. Animal cells typically possess an external coat of mucoprotein, consisting of collagen and high-polymeric polysaccharide closely

[1] L. Picken, "The Organization of the Cell," p. 441, Clarendon Press, Oxford, 1960.

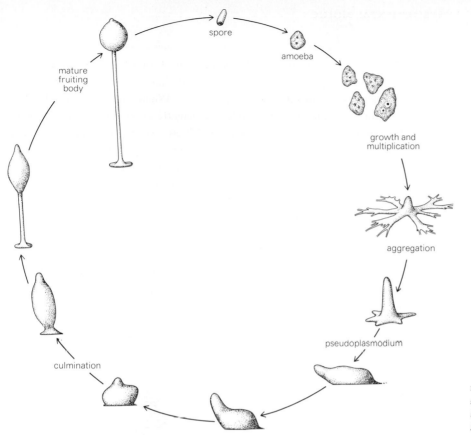

spore

amoeba

mature
fruiting
body

growth and
multiplication

aggregation

pseudoplasmodium

culmination

FIGURE 5.2 Life cycle of the slime mold, *Dictyostelium discoideum*. [*After Wright, copyright 1966 by the American Association for the Advancement of Science.*]

associated down to molecular dimensions. As a rule substances of this class are highly hydrated, so that they tend to form voluminous aqueous gels of a relatively low degree of orientation, rather than structures as compact as cellulose. Even where the crystallite structure of the extracellular polysaccharide is compact, as in the widespread cuticular chitin, the formative animal cells (unlike plant cells) have a greater affinity for each other than for the product, so that they retain an epithelial arrangement. Presumably, in plants, the affinity of the protoplasmic surface for the cellulose fabric is greater than that of the protoplasmic surfaces of adjacent cells for one another, and consequently each cell tends to surround itself with a sheath of cellulose that is more or less discrete because of its high degree of crystallinity. Various phenomena and problems relating to multicellularity, including the role and nature of the extracellular secretion, are well illustrated in the life cycle of the cellular slime molds, of which the species *Dictyostelium discoideum* has received the most attention.

MORPHOGENESIS IN CELLULAR SLIME MOLDS

The cellular slime molds, like other fungi, reproduce through spores. The life cycle, in brief, is as follows: As spores germinate, each liberates a small amoeba which

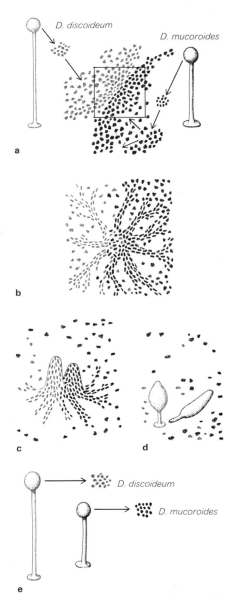

divides repeatedly by binary fission; the population of amoebae continue to divide as long as an adequate supply of bacteria (the colon bacillus, *Escherichia coli*, is employed in laboratory cultures) permits them to feed and grow. When the bacterial food supply is exhausted and the population of amoebae sufficiently dense, they begin to stream together to form cell masses, or aggregations. These assume a sluglike form and appear to move about for a while but eventually become upright. The anterior cells of the slug give rise to stalk cells, the posterior cells to spore cells which form an ovoid mass at the tip of the stalk, the whole being known as a *sporocarp*. The proportioning of stalk and spore cells varies with the species. In *Dictyostelium mucoroides*, for example, more cells contribute to stalk and relatively fewer become spores, so that the sporocarp shape is correspondingly different. In another form, *Polyspondylium violaceum*, also well studied, the stalk bears whorls of small branches, each of which carries a minute fruiting body at the end, in addition to the main fruiting body borne at the top of the main stalk. Such differences in form are helpful in analyzing the process of morphogenesis.

An organism with such a life cycle obviously lends itself to the study of many aspects of cell and developmental biology. Primarily three phases predominate:

1   The phase of *aggregation*, whereby a local population of free amoebae congregate at a center and become an integrated mass
2   The phase of the slug, or pseudoplasmodium, with the degree of *differentiation* that its cells exhibit
3   The phase of morphogenesis, whereby a slug transforms into a mature fruiting body, the sporocarp, a process known as *culmination*

Each phase presents problems of its own, though all relate to the attainment of the multicellular state and its subsequent self-assembly into an organized and differentiated reproductive structure.

Aggregation

Growth and cell division proceed as long as the amoebae have available food. As long as they are able to feed voraciously on bacteria in the surrounding medium, they move about and actively ingest bacteria by means of the pseudopodia which continually form from the amoeba cell surface. In this circumstance, in other words, growth and division proceed at maximum rates. The cell generation time at room temperature is about 3 hours, and the cell surface exhibits mobility. Neither activity is readily compatible with a stable multicellular system, any more than a crowd running wild is likely to deliberate the affairs of state. When the food supply becomes exhausted, growth ceases, cell mobility lessens, cell surfaces become more stable, and cell divisions become greatly reduced though not entirely ended. Only then does aggregation begin. It is as though a cell cannot be a rampant, gluttonous, self-replicating individual and also be a responsive contributive member of a cell society at

FIGURE 5.3 Sequential stages of development following the intermixture of amoebae of *Dictyostelium discoideum* with amoebae of *Dictyostelium mucoroides*, showing (a) intermingling, (b) aggregation, (c) species-specific segregation, (d) separation of pseudoplasmodia, and (e) formation of separate fruiting bodies. [*After Holtfreter.*]

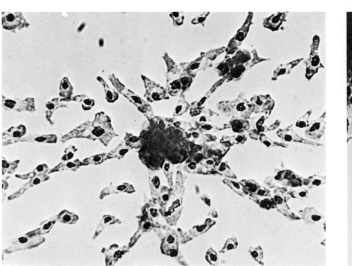

a

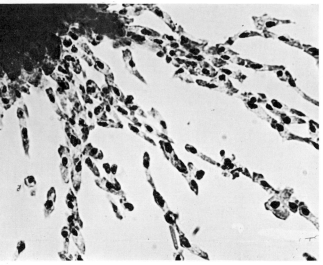

b

FIGURE 5.4 (a) Early and (b) late stages in aggregation of amoebae at a center showing alignment and head-tail contacts of amoebae in migration. [*Courtesy of J. T. Bonner.*]

the same time. Lack of food and the consequent suppression of growth, cell division, and random movement permit other forms of cell behavior to be expressed. Aggregation depends on the capacity of cells to adhere to one another and also on some directive agent which causes them to migrate toward a common center, i.e., they must come together before they can cohere.

If a young aggregation center is repeatedly dispersed by mechanical means, the cells may reaggregate radially around a so-called founder cell a hundred times or more, at least in *D. discoideum.* Such a founder cell was therefore thought to secrete a specific attractor substance, which was called *acrasin.* This, it was thought, formed a diffusion gradient extending outward from the source, permitting cells within a certain distance to orient and move in the direction of higher concentration, i.e., by chemotaxis. The existence of such an orienting substance, though not its identity, was shown when it was isolated in vitro, reintroduced into a culture, and seen to orient the sensitive amoebae. Twenty years of intensive study in several laboratories elapsed between the time when J. T. Bonner named the attractive substance acrasin and its final identification as a specific chemical compound secreted by certain of the amoeboid cells.

Acrasin was shown to be destroyed, presumably by an enzyme, soon after its secretion into the medium surrounding the cells. The system has the general features of chemical control systems, as in, for instance, the neuromuscular synapse, where a transmitter chemical, acetylcholine, is released by the nerve endings but is limited in action by the extracellular enzyme cholinesterase, which destroys it. It is now known that:

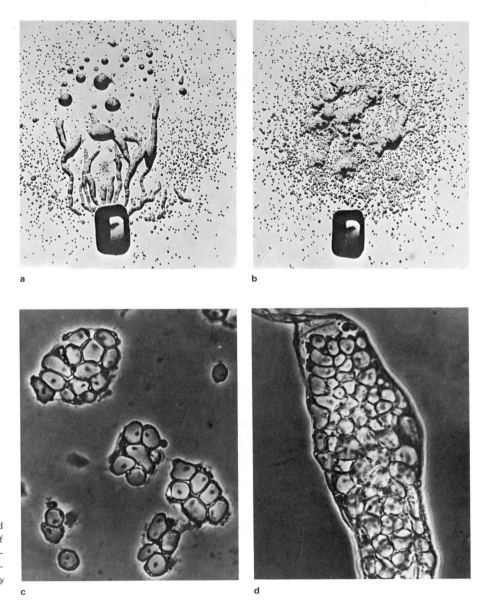

FIGURE 5.5 (**a,b**) Aggregation of amoebae toward diffusion center of cyclic AMP. The small block of agar at bottom of each photo contained the attractant. (**c,d**) Direct transformation of unaggregated cells into stalk cells by cyclic AMP. [*Courtesy of J. T. Bonner.*]

1 Slime-mold amoebae aggregate in typical manner when a substance widely involved in numerous hormone reactions in mammals, namely, cyclic 3',5'-adenosine monophosphate (3',5'-AMP), is added locally to a culture, and this substance is given off by slime-mold amoebae.

2 The amoebae secrete the enzyme phosphodiesterase, which breaks down cyclic AMP.

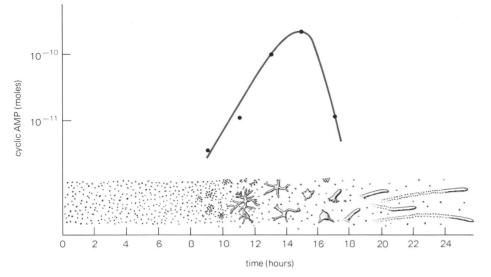

cyclic AMP (moles)

10^{-10}

10^{-11}

time (hours)

FIGURE 5.6 Concentration of cyclic AMP in medium at successive stages of life cycle. [*After Bonner, 1969.*]

It is significant that both the attractant and the agent that destroys the attractant are almost entirely extracellular, which is reasonable, since these two substances must be partly responsible for the aggregation process that requires communication between cells that are initially separate from one another. Aggregation masses may consist of more than 100,000 cells or, in extreme cases, of little more than a dozen or so.

Aggregation accordingly appears to be a response to several factors working together:

1   The cessation of feeding and growth, which is associated with sensitivity to the orienting attractant substance

2   The density of the amoeba population, which determines how many amoebae are within a certain distance of one or several founder cells

3   The combined diffusion gradients of the attractant and its degrader, which define the area of influence of a particular center.

It remains to be seen to what extent such chemical gradients serve as guidelines in developing systems generally.

Differentiation

The process whereby an aggregate becomes an erect fruiting body, although divided for convenience into a so-called slug phase, or pseudoplasmodium, and a culminating phase, is a continuous process of differentiation and morphogenesis. That the slug presents problems peculiar to this phase, particularly the means by which it appears to migrate over the substratum, should not distract attention from the continuity of the overall differentiation and assembly process.

 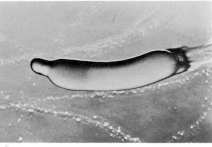

a   b

FIGURE 5.7 (a) Side and (b) top views of slug (pseudoplasmodium) near close of migrating period, showing tip lifting from substratum and a trail of mucus behind. [*Courtesy of David Francis.*]

A change takes place in the cells of a colony even before aggregation is accomplished, for they change from being amoebae with active, blunt pseudopodia to elongate, rectangular cells oriented in the line of march. Aggregation, moreover, is a progressive process. Cells pile up at the center while others are still streaming in toward the periphery. When the process is complete, therefore, the central cells have already been part of the aggregation mass for a considerably longer time than those close to the periphery, so that both in location and in the length of exposure to new circumstances the mass as a whole is regionally differentiated. Cells moving inward from peripheral areas inevitably tend to pile up at the center, and, not surprisingly, an aggregate acquires a hemispherical center. This normally elongates upwards as late-coming cells move into the center from below, until the whole assembly falls on its side. In most species of slime molds it begins to form a fruiting body at once, but in two species, including *D. discoides,* the fallen aggregation becomes a migrating pseudoplasmodium, moving as a unit and secreting a slime sheath as it goes. The mechanics of movement poses a special problem; it is enough here to say that although the amoebae within the sheath vigorously move their pseudopodia in a coordinated manner and in some way move as a mass with the original tip forward, their traction is with the inside of the secreted sheath, not directly with the ground.

From the beginning of aggregation to the early stage of culmination following the slug phase, whatever processes of differentiation have occurred are reversible. Dispersed cells of an aggregate will reaggregate. Dispersed cells of a slug can re-form a slug. Cells derived from any point of the life cycle short of culmination, when cultured in isolation, give rise to cultures that undergo the normal fruiting cycle. In this sense the cells are equipotential; this does not imply that differentiation is not taking place, only that it is reversible.

That some differentiation already exists is shown by experiments in which a colored anterior portion of a slug is grafted into the posterior region of an intact migrating slug. The vitally stained portion thereupon moves forward through the mass until it assumes the anterior position in the new combination slug, indicating that anterior cells are in some way already different from posterior cells. On the other hand if a normal slug is divided transversely into two parts, each part exhibits the properties of a whole and transforms into normal fruiting bodies, though of reduced size. Yet left alone in the intact original slug, each part gives rise to a particular component of the whole, i.e., stalk or spores.

As the name implies, slime-mold cells secrete slime, or mucopolysaccharide, which is very evident as the containing sheath of the slug and as a trail attached to the substratum, indicating the path of slug migration. During the culminating stages of fruiting-body formation such extracellular material assumes even greater importance.

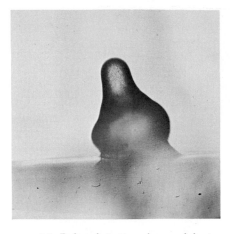

FIGURE 5.8 Early culmination phase and beginning of stalk formation in *Dictyostelium discoideum*. [*Courtesy of K. B. Raper.*]

Culmination

At the end of the migration period the tip of the slug becomes stationary while the rear portion continues to push forward beneath it, causing the original front end to rise up. The process has been followed by various means, including grafting the tip of a vitally stained slug onto a decapitated unstained slug, thus making relative cell movements more easily followed.

1   The tip cells begin to form a stalk, by laying down an internal cylindrical sheath, within which cells become vacuolated stalk cells.

2   Cells on the outside move upward and extend the sheath, become trapped inside it when they reach the apex, and become additional stalk cells. This process has been called a *reverse fountain*. The effect is as though a stalk (sheath plus contained vacuolating cells) is being pushed downward through the cell mass as a whole. Yet this is only a relative movement, and in reality the stalk as it is laid down remains almost stationary while the more posterior cells of the slug phase, crowding around the base, climb higher and higher as additional stalk is added at the top. Moreover, the stalk cells within the sheath, during vacuolation, increase about five times in volume, and push upwards accordingly. A tall stalk thus forms that is dependent on the continual laying down of an extracellular sheath and the expansion of the contained stalk cells. The rigidity of the confining sheath causes the expansion to be expressed mainly as linear growth. Without the sheath no stalk could form.

3   Finally the cells of the posterior part of the slug overtake the stalk-forming cells and accumulate at the top to form a more or less globular mass of spore cells contained within widely extended apical sheath material.

Questions which arise concern the agency that causes some cells of an aggregate to differentiate toward stalk cells and others toward spore cells; the processes involved in the production of the sheath substance; and the levels of control of morphogenesis.

Is there a genetic basis for the regional differentiation? Apart from mutations,

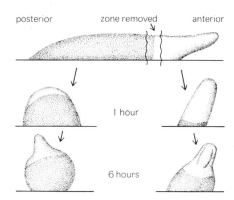

FIGURE 5.9 Partially differentiated migrating cell mass, stained with Nile blue sulfate for polysaccharides, is divided into an anterior nonstained part and a posterior stained part. Reconstitution of a whole system occurs in both parts, giving rise in each case to a prespore and a prestalk region. [*After Bonner, 1967.*]

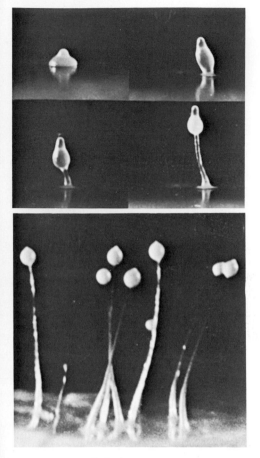

FIGURE 5.10 Stalk formation and sorocarp development of *Dictyostelium discoideum*. [*Courtesy of J. T. Bonner.*]

all the cells forming an aggregate are genetically alike. The same characteristic development occurs whether an aggregating culture originates as a single amoeba clone or as a mixed assembly from several sources. The location of cells within an aggregate or a slug is related to its destiny in the mature system; yet the cells of each part of an aggregate or a slug contain all the information necessary for development, since an isolated part can give rise to a whole.

Certain mutant forms show that a genetic basis exists, although not how it operates. In the mutant "aggregateless," the cells feed, grow, and divide in a normal manner but fail to aggregate when the food is exhausted. In the mutant "fruitless," they feed, grow, divide, and aggregate, but go no further. In other mutants they may form large numbers of small slugs that produce irregular fruiting bodies. But if the mutant "fruitless" cells, which form loose aggregations, are mixed with "aggregateless" cells, which never aggregate, together they produce normal mature fruiting bodies. These and comparable experiments *suggest that cells communicate by means of a chemical passed from cell to cell across cell membranes,* somewhat as some cells secrete acrasin and thereby communicate with neighboring cells and direct their movement and orientation. This is a plausible hypothesis of general importance.

Although some evidence exists that gene products respectively responsible for the sequential events of development may be produced in sequence, the gene-determined information is probably present from the beginning in all the amoeboid cells, and development simply proceeds in a more or less normal manner until some important component is found to be missing, when development stops. The question remains: How do two different deficient types of cells compensate for each other's deficiencies, as though the two types conjugated in pairs to produce normal cells? There is no evidence that any such cell fusions take place, and we are left with the problem of how two kinds of abnormal cells can create a normal structure. Some mutual interchange of substance or signals seems to be required, or perhaps complementary contributions to an essential intercellular complex.

However this may be, the later stages of development, at least, appear to be closely related to the secretion and properties of the extracellular sheath material. It would be a mistake to look on any such extracellular structure as the one agency responsible for the determination of multicellular organization or pattern, but neither should its possible importance be ignored. Differentiation is a complicated phenomenon which may well be brought about by varied and independent forces, all of which are required. Changes in enzyme activity, RNA metabolism, gene activation, levels of specific substrates or inhibitors may all influence a reaction critical to morphogenesis. Studies of the production of sheath substance in the slime mold have been made with this concept in mind.

The sheath material, of both stalk and spore coats of the fruiting body, is composed of an insoluble polysaccharide complex of cellulose and glycogen. It represents 5 percent of the dry weight of the fruiting body and is present in insignificant amounts at the beginning of differentiation. Since the beginning of differentiation coincides with cell starvation, competition must occur within and among the cells for sources of energy and for precursors of the polysaccharide complex. Polymerization of glucose to form cellulose and its polymerization to form glycogen appear to be competitive processes. Any changes affecting the synthesis of one polymer, such as changes in concentrations of substrates, primer substances, effector or inhibitor agents, and enzymes, will automatically be reflected in the synthesis of the other. The control of the synthesis of the two polymers must therefore be interdependent. Specific enzymes are present long before the initiation of the differentiation process, but conditions or precursors necessary to their function appear at different times. We are compelled to consider the timing of events at the molecular level and the complexity of the control of time as vital to further understanding of the differentiation process.

As we have seen, the sequence of developmental changes occurs in an invariant order. Morphogenetic activities are accompanied by a sequential appearance of a variety of specific materials. One is an acid mucopolysaccharide composed of galactose,

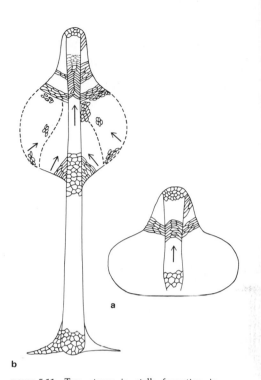

FIGURE 5.11 Two stages in stalk formation in *Dictyostelium discoideum*. **(a)** Sectional view of young sorocarp, showing that young stalk extends almost to substratum and that cells in basal area are beginning to vacuolate, creating an upward pressure which displaces the still-plastic cells above. **(b)** Sectional view of later stage, showing preorientation of prestalk cells, and differentiation of spores from periphery inward. [*After Raper and Fennell, 1952.*]

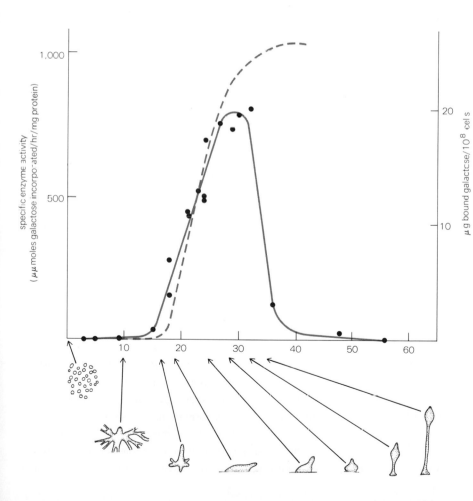

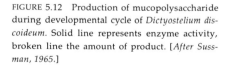

FIGURE 5.12 Production of mucopolysaccharide during developmental cycle of *Dictyostelium discoideum*. Solid line represents enzyme activity, broken line the amount of product. [*After Sussman, 1965.*]

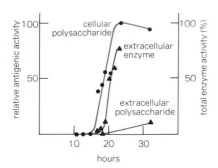

FIGURE 5.13 Relation between mucopolysaccharide synthesis and enzyme release. [*After Sussman, 1965.*]

galactosamine, and galacturonic acid. According to Sussman it first appears when the completed aggregates are transformed into pseudoplasmodia. It accumulates to a level of about 2 percent dry weight by the end of fruiting-body construction. All the polysaccharide is sequestered in the spore mass, none in the stalk. Several mutants whose development is blocked at a stage prior to the start of fruit construction do not synthesize the polysaccharide, while one mutant whose temporal controls are deranged displays a consonant variation in polysaccharide synthesis. Thus, as Sussman states, *the formation of the mucopolysaccharide appears to be regulated temporally, spatially, and quantitatively by the overall developmental program.* Incorporation of galactose is apparently brought about by the enzyme UDP-Gal polysaccharide transferase.

Specific enzyme activity first appears in *D. discoideum* after the completion of cell aggregation, just before the onset of mucopolysaccharide synthesis. It reaches a peak during the terminal phases of morphogenesis, when synthesis stops. The bulk of the enzyme is then released by the cells during the next few hours and later disappears. This preferential release of the transferase at the close of morphogenesis, and the subsequent restriction of the polysaccharide product to the spore component of the fruiting body indicate spatial regulation and restriction.

CELL ADHESION AND COMMUNICATION

Many of the phenomena and problems in the life cycle of the cellular slime molds are encountered in the study of tissue cultures of metazoan organisms—i.e., of cells or tissues cultured in artificial media outside the body—initiated by Harrison. At the moment we are concerned with the means by which normally stationary cells move about when in culture, with the cause of their immobilization when they come together, and with the nature of mutual cell adhesion in general. These are all properties of great importance in the development and maintenance of multicellular organisms.

Cell Movement

The studies of individual cells in tissue culture, particularly those of Weiss and his colleagues over many years, have shown that cells in suitable culture medium, separated from their neighbors, usually move continuously and nondirectionally over a glass surface. Contact of this type, i.e., contact with a surface which is not that of another cell, does not inhibit movement; movement continues but in a way that preserves contact. Yet the cells do not move by amoeboid movement but flatten and glide over the substratum without pseudopodial formation; nor is there any evidence of protoplasmic flow. Such cells move across a glass surface by making intermittent contacts resulting from ruffling, peristaltic movements in the whole plasma membrane. The ruffles are waves of large-scale expansions that travel along the cell membrane.

Phase and interference microscopy show that the leading edge of a moving fibroblast consists of an exceedingly thin, fanlike membrane, 5 to 10 μ wide and closely applied to the substratum. This membrane undergoes continual folding movements which beat inward, i.e., appear as ruffles. Cultures of isolated embryonic cells of amphibian embryos show wandering cells with the same kind of thin, advancing, fanlike edge.

Contact Inhibition

Cells in the body do not as a rule move about, but remain in stationary contact with neighboring cells. In culture, epithelial cells and mesenchymatous cells, whether fibroblasts or another type, tend to move freely about, much as malignant cells do both within and outside the body. Nonmalignant cells, however, even in culture, become stationary when they make contact with one another. When a ruffled membrane makes contact with a neighboring cell, an adhesion usually forms at the point of contact. The membrane ruffles then become stationary: a contact signal or reaction prevents further movement in the original direction. If only a few contacts have been made, the cell may move in a new direction and break the initial contacts with a snap, after considerable tension has been produced. If many contacts have been formed, the large-scale movement of the cell ceases entirely.

Movement and inhibition of movement of this sort undoubtedly play vital roles during embryonic development and in wound healing. Epithelial cells typically unite to form flat cohesive sheets, or epithelial tissue, one cell layer thick; and an epithelial sheet will invariably spread, providing it adheres to a substratum and has a free edge. The advance of the free edge is accompanied by a similar advance of the sheet behind it, in a fully integrated manner so that the sheet spreads as a unit. Even so, the activity of the individual cells appears to provide the momentum for the movement of the whole. At the onset of spreading by a wound epithelium, for instance, only the marginal cells show movement; the cells behind join in the advance at progressively later times. The delay in their mobilization increases with their distance from the wound, somewhat as in a line of traffic moving forward when the light turns green.

Spreading of an epithelial sheet is stopped by contact with another epithelial edge. Contact inhibition operates equally when isolated wandering cells meet or when spreading edges of adjacent epithelia come together. In either case the cells or tissue do not spread over cells or tissue of their own kind. The nature of the paralysis of movement brought about by contact is not known. A suggestion is that it results from some chemical or electrical interaction—which leaves the question wide open.

Cell Adhesion

Adhesion is the first step toward multicellularity. The means of adhesion of one cell to another is a much-debated topic, points of view varying according to whether

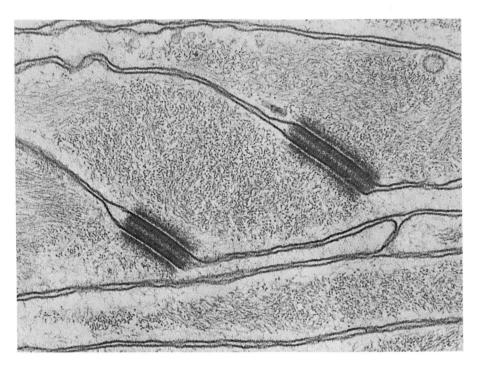

FIGURE 5.14 An intercellular junction (desmosome) in muscle tissue. From three to five bands of differing stain contrast are visible between the plasma membranes. The adjacent plasma membranes lie about 100 Å apart. Dense-staining material inside the cytoplasm forms part of the desmosome. [*Courtesy of D. Fawcett.*]

only what can be seen in electron micrographs is considered, or whether indirect evidence of intracellular material not visible by electron microscopy is accepted. Epithelial cells have been the most studied because of their characteristically firm adherence to one another. Special junctional attachment sites have been recognized for some time, and junctional areas are now categorized as follows:

Zona occludens   Tight junctions where the outer components of unit membranes of adjoining cells appear to fuse and to act as a seal, preventing passage of substances through the cell interface

Zona adherens   Area where the two plasma membranes of adjacent cells approach within an approximate distance of 200 Å, and the intervening space and the cortical cytoplasm appear to be of increased density and to represent continuous circumferential attachment regions between cells

Macula adherens, or desmosome   A small ellipsoid and very dense area between cells, consisting of parallel layers of dense material and extending from within the cytoplasm of one cell to within the cytoplasm of the other

Junctions similar to the zona adherens have been seen in connective tissue cells, but desmosomes and tight junctions have not been seen. This zone, or 100- to 200-Å gap, is generally considered as the principal adhesion zone, whatever the cementing materials within the space may be. That cementing materials may bind cells together is indicated by the effects of proteolytic enzymes in disaggregating tissues with their constituent cells.

Cell adhesion is selective: like cells adhere together, usually to form a monolayer over a substratum. Unlike cells tend to separate from one another, either as individual cells or as tissues. Much has been done in this field of study, in relation to morphogenetic processes and tissue differentiation, a topic discussed later in Chapter 16. Meanwhile we are concerned with the technique and phenomena connected with the experimental disassociation, or disaggregation, and subsequent reaggregation of tissue cells. Our particular interest is the chemical or physical-chemical nature of the binding mechanism.

The tissues of sponges have been used from first to last in analytical studies of this sort, although not exclusively. As multicellular organisms, which have probably attained to multicellularity independently from true metazoans, sponges are somewhat loosely organized in terms of cells and tissues and are readily disaggregated. They are mostly marine organisms; hence a chemical approach to the effects of manipulating their environmental medium presents little difficulty. Seawater is a very complex but normally very stable inorganic salt solution which forms a natural physiological medium for the cells of most marine organisms. Artificial seawater is readily made up, as is seawater deficient in this or that constituent. Many technical problems met with in the experimental study of the tissues and cells of the higher vertebrates are not therefore encountered.

The earliest experiments were made in 1907 and 1910 by H. V. Wilson; they were followed up later by Galtsoff (1925) and others, who separated sponge tissue into constituent cells by pressing the tissue through fine bolting silk. Upon settling, certain cell types, such as the amoeba-like cells of the interior and the cells of the sponge epidermis, flatten and exhibit amoeboid movement. Upon coming into contact with one another, the cells coalesce; the plasma membranes of opposing cells join in what seems to be a zipperlike manner, proceeding from the area of initial contact. Specificity is shown by the observation that cell mixtures of the red sponge *Microciona prolifera* and the yellow sponge *Cliona celata* produce cell aggregates which are species-specific, i.e., in a mixed suspension of red and yellow sponges, within a few hours one will first see aggregation, with islands of one type inside the other, but shortly afterwards the two kinds separate from each other.

The main interest in sponge disaggregation and reaggregation in the earlier phases of this continuing investigation has been in the capacity of aggregates to reconstitute new individual sponges. To perform successfully in this way aggregates need to be neither too small nor too large, namely, about 1.0 to 1.5 mm in diameter, and containing about 2,000 cells. Not all kinds of cells necessarily survive the disaggregation process, and some kinds appear to lose distinctive features and become unrecognizable. Yet the overall picture is directly comparable to what we have seen in the culmination phase of the cellular slime molds. Different cell types within an aggregate sort out and assume their characteristic topographical relationships with one another and, mainly through this process of reassembly, construct typical sponge organization. Selective adhesion of cell to cell appears to be the dominating factor in this event.

A concept of intercellular reaction and coherence put forward independently by Tyler and by Weiss is that contiguous cell surfaces are normally held together, at least partly, by forces like those between antigens and homologous antibodies. Adhesion, in other words, depends on the presence in the cell membranes of two intramolecular configurations with the reciprocal structural relationship of antigen and antibody. Putting this concept to the test, Spiegel (1954), using antiserums (made in rabbits) to cell suspensions of two sponge species and also to a mixture of cells of both species, found that reaggregation of dissociated cells was reversibly inhibited in the homologous antiserum; in an antiserum versus both species, large aggregates formed in which cells of both species were distributed at random throughout. The Tyler-Weiss hypothesis thus receives support, though the results refer more to the nature of the species-specificity property than to the mechanics of adhesion as such.

More recently a different technical and conceptual approach has been employed by Humphreys (1963), who used cold calcium- and magnesium-free seawater, i.e., seawater devoid of divalent cations, to dissociate sponge tissue into its constituent cells, and a rotatory agitating apparatus to accelerate reaggregation. Using these techniques, a specific requirement for the divalent calcium and magnesium cations was established for sponge-cell adhesion. However, low temperature was found to inhibit adhesion of such chemically dissociated cells even when the divalent cations were added back to the cells, whereas mechanically dissociated cells aggregated rapidly under the same conditions. This difference is shown to be due to a factor released into the supernatant fluid during chemical dissolution of sponge tissue. When this factor is added back to the chemically dissociated cells along with divalent cations, the cells adhere rapidly even at low temperatures. Moreover, the factor is species-specific, causing adhesion only of cells from the same species.

These results have been interpreted as indicating that sponge cell adhesion is composed of three components: cell surface (as represented by whole cells), divalent cations, and an intercellular material. Cellular aggregation appears to depend on intercellular protein integrity and on the presence of intact disulfide groups, for which calcium is necessary. The several components, separated during chemical dissociation of cells, are capable of spontaneously and species-specifically reassembling themselves to re-form an apparently normal cell adhesion. Thus sponge cells appear to be held together by an intercellular material bound to each cell surface by specific bonds involving divalent cations.

Cell Communication

Perhaps the greatest problem concerning the organization of cells to form tissues, organs, and the organism as a whole concerns the nature of cellular communication.

The unity of a tissue or of an organism is a dominant feature, and many attempts have been made to determine its physical basis; is it chemical gradient systems, or electrical fields, or more substantial supracellular matrices? As we have seen in comparing the behavior of tissue cells in isolation with the same cells in contact with one another, the individuality of the cell is differently expressed in the two circumstances. Contact inhibition of cell migration and membrane ruffling, however, is only one aspect of the changed relationship. Another is seen in cultures of heart cells. Isolated single beating cells, taken from the hearts of young rats, grow and multiply in an appropriate culture medium. As long as the cells in the culture remain separate from one another, each heart cell beats at its own individual rate, if at all, the beating of the cells varying from slow to fast and from intermittent to regular. As growth continues and the number of cells increases, the cells grow into physical contact and the beating becomes synchronous. Single rat heart cells, in other words, can aggregate into a beating sheet of cells without loss of function. Physical contact is essential to the establishment of a synchronous beat, and an extremely rapid form of communication of physical or chemical events must exist.

The present situation concerning cellular communication has been well expressed, and a new hypothesis offered, by W. R. Loewenstein (1967), on the basis of further study of sponge disaggregation and reaggregation. At first, when mechanical contact has just been made between sponge cells, there is no electrical communication between them; current directed into one cell fails to pass into a neighboring cell. At later stages of cell association, signs of electrical communication appear, until about 40 percent of the voltage originally produced in the first cell appears in the neighboring cell, the sum of the voltages remaining constant. Once communication is established between a cell pair, it persists as long as an aggregate remains cohesive.

Cells which are pressed together in seawater devoid of or poor in calcium and magnesium ions develop no such communication but do so if these cations are added to the medium. Further, in another experiment, sponge cells washed and maintained in calcium- and magnesium-free seawater at low temperature to eliminate their organic factor and then returned to normal seawater also fail to develop communication, but do so as soon as purified extract of the organic factor is added. Thus intercellular

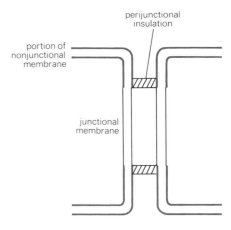

FIGURE 5.15 A junctional unit. [*After Loewenstein, 1968.*]

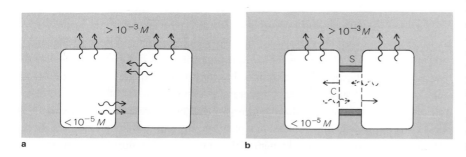

FIGURE 5.16 Permeability conversion during junctional formation. Intracellular calcium ion concentrations are kept low by energized extension of calcium ions (wiggled arrows). (**a**) In unconnected cells the entire cell membrane faces the extracellular compartment where calcium ions attach to membrane binding sites, creating an impermeable state. (**b**) In connected cells the portion of membrane *C* circumscribed by elements *S* is insulated from the extracellular compartment; calcium ions detach from the membrane, creating a permeable state. [*After Loewenstein, 1968.*]

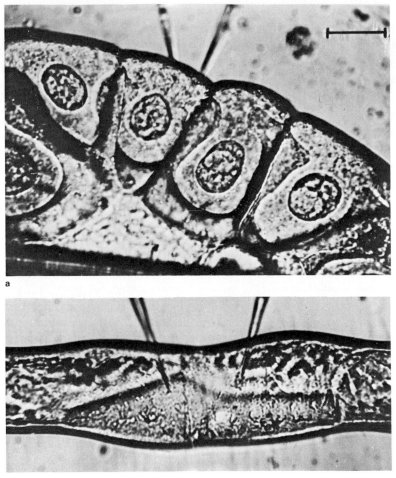

FIGURE 5.17 Measurement of intercellular electrical communication. Light-microscope views of living, unstained material showing microelectrodes inserted: (a) salivary gland epithelial cells of *Chironomus* larva; (b) cells of Malpighian tubule cells of prepupal stage of *Chironomus*. (Calibration = 50 μ.) [*Courtesy of W. R. Loewenstein.*]

communication, as well as intercellular adhesion, depends on the presence of both the organic factor and the divalent cations.

The active constituents of the species-specific factor have been termed *cell ligands* by Moscona, to indicate that they participate structurally in the linking of cells into multicellular systems. Ligand activity is apparently associated with unit particles from 20 to 25 Å in diameter, consisting mainly of glycoprotein. He suggests that the molecular features, differences or complementarities, and distribution patterns of ligands on and between cells are responsible for differential adhesiveness of the cells, cell "recognition," affinities, and preferential association. In other words, specific macromolecular products localized and organized dynamically at the cell surface and between cells (extracellular materials) play an important role in linking cells into histogenetic patterns and provide them with a framework for morphogenetic processes.

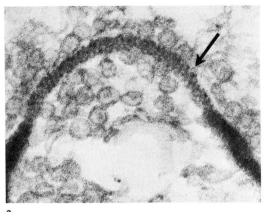

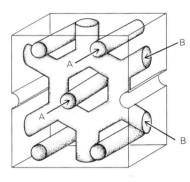

FIGURE 5.18 (a) Electron micrograph of section through a "gap junction." Very dense staining is seen in the extracellular space at the edge of the junction, and a repeating pattern is discernible within the junction (arrow). (b) The model shows possible distribution of intercellular channels (A) and extracellular channels (B) within the junctional membrane complex. [*Courtesy of B. W. Payton.*]

a   b

This is in keeping with the model of interactions between intercellular mucopolysaccharides and collagen proposed by Matthews, and with Grobstein's concept that the structural and physiological consequences of interactions between various cell-surface and extracellular components may be of prime significance to their ordering and thereby to the organization of cells into developmental patterns.

According to Loewenstein, a generalized scheme of a cell-membrane junction concentrates on a region of cell junction (junctional membranes, C) where the membranes are markedly different in permeability from the rest of the cell surfaces (nonjunctional membranes, O): "At these junctional membrane regions, permeability to small ions is more than 10,000 times greater than at the rest of the cell surface. The permeability at these regions is high even to relatively large molecules, 10^3 mol. wt., and, possibly, up to 10^4 mol. wt. The region of high permeability is completely surrounded by a diffusion barrier (perijunctional insulation, S) insulating the system from the exterior. The arrangement of many such connected cells in series forms thus an efficient communication system in which substances can flow rather freely from one cell interior to another without leaking to the exterior." Once the perijunctional insulation is formed, the membrane portion circumscribed by it is incorporated into the intracellular compartment; chemical gradients now favor calcium ion detachment from the membrane structure, and conversion from low to high permeability follows; while the driving force behind the permeability transformation and its maintenance is the metabolic force that keeps the calcium ion concentration low in the cytoplasm.

Several agencies are therefore involved in establishing and maintaining tissue organization, namely, species- and tissue-specific intercellular bonds, or cell ligands; chemical diffusion gradients, particularly from cell to cell; electrical potentials; and ionic conditions and stability of the cell and tissue environment.

CONCEPTS

The multicellular state can arise through the aggregation of a population of previously separate cells, or it may evolve or develop as a consequence of the failure of cells to separate following cell division.

Multicellularity makes possible the extension of cellular material as thin membrane, thereby maintaining necessary surface/volume ratios not otherwise possible.

The multicellular organism is an integrated system which exhibits a "wholeness" or organizational unity that is of a real but elusive nature.

Development, whether of a slime mold or of an egg, is a continuous process, arbitrarily divisible into three phases: a cell multiplication phase, a differentiation phase, and a morphogenetic phase.

Differentiation is to a degree incompatible with a maximal rate of cell multiplication.

Cell aggregation may be controlled by chemical diffusion gradients.

Processes of cell assembly and of early differentiation are to some extent reversible: an isolated part may develop as a whole.

During morphogenesis, or final assembly, cells assume position according to their specific character.

Formation of structural components at the molecular level is regulated temporally, spatially, and quantitatively by the overall developmental program.

Cells move primarily by a ruffling movement of the cell membrane and continue to move when not in contact with other cells.

Contact inhibition occurs when cells meet, i.e., locomotory cell movement ceases, whether as single cells or as cell sheets.

Cells in tissues form special functional attachment sites of several kinds. Disassociated tissue cells can reunite to form typical tissue organization.

Mutual cell adhesion depends on the presence of extracellular cell matrix containing species-specific bonds known as "cell ligands," associated with unit particles of a glycoprotein nature.

Communication between cells is effected through electrical current and chemical diffusion pathways established between apposing areas of cell surface bounded by junctional membrane.

READINGS

BONNER, J. T., 1970. Induction of Stalk Cell Differentiation by Cyclic AMP in the Cellular Slime Mold *Dictyostelium discoideum, Proc. Nat. Acad. Sci.,* **65**:110–113.

——, 1969. Hormones in Social Amoebae and Mammals, *Sci. Amer.,* June.

——, 1967. "The Cellular Slime Molds," 2d ed., Princeton University Press.

——, 1952. "Morphogenesis," Princeton University Press.

——, et al., 1969. Acrasin, Acrasinase, and the Sensitivity to Acrasin in *Dictyostelium discoideum, Develop. Biol.,* **20**:72–87.

CURTIS, A. S. G., 1967. "The Cell Surface," Academic.

GROBSTEIN, C., 1967. Mechanisms of Organogenetic Tissue Interaction, *Nat. Cancer Inst. Monogr.,* **26**:279–299.

HUMPHREYS, T., 1963. Chemical Dissolution and in Vitro Reconstruction of Sponge Cell Adhesions, *Develop. Biol.,* **8**:27–47.

LOEWENSTEIN, W. R., 1968. Communication through Cell Junctions, in M. Locke (ed.), "The Emergence of Order in Developing Systems," 27th Symposium, The Society of Developmental Biology, Academic.

——, 1967. On the Genesis of Cellular Communication, *Develop. Biol.,* **15**:503–520.

MATTHEWS, M. B., 1965. The Interaction of Collagen and Acid Mucopolysaccharides: A Model for Connective Tissue, *Biochem. J.,* **96**:710–716.

MOSCONA, A. A., 1968. Cell Aggregation: Properties of Specific Cell-ligands and Their Role in the Formation of Multicellular Systems, *Develop. Biol.,* **20**:250–277.

PICKEN, L., 1960. "The Organization of the Cell," Clarendon Press, Oxford.

RAPER, K. B., and D. I. FENNELL, 1952. Stalk Formation in *Dictyostelium, Torrey Bot. Club Bull.,* **79**:25–51.

RASMUSSEN, H., 1970. Cell Communication, Calcium Ion, and Cyclic Adenosine Monophosphate, *Science,* **170**: 404–412.

SHAFFER, J. T., 1962–1964. The Acrasina, *Advance. Morphogenesis,* **2**:109–182; **3**:301–322.

SPIEGEL, M., 1954. The Role of Surface Antigens in Cell Adhesion, I: The Reaggregation of Sponge Cells, *Biol. Bull.,* **107**:130–148.

SUSSMAN, M., 1965. Temporal, Spatial, and Quantitative Control of Enzyme Activity during Slime Mold Differentiation, in "Genetic Control of Differentiation," Brookhaven National Laboratory, Symposium in Biology, no. 18, pp. 66–76.

WRIGHT, B., 1966. Multiple Causes and Control of Differentiation, *Science,* **153**:830–837.

CHAPTER SIX

POLARITY, FORM, AND DIFFERENTIATION

Organisms, whether multicellular or unicellular, are typically polarized, with few exceptions. They have an apico-basal, or anteroposterior, axis and usually exhibit some asymmetry about that axis. Structure is to some extent differentiated according to its position along such an axis, generally showing greater complexity at the distal, or anterior, end.

CONTENTS
Axial gradients
Polarity
Individuality
Growth and form
Cell and tissue differentiation
Germ-cell determination
Reconstitution of form
Chemical diffusion theory
Concepts
Readings

AXIAL GRADIENTS

An axial gradient, a purely descriptive term, is commonly used to denote an observed gradation in degree of organization along an axis, as in flatworms and vertebrates, where head, trunk, and tail clearly reflect an axial structuring. Such sequential ordering of structure may have a basis in time as well as in space. This is seen in the development of embryos and in stages of regeneration or reconstitution, where new structure is added sequentially to preexisting structure. Accordingly *axial gradient* is a useful term referring to a commonly observed phenomenon, and is neutral in that it says little or nothing about underlying causes.

Many years ago C. M. Child developed the concept that metabolic gradients, specifically gradients in oxidative metabolism, underlie the manifest axial gradients. Regions of relatively great complexity such as head regions containing "brain" and sense organs, or regions of obviously greater metabolic activity such as the so-called animal pole of, for example, a frog egg, were considered to be regions of relatively high oxidative metabolism. Regions farther down the primary axis were considered to have relatively lower rates of oxidative metabolism, regions of high metabolic rate being dominant over regions with lower rates. The theory of metabolic gradients has been an integrating concept of great value and virtually the only one of its kind. Most of the supporting evidence presented during the first half of this century, however, has been indirect, mainly the demonstration of gradients in susceptibility to mildly toxic solutions or gradients in rate of reduction of various dyes; direct evidence for the presence of gradients in oxidative metabolism conforming to axial gradients in organization has been generally sparse or contradictory. Nevertheless the main concept stands, with chemical diffusion gradients, electrical fields, and graded potentials replacing the oxidative metabolism interpretation.

In this connection intensive studies have been made on the simpler of the more readily available and maintainable animals, namely, hydroids and flatworms. They exhibit comparatively low levels of tissue and body organization, and they have amazing capacities for reconstituting whole organisms from fragments. The investiga-

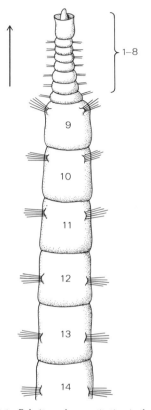

FIGURE 6.1   Polarity and reconstitution in the poly-chaete *Clymenella,* showing polarized (anteropos-terior axis) structure and reconstitution, anteriorly or posteriorly, of whatever number of segments are required to restore the original total of 22 segments. In this instance, segments 1–8 are re-generated from the anterior surface of segment 9.

tion of the nature of the multicellular organism has consisted mainly of the experi-mental study of reconstitutive processes and the relative roles of tissues and cells involved in those processes. The hydroids *Hydra* and *Tubularia* and various flatworms (planarians) have been subject to experimental investigation for well over a century. As early as 1814 Dalyell wrote that planarians are "almost to be called immortal under the edge of the knife." Since that time more than 500 papers have been published concerning the responses of planarians to operational procedures, and a comparable number on hydroids.

POLARITY

Most organisms exhibit polarity. It is evident in flatworms and annelid worms as head-tail structure and in hydroids such as *Tubularia* as hydranth-stem structure. The dynamic state of this polarity is seen in regeneration experiments. When a transverse piece is cut from the body of a worm or from the stem of a hydroid, a new head or hydranth grows from the anterior cut surface and new tail or stem from the posterior cut surface. Moreover, Lund (1921) showed that such polarity has an electrical basis, that there is an electrical potential difference between the two ends of a regenerating piece of hydroid and that this can be neutralized or reversed by an external field, with concomitant inhibition or reversal of the regeneration. Accordingly various hypotheses have been put forward concerning the underlying control system. All assume the existence of a gradient or of gradients of some sort, with high levels or concentrations being generally dominant over lower and with a critical difference between two levels being significant. Time itself seems to be either a factor or a feature of some importance, and a somewhat vague concept of time-graded regener-ation, or morphogenetic fields, has been developed in connection with the varying capacity of flatworms to regenerate a head according to the level of section along the animal's anteroposterior axis (Brønstedt, 1955, 1970).

However this may be, the question of regeneration of anterior or posterior structures from cut surfaces is strikingly affected both by the distance between the anterior and posterior cut surfaces and by the time interval between the making of the cuts. For example, a piece of tubularia stem several times as long as it is wide, a piece cut from the anterior half of a flatworm, about as long as it is wide, and a comparable piece cut from the abdominal region of the ascidian *Clavelina,* all regenerate according to rule: the anterior cut surface regenerates the missing anterior structure, and the posterior cut surface regenerates the missing posterior structure. If, however, in each of these examples the isolated piece is very short, i.e., the anterior and posterior cut surfaces are relatively close to one another, then an anterior, or head, structure is produced from each surface. Much of the strength of Child's thesis depended on observations of this sort, and any proposed theory must take this into consideration, in addition to the graded aspect of regenerative or reconstitutive capaci-ties generally.

*Con*d In the case of the ascidian, Brien demonstrated that the timing of cuts is as important as the distance between them. If short pieces are isolated by cuts made almost simultaneously, anterior (thoracic) structure forms from both surfaces. If the posterior cut is made an hour, more or less, *after* the first cut, typical anterior and posterior regeneration occurs at the two surfaces respectively; i.e., the original polarity of the tissue persists. The *orientation,* presumably of macromolecular components, may be more significant than concentration gradients of either ions or molecules in determining and maintaining tissue polarity.

INDIVIDUALITY

A hydra, as we shall see, is a strongly polarized multicellular organism that is maintained by an uninterrupted production and flow of cellular elements, combined with a certain rate of differentiation and aging of the cells. The system has to be recognized as the direct consequence of a series of interacting and balanced processes constituting a steady state rather than a fixed, structural pattern. Not only form but also size is a property of the specific state characteristic of this organism. When, for example, the heads and bases of hydras are amputated and the trunks are grafted together in series, the giant tube thus formed, although regenerating a new head with tentacles, does not remain a single giant individual; each portion asserts its individuality, regenerates head and stalk, and separates. It is significant that individual separation occurs sequentially, beginning with the unit farthest from the regenerated head.

In contrast to the amorphous individuality of slime molds and sponges, the full multicellular organismal state is seen in almost idealized simple form in *Hydra.* Moreover the fact that hydras continually and almost irrepressibly reproduce asexually by budding makes possible the maintenance and experimental use of clones derived from successive buds of single somatic parentage.

It is no accident that, among animals, *Hydra* has been so intensively studied over the years. It is simple in gross structure, being essentially a tube with a mouth (an oral cone, or hypostome) and a ring of hypostomial tentacles at one end, and an adhesive foot, or basal disc, at the other. It is simple in tissue structure, consisting of two epithelial layers, the epidermis and gastrodermis, separated by a thin layer of collagenlike, noncellular mesoglea. It is small, about 10 mm long and 1 mm wide when extended; yet it is large enough to be readily observed and amenable to operation. It is composed of about 100,000 cells, a seemingly large number, yet relatively small. There are only 17 distinguishable kinds of cells, representing a much smaller number of basic types. Yet despite its comparative simplicity of structure, the organism presents most of the familiar problems concerning the maintenance of form and the diversification of tissues and cells encountered in the study of multicellular organisms generally. Accordingly *Hydra* is introduced here for the perspective it offers with regard to the entire field of developmental biology. The establishment and maintenance of

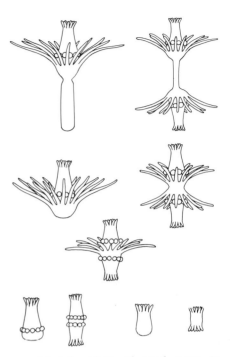

FIGURE 6.2 Independence of apical region in hydroid *Tubularia.* Partial regeneration occurs in short-stemmed pieces of stalk, which may be uniaxial or biaxial; it gives rise to as much of the rest of the organism as can be formed from the available material. [*After Child, 1915.*]

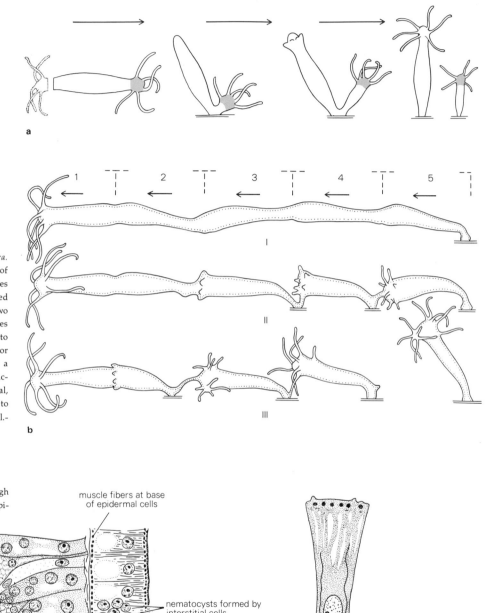

FIGURE 6.3 Polarity and individuality in *Hydra*. (a) Head of hydra (shaded) grafted to foot of decapitated hydra: decapitated hydra regenerates new head from its anterior cut surface, grafted head grows new column of its own, and the two individual hydras eventually separate. (b) A series of five decapitated hydras are grafted tandem to form a single long column. The original anterior region of each decapitated hydra reconstitutes a new head (hypostome and tentacles), commencing with the most posterior original individual, the process proceeding anteriorly according to their original polarity. [*After Tardent*, Arch. Entwickl.-mech., **146**:640(1954).]

FIGURE 6.4 Histology of *Hydra*. (a) Section through the gastric region of the body wall. (b) An epitheliomuscular cell. [*After Hyman, 1940.*]

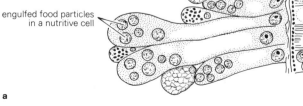

secretory cells in the endodermis

engulfed food particles in a nutritive cell

muscle fibers at base of epidermal cells

nematocysts formed by interstitial cells

a

b

the organism as a whole are paramount; in this connection the simplicity of structure and plasticity of response in *Hydra* are challenging to an extreme. At the same time one should remember that *Hydra* is but one of a large phylum, the Coelenterata, or Cnidaria, in which the same basic organizational plan is present throughout but is expressed in many shapes, sizes, and degrees.

Hydra became an experimental organism of complexity almost from the moment of its discovery more than two centuries ago. The Abbé Trembley in Holland observed and experimented on hydras, employing very simple means. He saw the process of reproduction by budding; he succeeded in cutting the small animal in two and observed the regeneration of the amputated part; and he managed to insert a hair through the mouth of a hydra and pull the creature inside out, subsequently observing the process of recovery to the normal state. Such was the beginning. During the present century extensive experimental studies have been made in attempts to comprehend the underlying nature of these and other developmental or reconstitutive events, for it has long been evident that the mature organism is not static. It has to be dynamically maintained not only metabolically but in terms of cells and tissues. The primary questions concern, therefore, (1) the basis, maintenance, and reconstitution of form, and (2) the determination or control of cell differentiation with regard to both diversification and distribution.

GROWTH AND FORM

Hydra exists in three states: (1) the nonreproductive state, which may be juvenile or fully grown; (2) the nonsexual reproductive state, in which buds are successively produced from the lower part of the body column; and (3) the sexual state, in which testes, ovaries, or both differentiate within the epidermis. Apart from the production of buds or gonads, a hydra may maintain itself for months and even years without significantly changing its size and form. Yet its cells are forever undergoing replacement, and its tissues are in continual flux. As in a candle flame, shape and certain dimensions are maintained in spite of constant turnover of constituents, although the movement of cells and tissues is not immediately visible. Much effort has gone into devising means of making the invisible apparent.

The pioneering work of Brien and Reniers-Decoen more than 20 years ago laid the foundation for the present analysis of the subtle processes of growth and maintenance in hydras. They grafted various portions, such as hypostome and the upper part of the body column, of individuals which had been stained with a vital dye—neutral red, methylene blue, or Nile blue sulfate—into various levels of the body column of unstained individuals. Such stained material migrates toward the basal disc, where it eventually disappears. When grafted immediately beneath the tentacle ring, stained tissue migrates toward the tentacle tips. The interpretation, which has been widely accepted, is that hydras possess a growth region, or special growth zone,

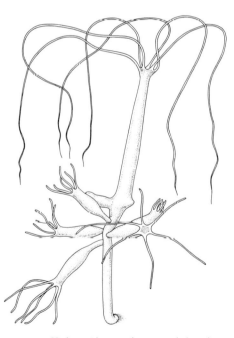

FIGURE 6.5 Hydra with tentacles expanded and with a number of buds forming on the body column. Successively forming buds arise in helical arrangement, seemingly according to available space on the narrow column in the budding region. [*After Brien and Decoen, 1949.*]

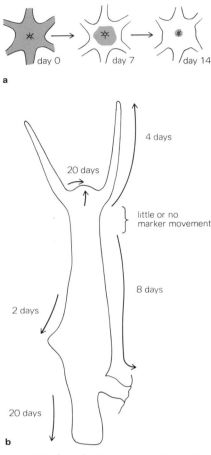

FIGURE 6.6 General tissue-movement pattern, based on movement of tissue markers. Numbers indicate the approximate time (days) for a marker to move along the path shown by the arrow. (a) Movement along the hypostome. (b) Movement distally and proximally from the stationary zone below the tentacles. [*After Campbell, 1967.*]

just below the level of the tentacles, where relatively rapid cell division is supposed to occur in both epidermis and gastrodermis. Only the hypostome, i.e., the conical and expansible tissue between mouth and tentacles, appears to be stable.

According to this interpretation, supported by Burnett, a region of maximal growth appears to exist as a subhypostomal growth zone beneath the ring of tentacles. As cells multiply, new tissue forms most rapidly in this region and continuously moves not only toward the base of the organism but also distally into the tentacles, displacing older tissue until the aging tissue reaches the extremities, where it is sloughed off. The length of the column and of the tentacles would then be determined by the rate of production of new tissue in the growth zone and the durability of the cells produced. Cessation of cell proliferation in the growth zone would be followed by a shortening of the tentacles and of the column as cells are lost without replacement. This phenomenon of regression is often observed in hydra cultures.

The validity of this view of hydra growth, plausible and suggestive as it is, depends in part on the validity of the assumption that the cells which have been vitally stained with the dyes just mentioned are entirely normal except for color. Supporting observations have been made by exploiting an old discovery that green hydras placed in a 0.5 percent glycerin solution get rid of their green symbiotic algae within a few days. By grafting pieces of such individuals to various regions of normal green individuals, chimaeras are formed, some of which, or their buds, will have columns that are white with a few green patches. Such green patches are seen to move from the growth zone to the base of the column in from 2 to 3 weeks, and to the tips of the tentacles in about 5 days. Yet, for all this, the issue does not rest, for the picture seems to be too sharply drawn.

There is still agreement that a band of column material which appears stationary exists just below the tentacles. From this region tissues move distally along the hypostome, outwards along the tentacles, and proximally down the column to the buds or onto the stalk. Yet studies by Campbell of the distribution and frequency of cells in division along a hydra column, together with related studies of the incorporation by cell nuclei of the radioactive nucleotide tritiated thymidine, suggest that cell growth and division are much more widespread through the column than has been supposed. The existence of a restricted and exclusive growth zone is in doubt; the *body column of a hydra is better viewed as an expanding cylinder whose elongation is balanced by tissue loss at the two ends.* Since tissues at the two ends of the column are moving in opposite directions, there has to be one intermediate column level where tissue does not move with respect to the animal's structure. *Therefore a stationary region must be present regardless of how the growth is distributed along the column.*

The movement of cells is of two kinds; some move as individual cells, some only as constituents of an epithelium. Nematocysts specifically labeled with methylene blue and with $^{14}CO_2$, respectively, are seen to migrate in amoeboid manner from the column to the tentacles; interstitial cells, also isotopically labeled, migrate from the lower column to the distal column. On the other hand the epidermis and gastrodermis move as coherent epithelia, although at somewhat different rates and therefore to some extent independently of each other.

When a distal half of a hydra in which all interstitial cells have been destroyed by irradiation is grafted onto a complementary proximal half which has previously been labeled with radioactive isotope, labeled interstitial cells and cnidoblasts invade the distal half; but the epitheliomuscular cells do not participate in such a movement. In keeping with this difference in cell behavior is the observation from electron micrographs that the epitheliomuscular cells of the epidermis and of the gastrodermis are linked by desmosomes, whereas interstitial cells adjoining one another show no such connections.

The problem of form in a fully grown and well-fed hydra necessarily includes the initiation and growth of buds from a level about two-thirds down the body column, since the buds are an integral part of the individual until the time of their separation. Furthermore, the growth and development of the hydra form begins with the formation of the bud rudiment from which a hydra is usually derived. In fact, the rate of lengthening of a hydra is ten times greater during its development as a bud than

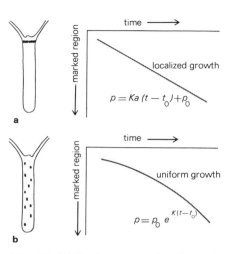

FIGURE 6.7 Relation between growth pattern and tissue movement. Graphs show expected position, p, of a marked region as a function of time. (a) Model of growth (black) localized below tentacles. Tissue should move linearly with time. (b) Model of growth distributed uniformly along column. Tissue should move at increasing rates as it advances down column. A marker's distance from the stationary region should increase exponentially with time. [After Campbell, 1967.]

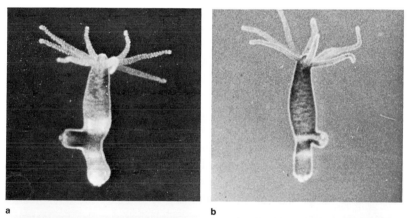

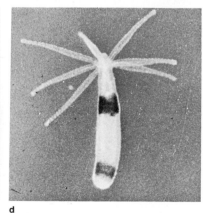

FIGURE 6.8 Starved (a,b) budding and (c,d) non-budding *Hydra* individuals consisting of portions of green and white hydras fused together. (a) A green animal bearing two white rings. (b) During 4 days the thickness of the distal zone and of the green zone between the rings became considerably reduced. (c) A white animal bearing two green rings. (d) During 4 days there was no conspicuous change in the distance between the rings, the dimensions of the rings, or the relative position of the rings. [Courtesy of S. Shostak.]

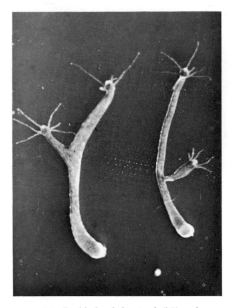

FIGURE 6.9 Double-headed animal (*left*) and normal animal with advanced bud (*right*). Secondary head is more broadly attached to remainder of parent animal than the bud which has begun to form a base, as a result of treatment with colcemid. [*Courtesy of S. Shostak.*]

following detachment from the parent; and buds begin to form on a new individual only after the parental rate of growth is already decreasing. During the process of bud formation both parental epidermal and gastrodermal tissue extends outward as the body column of a bud, so that much of the growth of the parental body column as a whole becomes channeled into the growth of buds. Well-fed hydras bud continually; underfed hydras may not bud at all.

Recent analyses by Shostak show that in the gastrodermis, at least, there are four separate self-generating cell populations: a population at each end of the body column supports the tentacles and the basal disc, respectively, and two populations meet at the lower edge of the budding region and support the development of buds, although they can also expand and produce lengthening of the body. Altogether this is a different picture of tissue maintenance from that of a subtentacular growth zone. Yet there is no doubt of a general movement of the epidermal (epitheliomuscular) and gastrodermal (digestive) cells down the length of the body column. Since the pattern and rates of cell division do not satisfactorily explain it, what does? Does the mesoglea play a role? This has been suggested. It is probably significant that mesogleal strands, bands of interstitial cells, and the succession of bud initiations all take a helical course in the column.

Hydra may be regarded as essentially a single cell layer, or epithelium, adhering to each side of a cylindrical collagenous endoskeleton. Cells move proximally over this endoskeleton, the two layers tending to move at different rates. The completely isolated mesoglea retains the shape of the animal and remains elastic. New epidermis slipped over denuded mesoglea adheres and migrates. At present too little is still known concerning the production of mesogleal collagen, its stability, and its structural properties; but it is possible, for instance, that the process of cell-sloughing at the column base or at the tips of tentacles may result from local aging of the mesoglea rather than from the death of cells. In any case the mesoglea permits cells to migrate over it, and does not act as a cement that prohibits cell migration. This concept relates to the broader question of the relationship of cell growth, differentiation, and life span to the presence and nature of extracellular macromolecular material generally.

CELL AND TISSUE DIFFERENTIATION

Although in *Hydra* the two epithelial sheets of tissue—the epidermis and gastrodermis—represent the primary cellular structure, according to Burnett 17 kinds of cells are distinguishable. They fall into fewer categories:

1 Epitheliomuscular cells, which are the large cells comprising the epidermal epithelium

2 Interstitial cells, also called neoblasts, which lie at the base of the large epidermal cells and are small, unspecialized reserve cells

3   Three kinds of nerve cells, all epidermal—sensory cells, ganglion cells, and neuro-secretory cells

4   Four kinds of sting cells (cnidoblasts), all epidermal

5   Reproductive cells, i.e., ova and sperm cells, both epidermal, though only seasonally present

6   Digestive cells, or the large cells comprising most of the gastrodermis

7   Two kinds of gland cells, both gastrodermal, of the hypostome and gastric column, respectively

8   Three kinds of mucous cells, those of the hypostome, of the gastric column (gastrodermal), and of the basal disc (epidermal).

These cell types represent different degrees of differentiation.

1   Interstitial cells not only are small and without specialized visible structure but also are multipotent, in that they are able to differentiate into cnidoblasts, nerve cells, or ova or sperm cells, depending on circumstances.

2   Cnidocytes (i.e., fully differentiated cnidoblasts), nerve cells, and sperm cells are highly specialized, highly structured cells that are unable to undergo cell division and can only be formed anew from reserves of interstitial cells.

3   The large epitheliomuscular cells of the epidermis and the large digestive and glandular cells of the gastrodermis are capable of division and constitute essentially self-maintaining epithelia; their high degrees of differentiation must be regarded as elaboration of general cell properties rather than exclusive forms of cell speciali-zation. The total cell population is maintained, therefore, through cell division of only four kinds. Yet even these appear to be intertransformable to some extent.

The epidermal and gastrodermal layers of hydras can be separated by a perfusion technique so that each layer is completely free from contamination by the other layer. Each layer is capable of reconstituting a whole hydra. Thus isolated gastrodermis consisting only of digestive and gland cells is capable of giving rise to a complete hydra containing epidermis, mesoglea, and gastrodermis.

1   Gland cells lose their characteristic granular endoplasmic reticulum, dedifferentiate into interstitial cells packed with free ribosomes and without endoplasmic reticu-lum, and subsequently divide and redifferentiate into nests of cnidoblasts or, alternatively, into nerve or sperm cells.

2   Digestive cells transform to epitheliomuscular cells directly, through breakdown of their algal cells and by secreting a mucoprotein border and forming myofibrils (myonemes). These are necessary for contraction of the cell bases in the longi-tudinal axis of the animal.

Burnett considers that basically there are only two types of cell in a hydra—the epitheliomuscular cell and the interstitial cell—and that the different fates of these two kinds of cells depend on whether they occupy positions in the epidermis or gastrodermis and on the position they occupy in a chemical gradient system. That is, the epidermal epitheliomuscular cell and the digestive gastrodermal cell represent

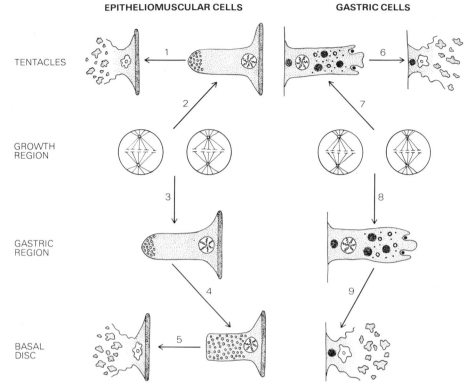

EPITHELIOMUSCULAR CELLS   **GASTRIC CELLS**

TENTACLES

GROWTH
REGION

GASTRIC
REGION

BASAL
DISC

FIGURE 6.10 Fate of epitheliomuscular cells of *Hydra* in tentacles, growth region, gastric region, and basal disc. They die at tips of tentacles and are sloughed, they divide in growth regions, they may differentiate a mucus border and a myoneme, or, before death at the level of the basal disc, they secrete a specific acid-mucopolysaccharide not found elsewhere. Gastric cells, moving distally from the growth region, engulf food, cease division, form an intracellular myoneme, and are then sloughed off into the enteron at the extremities. Those moving proximally undergo a similar cycle. [*After Burnett, 1966.*]

a single cell type expressed differently in two environments. An interstitial cell in the epidermis differentiates into a nerve cell, a cnidoblast, or a sperm cell, or else grows and divides without differentiating. On the other hand, one in the gastrodermis never differentiates into a cnidoblast but rather into either a mucous cell or a gland cell. Even this distinction is not absolute, since in small, rounded-up fragments of isolated masses of epidermis or of gastrodermis, according to Haynes and Burnett, epidermal, epitheliomuscular cells can arise directly from gastrodermal cells which lose their enclosed algae and food droplets and begin mucous secretion. Interstitial cells appearing in the mass arise not from preexisting interstitial cells but from digestive cells, gland cells, or mucous cells which void their internal secretions. Therefore, the interstitial cells do not represent a "modulated" form of a gastrodermal cell, since they are capable of differentiating into cnidoblasts containing mature nematocysts; neither do they represent a persistent embryonic stock which is maintained solely by the division of interstitial cells. Evidently, certain specialized cells in the hydra are not "end points" of development but can acquire a new potency if properly stimulated.

Interstitial cells provide exceptional opportunities for investigating the factors influencing the selection of one of several particular paths of differentiation which

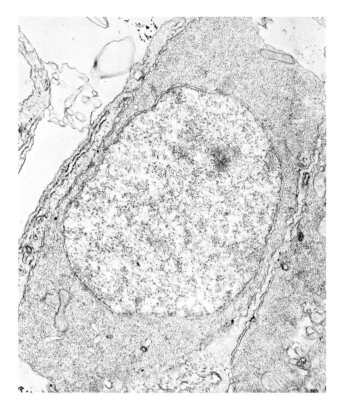

FIGURE 6.11 Photo of an interstitial cell, showing large size of nucleus relative to cell as a whole, and comparatively undifferentiated state of the cytoplasm. [*Courtesy of T. Lentz.*]

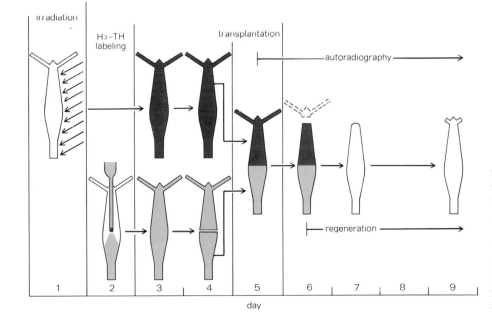

FIGURE 6.12 Experimental investigation of regeneration in *Hydra*: first day, irradiation to kill interstitial cells; second day, in another hydra, tritiated thymine injected into body cavity; third and fourth days, individuals bisected and anterior irradiated half grafted to tritiated posterior half (fifth day); sixth day, composite individual decapitated; seventh to ninth days, labeled interstitial cells migrate from posterior half into irradiated part and regeneration proceeds. [*After Tardent and Morganthaler, 1966.*]

are open to an initially indifferent or unspecialized cell. As interstitial or reserve cells, like comparable cell types in many other organisms, they are small, more or less round, with a relatively large vesicular nucleus and a relatively small cytoplasmic volume. As a rule they give rise either to cnidoblasts or to germ cells, male or female. Otherwise, except for reconstitutive emergencies, they remain at rest and unspecialized. If they are activated, however, the response varies according to circumstances.

Several types of nematocysts are formed by cnidoblasts in hydras, principally stenoteles, desmonemes, and glutinants. Stenoteles may be present in both large and small forms (e.g., *Hydra vulgaris attenuata*) or only as small forms (e.g., *Hydra oligactis*). Cnidoblasts also vary in the number which are seen to differentiate together as a "nest." *Each nest,* whatever the number of cells, *arises from a single interstitial cell* near the base of the epidermis, called the *lineage cell.* The size and number of cnidoblasts in a nest depend on the number of synchronous cleavages that the original interstitial cell undergoes, and the total volume of a nest does not vary appreciably with the number of cleavages. Evidently the growth potential for an interstitial cell entering the path of cnidoblast differentiation is determined even before the number of cleavages has been settled, i.e., it is independent of other aspects of differentiation.

The sister cells in a nest remain united by cytoplasmic bridges, at least while they constitute a group, thus forming a temporary syncytium of the type commonly seen in mitotic divisions of germ cells and some somatic cells. The number of divisions occurring during cnidoblast differentiation ranges from two to five. The types of nematocysts formed by cnidoblasts resulting from nests of various sizes are indicated in the table below.

Since the total growth of the initial cell and its descendants destined to form a nest appears to be the same in all cases, we are left with three possible variants responsible for the different outcomes: rate of growth, rate of divisions, and rate of the differentiation process. There is no clear evidence as to which is the responsible variant, but there is remarkable opportunity here for investigation of the inter-relationship of these features.

The course of differentiation of an individual cnidoblast has been closely studied. In stained preparations examined with the light microscope, the earliest rudiment of the future nematocyst is seen to be a small vacuole. Cells in tissue culture have been seen to divide after a vacuole has appeared. With the electron microscope the vacuole is seen to be pear-shaped and surrounded by a thick protocapsule which becomes thin over the narrow end of the vacuole. Stacks of Golgi membranes are seen in this region of the cytoplasm; later the rough endoplasmic reticulum becomes increasingly abundant and the mitochondria enlarge, all signs indicative of intense protein synthesis. Early capsules show the nematocyst thread (tube) rudiment either invaginated within or evaginated from the capsule. Once formed, cnidoblasts actively migrate from the trunk toward the distal ends of tentacles, as shown both by their rate of translocation relative to rate of tentacle growth and by the migration of the large stenotele cnidoblasts of *Pelmatohydra oligactis* along the tentacles of another species following grafting.

Number of Cells per Nest	Type of Nematocysts
4–8	Large stenoteles
8–16	Small stenoteles
16	Glutinants
16–32	Desmonemes

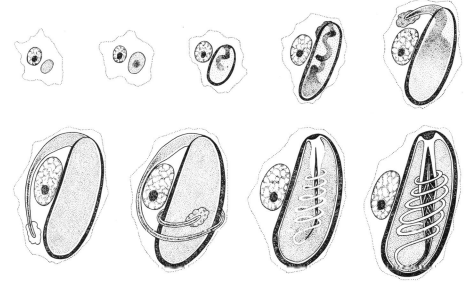

FIGURE 6.13   Differentiation of a cnidoblast and formation of nematocyst (description in text). [*After Mergner*, Natur Museum, **94**:22(1964).]

GERM-CELL DETERMINATION

The most dramatic constitutional change during the life cycle of an individual hydra concerns the transformation from the nonsexual state, in which hydras normally produce buds in rapid succession low down on the body column, to the sexual state in which testes or ovaries (or in some species both testes and ovaries) differentiate in the column epidermis and, in most species, the budding process ceases. A monumental amount of work has been done at the cellular level in efforts to interpret the transformation from the rapidly growing, budding state to the formation of the differentiated male and female condition, and to determine what factors in the external or internal environment of the hydra are responsible for the initiation of sexuality.

The onset of the sexual phase is marked by several events. In most species of *Hydra*, at least, the linear growth of the body column seems to be arrested. Throughout most of the column the interstitial cells cease to differentiate into cnidoblasts; instead they continue to divide and form cell masses leading to the production of either testes or ovaries, depending on the species. In the species where both testes and ovaries are produced in the same individual, the location of the interstitial cells in the column controls their fate. As a rule bud formation ceases during sexual reproduction, apparently because the budding zone becomes involved in the sexual differentiation process and cannot proceed both ways at the same time, except in species where neither process takes up all the available tissue resources. Testes are usually numerous and form as helically arranged plates of cells in the upper and middle part of the column epidermis. Each develops into a number of cysts, or testicular

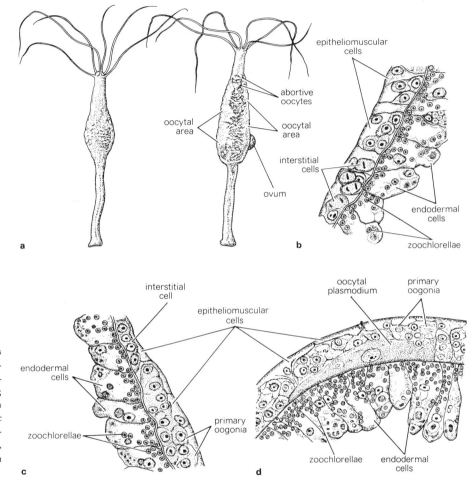

FIGURE 6.14 Ovogenesis in *Hydra*. (a) Individuals at early and late stage of egg formation (from epidermis). (b) Onset of ovogenesis showing interstitial cells proliferating. (c) Interstitial cells giving rise to primary oogonia resting against mesoglea at base of epitheliomuscular cells, and symplasmic union of oogonia. (d) Transformation of symplasmic mass into unicellular ovocytal plasmodium, other nuclei having been absorbed. [*After Brien and Decoen, 1949.*]

follicles, each of which exhibits a gradient in spermatogenesis, with the sperm mother cells, or spermatogonia, at the base of each follicle, and with the spermatozoa at the apex of each. During ovogenesis, which typically occurs more in the middle region of the column, the epidermis becomes profoundly thickened by excessive interstitial-cell proliferation. This coincides with a total and permanent disappearance of cnidoblasts from the area of proliferation. The thickened tissue forms several ovocytal areas, in each of which the interstitial cells in contact with the mesoglea fuse as a mass, within which only one of the contained cell nuclei survives. Each fused mass becomes a single ovum with a single nucleus.

Two extremely important concepts emerge from this:

1   The sexual reproductive cells, or germ cells, differentiate from unspecialized somatic cells, the interstitial cells, which may in turn have been produced by more

specialized somatic cells, the gland cells. *There is no so-called germ line, continuous from generation to generation, from which the soma, or body cell assembly, is produced.* Somatic tissue gives rise to sexual reproductive cells, just as it gives rise to other kinds of highly differentiated cells.

2 *Factors external to cells determine the direction of cell differentiation.*

The nature of the controlling agency remains obscure, in spite of intensive efforts to identify it. The switch from the nonsexual, budding state to the sexual state is reversible and is clearly under the influence of environmental conditions. Temperature level and change, nutrient level, and carbon dioxide tension have all been implicated, yet none, separately or in combination, appears to be the exclusive factor. Both Berrill (1961) and Burnett and Diehl (1964) emphasize that epidermal interstitial cells differentiate as germ cells when the body growth of a hydra ceases. In normal periods of growth the interstitial cells continually differentiate into cnidoblasts. When normal growth ceases, no more cnidoblasts are formed and the interstitial cells take the other main path of differentiation, i.e., the paths are mutually exclusive.

RECONSTITUTION OF FORM

The problem of form in *Hydra* relates not only to the growth and maintenance processes seen in the normal cycle. A hydra can regenerate a hypostome and tentacles rapidly and repeatedly after all interstitial cells have been destroyed by exposure of the hydra to nitrogen mustard. Normal proportions can be retained or reconstituted apparently even without any cell division at all, e.g., a starved hydra keeps its shape while slowly diminishing in size; while a minced hydra rapidly reorganizes into a complete animal or animals, mainly through reassembly of the constituent cells. A decapitated column regenerates a new hypostome and tentacle ring, though the time necessary for initial hypostome determination is less when the cut is made distally and becomes progressively greater as the cut is made more proximally. Amputation of the basal disc is followed by regeneration of a new basal disc. A hypostome grafted into the side of the body induces the outgrowth of a new column and tentacles. Hydras partially divided lengthwise and fused together longitudinally, thereby forming individuals with an excess of tentacles, subsequently reduce the tentacle number to normal, and so on. In most cases the key event seems to be the initiation or formation of a new hypostome, from which all else follows. This was well demonstrated by Chalkley (1945), who found that in masses of minced body sections new organization centers first appear as hypostomal outgrowths, and also that the number of such centers varies directly with the volume of tissue in the mass.

The crucial question is: How is a distinction made between the hypostome and the rest of the form? Burnett suggests that the differentiation and maintenance of a polarized cellular form in hydras is at least in part controlled chemically by a diffusible factor concentrated in and probably produced by nerves of the hypostome,

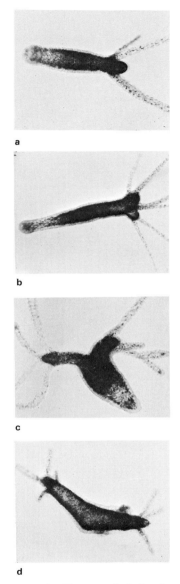

a

b

c

d

FIGURE 6.15   A midsegment of a hydra allowed to regenerate normally for 3 days is compared with three others exposed to a tissue-extract fraction (F_3) containing neurosecretory granules. The normal regenerate (**a**) possesses a distal head (hypostome and tentacles) and a proximal base; those exposed to F_3 are abnormal, having (**b**) two distal heads, (**c**) with additional head protruding from the body, and (**d**) heads protruding both proximally and distally. [*Courtesy of T. L. Lentz.*]

and that this substance directs the differentiation of interstitial cells into nerve cells. These in turn control the fate of the remaining interstitial cells.

In brief, the hypothesis is that in *Hydra* there is a *single* inducer of all cell differentiation. The inducer controls the *qualitative* cellular effects, i.e., the differentiation of the basic cells into diverse cell types, by varying *quantitatively* along a gradient extending from apex to base of the organism. Furthermore, according to this view, this inducer is also the growth-stimulating agent responsible for polarity in the organism.

On the other hand, an inhibitor of the inducer is assumed to be produced by dividing cells, so that the actual controlling condition for local differentiation is always the ratio between inducer and inhibitor in the region in question. Conditions in the epidermis and gastrodermis would be different, since an inhibitor substance produced by dividing epidermal cells diffuses into the external medium, whereas inhibitor from gastrodermal cells diffuses into the digestive cavity and accumulates there. Such a hypothesis is essentially a plausible speculation; in other words, it is a working hypothesis that serves as a model accounting for some of the observed phenomena.

Some evidence exists that a diffusible substance, probably a polypeptide, is present in the hydra hypostome, and also in *Tubularia*, that can control the polarity of the organism and possibly cell division. This substance may be associated with neurosecretory granules, since neurosecretory granules are released at sites where new heads are regenerated. Furthermore, extra heads are produced where isolated neurosecretory granules are administered to the hydra tissue. The gradient theory of control as operating primarily through the regulation of hypostome differentiation has been tested in another manner, however, by studying the regeneration of the basal disc at the far end of the body from the hypostome.

The disc is an organized structure by which hydras adhere to the substrate. It consists of a compact group of mucous cells, of a kind found only in the disc, that secrete a mucopolysaccharide. Regeneration of a new disc is possible, however, from any level of the peduncle, i.e., at any level of the body column below the budding zone. Epidermal cells of the peduncle continually move basally and, in the intact animal, replace the disc cells that are steadily sloughed off. In regeneration a new disc forms from the epidermal cells that would otherwise have moved to the base as replacement cells. The question is: What local agents may be active in controlling the event? Experiments in which amputated basal discs were grafted to the cut peduncles of host hydras at various intervals after hosts had begun their own disc regeneration process show that a grafted basal disc can inhibit the development of another disc if the latter is present during a period of induction. During this period, less than 5 hours, removal of the grafted disc permits host regeneration to proceed, i.e., the inhibition is removed. The possible explanation is that a basal disc normally produces, as it grows, an inhibitory substance which serves as a negative feedback

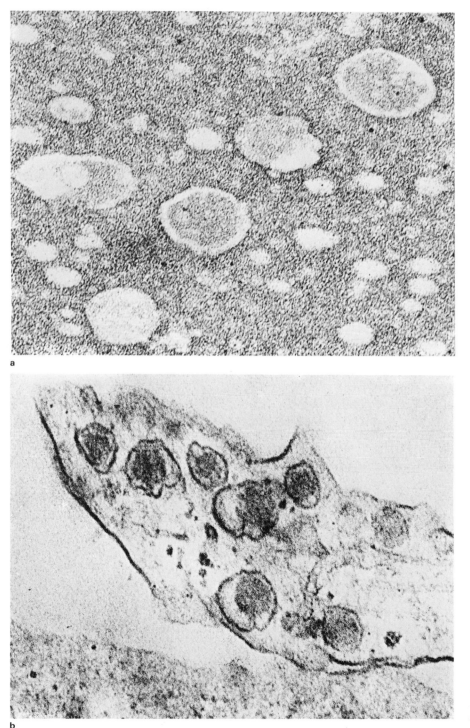

FIGURE 6.16   (a) Electron micrograph of a suspension of fraction F_3 of homogenate of whole hydras, showing membrane-bounded particles. This fraction can still induce extra heads anywhere along the body column after cold storage of several months. (b) Electron micrograph of a neurosecretory cell neurite containing dense, membrane-bounded granules. [*Courtesy of T. L. Lentz.*]

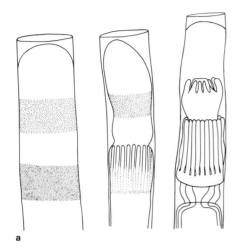

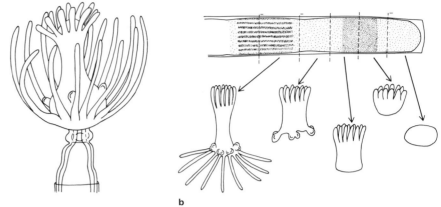

FIGURE 6.17 (a) Reconstitution of a new head (hydranth) from the distal part of a piece of *Tubularia* stem. Four stages, from left to right: pigment-band stage representing future tentacle zones, later stage with tentacle ridges in proximal but not in distal zone, completed hydranth still in original position within cuticle, and functional hydranth extended beyond cuticle. [*After Berrill, 1948.*] (b) Prospective development of parts isolated from reconstituting *Tubularia* stem. [*After Davidson and Berrill, J. Exp. Zool., 107:473(1948).*]

in the intact animal, thereby regulating the size of the basal disc in normal growth, maintenance, and regeneration. At the present time the interpretation of the experimental data is speculative, but this explanation is in keeping with current ideas concerning the control systems for organ size in general in animal organisms.

CHEMICAL DIFFUSION THEORY

Control systems generally require a stimulating, or triggering, agent and an inhibitory, or removal, agent. A nerve axon discharges acetylcholine; acetylcholinesterase destroys it. Slime-mold cells secrete cyclic adenosine monophosphate, and a secreted enzyme destroys it. Any axial gradient system in which a morphogenetic stimulatory activity falls off from a high point toward lower values necessarily implies that there must also be some gradient in inhibitory actions, since otherwise the final result would be that the whole mass of tissue would eventually turn into a product characteristic of the top end of the gradient, even though the lower end might take longer to attain it. In hydras a hypostome does indeed exert an inhibitory action which falls off with distance, tending to prevent the formation of another hypostome. Moreover this gradient of inhibition must be maintained by the continued activity of the hypostome, since the level of inhibition begins to fall as soon as the hypostome is removed. This concept has been applied to axial gradients generally, particularly to other hydroids and to flatworms.

In *Tubularia,* when rather long pieces of the stem, e.g., 10-mm pieces, are isolated, the original anterior end (or sometimes both ends) will transform into a hydranth, whereas very short pieces, less than 1 mm in length, which are too short to form a whole hydranth, form only part of a hydranth—always the anterior part. Therefore every small part contains within itself the information required for hydranth formation, the only stimulus needed for transformation being isolation.

According to Rose, who has worked extensively with this organism, "If every small part can become an anterior part of a hydranth how do posterior parts arise? This has been experimented upon. If we start with two pieces of stem and let them proceed almost to the point of producing hydranths and then fuse the two primordia, the anterior member will continue its development and become the anterior part of a hydranth. The more posterior positioned member cannot continue toward the production of anterior structures. Instead, its more anterior partner forces it to lose its partially formed anterior structure and transform to posterior structures. . . . The same relationship has been demonstrated in *Hydra*.

"Another feature of all differentiating systems is that they are polarized. In the case of *Tubularia* this can be demonstrated by a simple experiment. As noted above, an anterior region prevents more posterior regions from producing anterior structures. If the same combination of two potential anterior pieces is made but they are combined anterior to anterior neither one controls the other. Both produce anterior structures. We learn from this and many such experiments that the controlling information passes in one direction only—from anterior to posterior. . . . The situation in plants, and *Stentor*, and amphibian limbs is essentially the same. This kind of polarized inhibitory control is probably a universal phenomenon.

"The basic question is a double one. What kind of a transmission system is there during polarized differentiation and what is the nature of the information passing along it? For a long time it has been known that the site and the direction of differentiation can be determined by an imposed electric current. Could the ubiquitous bioelectrical potentials cause something to move from one cell to the next and not in the reverse direction?"[1]

Many experiments have been made, on a variety of organisms, concerning the existence of inhibitory substances, in relation to regeneration of anterior or posterior parts of *Tubularia*, or of the nemertean worm *Lineus*, the annelid *Clymenella*, and various flatworms, and in relation to regional suppression of developing parts in vertebrate embryos.

In all such cases an aqueous extract of tissue from an anterior part of an organism, when added to the culture medium in which regenerating is taking place, tends to prevent the formation of the anterior part, while allowing posterior regeneration to proceed. Extracts of posterior tissue blocked posterior regeneration but allowed anterior regeneration. In flatworms, if a brain is grafted to a headless piece, the latter regenerates all parts except the brain, while pharyngeal extract inhibits reconstitution of a new pharynx but permits other regeneration to proceed. In chick embryos, extracts of anterior brain suppress the development of that part of the brain; midbrain extracts inhibit development of midbrain and anterior brain, but not of posterior brain or

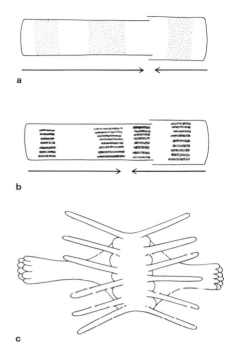

FIGURE 6.18 Induction in *Tubularia*. (**a**) Two portions of *Tubularia* stems showing regeneration and transformation when hydranth primordia are combined. The left component already has two pigmented prospective tentacle areas and represents a potential whole new hydranth; the right component possesses only one (anterior) pigment area and gives rise only to the anterior part of a hydranth if isolated. (**b**) In combination the host develops two rows of posterior tentacles, of which the second row, although in host territory, develops in conformity with the graft and is presumably induced by the graft. (**c**) The doubled hydranth with the secondary posterior tentacles oriented as part of the graft. [*After Rose, 1957.*]

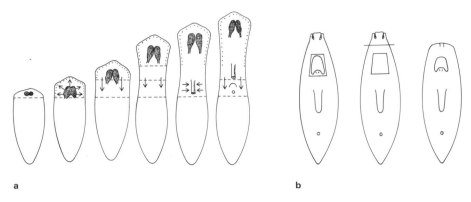

FIGURE 6.19   (a) Sequence of induction in regenerating flatworm, from left to right: brain first differentiates, which induces differentiation of eyes along anterior and anterolateral margin, to constitute head structure; head induces prepharyngeal region and then the pharyngeal region posteriorly, which induces reproductive organs posteriorly. [*After Wolff, 1962.*] (b) Window is cut out of forepart of flatworm, and head regenerates within it, whether or not original head is amputated. [*After Bronstedt, 1955.*]

a   b

spinal cord; and spinal cord extracts suppress spinal cord and all brain development anterior to it. That is, the whole central nervous system can be suppressed by extracts of the most posterior unit. Accordingly there is good evidence that specific repressors of differentiation exist in tissues, although it should be noted that differentiation is a term applied here to form and structure differentiation rather than to cell-type differentiation.

In addition to repression of differentiation, specific inhibitors are known which prevent mitosis in the region homologous to the donor region. The existence of specific chemical messengers that control the normal process of cell division in living tissue has been established by a group of British and Norwegian investigators, principally Bullough and Iverson, respectively. Each of the messengers, known as *chalones*, appears to act chiefly on the cells of a specific organ or part of an organ, e.g., they affect the cells of the mammalian epidermis but not the hair bulb. Each chalone, however, influences its target organ in many kinds of animals. Chalones extracted from the epidermis of pigs, rabbits, guinea pigs, human beings, and cod all depressed cell division in mouse epidermis. Chalones with comparable properties have also been extracted from liver, kidney, lens, white blood cells, and hair bulbs—each specific for the type of its tissue of origin but able to affect corresponding tissues in other classes of vertebrates.

Is differentiation, in the broader sense, the result of the cell-to-cell transport of charged specific repressors in bioelectric fields? Loewenstein's proposals for cell communication, discussed in the preceding chapter, suggest the pathway. *Tubularia* again serves as a test system because of the linear sequence of the parts under consideration. Again the work of Rose involving grafts of an anterior part of a regeneration primordium to a whole primordium serves as a test. This time, in the few hours when the graft is known to affect the host, the combinations were placed in an electrical field, facing either the positive pole or the negative pole. When they faced the positive pole, the anterior graft prevented anterior differentiation in the

CH

**TH
OF
OF

Sin
is
me
or
pop

REF

Sin
circ
bec
cell
self
and
of
nutr
repr
relat

thro
fere
of c
by c
prod

sivel
as a
amo
chro
and
sexu
Non
the
adult

Germ cells
into sp
in coel

Each region o

In total recon
compo

Gradient syst
ganism

Bioelectric fie
totipote

READINGS

BERRILL, N. J.,

BRIEN, P., 1960

—— and R.
    Bull. Bio

BRØNSTEDT, H.

BULLOUGH, W.
    the Cha

BURNETT, A. L.
    100:165-

CAMPBELL, R. D
    Cell Di
    121:19-2

CHALKLEY, H. W.,
    Tissue V
    Inst. J., **6**

CHILD, C. M., 19

HYMAN, L. B., 1

LENHOFF, H. M.,

LENIQUE, P., and
    of the H

LENTZ, T. H., 196
    Science, **1**

LUND, E. J., 192
    of Electr
    34:471-49

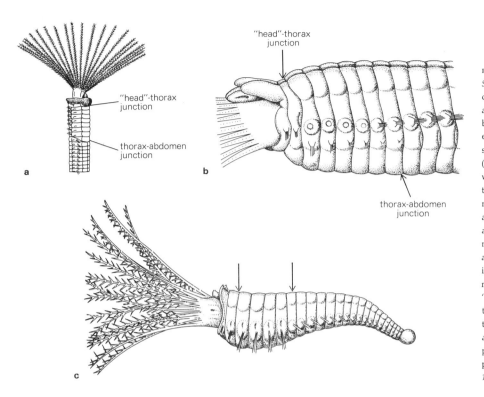

host. When they faced the negative pole, the graft formed anterior structures as expected but had no effect on the host: the host also produced anterior structures.

It appears that the current blocks the flow of morphogenetic information, an imposing name for a near-unknown, in one direction but not in the other. The hypothesis is that specific inhibitors are positively charged and are free to move accordingly. There is already evidence that inhibitors are protein or polypeptide, which carry a positive charge, and also that regional extracts move in the expected way in an electrical field. Histones have been suggested as the effective agents. According to Rose, the bioelectric fields determine the direction of flow of specific inhibitors, thus changing a totipotent system into a differentiated one.

Apart from the identity of inhibitors, there is the question of whether differences among them are quantitative or qualitative. Over the years there has been a tendency to think that within a given part there is one inhibitor but that its quantity may vary, in which case the level of differentiation would depend on the balance between stimulation and inhibition (Burnett) or between quantity and threshold of inhibition (Webster and Wolpert). Rose has considered the differences to be primarily qualitative. Yet if an extract contains both inhibitors and stimulators, changing the concentration would not change the proportions. Reports that stimulators and inhibitors of regener-

cell itself. The capacity may be lost, as in cells that have exclusively taken narrow paths of specialization, even though the capacity for mitotic division is retained; or it may be inexpressible except under certain sustaining circumstances.

Totipotent Somatic Cells

Tissues from many plants have been successfully cultured in vitro, i.e., as tissue cultures, under suitable physical and chemical conditions. Typically they form proliferating masses of undifferentiated parenchyma cells showing no organization. As such they can be propagated indefinitely as callus tissue culture. Cells obtained from such cultures may give rise to complete plants; for example, callus tissue derived from the hypocotyl, or root of the wild carrot, cultured in a coconut milk medium on agar and then subcultured under different conditions, spontaneously differentiates into bipolar embryos which even appear to pass through the distinctive developmental stages typical of wild-carrot embryos.

Suspensions of cells settled on agar plates can form thousands of such carrot pseudoembryos, some of them arising from single cells. Larger numbers of single cells are obtained by maintaining cultures in media kept in motion by shaking or aeration mechanisms. In this way many of the new cells produced by division in tissues grown under these conditions separate as single cells from the original masses

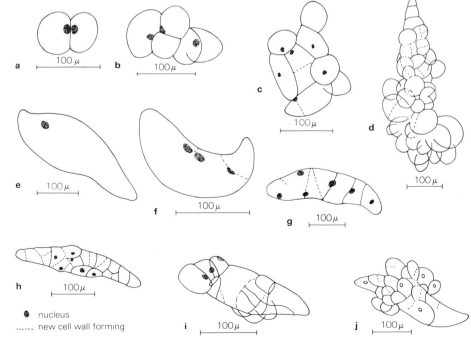

FIGURE 7.1 Development of masses of cells from single carrot-root cells in microculture. [*After Steward, Mapes, and Smith,* Amer. J. Bot., **45**:705, (1958).]

and float freely in the liquid medium. Some of these then divide to form small masses of cells from which differentiated plant tissues and shoot and root apices form. The essential feature in initiation of such independent development seems to be the isolation of the cell from intimate association with its neighboring cells, although isolation alone is not enough.

Once such a cell has been stimulated, it undergoes both cell division and growth. If the cell is of minimal size to begin with, growth and cell division go hand in hand and soon form a small mass of cells known as a *moruloid*. Some cells, however, may be giant vacuolated cells, as large as $300 \times 150 \mu$, and these undergo repeated internal divisions without growth until a compact mass of small nonvacuolated cells has been produced. Thus the starting point for further development is much the same in the two cases.

Earlier reports emphasized the importance of culture media such as coconut milk, which contains specific growth stimulators as well as many other supportive substances; later certain well-defined synthetic media were successfully substituted. All such experiments demonstrated that the cells of most plant tissues are totipotent and are capable of undergoing morphogenesis to form plants complete with roots, shoots, and usually mature flowers. Yet single cells of flowering plants removed from association with other cells and cultured in complete isolation have failed to give rise to entire plants; "nurse" tissue or a "conditioned medium" has been necessary.

TOTIPOTENCY OF DIFFERENTIATED PLANT CELLS

Species	Organ	Cell Type
Daucus carota (carrot)	Root	Cells in liquid culture
Nicotiana tabacum (tobacco)	Tumor on stem Stem pith	Single tumor cell
Begonia rex (begonia)	Leaf blade	Epidermis
Linum usitatissum (flax)	Hypocotyl	Epidermis
Cyclamen persicum (cyclamen)	Tuber	
Sphagnum cymbiogolium (sphagnum moss)	Leaf	Chlorophyllose and hyaline cells
Cichorium intybus (chicory)	Root	Phloem parenchyma
Taraxacum laevigatum (red-seeded dandelion)	Root	Phloem parenchyma
Saintpaulia coriantha (African violet)	Petiole Leaf blade	Epidermis

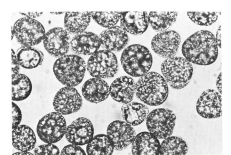

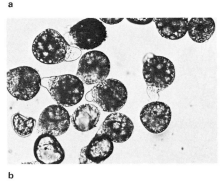

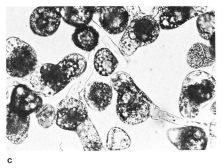

FIGURE 7.5 Development of spore cell of fern *Pteridium.* (a) Dry spores. (b,c) Protonemas in filamentous or one-dimensional stage, after 60 and 84 hours of development. [*Courtesy of A. E. DeMaggio.*]

FIGUR
of br
loid
Meio
Austi
Prenti

as a cell and its development *as a particular kind of cell,* namely, an egg. In general, the eggs of animals, with some exceptions such as those of mammals, are adequately endowed as cells to undergo development completely sustained by their own substantial contents. In plants, on the other hand, growth and cell division are sustained either through photosynthesis (as in spores) or by nutrients supplied by external supporting tissues (as in fertilized eggs of terrestrial plants); in either case the cell at the onset of development as a rule is not significantly larger than most somatic cells of plants. The early development of plants accordingly offers opportunities for comparison of development of apparently unspecialized totipotent cells under different circumstances.

This is most evident in the so-called alternation of generations in plants. We have already seen that the stage of the total life cycle at which meiosis occurs varies independently of the actual setting aside and differentiation of reproductive cells, and also that the difference in size and structure between the two generations is not related to whether the constituent cells are haploid or diploid. Yet the differences between the two alternating generations are generally striking and persisting. Therefore the questions are: When, how, and why do these differences arise during the development of the cell?

This problem has been studied mostly in the fern, and as long ago as 1909 Lang invoked environmental influences to account for the different developmental courses of spore and fertilized egg, respectively: The free spore, liberated from the sori of the mature sporophyte and settling upon damp soil, grows without restraint and produces a green filament, the protonema, which ordinarily develops into a heart-shaped prothallus bearing sex organs, i.e., antheridia and archegonia. The fertilized egg is not exposed as a free cell but is enclosed in the chamber (ventor) of an archegonium, which in turn has developed in the cushion of the thallus of the gametophyte, where there is some physical restriction and where chemical gradients almost certainly exist. Subsequently, part of the embryo becomes specialized as the foot, an organ which absorbs nutrients from the gametophyte, while other areas of the embryo begin to develop as the first leaf, root, and shoot apex.

The embryo develops three-dimensionally; the spore develops as a two-dimensional thallus. Are the differing circumstances responsible for the differences in development? This would be quickly answered if it were possible to place a spore cell within a confined space such as an archegonium, and to remove the fertilized egg from an archegonium and allow it to develop free of constraints or other adjoining tissue influences. The first such experiment (confining a spore) has not yet been accomplished, although spores lend themselves readily to experimental treatment in other ways, because of their numbers and accessibility. Embryos, however, have been extracted from the constraint of archegonia at progressively earlier stages and allowed to develop in liquid culture media supplied with inorganic salts, vitamins, and sucrose. In this circumstance it is found that the younger the embryo at removal, the more the early pattern of development deviates from the normal and slows down. One-celled fertilized eggs become prothalloid embryos, and none attains the structure and dimensions of a sporophyte.

Such evidence suggests that a cell (fertilized egg) develops into the large, three-dimensional sporophyte fern in consequence of its particular situation in and early nourishment by the gametophyte tissue. The "natural" course of development of such a cell if it is free from constraint would be one leading to the heart-shaped two-dimensional plant which subsequently differentiates the male and female sexual tissues. This conclusion is further supported by experiments where leaves from young fern plants are placed on the surface of an agar–mineral salt medium with little or no sucrose; after 2 or 3 months, prothalloid filaments develop from marginal or surface cells, which, if illuminated, become heart-shaped and bear sex organs and rhizoids, i.e., they become typical gametophytes. However, older embryos treated in the same way develop normally and grow into full-sized fern sporophytes. Thus, according to Wetmore, who has pioneered in this field, "It would appear that the determination of what might be called the *prothallial* or the *sporophyte* pattern is not a commitment because of hereditary or DNA differences in spore and zygote. . . . The unenclosed spore grows into a filament if and when exposed to formative blue light which somehow influences the planes of its cell divisions so that it gives rise to a two-dimensional prothallus with its rhizoids and sex organs. By contrast the fertilized egg is enclosed within its jacket of turgid archegonial cells. These cells collectively may well exert a determining influence on planes of cell division of the zygote, divisions which follow a consistently regular pattern. This enclosed embryo becomes multidimensional, adequately massive to acquire its own gradients. Gradients produce differential distributions in cell masses with the evident effects of such differentials upon inductions and inhibitions. At least these entities early acquire a leaf, a stem and a root, each with its apical meristem and capacity to extend itself. . . . Morphologically and physiologically, single cells or small cell groups, which can give rise to cell aggregates by regularized pattern or by statistically haphazard cell division, all produce sporophytic or multidimensional types of plants. . . . Whenever cell development is free on a surface which supplies moisture and nutrients, development is mainly two-dimensional and the prothallus plant is formed. Once initiated, neither of these two types is easily changed; either can be varied or may even assume characters of the other type, or may give origin to plants of the other type. But each tends to remain in its own pattern."[1]

In flowering plants the difference between the two generations has become more extreme. The single-celled fertilized egg or embryo, if we can call it that, is buried within the multilayered structure of the ovule, complete with integuments, nucellus, and embryo sac. At the time of fertilization one of the two nuclei introduced

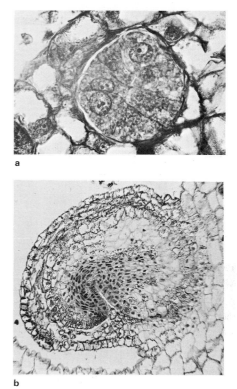

FIGURE 7.6   (a) Early cleavage stage and (b) embryo of fern embedded in parental tissue (i.e., development of zygote). [*Courtesy of A. E. DeMaggio.*]

[1] R. H. Wetmore, A. E. DeMaggio, and G. Morel, *India Bot. Soc. J.,* **42A**:316 (1964).

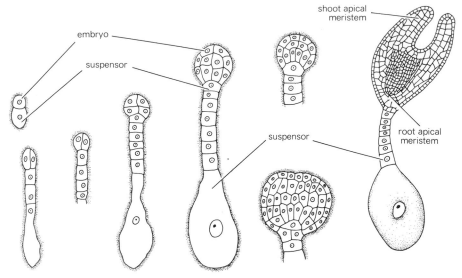

FIGURE 7.7 Early development of zygote of a dicotyledonous plant.

FIGURE 7.8 Complete plants growing from pollen grains. [*Courtesy of J. P. Nitsch and C. Nitsch.*]

by the pollen grain fuses with the egg nucleus proper to form the zygote, the other fuses with the nucleus of the embryo sac to form the endosperm nucleus, the endosperm subsequently providing the elaborate nutrient materials which the embryo needs for its development. Once development begins, cell divisions and cell enlargement produce a multicellular embryo consisting of thousands of cells. The embryo passes through a succession of distinctive morphological stages, absorbing the entire endosperm and surrounding tissues of the ovule. The ovule, in fact, is the female gametophyte formed and retained by the large sporophytic plant. Pollen grains represent the male gametophytes, set free but comparably reduced; part of the pollen grain grows out at the time of pollination as the pollen tube. Yet much the same situation seems to hold, namely, that the diverse nature of gametophytic and sporophytic courses of development is based on circumstance and is not directly genetically determined.

This is shown by the discovery that full-grown plants of various species of tobacco can be raised from pollen grains alone. They are haploid and derived from a cell normally destined to form a pollen tube and fertilize an egg. Stamens are removed from young, unopened flower buds, at a stage when pollen grains are individually present but are still uninucleate and free of starch. When grown on a simple medium, including sucrose, in the presence of light, many pollen grains develop into plantlets, which mature and flower profusely, although they do not set seed. However, mature pollen grains given the same treatment produce normal pollen tubes but no plantlets. In other words a presumably unspecialized cell (the immature, uninucleate pollen grain) can develop into a full-grown flowering plant or into minute gametophytes, depending on conditions external to the cell.

CONDITIONAL AND MULTIPLE DEVELOPMENT: ANIMALS

Certain groups of animals, as already mentioned, also undergo asexual as well as sexual reproduction. Asexual reproduction may be concerned with the formation of colonial systems consisting of many individuals associated in various degrees of organic or physical intimacy to form in some cases a true superorganism in which the life and function of the individual are subordinated to the life and function of the whole; or it may be concerned with the production and liberation of large numbers of free individuals in order to build up enormous populations while circumstances are favorable, although with minimal genetic variation. The contrast between the sexual and asexual reproductive processes is comparable to that between the development of a plant from a zygote and development from a group of unspecialized cells of cultured callus tissue. In all cases the developing multicellular system is diploid.

When the originating masses consist of few cells, they necessarily remain organically related to the parental organism, since nutritive support for growth and differentiation cannot be otherwise supplied. Large fragments of an animal organism, however, may contain sufficient reserves to sustain regeneration and reorganization processes which are also truly developmental.

Certain features of asexual development in animals need to be emphasized here, particularly features evident in certain coelenterates.

Polymorphic development A hydroid colony commonly produces new individuals from various sites along its system of stems and connecting basal stolons. New growths arise as outgrowths from stem or stolon, resembling an early stage of bud outgrowth in hydras. The terminal part of such an outgrowth typically develops into a feeding polyp, or hydranth, which consists essentially of rings of tentacles surrounding a gastric column. Alternatively, an outgrowth may develop into a medusa of different form and size from a hydranth and complete with internal canals and marginal sense organs. Eventually the medusa is set free to grow to sexual maturity. The same parental tissues are involved in the production of both types of individual.

What determines which course of development will be taken is far from being understood, although temperature and light are among external factors, and nutrition and cell relationships are among internal factors. However this may be, the important fact is that one of two or more very different paths of development, leading to the formation of very different kinds of individuals, may be taken by groups of cells with the same genetic constitution and from adjacent regions of the same parental structure.

Alternative differentiation Depending on circumstances, the same tissue may give rise to either a newly developed individual (asexual reproduction) or to differentiated sexual tissue, male or female. This is evident in hydras; the middle region of the column may

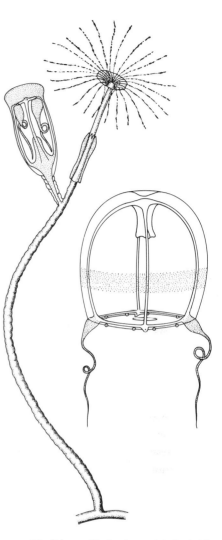

FIGURE 7.9 Polymorphic development in hydroid, showing fully formed hydranth with a fully developed medusa growing from stalk, and a liberated, expanded medusa. [*After Rees,* J. Mar. Biol. Assoc. U.K., **23***:1, (1939).*]

be heavily involved in forming bud outgrowths which successively develop and detach, with both the gastrodermis and epidermis taking part; under certain other conditions, the epidermal cells of the same region differentiate into testes or/and ovaries, when budding usually ceases.

Temperature-dependent development   An even more striking case is seen in certain small jellyfish (the medusae of *Rathkea,* a marine hydrozoan). Medusae are formed and set free from the small hydralike polyp, but instead of growing directly to a sexually mature state they proceed to form exclusively epidermal buds on the manubrium, which develop directly into new medusae. These are set free successively, to grow and produce similar medusae in turn, until enormous populations are established. The procedure continues as long as the ocean temperatures remain below about 7°C. When the temperature rises above this level, the budding process stops and all medusae differentiate testes or ovaries from the tissue which had previously been involved in budding. In this case only the epidermis is involved in the process of bud production. It is the same epidermal tissue which becomes sexually differentiated. The two

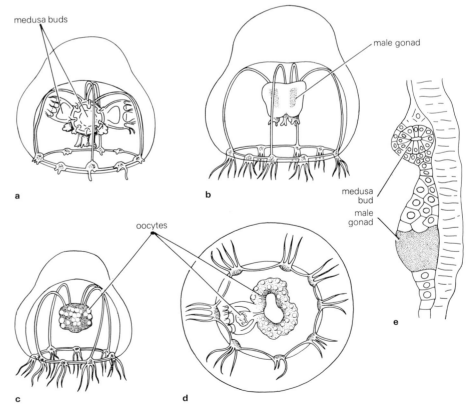

FIGURE 7.10 Conversion from medusa-bud production to sexual reproduction in the medusa of *Rathkea octopunctata.* (a) At low temperatures medusa buds form successively from the ectoderm of the manubrium. (b,c) At higher temperatures medusa-bud production ceases and the same ectodermal tissue differentiates into either male or female gonad, according to the genetic sex of the parental individual. (d) Sexually differentiating medusa produces sexually differentiating medusa bud. (e) Section of manubrium wall shows simultaneous origination of medusa bud and a male gonad from the ectoderm. [*After Berrill, 1952.*]

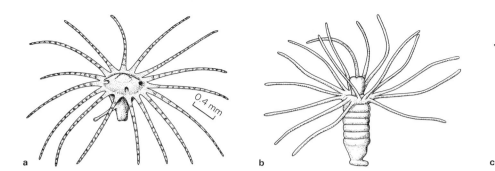

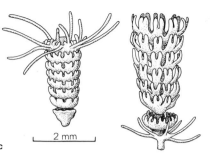

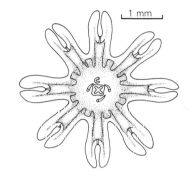

processes may well be competitive. Yet there is clearly a temperature-operated switch, at least with regard to the initiation of sexual differentiation.

Thus somatic tissue in coelenterates and at least some other kinds of animals, often initially consisting of a very few cells indeed, can readily develop asexually into complete new organisms capable of becoming sexually mature individuals, or they can produce sex cells. Furthermore, somatic development may take different paths, leading to different individual forms, depending on circumstances, even though the constituent cells have the same genetic constitution.

Transformative development The egg of the common jellyfish *Aurelia*, like that of virtually all hydrozoan and scyphozoan coelenterates, hydra included, develops into a small elongate, cilia-covered planula larva, which attaches by one end and develops a mouth and ring of tentacles at the other, i.e., it becomes a hydralike polyp, or scyphistoma. The polyp grows and produces buds in essentially the same way as hydras do, the buds separating and becoming attached alongside the parent polyps which they replicate. Extensive polyp colonies are thus produced, as in hydras, that persist throughout most of the year. Each polyp grows to a certain maximum size, with growth otherwise being siphoned off, so to speak, through the process of budding.

In nature, during spring months, polyps undergo a segmentation, or strobilation, process, i.e., transverse epidermal constrictions mark off a series of segments of the trunk, forming in sequence beginning at the distal end and eventually consuming all but the basal part of the polyp. During the process of constriction each segment develops the basic organization of a medusa, complete with radial symmetry, gastric filaments, statocysts, and pulsating musculature. The young medusa, set free as an ephyra, grows massively, to become a sexually mature aurelia. This life cycle is of course well known, but it exemplifies again that diverse forms—polyp and medusa—can be expressed by the same genome under different circumstances, and also that here there is an actual transformation of polyp (hydranth) into multiple medusae. Segmentation initiates the process and is followed in each segment by a true metamorphosis.

FIGURE 7.11 Strobilation in *Aurelia*. (**a**) Young scyphistoma stage with 16 tentacles. (**b,c**) Three stages in strobilation of scyphistoma column. (**d**) Expanded medusa (ephyra) stage liberated from top of strobilating column. [*After Uchida and Nagao, 1963.*]

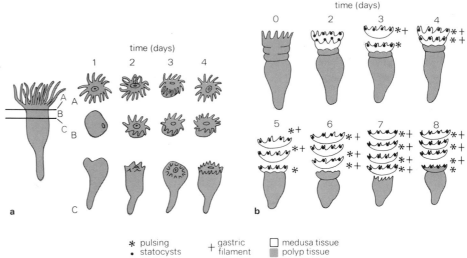

FIGURE 7.12 Strobilation in *Aurelia*. (a) Regeneration of segments of polyp stage, to reconstitute new polyp. (b) Time sequence of medusa (ephyra) development and liberation. [*After Spangenberg, 1967.*]

* pulsing
• statocysts

+ gastric filament

☐ medusa tissue
▨ polyp tissue

FIGURE 7.13 Strobilating scyphistomas of *Aurelia*, both polydisc and monodisc. [*Photo by Clarence Flaten, courtesy of Dorothy Spangenberg.*]

Many attempts have been made over the years to identify the environmental factor or factors responsible for triggering the strobilation process. Several have been implicated, particularly temperature; a preconditioning by exposure to comparatively low temperatures appears to be generally necessary. The only dependable effective agent so far discovered, however (by Spangenberg in 1967), is iodine. It is effective as iodide or as thyroxine when added to artificial seawater used as a culture medium for laboratory-maintained *Aurelia* polyps. Natural seawater normally contains iodine in trace amounts, and since the iodine content undoubtedly fluctuates with the seasonal growth and decay of iodine-binding seaweeds, iodine may well be the trigger in nature as well.

In any case it is striking that iodine and thyroxine are capable of initiating a profound metamorphosis in such unrelated organisms as jellyfish and amphibians. What the significant change may be that iodine brings about in the tissue metabolism of the responsive organisms remains a question for the future.

MERISTEMS

Production in hydras and in those medusae that produce medusa buds from the manubrium follows a characteristically sequential course. New buds are continually produced at a critical distance from the distal end of the organism or manubrium, as the case may be. Older buds and their adjoining parental tissue shift proximally accordingly. Bud rudiments appear initially to be of essentially the same size, and the number of buds present at any one time depends on the circumference and extent of the total area involved. In other words, irrespective of the details of the budding process in any particular case, buds are initiated and exist in direct relation to available space. In essence the procedure is meristematic and may be compared to the situation in the vegetative and floral meristems of the higher plants.

CONCEPTS

Reproduction in multicellular organisms is generally a function of single cells.

Totipotent somatic cells, singly or in groups, may undergo direct development to form sexually mature multicellular organisms. The potential for development is therefore a property of the unspecialized cell and is not uniquely that of designated reproductive cells.

Gametes are divergently differentiated reproductive cells designed to fuse (conjugation or fertilization) in pairs to form zygotes. Conjugation relates primarily to chromosomal recombination and is not necessary to the developmental process as such.

Reduction division (meiosis) and cell fusion (nuclear conjugation) involve a successive halving and doubling of the number of sets of chromosomes (ploidy) necessary for generational stability. Meiosis may be widely divorced from other aspects of reproduction and is not essential either to development or even to the differentiation of gametes.

Single reproductive cells, whether eggs, spores, or somatic cells, may develop into essentially normal multicellular organisms irrespective of ploidy or number of sets of chromosomes present. A single set of chromosomes can sustain development.

Cells with identical genetic constitution may follow different paths of development and form very different types of organisms, depending on external factors and physiological conditions.

READINGS

ALLSOPP, A., 1964. Shoot Morphogenesis, *Ann. Rev. Plant Physiol.*, **15**:225–254.

AUSTIN, C. R., 1965. "Fertilization," Prentice-Hall.

BELL, P. R., 1959. The Experimental Investigation of the Pteridophyte Life Cycle, *J. Linn. Soc. (Bot.)*, **56**:188–203.

BERRILL, N. J., 1961. "Growth, Development and Pattern," chaps. 16, 17, Freeman.

——, 1949. Developmental Analysis of Scyphomedusae, *Biol. Rev.*, **24**:393–409.

DEMAGGIO, A. E., 1963. Morphogenetic Factors Influencing the Development of Fern Embryos, *J. Linn. Soc. (Bot.)*, **48**:361–376.

—— and R. H. WETMORE, 1961. Morphogenetic Studies on the Fern *Todea barbara*, III: Experimental Embryology, *Amer. J. Bot.*, **48**:551–565.

HYMAN, L. H., 1940. "The Invertebrates," vol. I, "Protozoa through Ctenophora," McGraw-Hill.

NITSCH, J. P., and C. NITSCH, 1969. Haploid Plants from Pollen Grains, *Science*, **163**:85–87.

SPANGENBERG, D., 1967. Iodine Induction of Metamorphosis in *Aurelia*, *J. Exp. Zool.*, **165**:441–450.

STEWARD, F. C., 1970. From Cultured Cells to Whole Plants: The Induction and Control of Their Growth and Morphogenesis, *Proc. Roy. Soc. (London), Ser. B*, **175**:1–30.

—— and H. Y. MOHAN RAM, 1961. Determining Factors in Cell Growth: Some Implications for Morphogenesis in Plants, *Advance. Morphogenesis*, **1**:189–265.

SUSSEX, I. M., 1962. Plant Morphogenesis, in W. H. Johnson and W. C. Steere (eds.), "This Is Life," Holt, Rinehart and Winston.

TORREY, J. G., 1967. "Development in Flowering Plants," Macmillan.

——, 1966. The Initiation of Organized Development in Plants, *Advance. Morphogenesis*, **5**:39–92.

UCHIDA, T., and Z. NAGAO, 1963. The Metamorphosis of the Scyphomedusa *Aurelia limbata*, *Annot. Zool. Japan.*, **36**.

VANABLE, J. W., and J. H. CLARK, 1968. Early Development of the Fern Gametophyte, chap. 4; Shoot Apex Study, chap. 5; Development of Excised Fern Leaf Rudiments in Sterile Culture, chap. 6, in "Developmental Biology" (a laboratory manual), Burgess.

VASIL, V., and A. C. HILDEBRANDT, 1968. Differentiation of Tobacco Plants from Single, Isolated Cells in Microcultures, in J. R. Whittaker (ed.), "Cellular Differentiation," Dickenson.

WARDLAW, C. W., 1966. Leaves and Buds: Mechanisms of Local Induction in Plant Growth, in W. Beermann (ed.), "Cell Differentiation and Morphogenesis," North-Holland Publishing Company, Amsterdam.

——, 1965. The Morphogenetic Role of Plant Apical Meristems, *Encycl. Plant Physiol.*, **15**:443–451, 966–1076.

WETMORE, R. H., 1959. Morphogenesis in Plants: A New Approach, *Amer. Sci.*, **47**:326–340.

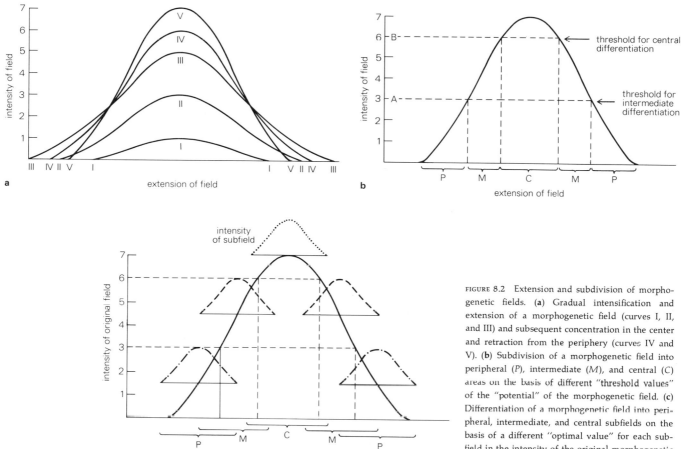

FIGURE 8.2 Extension and subdivision of morphogenetic fields. **(a)** Gradual intensification and extension of a morphogenetic field (curves I, II, and III) and subsequent concentration in the center and retraction from the periphery (curves IV and V). **(b)** Subdivision of a morphogenetic field into peripheral (*P*), intermediate (*M*), and central (*C*) areas on the basis of different "threshold values" of the "potential" of the morphogenetic field. **(c)** Differentiation of a morphogenetic field into peripheral, intermediate, and central subfields on the basis of a different "optimal value" for each subfield in the intensity of the original morphogenetic field. [*After Nieuwkoop, 1967.*]

regular change in shape correlated with the initiation of each secondary center. During this growth the apex passes from a minimum to a maximum volume. Each secondary center initiated successively from the apical region, whatever its manner of formation may be, represents a local region of increased growth rate relative to the rate of growth of its immediate surroundings. A rhythm in the frequency of mitosis is evident in the meristematic zone immediately distal to the zone of initiation of the primordia.

INITIATION OF PRIMORDIA

Most of the experimental work conducted on shoot meristems has concerned leaf primordia, rather than bud or reproductive primordia. The primary event is obviously

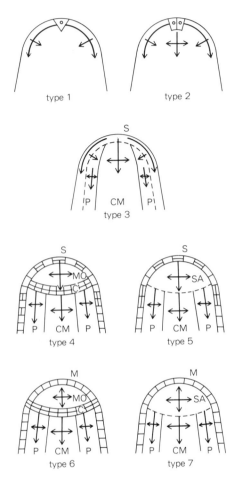

FIGURE 8.3 Diagrams illustrating seven principal types of shoot apex organization found among vascular plants. *S*, surface meristem; *M*, mantle; *MO*, central mother cells; *C*, cambiumlike zone; *SA*, subapical initials; *CM*, central meristem; *P*, peripheral meristem. [*After Popham, 1964.*]

the growth of the apical meristem, whether it consists of one single large cell (as in ferns and conifers) or of a group of cells (as in flowering plants). The regulatory system in the two cases, however, may not necessarily be the same. In the fern, for instance, destroying the large apical cell by puncturing it with a very fine needle brings axial growth to a close. Puncturing the group of apical cells of a flowering plant is less effective, presumably because as long as one or more cells survive intact, the apex can be restored. The apical meristem is gradually used up in the initiation of new leaves, the last of which may be close to the punctured tip.

The growing apical cell is evidently necessary as the source of new material continually added to what has already been produced. On the other hand the apical cell is just as evidently not necessary for the initiation of new leaf centers in meristem already formed. In fact active apical growth is not a condition for the establishment of secondary growth centers. The embryo of species of pine, for instance, with its massive apical meristem, in the absence of older primordia, gives rise simultaneously to from six to sixteen primordia. As Wardlaw (1966) has pointed out, the organization and reactivity of shoot apical meristems are such that they give rise to regularly spaced growth centers. The situation is comparable to the initiation of multiple hypostomal growth centers from the surface of a mass of minced hydras.

Many operative experiments have been conducted on the shoot apex of ferns and cycads, with particular concern for two problems: (1) Why do leaf primordia appear at all? (2) Why do they appear in a precisely ordered sequence, i.e., in a specific phyllotactic pattern? So far little has emerged from operative or other experimental approaches to the first question. With regard to the second, one theory is that a specific stimulus activates a new site of a leaf primordium, which moves along a spiral—a purely theoretical approach. The other, proposed by the Snows, is that a new leaf primordium arises in "the next available space," namely, the area of the apex least occupied by previously formed primordia. This might be simply a matter of physical space available which allows the primordium to develop or, alternatively, preexisting primordia may produce chemical inhibitors which diffuse radially outward and inhibit new centers from forming within a certain distance. Isolation experiments performed by Wardlaw, essentially a series of precisely made cuts in the apical meristem, tend to support the available-space theory rather than the repulsion effect of chemical inhibitors. The question, however, is not settled.

Careful observations of a shoot apex during the period that a new leaf primordium is being formed show that in cases of a spiral sequence, the new primordium arises on the side of the meristem some distance down from the summit and between two slightly older primordia. These three points may determine where a new primordium emerges; it may be that some critical distance must exist between them. However, if one of the adjacent leaf primordia is surgically removed, the new center arises closer to the position of the removed primordium than would normally be the case. Removal of the primordium appears to remove or reduce some inhibiting factor, although this may not be a diffusing chemical.

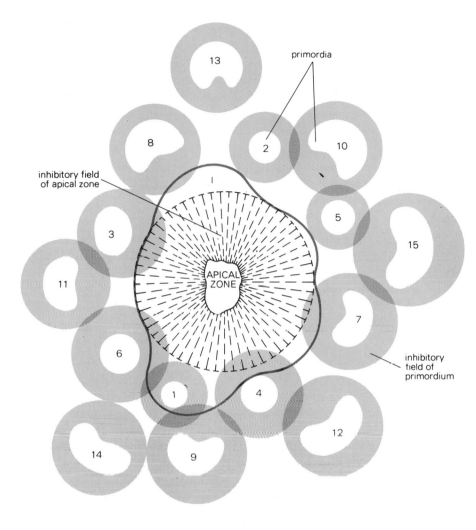

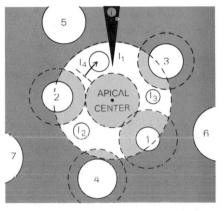

FIGURE 8.4 Shoot apex seen from above, showing apical zone and its inhibitory field (circular), together with successively formed secondary primordia, each with its own inhibitory field. New primordia arise only from the apical meristem at a certain distance from the apex, and only in a region relatively free from the inhibitory influence of adjacent primordium fields. Numbers refer to sequence of primordia already formed. I represents the next primordium to be established. [*Based on Wardlaw, 1949.*]

FIGURE 8.5 Apical meristem of fern *Dryopteris.* Diagrammatic representation of the expected result when the I_1 position is incised, i.e., before the field was developed: I_2 and I_3 will occupy normal positions for these primordia, but I_4 will tend to arise closer than normal to the I_1 position. [*After Wardlaw, 1949.*]

Experiments of this sort suggest that each primordium represents a growth center which is surrounded by a physiological field of some sort which prevents the establishment of another primordium within its territory. This "field" concept has been employed extensively in animal development biology, and there is no doubt that such "morphogenetic fields" exist. What is yet to be discovered is the actual basis of such fields—whether they are primarily chemical or electrical, or have some other sort of basis.

The first signs that a primordium will develop into a leaf are the change from an initial radial shape to a bilateral pattern of growth and, later, the differentiation of tissues in a dorsoventral pattern. This inevitably leads to the formation of the bladelike structure of a leaf. Wardlaw found that if an initial primordium is isolated on the shoot meristems by means of small incisions on every side, the structure which grows from the isolated site frequently develops as a shoot apex, radially symmetrical and able to produce a succession of leaves itself. Similarly when fern-leaf primordia are actually excised and grown in sterile culture, they develop as shoots. Evidently the initial primordium is a radial field with the potential for developing into an entire shoot, but when it is undisturbed on the side of a meristem it comes under influences that transform it into a primordium with more restricted potential.

Leaf primordia arise relatively close together near the summit in most flowering plants but form as centers relatively widely separated much further down the flank in ferns. According to Wardlaw, a young fern primordium, situated above and between two older ones, occupies only a small portion of the cone. While the primordium is still part of the cone, its growth seems relatively slow; but once it is in the subapical region, growth and morphological development are relatively rapid. We need to remember, however, that the meristem as a whole, from apex to base (disregarding the local outgrowths along its side), is a system with maximal growth rate at the apex and minimal at the base.

Consequently a new, secondary center of growth needs to have a very high rate of growth to make an appearance near the apex, but it can appear readily at lower levels down the primary growth gradient. Inspired by Child's theory of metabolic gradients, Wardlaw suggests, in line with Child's theory, that a leaf primordium is formed as a growth center of special metabolism and that the position of the center is apparently controlled by diffusion of substances from the apical cell or cell group at the shoot apex and by the two adjacent, youngest primordia. However this may be, a gradient of some sort, indicated by the growth-rate gradient of the entire meristem and by its electrical polarity, seems to exert an overriding effect on the initially radially patterned primordium growth centers, imposing a dorsoventral property, i.e., its own polarity, on the potentially independent center. Incisions around such a center remove the dominance of the apex and permit the primordium to retain its original radial and equipotential character.

MERISTEM GROWTH

Obviously the events occurring in the meristem, with regard to growth especially, are of prime importance, and so are the even more elusive procedures leading to the sequential and patterned establishment of the secondary growth centers. Since there is an essential similarity in the overall picture of meristem growth and sequential leaf-primordia formation in ferns and certain conifers (with their single large apical cell) and in flowering plants and others (with a cell population occupying the same position of eminence), we may assume that the growth or other activities that we are probing for are essentially the same whether they are incorporated in one cell or in many cells.

It is often stated that the shoot apex is a dynamic system, but the profundity of this statement is only recently becoming fully appreciated. This dynamic quality is particularly evident from experiments on the multicellular apex of flowering plants. The difficulty is not so much to understand the processes involved as to see what is going on. Consequently marking experiments have been employed, of the kind widely used during the period of so-called classical embryology, for the purpose of tracing cell movements in the summit region. How do the summit cells give rise to the flank cells? Is there a kind of permanent though proliferating population of cells residing at the summit and continually adding peripheral cells to what is already present below, thereby surmounting a base of increasing dimensions, or is there a dynamic movement and shift within the summit population itself? If the latter is the case, then does something comparable take place within the confines or in the cortex of the giant single apical cell of the other type of meristem?

These are difficult questions to answer, and this is a field of inquiry only now opening up. Preliminary marking experiments have been performed with carbon particles which are made to adhere to the apical cell or cells, with the fern by Wardlaw (1949) and with lupin by Soma and Ball (1964). In the fern, after a few weeks most of the particles have shifted to lower levels. In the flowering plant the carbon spots on the dome also are shifted toward the flank and, like shifts in position of punctured apical cells, indicate considerable cell movement. Apparently the centrally located cells or their derivates shift toward the sector which includes the most recently initiated leaf primordium at its center. It is probable that these movements of the carbon particles indicate directions of growth involved in supplying cells to these areas. With the passage of successive leaf-initiating growth periods, or plastochrons, this direction of growth is changed and follows a genetic spiral, at least in forms exhibiting spiral phyllotaxis. In others, where whorls of primordia appear rather than a helix of primordia, the central cells appear to shift in straight lines to the flanks.

It becomes increasingly evident from the studies of the plant meristem and equally from studies of the development of eggs and other developmental units that the great technical problem arises from the fact that in all cases the most important developmental events occur in the beginning. This of course is far from surprising; yet it means that the primary analysis concerns events taking place either in single cells (though they may be large, as in the case of the apical cell of a meristem and all animal eggs) or in a group of apical meristem cells and the buds of colonial animals. So much must go on in so little space, or in so little a mass of living material, that analytical progress has been severely retarded by technological difficulties. The need, with regard to experimental analysis, is to isolate the growing systems so that they can be studied under various conditions in the laboratory. Animal eggs, of course, are ready-made for such a procedure and have therefore been exploited in this way for nearly a century. The asexual developmental units of plants, excluding spores, and of those animals that exhibit such development comprise a different situation, since the developmental units are usually not only initially much smaller than most animal eggs but are typically dependent on tissue continuity with the parent organism if they are to grow and develop.

Recently, however, Romberger and his coworkers (1970) have evolved a tech-

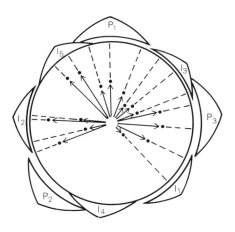

FIGURE 8.6 Diagram of a shoot apex, showing recorded positions of carbon spots after an original spot had been applied to the center of a shoot apex in the minimal area. The carbon spots were found in a wide sector of which P_1 is the center, and no spot was found opposite the position of P_1 adjacent to I_4 and P_2 and between I_1 and I_2. [*After Soma and Ball, 1964.*]

FIGURE 8.7 Lateral view of whole-shoot meristem from a dormant vegetative bud of spruce, showing domelike apical meristem at the top, and helical whorls of primordia (secondary fields). [*U.S. Forest Service Photograph No. 520097, by J. A. Romberger, reproduced by permission.*]

FIGURE 8.8 Activated shoot meristem of spruce bud showing apical dome within whorls of youngest leaves. [*U.S. Forest Service Photograph No. 520096, by J. A. Romberger, reproduced by permission.*]

nique for maintaining excised shoot meristems of Norway spruce and other plants under laboratory conditions, thus making the meristematic system, normally hidden and almost inaccessible, available for experimental study. A comparable situation has prevailed in the study and experimental approach to the early development of the eggs of mammals, which are minute and are of exceptional interest but normally develop out of sight and out of reach within the mammalian womb; very recently techniques have been evolved for maintaining and experimenting with early developmental stages of mammalian eggs in vitro, an account of which is included in the last two chapters of this book.

The domelike meristem of the spruce is about 0.3 mm in diameter and somewhat less in height. To culture it one must, while maintaining sterility, expose it to view, cut it off with a microtool, causing as little damage as possible, and transfer it to a suitable nutrient medium. The culture must then be kept sterile and be nurtured under a controlled environment while its growth response to various treatments is periodically observed. The practical problems are legion. For plant scientists the problem is comparable to that of growing test-tube mammalian embryos, though the difficulties are not as great.

According to Romberger, once a basal medium has been designed, it is possible to test the effects, in relation to growth limitation, of known substances added to it. These may be metabolites known to be, or suspected of being, normally supplied to the meristem by older tissues; or they may be "growth regulators," the mode and locus of action of which is not yet known. In this way we can begin to learn how, under a given set of environment parameters, the growth and development of the meristem and the primordia arising from it are affected by known changes in the biochemical environment. Likewise the biochemical environment can be held constant while physical parameters are being varied. These domelike explants, initially including no primordia and having a fresh weight of about 15 μg (micrograms), survive and make good growth on a medium consisting of water, mineral salts, sucrose, urea, inositol, thiamine, and purified agar. No coconut milk, yeast extract, or hormonal preparation is needed. On this medium an initially smooth-surfaced dome will initiate as many as 10 new primordia during its first week in culture. Initiation and growth of these primordia can be observed under the microscope. A freshly excised dome is almost colorless, but the growing culture becomes quite green. During a 3-week period it typically shows as much as a thirtyfold dry-weight increase. Once we can control growth and development of the apical meristem in culture, new insight will be possible into such problems as the induction or inhibition of flowering, the regulation of apical dominance, and the prolongation of the juvenile phase of growth.

In terms of their cellular constitution there are three general types of plant meristem:

1   In certain ferns and some conifers there is a single large terminal apical cell which, through a polarized pattern of cell division, gives rise to subapical cells; these in turn repeatedly divide to produce all the cells of the shoot.

2 In many conifers there is a small cluster of cells, radially oriented about the apical center, which can divide both parallel and perpendicular to the apical surface.

3 In flowering plants, the commonest type of apical cell population shows a striking formation into distinct cell layers. This arrangement of tissues is customarily described in terms of the tunica-corpus concept, which is useful in somewhat the same way that the concept of the three primary germ layers still has useful descriptive value in animal embryology.

The technological advance just described for the pine opens the door for experiments concerning the single, giant apical cell type of meristem. The multicellular type characteristic of flowering plants has also been cultured, although in most cases at least one leaf must remain attached if the meristem is to grow. However, Morel (1964) found that isolated apical meristems of certain orchids with one or two leaf primordia attached can be grown in a nutrient medium and will give rise in a few weeks to a tissue mass, or protocorm, which can be quartered. Each sector produces either a whole plant or, if further successive quartering is performed, an entire clone of genetically identical plants. The same technique has since been successfully applied to the asparagus and carnation.

The tunica-corpus theory relates to the planes of cell division and therefore directs attention to the formation and organization of the whole shoot structure. Yet the organization of the shoot apex is by no means fixed, and the number of cell layers may change with time in the same apex. In any case, the outer tunica layer grows by continued anticlinal divisions, i.e., divisions that extend the area of the outer layer of cells rather than periclinal divisions which result in the formation of a deeper layer, and this sheet of cells becomes the epidermis of stem and leaves. The inner tunica layer, or layers, gives rise to certain cell regions of the leaf and stem, and these cells also divide anticlinally. The innermost layers, comprising the corpus, give rise to most of the stem and in part to the leaf tissues. The corpus cells undergo randomly oriented cell divisions centrally, but with greater tendency to anticlinal divisions toward the outside. In other words, a gradient exists from the surface to the interior, the outer cell layers extending as two-dimensional sheets, the inner cell population as a three-dimensional mass. These tissues are sensitive to hormones and metabolites of various kinds.

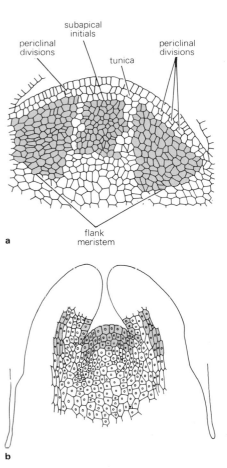

FIGURE 8.9 (a) Vegetative apex of flowering plant *Succinia pratensis*, showing various zones. [*After Philipson*, Ann. Bot., **10**:257 (1947).] (b) Growing point of *Vinca minor* in longitudinal section. Meristem cells (shaded) cover the apex of the shoot and the lateral folds which form the youngest pair of leaves. [*After Priestley*, New Phytol., **28**:54 (1929).]

LEAF DETERMINATION

Leaf initiation seems at first to be associated with the outer layers of meristem tissue; almost immediately thereafter it involves the inner tunica and even corpus tissue at a deeper level. The young primordium develops from a superficial mound on the side of the promeristem, i.e., the rapidly growing region near the summit of the meristem. It develops into a radially symmetrical outgrowth as the result of rapid cell division at the tip of the primordium and subsequent cell elongation of the cells

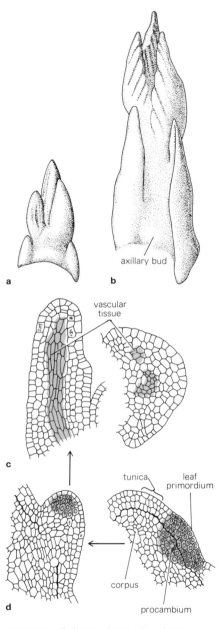

FIGURE 8.10 Embryonic leaves of raspberry, *Rubus idaeus.* (a) Leaf base together with stipule rudiments and three-lobed leaf embryo. (b) Later stage showing axillary bud within leaf base, extensive growth of stipule rudiments, and differentiation of three-lobed leaf embryo. (c) Longitudinal and transverse section of leaf embryo showing distribution of vascular tissue. (d) Origin and development of leaf rudiment in *Acacia* (heavy lines indicate delimitation of tunica and its derivatives from corpus and its derivatives). [(a–c) *after Engard, 1944;* (d) *after Boke, 1947.*]

thus formed. The growing leaf primordium may in fact be visualized as a circular or ellipsoid mound of tissue on the flank of the shoot apex. It grows as if it were pushed somewhat parallel to the shoot axis by deeper tissue arising from a somewhat lower level down the meristem, as an arc with its convex side outermost. The general form of the leaf, in this concept, is determined primarily by the pattern and growth orientation of the inner component, that is, the procambial or presumptive vascular tissue. In other words, a more or less circular secondary growth center, or field, becomes established on the sloping flank of the meristem, but the direction of outgrowth is determined by the axis of growth of the deeper, corpus prevascular tissue activated by the primordium, which in this case is more or less in line with the primary axis of the meristem: hence the bilateral, bladelike, distally growing structure which becomes a leaf.

When incisions are made around a leaf primordium, the associated corpus tissue is inevitably isolated on all sides from the polarized meristematic tissue of which it was a part, except for continuity with tissue directly inward from its base. If in this circumstance the presumptive vascular tissue then grows directly outward, rather than upward, the primordium grows symmetrically about a center, with a radial orientation of its own, i.e., with the complete organization of a new meristem, as a bud giving rise to a new shoot. This is in keeping with the long-standing concept of a leaf as a modified shoot.

Accordingly, if, as Wardlaw's experiments indicate, a leaf primordium is to be regarded as initially a bud or a potential shoot, which develops as a leaf owing to the polarized, axially directed growth of a crescent of subjacent, presumptive vascular tissue, then it is obvious, on territorial and mechanical grounds alone, that primordial tissue which is off-center from the concave side of the arc will not be involved in leaf formation. This concave area becomes the axillary bud and may be regarded as an integral but undeveloped part of the primordium proper. In effect, then, a leaf is a shoot with an extremely off-center apex (the axillary bud) and one large foliar appendage (the leaf blade and petiole). All variations are to be found, from a large leaf with a minute or even invisibile axillary bud to a very small leaf with a prominent axillary bud, depending on the extent or force of growth of the presumptive vascular tissue, and perhaps depending in some degree on the orientation of outgrowth relative to the meristem or stem axis.

It seems clear, therefore, that the particular direction, extent, and pattern of growth, whether they result in a leaf or a bud, depend directly on the orientation of planes of cell division, the number of cell divisions, and the degree of cell elongation within a primordium and its immediately subjacent corpus tissue. There can be little doubt that the whole developmental event can be described precisely in these terms. In fact any process of growth of a multicellular tissue may be expressed as the product of cell multiplication, planes of division, and individual cell enlargement, and a change in any of these factors must result in a related change in the outcome. These several aspects of formative growth have been invoked by Stebbins as the operative controlling agency responsible for the development of the particular character of leaf or bud or floral unit, as the case may be. Such an interpretation, however, omits consideration of the guidelines that necessarily determine and correlate these respective growth activities.

Nevertheless, cell division and cell enlargement are subject to the influence of external agents (environmental, nutritional, or hormonal), although these must be regarded as affecting equally a system represented either by the primary apical meristem and its subjacent territory or by a secondary primordium with an initially equipotential character.

HORMONES AND GROWTH REGULATION

Developing primordia respond to various external influences. Hormones especially have been employed in this connection, since they are directly involved. Three general kinds of hormones are now known in plants—auxins, cytokinins, and gibberellins. Their role has been interpreted in two ways:

1   These hormones are all stimulants to cell division and cell enlargement, particularly in plant cell and tissue cultures in vitro, and they make it possible for the cell population to proceed with its intrinsic developmental program. If this interpretation is correct then, as stated before, there must exist an as yet undiscovered built-in control mechanism for pattern determination, independent of externally supplied substances, although affected by them.
2   These hormones, the time of their specific release in the intact plant, and their reaching of the target cells in vivo are generally considered to constitute an integral and essential part of the genetically controlled program of development for organizing cell populations. If this interpretation is valid, then hormone levels must themselves be under the genetic control of plant cells that are determining their own patterns of development.

How the various hormones are moved through a plant system and where they are manufactured are physiological problems of seemingly little concern to development, any more than the circulation and local production of hormones in the animal body relate to the nature of animal development. In both cases the hormonal mechanism plays a significant role only in relatively well-developed organismal systems. Nor do the specific molecular configurations of the many kinds of hormones which are known relate in any enlightening way to their respective influences on the target tissue. They are intermediaries with a modifying influence, produced by an established, differentiated system and affecting developing systems that are within their reach. Nonetheless, hormones play a dominant role in regulating the overall growth phenomena within the composite system of a plant.

Auxins, gibberellins, and cytokinins, by themselves or in various combinations, can cause many different responses in plants; they are much less specific than most

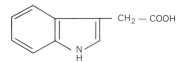

auxin (indoleacetic acid)

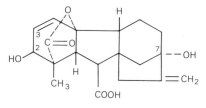

gibberellin 3

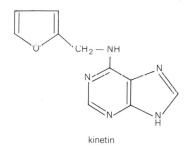

kinetin

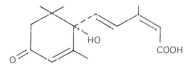

abscisin II, or dormin

FIGURE 8.11 Plant hormones. All are polycyclic hydrocarbons. A number of varieties of each type are known.

animal hormones. More than one hormone may be involved in regulating a certain process. Cytokinins, for instance, induce cell division in meristems and elsewhere but do not stimulate division in cambial tissue. Auxin also is involved. In general, the control mechanism of a particular growth process appears to be the same in different plants, but the limiting hormonal factor or factors may be different.

For a substance to be classified as an auxin it must cause enlargement of shoot cells. In addition, however, it usually induces other activities, such as cambium growth, callus formation, xylem differentiation, root initiation, delayed abscission, etc. Auxin fails to act under conditions where messenger RNA and protein synthesis are almost completely inhibited. This and other data have led to the conclusions that auxin brings about cell enlargement through the synthesis of mRNA, followed by formation of new enzymes which increase cell-wall plasticity.

Cytokinins usually exert their effect in combination with auxins, and for cell division to occur, at least in tissue cultures, both auxin and kinin must be present in the medium. A high ratio of auxin to kinin favors root formation; a low ratio favors that of shoots. On the other hand, cytokinins, auxins, and gibberellins together play a regulatory role in the growth and development of fruit. Gibberellin has been shown to activate hydrolytic enzymes, and in this way it may exert its control over the growth and developmental processes in plants.

It is increasingly evident, therefore, that hormones act not in isolated systems but in an interrelated manner in the plant as a whole. Accordingly the varying ratios

Hormones	Chemical Nature	Production Site	Site of Action	Action
Auxins	Indoleacetic acid (IAA)	Shoot apex Young leaves Young fruits	Shoots	Cell division and enlargement
Root hormones	B vitamins	Leaves	Root	Growth
Cytokinins	Purine derivatives	Young fruits Roots	Meristems	Cell division
Gibberellins	Diterpinoids	Shoot apex Young leaves Young fruits Roots	Shoots	Cell division and enlargement
Dormin, the same as abscission II	Sesquiterpene	Leaves Fruit	Shoot apex	Inhibition of cell division
Florigen	?	Leaves	Shoot apex	Initiation of floral primordia

of the different hormones present at one time may vastly affect the growth rate and subsequent differentiation patterns of the tissues in the whole organism. Moreover, the stimulatory and inhibitory hormones together permit a precise control of many developmental activities. Like the action of animal hormones, the action of plant hormones appears to lie in the control of the mechanism by which enzymes are made in the cell.

In summing up the contemporary situation, Galston (1969) states that hormonal action, even of the naturally occurring inhibitor dormin, or abscissic acid, appears to be connected with the control of nucleic acid metabolism. Thus, several hormones appear to control development by regulating the ultimate expression of genes, i.e., the coding for enzymes which modulate the different phases of plant growth. But as the relation between hormones and the genetic control of enzyme formation becomes more firmly established, it becomes equally clear that this mechanism does not supply a complete answer to the problem. The rapid effects of auxin on growth indicate that we still have to search for an alternative primary mode of action of this class of hormones, the interaction with the nucleic acids might then account for the continuation of the response.

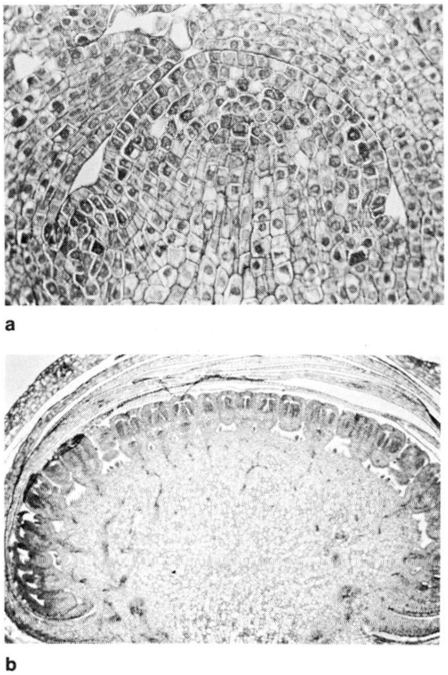

TRANSITION TO FLOWERING

Flower primordia may arise on the apical region of a leaf-shoot meristem in company with preexisting developing leaves. In such cases, modified forms of leaves typically arise from primordia lying between the earlier, normal, leaf primordia and the primordia which are fully determined for floral development. In other cases a meristem may from the beginning be almost entirely of the floral type and no such transition may be evident. Whatever the situation, the initiation of floral primordia is controlled in many plants by low temperature (vernalization) or day length (photoperiodism) or by both. Various experiments, including control of the photoperiod, leaf removal, and grafting, show that a floral hormone called *florigen* is produced in leaves and is transported to the meristems. Like other plant hormones, it is not species-specific. Its chemical nature is not known, although it may be related to the gibberellins.

Whatever the flowering hormone may be, it is transported to the apical meristem, where it initiates differentiation of floral primordia. Since this involves a new developmental pattern, it seems logical to assume that the action of floral stimulus in the apex consists of the activation of new genetic information, although this has yet to be proved. In any event, nucleic acid and protein synthesis are required for floral initiation, although the same might be said for any new developmental event. The production of new species of RNA and protein in the floral apex has not been demonstrated.

The most obvious difference between the vegetative and floral meristems is one of shape. The apical dome, whatever its original shape as a vegetative meristem,

FIGURE 8.12 Longitudinal sections through chrysanthemum apex (**a**) transforming to floral type and (**b**) showing differentiation of floral units. [*Courtesy of R. A. Popham.*]

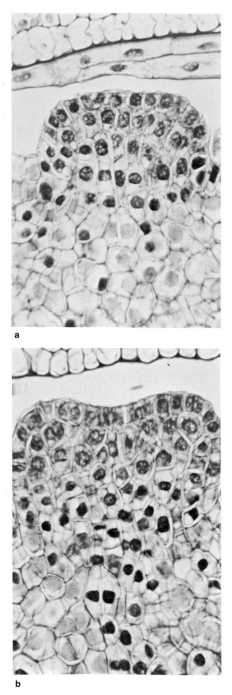

a

b

FIGURE 8.13 Two stages in development of a single floret. [*Courtesy of R. A. Popham.*]

becomes larger and more domed. The transformation, whether slow or sudden, represents a progressive developmental change in apical growth and constitution. According to Boke (1947), in *Vinca rosea*, three types of apex are produced during growth, all with a two-layered tunica and an active corpus: **(1)** a juvenile apex, which forms only leaf primordia; **(2)** an adult apex, which bears both leaf and flower primordia; and **(3)** a floral apex, which forms only flower primordia.

Within the presumptive floral apex, two changes are paramount:

1   Cessation of growth of the central initial zone, or apical mother cells, with a corresponding disappearance of the underlying stem meristem
2   Progressive extension of peripheral meristem toward what until this time has been the tip of the apex

These changes coincide with a decrease in auxin content of the apex. Growth of the corpus tissue continues, although no longer with a predominantly axial orientation, and gives rise to a fairly massive, dome-shaped apex. The whole base of operation has been radically altered, although what the regulatory control may be is utterly unknown. The new apex may give rise to a single or multiple flower.

The transformation from the vegetative to the floral apex has both positive and negative correlations. As a result of the diminishing activity of the corpus initial cells, less and less meristem is produced in a given time. In the initiation of leaf primordia, the incorporated corpus tissue is already incipiently provascular and exhibits a spectacular capacity for polarized growth. In floral primordia the corpus material is much less determined and lacks the axial growth tendency. The floral primordium, in fact, does not lose its initial radial symmetry and may be compared with a bud primordium, except that, unlike a bud primordium, with its prospective provascular core of vascular tissue, the floral primordium acquires a distinctive pattern of its own.

The specific floral pattern, unexpressed but latent in the floral primordium, can be accounted for more readily in negative terms than in positive ones. The growth circumstances of the vegetative meristem are such that every primordium that is formed is forced by the polarized growth of provascular corpus tissue to become either a leaf or another vegetative meristem of the same character as the one of which it is a part. In either event, there is no room at the top, or there are no conditions at the top, which permit an initial primordium to express the complete differentiation or pattern potential of the genome.

THE FLORAL PATTERN

The incipient floral primordium as a whole is a circular territory which may occupy the whole summit of the floral meristem and probably corresponds to an independent morphogenetic field. The general sequence of secondary primordia is as follows:

1   During its development the circumferential territory, i.e., the presumptive perianth, of the floral apex gives rise to two whorls of primordia which commonly differentiate as sepals and petals, respectively. Both whorls or primordia grow in much the same manner as do those of the vegetative shoot that become leaves. They incorporate provascular tissue, which is typically arc-shaped at its base.

2   The axillary buds of both sepals and petals develop into stamens, although additional whorls of primordia inward from the peripheral whorls may also develop into stamens without association with the leaflike sepals or petals. Stamens develop from primordia which, in their early growth stages, elongate perpendicularly to the surface of the initiating tissue, not obliquely as do all foliar appendages.

3   Carpels form from primordia on the more central apical meristem. They differ from stamen primordia not only in their relative position on the floral apex but also in their relative initial size. Small primordia can develop into stamens but not into carpels.

   The flower as a whole is consequently the composite of the various rings of primordia forming in concentric circles from the peripheral territory of the whole field toward its center.

   The subject, however, is as vast and diversified as the flowering kingdom in its entirety, and it awaits investigation on a scale far greater than has so far been attempted. The one outstanding variable is the obvious fact that the floral apex may become organized as a large single flower unit, composite though it may be, or many small fields may become established on the dome, each giving rise to a relatively miniature flower, complete or incomplete as the case may be.

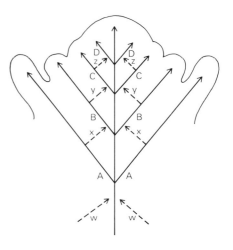

FIGURE 8.14 Model for a differentiating floral apex based upon the idea of control through a relay system. A hormonal stimulus, florigen (w), initiates the transition from vegetative growth to flowering. That is, an early gene product forms an inducer (x) which diffuses to the next group of primordia, where it activates gene complex B, and so on toward the apex. [*After Heslop-Harrison, 1957.*]

CONCEPTS

The apical meristem proper, i.e., the extreme tip, is a continually growing center, consisting of a single large cell which through division successively adds new cells to the territory beneath (as in ferns), or it is a terminal group of cells (in flowering plants) which proliferate with essentially the same effect, i.e., maintenance of the terminal growth center and addition of new tissue below it.

At a short distance from the apical growth center, developmental centers arise in sequence and according to available space, either as a spiral or helical sequence, or as a succession of whorls. Each developmental center is a unit of a certain extent and shape, separated from adjacent units by neutral or uncommitted territory.

Each such unit represents a complete morphogenetic field, as does each newly initiated rudiment of a hydra bud. The course of development, however, depends on circumstances, at least in the higher plants, such as area, shape, relative growth axes, external environmental conditions, and hormonal substances reaching the meristem from other regions of the plant.

Higher plants are essentially compound organisms resulting from sequential or simultaneous development of many semi-independent developmental units capable of various conditional forms of development, conjoined and nourished by a common vascular system.

READINGS

BERRILL, N. J., 1963. Morphogenetic Fields, Their Growth and Development, *Develop. Biol.,* **7**:342–347.

——, 1961. "Growth, Development and Pattern," part III, Freeman.

BOKE, N. J., 1947. Development of the Shoot Apex and Floral Initiation in *Vinca rosea* L., *Amer. J. Bot.,* **37**:117–136.

Brookhaven National Laboratory, Symposium in Biology, no. 16, 1964. "Meristems and Differentiation."

CLOWES, F. A. L., 1961. "Apical Meristems," Oxford.

ENGARD, C. J., 1944. Organogenesis in *Rubus, Univ. Hawaii Res. Publ.,* **21**:1–233.

EVANS, L. T., 1969. "The Induction of Flowering," Cornell.

GALSTON, A. W., and P. J. DAVIES, 1969. Hormonal Regulation in Higher Plants, *Science,* **163**:1288–1297.

GAUTHIER, R. J., 1966. Factors Affecting Differentiation of Plant Tissues Grown *in vitro*, in W. Beermann (ed.), "Differentiation and Morphogenesis," North-Holland Publishing Company, Amsterdam.

GIFFORD, E. M., 1964. Developmental Studies of Vegetative and Floral Meristems, in "Meristems and Differentiation," Brookhaven National Laboratory, Symposium in Biology, no. 16.

——, 1954. The Shoot Apex in Angiosperms, *Bot. Rev.,* **20**:477–529.

GREEN, P. B., 1964. Cell Walls and the Geometry of Plant Growth, in "Meristems and Differentiation," Brookhaven National Laboratory, Symposium in Biology, no. 16.

HESLOP-HARRISON, J., 1964. Sex Expression in Flowering Plants, in "Meristems and Differentiation," Brookhaven National Laboratory, Symposium in Biology, no. 16.

——, 1957. The Experimental Modification of Sex Expression in Flowering Plants, *Biol. Rev.,* **32**:38–90.

LAETSCH, W. M., 1967. Ferns, in "Methods in Developmental Biology," Crowell.

MOMENT, G., 1952. A Theory of Differentiation, *Growth,* **16**:231–242.

NIEUWKOOP, P. D., 1967. The "Organization Centre," II: Field Phenomena, Their Origin and Significance; III: Segregation and Pattern Formation in Morphogenetic Fields, *Acta Biotheor.,* **17**:151–177, 178–194.

POPHAM, R. A., 1964. Developmental Studies of Flowering, "Meristems and Differentiation," Brookhaven National Laboratory, Symposium in Biology, no. 16.

RICHARDS, F. J., 1951. Phyllotaxis: Its Quantitative Expression and Relation to Growth in the Apex, *Phil. Trans. Roy. Soc. London, Ser. B*, **235**:509–564.

ROMBERGER, J. A., 1963. Meristems, Growth, and Development in Woody Plants, *Forest Service Tech. Bull.* 1293, U.S. Department of Agriculture.

—— and C. A. TABOR, 1970. Culture of Apical Meristems and Embryonic Shoots of *Pica abies:* Approach and Technique, *Tech. Bull.,* U.S. Department of Agriculture.

SNOW, M., and R. SNOW, 1955. Regulation of Sizes of Leaf Primordia by Growing-point of Stem Apex, *Proc. Roy. Soc. London, Ser. B*, **14**:222–234.

SOMA, K., and E. BALL, 1964. Studies of the Surface Growth of the Shoot Apex of *Lupinus albus,* in "Meristems and Differentiation," Brookhaven National Laboratory, Symposium in Biology, no. 16.

STEBBINS, G. L., 1967. Gene Action, Mitotic Frequency, and Morphogenesis in Higher Plants, 24th Growth Symposium, *Develop. Biol.,* suppl., **1**:113–135.

STEEVES, T. A., and R. H. WETMORE, 1953. Morphogenetic Studies in *Osmunda cinnamomeal:* Some Aspects of General Morphology, *Phytomorphology,* **3**:339–354.

STEWARD, F. C., 1970. From Cultured Cells to Whole Plants: The Induction and Control of Their Growth and Morphogenesis, *Proc. Roy. Soc. London, Ser. B*, **175**:1–30.

SUSSEX, I. M., 1964. The Permanence of Meristems: Developmental Organizers or Reactors to Exogenous Stimuli, in "Meristems and Differentiation," Brookhaven National Laboratory, Symposium in Biology, no. 16.

TORREY, J. G., 1967. "Development in Flowering Plants," Macmillan.

——, 1966. The Initiation of Organized Development in Plants, *Advance. Morphogenesis,* **5**:39–91.

WARDLAW, C. W., 1966. Leaves and Buds: Mechanisms of Local Induction in Plant Growth, in W. Beermann (ed.), "Cell Differentiation and Morphogenesis," North-Holland Publishing Company, Amsterdam.

——, 1949. Phyllotaxis and Organogenesis in Ferns, *Growth,* **9**:92–132.

WETMORE, R. H., 1956. Growth and Development in the Shoot System, in D. Rudnick (ed.), "Cellular Mechanisms in Differentiation and Growth," 14th Growth Symposium, Society of Development and Growth, Princeton.

ZEEVART, A. D., 1966. Hormonal Regulation of Plant Development, in W. Beermann (ed.), "Cell Differentiation and Morphogenesis," North-Holland Publishing Company, Amsterdam.

PART TWO
THE NATURE OF ANIMAL DEVELOPMENT

In studying the development of animal eggs it is important to remember that the egg is not necessarily the only road to the attainment of typical adult structure. The egg is essentially a single cell capable of undergoing development to form a larger and more complex organism at the expense of special reserves which compensate for the fact that it is not autotrophic. The development of the animal egg is therefore something which calls for analysis in terms of the processes and phenomena involved and also for explanation in terms of adaptation and evolution responsible for its existence and character.

UNITY OF DEVELOPMENT

Following activation, an animal egg undergoes successive cleavages to become a multicellular organism consisting of a multitude of small more or less differentiated cells cohering and functioning together as a whole. Such an organism must be viable, i.e., capable of self-maintenance, by the time the various processes released during activation have run their course. The fertilized egg has an obvious unity to begin with, as a single cell with but one zygote nucleus and with a single unbroken surface layer of plasma membrane and cortical material. The division of the egg into cells does not alter the fact that the developing organism remains an integral protoplasmic system. This needs to be kept in mind, in spite of the elusive nature of the integration, particularly since analytical studies tend to break the whole developmental process into separate components that appear to be independent.

An egg, any egg, at the moment of activation is awesome in its potential. It is a single cell in the process of self-creatively becoming a comparatively large, complex, animated, and perhaps thoughtful being. Small eggs may develop into prodigious organisms, particularly in mammals, although only with major nutritive assistance from the maternal organism. Large eggs, enormously rich in yolk, may develop independently and directly into near replicas of the adult, as in birds, reptiles, sharks, and some others. Very small eggs, entirely self-contained and unassisted, develop into adult-type creatures only when the adult itself is of minute dimensions, as in rotifers where the egg divides into about a thousand cells and that number suffices to make a rotifer; otherwise such development is indirect, the small egg becoming only what is possible with the material it is endowed with, and attaining adult size and character by devious and diverse means at later times. Yet whatever the course of development or the final destiny, all eggs undergo various progressive changes simultaneously in a strictly coordinated manner.

If we say that the whole of development consists of all that occurs between the moment of activation and the attainment of full size and reproductive maturity, then at least for the purpose of the present discussion we must recognize two phases. In all eggs which are shed or in any way are effectively separated from the parental body, and cleave completely from the first, there is a phase which lasts from the moment of activation until the egg has undergone all the cell divisions possible utilizing only its original resources. The process of cleavage and subsequent cell divisions in the differentiating tissues eventually come to an end as the cells attain the small dimensions characteristic of mature body cells and when all contained yolk platelets have been transformed and utilized. At this time the developing egg has become an organism normally capable of feeding and otherwise actively maintaining itself, and it will die if an extraneous food supply is not available. Given suitable external circumstances it feeds and grows, and its component cells grow and continue to multiply in pace with such growth. The second phase of development accordingly begins with the onset of feeding and continues through all subsequent existence, either as clearly recognizable growth and differentiation, or later as the more subtle form associated with maintenance and replacement.

A sharp distinction between the two phases is evident in the development of the majority of animal species, particularly when the egg is comparatively small and not excessively yolky, and divides totally during the early cleavage stages, i.e., when it is *holoblastic.* In this type of egg, every cell inherits a modicum of yolk reserves, although as a rule some cells receive much more than others. The distinction is much less clear, however, in the case of so-called *meroblastic* eggs, eggs which are typically larger and densely packed with yolk, and in which cleavage is confined to a cytoplasmic cap or cortical layer more or less separated from a mass of nondividing yolk. In eggs such as these, including the small mammalian egg which is a derivative of this type, the dividing cells receive nutrients and actively grow, commencing at a relatively very early stage of development. The crisis when a viable organism is required to be fully functional and to sustain itself in the external environment is thus postponed to a much later stage in the growth cycle as a whole.

ANALYSIS OF DEVELOPMENT

The main problems of development were seen as clearly by the pioneering embryologists in the late nineteenth century and the early part of the twentieth century as they are today. E. G. Conklin defined development as progressive differentiation, coordinated in space and time, and leading to specific ends. He went on to state that the chief problems of development may be grouped around the three phases of this definition, namely:

1　When and how do progressive differentiations arise?

2   How are coordination, orientation, and regulation brought about?

3   How can we explain the teleological character of development, in which the end seems in view from the beginning?

He commented that at the time (1929) only the first of these general problems had received adequate treatment; that an important beginning had been made on problems of coordination and regulation; but that with regard to the teleological character of development biologists were still almost completely in the dark, and that many regarded this aspect of the problem as no problem at all—in other words, they explained it by explaining it away.

During the decades since these statements were made, we have discovered more about progressive differentiation, have learned a little about coordination and regulation, and have continued to ignore the teleological appearance of development, being concerned with the what, how, and why of development in that order of priority. Conklin, however, went on to say: "The chain of cause and effect is endless and every cause discovered leads to inquiries as to the cause of this cause. We trace differentiation to inductions and these to earlier formations and localizations of formative materials and these to the promorphology of the egg, and all of these to the genes—only to be met by the eternal question of the cause of this last link. We find mechanisms only to inquire as to the causes of these mechanisms. We find orientations, regulations and teleology only to be mystified by the immensity and complexity of these problems of development. But this is the method and these are the limitations of science, for nature is infinite and our science only touches the hem of her garment. But so far from discouraging research it should stimulate us to know that we are working in a field which has no limits and that our explorations will never end. . . ."[1]

The development of an egg, i.e., ontogenesis, or of any other kind of cell, is a multidimensional event of great and subtle complexity, seemingly beyond the capacity of the mind to embrace in its totality. Various events proceed simultaneously, and it has been expedient to abstract these events for separate study. Consequently a view of development has evolved which regards the whole process basically as several more or less independent processes in some way geared together and directed to a common end. This view, which has considerable observational and experimental evidence to support it, almost inevitably calls for some supervisory, coordinative, or regulatory agency, responsible for the wholeness, or integration, manifest in development. Circumstantial evidence for the existence of such an agency is strong, direct proof is lacking, and its hypothetical nature is a subject of controversy. In fact some biologists explain it by saying that the developing system of an egg is essentially a cooperative, progressing, cell community, that multiple causes are at work in development, and that their effects and interactions give rise to and maintain the product without any overall direction at any stage. In other words, there is no need for a guiding agency of any kind. It

[1] E. G. Conklin, Problems of Development, *Amer. Natur.*, **63**:34 (1929).

is evident, therefore, that there can be more than one approach to the study of development and that, in our present state of informed ignorance, one approach does not necessarily invalidate another.

DISSOCIABILITY OF DEVELOPMENTAL PROCESSES

During the development of an egg, four processes occur which are at least somewhat independent of one another, in the sense that each may to some extent take place in the absence of the others. These are cleavage, differentiation, morphogenesis, and growth.

Cleavage An egg cleaves and continues to subdivide to become a multitude of small cohering cells by an orderly process involving the mechanism of mitotic cell division.

Differentiation, or histogenesis[1] An egg progressively undergoes change of substance as a result of localized chemical processes initiated or released during activation. Its contents become regionally differentiated, first at the macromolecular level and later, as the egg becomes progressively compartmentalized through cell divisions, at the cellular level as specific histo and cytodifferentiation in structure and in function.

Morphogenesis As the number of cells in the cleaving egg increases, progressive changes in form take place, commonly converting the typically spherical form of the egg and early cleavage stages into an elongate and usually involuted form with regionally separated parts.

Growth There is an increase in the mass of the developing embryo, or part thereof, intimately associated with metabolic activity.

Joseph Needham, in a well-known essay entitled "On the Dissociability of Fundamental Processes in Ontogenesis," had this to say: "The fundamental processes of ontogenesis are evidently not existentially dependent on the integrity of the whole, but as in the case of the non-segmenting Chaetopterus egg, they do not take a wholly normal course when the integration has been interfered with. It is as if each fundamental process represented a layshaft which may or may not be in gear with the primary shaft, and animal economy is obviously so constituted that more than one secondary gear can be engaged with the primary shaft at one time. . . . Now what is the primary shaft? It is probably not identifiable with one chemical reaction, but it may be defined as whatever reaction the cell can carry out which will provide it with the minimum amount of energy necessary to maintain itself as a going concern in the physical world. Whatever under the worst environmental conditions suffices for basal metabolism may be thought of as the primary shaft, or rather, the automotive unit to which the primary gearshaft is attached."[2] This may be a useful analogy insofar as it emphasizes the underlying

---

[1] Some confusion arises through the use of terms, particularly *differentiation*. This is often taken to mean the whole process of change in both form and substance during development.

[2] J. Needham, On the Dissociability of Fundamental Processes in Ontogenesis, *Biol. Rev.*, **8**:180 (1933).

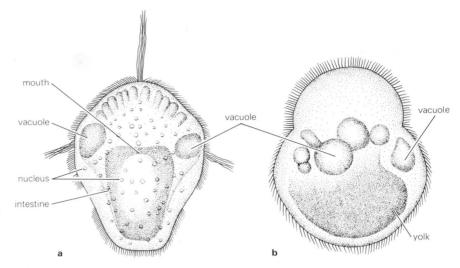

Differentiation without cleavage in the egg of the polychaete *Chaetopterus*. (a) Normal larva. (b) Unfertilized egg after treatment with dilute solution of potassium chloride in seawater for 24 hours. [*After Lillie*, Arch. Entwickl.-mech., **14**:477 (1902).]

oneness of the developing system; the semi-independence of the more obvious contributing processes; the possibility that among different kinds of related eggs different gearing ratios may exist with profound effect upon development; and that the study of development is above all a study of parts or processes in relation to the whole. These contributing processes, however, do require separate consideration, and they are accordingly examined in some detail in the chapters immediately following.

Meanwhile some properties relating to the wholeness of the developmental phenomenon call for comment. One of these is the extent and kind of control exhibited by the basal metabolism of the developing system, particularly in connection with the role of temperature in relation to rate of development.

RATE OF DEVELOPMENT

With the exception of birds and mammals, the temperature of any organism approximates that of its environment. The overall temperature range in aquatic environments is roughly from 0 to 30°C, the lower limit being about −1.8°C in seawater because of its high salt content. The total range, however, is considerably larger than the range for any particular region. The eggs of invertebrates and the lower vertebrates undergo normal development over a range of about 15°C, more or less. Above the higher end of this range development becomes abnormal in various ways, as a result of irreversible changes produced mainly in protein and fatty molecular organization by high temperatures. At the other end of the scale, low temperature may result in greatly increased viscosity or actual gelation of protoplasmic substance, which may bring all developmental activity to a standstill but is usually reversible and seemingly harmless. The extent of tolerated range for normal development is much the same for cold-, temperate-, and warm-water species, but the location of the range on the total temperature scale varies. Thus the eggs of most temperate-water species develop normally in temperatures from around 7 to 8°C to around 22 to 23°C, or a little higher. The eggs of cold-water marine fish may develop normally from less than −1.0 to no higher than 10 to 12°C. At the other extreme, eggs laid on land in the tropics may develop normally at temperatures close to 35°C, but may undergo temporary cold arrest at 25°C; in mammals and birds the upper limit for normal development has been pushed so high, close to 40°C, that no higher temperature can be tolerated, while developmental arrest occurs at temperatures only a few degrees lower than that to which such eggs have become accustomed. Natural selection has evidently operated at the macromolecular level of the egg and developing organism so that the temperature range for normal development has been shifted in each direction along the environmental temperature scale according to regional circumstances.

Within the range for normal development, wherever it may be on the overall scale, the effect of temperature change is not differential. The developmental process as a whole becomes accelerated or decelerated as the temperature rises or falls, the developmental rate increasing or decreasing about threefold for any 10° rise or fall in temperature. This is a somewhat higher temperature coefficient than that typical of simple chemical

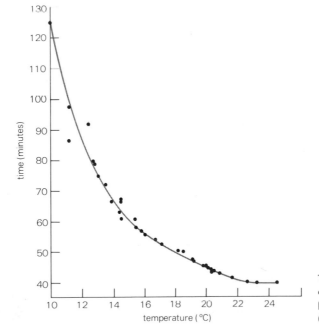

Temperature dependence of duration of the period of the first cleavage division of the sturgeon egg. [*After Dettlaff*, Advance. Morphogenesis, **3**:337 (*1964*).]

reactions, but it is of the same order, and it shows that the rate of the basic chemical reactions in the cell governs the rate of development as a whole. There is no significant difference, for instance, in frog tadpoles that have developed at 12, 17, and 22°C respectively.

This, however, may mean little more than that all the developmental processes are geared, as far as rate is concerned, to a common energy denominator, Needham's primary shaft (presumably adenosine triphosphate), derived from a variety of substances stored in the egg, and that insofar as the availability and utilization of this compound vary with the temperature, so will all the processes of maintenance, growth, and development that are dependent upon it.

DIRECT AND INDIRECT DEVELOPMENT

Since there is almost no limit to the adult size reached by some animals, at least when compared with cell sizes, the production problem faced by the egg becomes prodigious. The several means by which an initially microscopic oocyte can eventually give rise to a relatively giant organism have already been briefly mentioned. In the mammal the primary oocyte grows to form an egg of minimal size, whose development is sustained by nutrients from the maternal body until birth, which is generally an arbitrary stage in the course of growth as a whole. Extraordinarily large and yolky eggs, such as birds', etc., are necessarily few in number, are formed in the ovary, and are capable of producing juvenile organisms directly. Small eggs on the other hand typically develop into larval forms capable of feeding and growing but notably different in form, structure, function, and habits from the mature adult. Development in the first two categories is termed *direct,* in the last, *indirect.*

Indirect development has been the basis for much of the supposed support for the theory of recapitulation, mainly because transformations from one kind of organism to another during an individual life cycle have the appearance of a progressive evolutionary change. But because of the pressure of natural selection operating at all stages of development, the retention of a more or less ancestral form as a larval stage in development is only one of several possibilities. During the evolution of a particular group of animals, progressive change or adaptation may have consisted mainly in increase in the size of the fully adult stage, together with whatever internal adaptations may have been necessary to maintain the metabolic, functional, and structural efficiency of the organism. Or the adult state may have persisted essentially unchanged but new adaptive forms may have been interpolated at an earlier stage of the life cycle. Or little that is new has appeared either with regard to bigger and better end products or with regard to novelties at various earlier stages, but the whole developmental procedure whereby an egg develops and grows to form the adult organism may become streamlined, with shortcuts and increased efficiency in developmental processes occurring throughout.

DEVELOPMENTAL CYCLES

Echinoderm eggs and larvae Several classes of echinoderms, all marine, produce mostly small eggs in large numbers which develop into pelagic larvae of various types, each type being characteristic of a class. The diverse larval types are highly distinctive as larvae but are entirely unlike the respective juvenile and adult stages in either structure or habit. The weight of evidence is that each particular larval type has evolved within its respective class; each serves as an adaptive device which enables the developing organism to persist as an increasingly large pelagic stage that exploits the microplanktonic food supply and thereby increases the prospects of successful metamorphosis into the juvenile-adult pattern. In this group and among other marine invertebrates natural selection has clearly operated at both the larval and the final organismal levels, with regard to adaptation to and exploitation of two entirely different sets of environmental circumstances.

Spiralia eggs The eggs of flatworms, polychaete worms, and bivalve and gastropod mollusks exhibit a unique spiral course of cleavage during early development, culminating in the formation of larval organisms of a certain general type. The several animal phyla exhibiting this particular course of development are collectively called the *Spiralia*, even though the differences in their respective adult structures are of phylum magnitude. The implication is that the three phyla have had a common ancestry and that each has inherited essentially the same ancestral, specialized type of egg and embryo, although in each case further development leads to an adult organization of unique character. This retention of what is accordingly recognized as an ancient type of egg and embryonic development to some extent justifies a recapitulatory interpretation.

Insect eggs Insects are primarily terrestrial organisms that lay yolky eggs mostly on land and of an order of size larger than those of echinoderms. Primitively the development of the insect egg takes place on land and gives rise to a miniature of the adult form, except for body proportions and the absence of wings. Growth to the adult condition requires that the nonexpansible integument be cast off and renewed at intervals during juvenile existence. The greatest adaptive changes which have occurred within the group as a whole relate to the nature of the growth stages so that, variously, juvenile forms have evolved that can exploit food sources of kind and place beyond the reach of the adult; they may be adapted to aquatic life (e.g., the nymphs of dragonflies, etc.), or to a burrowing, boring existence (e.g., grubs), or to more exposed vegetarian life (e.g., caterpillars). Adaptive changes have been imposed or interpolated within an originally much more straightforward developmental cycle. All require some degree of later metamorphic change in order to become characteristic adults.

Chordate eggs Among the chordates, the eggs of the protochordates (tunicates and amphioxus) are of minimal size, i.e., about 100 μ in diameter, while the eggs of several relatively primitive kinds of vertebrates [such as the most primitive of living vertebrates (the lamprey), the most primitive of bony fishes (the lungfish, lobe-finned fish, and sturgeon), and the most primitive of terrestrial vertebrates (the salamanders)] are of moderate size, i.e., 1,000 to 2,000 μ in diameter. These have a remarkably uniform cytoplasmic organization and undergo a correspondingly similar course of development, giving rise to the basic chordate organization in each case.

The eggs of all the vertebrates listed above develop in freshwater into larvae with essentially the same features, complete with a tapering fish-shaped body, and, with the exception of the lamprey, external gills which function before internal gills are fully formed. Such a larva was almost certainly the kind typical of the ancestral freshwater vertebrates of long ago, and the particular character of the eggs and larvae of the surviving representatives can be reasonably regarded as an inheritance in common, an inheritance of a conservative type of egg. Here, at least, is a likely example of a recapitulation phenomenon, as in the case of the *Spiralia*, where the egg and larval type have not changed significantly during some 200 or 300 million years and have been retained as a common first step toward the attainment of diverse and larger ends. In each case, with continuing growth and greater size, natural selection has operated to diversify the end products, with little effect on the early phases of development. Nevertheless some specialization, developmentally prepared for, has differentiated, for instance, between the larvae of urodele and anuran amphibians: salamander larvae typically have functional jaws with true teeth, and a pair of transitory balancers on the side of the head; whereas frog and toad larvae have transitory horny teeth and transitory adhesive suckers—i.e., distinctive structures which have been well exploited in experimental developmental analysis.

Adaptation of life cycle to environment   Successful reproduction, as distinct from the mechanics of development, depends on adaptation of the life cycle to environmental circumstances, and it may or may not remain advantageous to a species to exploit, during the individual life of the organism, two such different environments as the planktonic and benthic world of the sea (or freshwater) and the dry land. With progressive evolution of the adult organism, with its implied change in habits and capacities, extreme conservatism in egg and larva may cease to have survival value except in terms of developmental mechanics as such. The production of young individuals that can live in the same habitat as the parent and feed in much the same way is clearly an evolutionary option that can have and has had great potential. Accordingly in virtually every group of animals we find kinds which, mainly through great increase in egg size and yolk content, have eliminated not only whatever functional larval form might once have been present but, necessarily, whatever embryonic and prepattern stages led to larva formation. In such cases traces or rudiments of vanished structures may persist; where they are recognizable they have at times been interpreted as evidence for a condensed recapitulation of the evolution of the species. But the probability is that such remnants have persisted by virtue of their importance in the developmental construction process.

DUALITY OF DEVELOPMENT

It is evident from these examples that the development of many kinds of eggs is directed toward at least two goals: (1) the development, growth, and maintenance of the young and mature adult in structure, function, and habit; and (2) in all cases where development is indirect, the development of transitory organizations differing in structure, function, and habit from the adult type. This duality (of adult and larval structure) is to be seen in striking form widespread through the animal kingdom, and it raises fundamental questions concerning the role of the genome and the preparation of the egg for its special course of development. In every case the adaptive need is great from the beginning to the end of the life cycle, and in one way or another the influence of the past is incorporated throughout development. There is a historical aspect to development, and though this view has little to do with the causal analysis of development (i.e., it does not relate to the "what" and "how" questions), it does illuminate the "why" questions.

It may be helpful to examine the relation of egg size to adult size a little further. The egg size of those marine invertebrates that produce eggs in very large numbers, together with that of the lower chordates, is rarely larger than 100 μ in diameter, and even this small size is very large in terms of other kinds of cells and is attained only through very special means. Such eggs subdivide to produce a few thousand standard-sized cells at the most, by which time the proliferating cell mass must have differentiated into a viable multicellular organism. In the early phases of metazoan evolution, when mature adults were also generally minute, such organisms may well have been clearly recognizable as juveniles. But insofar as the mature adult form of most animal types has evolved into progressively larger and more complex organisms, some means also has had to evolve whereby developing eggs can produce them.

M. V. Edds sums up the general situation as a series of questions: "In what sense, then, does the central problem of development remain unsolved if one can so often specify in detail the steps followed by an egg as it becomes an adult? The crux of the matter is that we can so rarely explain what we can so amply describe. An interminable array of examples could be listed, but let the reader ask any such questions as these:

"By what means does a sperm penetrate an egg surface?

"Why, generally, does only one sperm penetrate?

"What causes the streaming movements of cytoplasmic materials when an egg is activated?

"What directs the union of the pronuclei?

"Where do the fibrous components of the cleavage spindle come from?

"Do they move the chromosomes? How?

"What is the mechanism behind the deceptively simple subdivision of the cytoplasm during cleavage?

"What determines that blastomeres shall cleave synchronously? Or asynchronously?

"Why are some blastomeres larger than others, even in some eggs with little or no yolk to impede division?

"What factors lie behind the rearrangement of cells during gastrulation?

"How are the movements accomplished?

"What integrates the various movements throughout the embryo so that all parts share in a harmonious and orderly whole?

"By what means do the various cells of the gastrula start to become visibly different from one another?

"Do individual nuclei become different? Individual cytoplasms?

"Does the genome remain the same in all cells?

"What makes the neural plate fold; the notochord cells vacuolate; the neural crest cells migrate; the mesoderm segment; or the ear placode invaginate?

"What determines the conversion of mesenchymal cells so that some become muscle cells and start synthesizing myosin, while others become fibroblasts and make collagen or chondroitin sulfate?

"How do young nerve cells spin out their axons, some extending into the peripheral portion of the embryo, others remaining to course up or down the central nervous system?

"Consider a much-studied case, the formation of the eye. How are the several previously separated components brought together in an orderly spatial and temporal sequence with lateral forebrain, surface epidermis, head mesenchyme, blood vessels, and nerve fibers all sharing in the process?

"Once together, what cues do these components follow so that one forms lens, another iris, another sclera, another retina?

"What keeps the components in balance so that each becomes a proportional and appropriate part of the finished organ?

"Why does the entire organ grow only to some typical size, then enter a steady state in which positive and negative growth processes are equalized?"[1]

He goes on to say that these questions are representative of hundreds like them and that all have one thing in common, namely, that we really do not know their answers. In each case, it is possible to give partial answers, based sometimes on established facts, sometimes on calculated guesses. Many of the questions are interrelated; a partial answer to one may shed light on another. Seventy-five years of ingenious experimentation by hundreds of investigators permit answers framed in terms of embryonic inductors and fields, segregation of special cytoplasms, nucleocytoplasmic interactions, competence, metabolic gradients, and the like.

Recent evidence permits assertions that sound more sophisticated. They deal, for example, with the microvilli at the cell surface; the *de novo* synthesis of DNA or RNA; the changing molecular configurations in the mitotic spindles; the emergence of new macromolecular constituents; the correlation of new enzyme patterns with emergent structures or functions; the important role of both small and large molecules in the

---

[1] M. V. Edds, Animal Morphogenesis, in W. H. Johnson and W. C. Steere (eds.), "This Is Life," p. 272, Holt, Rinehart and Winston, 1962.

microenvironment of developing cells; the time course followed during the synthesis of specific proteins; and a host of others. But in each case, the new information serves as much to raise new questions as to clarify old ones.

CONCEPTS

An egg is a cell, and we have already seen that a cell, either as a single cell or as a small cluster of cells, may be able to grow and develop into the multicellular organism. This is commonplace among plants, and moreover the egg cells of plants are little different from other unspecialized cells, the developmental specialties being mainly nutritive devices external to the egg cell.

The primary oocyte of animals before growth has begun also appears to be a comparatively unspecialized kind of cell; and such a cell, if it is not already too much committed to differentiate into a maturing egg, may well already have a complete developmental capacity provided that adequate nutritive sustaining conditions could be supplied as in plants. There is evidence from certain kinds of tumors that this is so.

The small eggs of marine invertebrates that undergo indirect development and the larger but relatively small freshwater-type eggs of vertebrates that also undergo indirect development show evidence of precocious egg patterning before or immediately after the onset of development, which relates to the mechanics and special nature of the precocious larval development. In other words, such eggs are not merely large cells, but they are in some way preprogrammed for a course of development that is a detour from the direct path.

In the larger, yolky eggs with direct development there is no evidence of any such precocity or early determination of events, and the egg appears to be primarily a cell capable of giving rise to an extensive sheet (blastoderm) of small unspecialized cells, typically overlying the surface of a mass of undivided yolk which nourishes it; secondarily and consequently, it is a cell capable of establishing a developmental situation in which one or more embryos may form *ab initio* from a multicellular base, a phenomenon which may be compared with the development of one or more hydra heads, and eventually whole hydras, from the surface of a minced mass of hydra tissue, depending on the area. Eggs of this general type are accordingly specialized only with regard to their capacity, related to their size and yolk content, to establish multicellular bases for one or more embryonic developments.

READINGS

BERRILL, N. J., 1961. "Growth, Development and Pattern," part IV, "The Development of Eggs," Freeman.

BONNER, J. T., 1960. The Unsolved Problem of Development: An Appraisal of Where We Stand, *Amer. Sci.* **48**:514–527.

CONKLIN, E. G., 1929. Problems of Development, *Amer. Natur.,* **63**:5–36.

EDDS, M. V., 1962. Animal Morphogenesis, in W. H. Johnson and W. C. Steere (eds.), "This is Life," Holt, Rinehart and Winston.

NEEDHAM, J., 1933. On the Dissociability of Fundamental Processes in Ontogenesis, *Biol. Rev.,* **8**:180.

OPPENHEIMER, J., 1967. "Essays on the History of Embryology and Biology," M.I.T. Press.

CHAPTER NINE

GAMETES AND GAMETOGENESIS

The specialization of reproductive cells as male and female gametes represents a division of labor between two mating types. As already emphasized in the preceding section, the primeval function of gametes is to conjugate, following chromosomal reassortment at some earlier stage in the life cycle, in order to combine genomes of distinct parentage. In unicellular organisms, such as bacteria and most flagellates and ciliates, mating occurs between individuals that are apparently alike but differ in molecular pattern of surface structure or in substances associated with the surface. Even at the protistan level, however, differentiation into microgametes and macrogametes is a common phenomenon. This distinction into motile microscopic male gametes and more or less macroscopic, nonmotile female gametes has become the general condition throughout all the animal kingdom and most of the plant kingdom, although independently in the two groups. It may be regarded as a division of labor representing motility and storage, respectively.

CONTENTS

Male gametes
   Spermatogenesis
Female gametes
   The structure and function of eggs
   Oogenesis
The growth of oocytes
Yolk distribution
Pinocytosis and accessory cells
Nucleic acid accumulation and gene amplification
Egg envelopes
Concepts
Readings

MALE GAMETES

Except in the more specialized cases of the higher plants and, among animals, in many arthropods and the nemathelminths, the male gametes characteristically propel themselves by means of flagella. In plants the sperm cells of algae, mosses, and liverworts typically have two flagella; those of ferns and cycads have numerous flagella. The sperm cells of animals typically have but one flagellum, with rare exceptions such as the toadfish, which has two. In other words, the sperm cell, or male gamete, seems to be a relict retained in recognizable form in most of the animal and much of the plant kingdom as a single motile, flagellate cell primarily concerned with the introduction of a set of chromosomes (or haploid nucleus) from one parent into a haploid reproductive cell (the female gamete) of another. Superimposed on this primary function are special properties which have evolved in the sperm cells which relate to the process of fusion with the mature female gamete, or ovum, and to the stimulation of the ovum to commence development.

Spermatozoa (the male gametes of animals) show a remarkable uniformity of structure, apart from the exceptions already noted, from lower invertebrates to higher vertebrates. The main structural subdivisions are head, midpiece, and tail, the head being further divisible into acrosome and nucleus. All are contained, as in living cells generally, by a continuous plasma membrane. The whole cell is streamlined

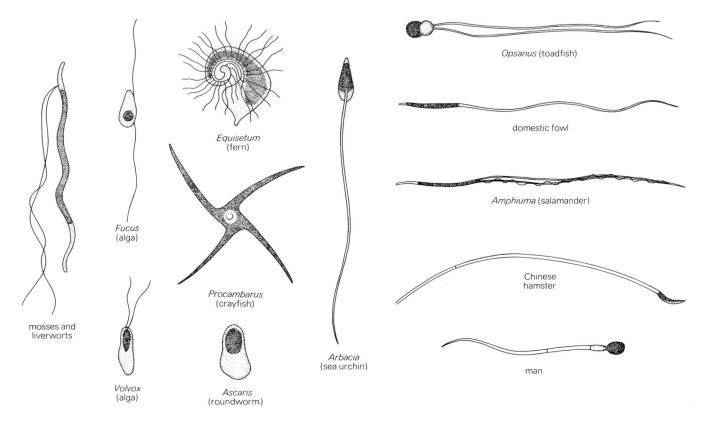

Opsanus (toadfish)

domestic fowl

Amphiuma (salamander)

Chinese hamster

man

Fucus
(alga)

Equisetum
(fern)

Procambarus
(crayfish)

mosses and
liverworts

Volvox
(alga)

Ascaris
(roundworm)

Arbacia
(sea urchin)

FIGURE 9.1 Forms of spermatozoa.

and pared down for action of a special sort and of limited duration, namely, to swim and meet an egg, to fuse with the cortex of an egg, and to introduce its nucleus and centriole into the egg interior. The acrosome is a caplike structure, varying greatly in size and shape among different species. Its universal occurrence on the leading surface of the cell has long suggested that it plays a vital role in the penetration of barriers which surround an egg. How this is accomplished is told in the following chapter on fertilization.

The sperm nucleus consists almost entirely of DNA protein, transcriptionally quiescent, but all so densely packed that little structural pattern is discernible. It is, in a sense, the burden of the spermatozoon which has to be conveyed to and passed into the cytoplasm of an egg, the sperm tail and acrosome representing the means of propulsion and penetration, respectively.

The tail flagellum exhibits typical flagellar structure consisting of nine double tubules around a pair of central tubules, all enclosed by the plasma membrane. In mammals the 9 + 2 array is known as the axial filament, for in this group there

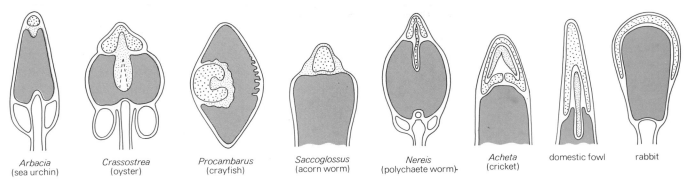

| Arbacia (sea urchin) | Crassostrea (oyster) | Procambarus (crayfish) | Saccoglossus (acorn worm) | Nereis (polychaete worm) | Acheta (cricket) | domestic fowl | rabbit |

FIGURE 9.2 Acrosomes (light-shaded) and nucleus (dark-shaded) in various spermatozoa. [After C. R. Austin, "Fertilization," © 1965. By permission of Prentice-Hall, Inc., Englewood Cliffs, N.J.]

is an additional outer ring of nine more fibers of a much thicker nature. It is not known why mammals require this extra set of fibers, if indeed they are truly necessary. Curiously, an outer ring of nine fibers is also present in some insects, a snail, a sparrow, and a snake, and possibly in many other creatures so far unexamined. The problem of locomotion is that of flagellar action generally.

The neck of the tail, which is inserted into an indentation of the posterior surface of the nucleus, is associated with a pair of centrioles. The proximal centriole represents the origin of the fibrous elements of the tail. The distal centriole, which lies at right angles to the other and probably was produced by it at an earlier stage of sperm-cell differentiation, appears to have no active function in the spermatozoon but is a potential activist within an egg during the later phases of fertilization.

The midpiece varies greatly among different species. In rabbit spermatozoa, for example, it consists mainly of a tightly coiled spiral of elongated mitochondria surrounding the anterior part of the tail; in the sea urchin a single mitochondrion wraps around the posterior part of the nucleus and adjacent part of the tail. Whatever the mitochondrial apparatus may be, it almost certainly supplies the energy required for the motility of the tail. This energy is limited and, once expended, apparently cannot be renewed. Spermatozoa are launched, so to speak, like torpedos with limited range, and are lost forever if the target is missed. Only in mammals and perhaps in other forms where spermatozoa are active only within the maternal body are energy sources available outside the spermatozoa themselves.

Because of the improbability that a spermatozoon will ever get close to an egg, let alone penetrate it, spermatozoa need to be produced in enormous numbers and as a rule to be liberated at times and places that improve the chances immeasurably; hence such devices as copulation in terrestrial and many aquatic creatures. In any case the production and liberation of spermatozoa are precisely controlled, either environmentally or physiologically. Temperature, light, hormones, and psychological state play a part in some degree, depending on the organism.

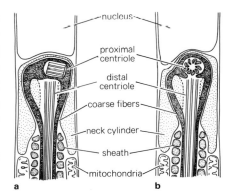

nucleus

proximal centriole

distal centriole

coarse fibers

neck cylinder

sheath

mitochondria

a   b

FIGURE 9.3 Vertical sections of snake spermatozoon, (a) parallel and (b) at right angles to the long axis of the proximal centriole. [After C. R. Austin, "Fertilization," © 1965. By permission of Prentice-Hall, Inc., Englewood Cliffs, N.J.]

Spermatogenesis

During the growth and differentiation of the testis the germ cells multiply and give rise to spermatogonia, or sperm mother cells. After further multiplication the spermatogonia become primary spermatocytes, which then undergo the first meiotic division to form secondary spermatocytes, and then the second meiotic division to form spermatids, the entire process being known as the *maturation* of the germ cells. Each primary spermatocyte thus gives rise to four haploid spermatids. Finally the spermatids, without further division, begin to differentiate into spermatozoa, a process known as *spermioteliosis*. This they do by a series of changes remarkably alike throughout multicellular animals. Apart from the preceding involvement of meiosis, the process is simply an example of cell differentiation leading to extreme specialization and as such is comparable to other examples such as the differentiation of nematocysts and neurosensory cells, i.e., of great interest but mainly as part of the general problem of special cell differentiation. It is, however, an example that has been broadly and intensively studied.

Cell divisions cease with the completion of the meiotic divisions, and each daughter cell then starts on its special course of differentiation. Small vesicles each containing a granule develop within the Golgi apparatus in the spermatid cytoplasm. The vesicles enlarge and fuse to form a single vesicle and granule which becomes attached to the surface of the nucleus and is the forerunner of the acrosome. The acrosome vesicle and its granule may remain projecting from the nuclear surface or may become spread out over the surface, depending on how closely the nucleus moves toward the cell membrane. At the same time the two centrioles of the spermatid move toward the cell surface and one of them gives rise to a small ciliumlike outgrowth which will become the axial filament of the future tail. The centrioles move inward again close to the nucleus, drawing the filament and a pocket of the adjacent cell membrane with them, so that the filament in reality becomes external to the cell. Subsequently the active centriole returns to the cell membrane, carrying the membrane fold with it. Mitochondria develop in association with the axial filament to form the midpiece. In mammals, the outer ring of nine coarse fibers grows out from the other centriole to thicken the tail. Finally the nucleus protrudes from the bulk of the cytoplasm to form the sperm head, with terminal acrosome and enclosing cell membrane; most of the remaining cytoplasm is discarded.

In plants the final differentiation of the motile, male gamete is somewhat different. There is no acrosome, for the sperm cell does not have to make its way through specialized egg membranes. More striking, in ferns and cycads, is the development of a ciliated band around the cell: in the spermatid the centrioles multiply until a large number are formed; these become arranged in rows, and each then produces a cilium. As in animal sperm differentiation, however, the residual cytoplasm is shed after the general structure of the mature cell has been formed.

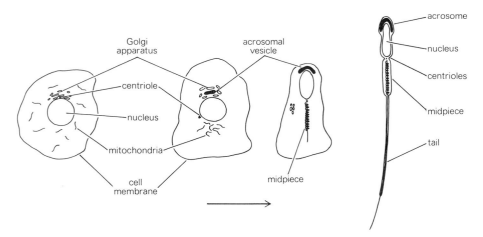

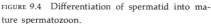

FIGURE 9.4 Differentiation of spermatid into mature spermatozoon.

Summarizing, the differentiation of spermatozoon:

1 Represents a very special type of cell differentiate.

2 Begins only after the cessation of cell division.

3 Involves a special activity, including:

Activity involving the Golgi apparatus (acrosome)

Special activity by one or both centrioles, both in filament formation and in movement to and from nucleus and cell membrane

Localization and growth of mitochondria relative to filament axis

Nuclear condensation, or compaction

Shaping by or of the cell membrane

Discard of residual cytoplasm

Although in animals meiosis always precedes differentiation of spermatozoa, in plants, as we have seen, it may be far removed in time and phase. Accordingly we may conclude, provisionally at least, that although spermioteliosis follows meiosis during the production of spermatozoa, the two phenomena are essentially independent and that meiosis as such has no causal relationship to sperm differentiation.

FEMALE GAMETES

In plants the macrogametes, or egg cells, are no larger than many somatic cells, are without protective membranes, although generally surrounded by nutritive tissue, and have no obvious specialized characteristics. They are essentially unspecialized, little-differentiated cells. As such they are capable of development into large multicellular organisms under certain conditions. All animal eggs, in contrast, are large compared

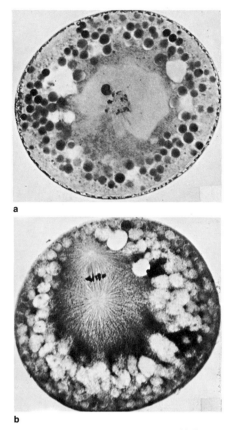

FIGURE 9.5 Maturation of *Nereis* egg. (a) Beginning of breakdown of germinal vesicle (shortly after fertilization). (b) First maturation division. [*Courtesy of D. P. Costello.*]

axis. As the cell prepares for the first maturation division, the nucleus and its associated centrioles move further toward the so-called animal pole. When the first maturation division begins, the egg nucleus and the mitotic spindle are close to the pole, with the mitotic apparatus as a whole vertical to the cell surface. When the cleavage furrow forms and cuts across the middle of the spindle in the usual way, a small cell is inevitably separated from a large cell. Insofar as the two centrioles and their respective asters may be responsible for initiating and locating the cleavage furrow, the distal centriole has but a very small region of cytoplasm available for astral organization, whereas the proximal centriole has almost all of the egg interior. The inequality of the cell division is thus accountable in terms of cell mechanics, which in turn is dependent on the positioning and orientation of the mitotic apparatus. In many eggs the meiotic spindles are oriented paratangentially (i.e., close to and parallel to the cell surface) during nuclear division, but rotate through 90° to become radially oriented more or less along the polar axis.

Experimental confirmation of the importance of the positioning of the egg nucleus plus associated centrioles was made long ago by A. C. Clement, who centrifuged the oocytes of the gastropod *Ilyanassa* at the onset of maturation and thereby caused the nucleus and division apparatus to move toward the middle of the cell or even more toward the opposite, so-called vegetal pole. In these circumstances the two centrioles organized roughly equivalent territories of the cell cytoplasm, and cleavage resulted in cells of approximately equal size.

After normal polar-body formation has been completed, the residual haploid egg nucleus, known as the *female pronucleus,* returns to its original position more or less central in the cell, though typically somewhat nearer the animal than the vegetal pole. Assuming that the locating of the division apparatus close to the cell surface and its orientation along a radius, hence vertical to the surface, are directly responsible for the extreme inequality of cell division, the question immediately arises concerning what agents are responsible for the outward and inward migrations of the oocyte nucleus and the actual orientation of the spindle preceding cleavage. As yet, there is no answer to this question for it has hardly been considered. Even so, when an answer finally is forthcoming, the further question appears: What determines the timing and nature of the controlling factors?

In vertebrates the growth and maturation of germ cells, together with reproduction as a whole, have long been known to be under the influence of hormones. This has been demonstrated specifically in oocyte maturation in the frog, where the early events of maturation, i.e., germinal vesicle dissolution and the procedure of meiosis to the metaphase of the second division, have been induced by the application of progesterone to full-grown oocytes outside the body. Since actinomycin D does not prevent progesterone-induced maturation but puromycin does, and since both nucleated and enucleated eggs increase protein synthesis to the same extent in response to progesterone treatment, it is suggested that the site of action of the hormone in inducing maturation is extranuclear proteins, possibly at the cell surface.

THE GROWTH OF OOCYTES

An egg may be regarded as a cell large enough to undergo a series of cell divisions without intervening growth, thus producing a multicellular mass of cells of typical somatic cell size. This clearly is a universal and essential feature of egg development and, without going any further, is one that immediately raises several questions:

1   What prevents the growing oocyte from undergoing cell division when it increases in size, as is customary in most other kinds of cells?
2   How does an oocyte accumulate its reserve substances?
3   What determines the extent of such accumulation?

Growth of the oocyte is at least a twofold process. It grows as a cell, and it accumulates nutritive reserves. The two phenomena are associated but are distinct. Eggs of closely related species may range considerably in size, yet contain virtually the same proportion of yolk to cytoplasm. Conversely, the yolk/cytoplasm ratio may vary greatly among eggs of the same size. Increase in egg size and increase in relative yolk content do, however, serve the same end, namely, to augment the number of standard-sized cells the egg is capable of producing from its own resources.

The fundamental property appears to be the capacity of the primary oocyte to grow beyond certain limits without dividing. How this is accomplished is clearly related to the question of what causes cells generally to divide when their volume becomes approximately twice what it was immediately following the previous division. The factors determining the onset of normal cell division are imperfectly understood, so that a clear understanding of what inhibits cell division in the growing oocyte either awaits or leads to better comprehension of the more general problem.

The ratio of surface area to cell volume, however, is one factor that should not be ignored. Gaseous exchange necessary to cell metabolism takes place through the cell membrane, a two-dimensional surface, whereas the associated metabolic activity is that of the three-dimensional cytoplasm. Consequently as a cell becomes larger, the available surface per unit volume or mass of cytoplasm becomes smaller. At some point in cell enlargement the surface area/volume ratio inevitably becomes critical and the rate of basal metabolism decreases. The surface area/volume ratio increases when a cell divides into two smaller bodies of the same shape, when additional cell membrane is actively formed. When this compensatory process fails to take place, whether by inhibition or from lack of stimulation, cell metabolism is necessarily affected. The later history of oocytes shows that this is the case, even though they appear to contain normal pathways for glycolysis, the citric acid cycle, and respiration. Most eggs, in fact, bog down, so to speak, before they can complete maturation. The oocytes grow rapidly for awhile, then progressively more slowly until a characteristic terminal size is reached. The maturation divisions follow, but the process generally becomes arrested while in the metaphase of either the first or the second division. Completion of the process then awaits the fertilization of the egg, with its attendant change in rate of metabolism. Fully grown but incompletely

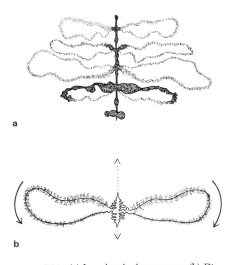

a

b

FIGURE 9.14 **(a)** Lampbrush chromosome. **(b)** Diagram of suggested mode of action. A pair of loops extend from the chromosome axis, the chromomere uncoiling at one end and rewinding at the other, with synthesis taking place along the loop axis as it unwinds. [*After Gall, 1963, and Wischnitzer, 1967.*]

although not attached to chromosomes. Evidence indicates that each is an autonomous site for synthesis of rRNA.

Why are so many copies of rDNA required? The answer is not clear, because synthesis of compounds in the oocyte relates both to the growth of the oocyte and to the utilization of synthesized substances at various stages during subsequent development. In any case, the mature oocyte contains a large pool of DNA precursors, e.g., thymidine, and a large amount of cytoplasmic DNA, of which 65 percent is bound to the yolk platelets. The ratio of precursor-pool content to that of bound DNA, however, varies from species to species. At least in certain insects the highly polyploid follicles, or nurse cells, actively incorporate thymidine and other nucleic acid precursors from the blood, and their nuclei show intense synthesis of rRNA which enters the cytoplasm and crosses by way of fusomes into the oocyte, the oocyte nucleus being more or less inactive in this respect.

Altogether we have a somewhat confusing picture of the build-up and storage of nucleic acids, ribosomes, yolk, and other reserves in the growing oocytes of the various species of animals. There is little doubt that the reason is that while all species are concerned with much the same project, namely, to prepare the egg for its developmental journey through time and space, and accordingly must prepare for an initial boost and for sustained action, there are many ways by which this can be done and few of them have been overlooked. In this connection, oocytes, follicle cells, and surrounding medium are to be seen as an integrated complex. In general, however, the large size of the oocyte as a cell, irrespective of any packaged yolk reserves, is itself a vital feature, analogous to a tethered balloon ready to rise the moment it is released. By contrast the yolk which has so elaborately been put together is utilized mainly late in the developmental process when all other sources of energy and substance may be nearing exhaustion.

EGG ENVELOPES

Animal eggs are characteristically enveloped by one or more coverings of some sort. Although some eggs are described as naked, all have a cell membrane, or plasmalemma, and most of them have a delicate vitelline membrane, or so-called fertilization membrane, closely applied to the periphery. This membrane is often surrounded by other envelopes which are of great variety of structure. Egg envelopes fall into three categories:

1   Primary envelopes, formed by the ovum itself. These consist of:

     A vitelline membrane

     A zona radiata, which represents the degraded microvilli of the growing oocyte

     A much thicker, structureless jelly surrounding the egg as a whole, as in echinoderm and many other eggs of marine invertebrates

2   Secondary envelopes, formed as a basement membrane by a layer of follicle cells surrounding the ovum, and called the *chorion,* as in insects, ascidians, etc.
3   Tertiary envelopes, formed by the oviduct or other maternal tissues not immediately connected with either ovum or ovary, such as the albumen of amphibian, reptile, and bird eggs, the shell membranes of reptile and bird eggs, and the egg capsules of mollusks and selachians

The distinction is not always simple. (The chorion of teleost fish eggs arises entirely from the oocyte and contains protein and polysaccharide.) In human eggs the zona pellucida is produced mainly by the follicle cells, together with some material produced by the oocyte. Whatever they are, egg envelopes appear to be initially secretions of either the oocyte or the surrounding follicle cells, mainly as by-products of cell activity or as a basement membrane typical of epithelial tissue as such, each chemically peculiar to the tissue and species concerned.

Any material or structure once present may be exploited secondarily to serve other functions. Thus material from either the oocyte or the surrounding follicle cells readily serves the egg as a protective structure, and it is commonly augmented for this purpose, particularly in relatively large, yolky eggs and in all eggs that are laid out of water. Even in the mammalian egg—small, almost yolkless, and existing in a nutrient liquid medium when shed from the ovary—the zona pellucida surrounding the ovum has been considered (1) as a protective device to prevent polyspermy, i.e., fertilization by more than one spermatozoon; (2) as a means of preventing egg fusion; and (3) as a necessary means of maintaining normal cleavage of the egg following fertilization. In ascidians, particularly solitary ascidians, not only does the basement membrane of the follicle cells survive as the chorion, which plays a role in preventing self-fertilization of the egg, but the follicle cells themselves persist attached to the chorion and, by virtue of vacuoles in their cytoplasm, function collectively to increase buoyancy.

Whatever use may be made of external membranes, jellies, or cell layers, however, these structures present barriers to penetration of the ova by spermatozoa, and much of the biology of fertilization of eggs and much of the specialization of the spermatozoon itself relate to this circumstance.

CONCEPTS

Sperm and egg represent two diverse, highly differentiated cells, comparable to somatic cell specializations.

Sperm and egg nuclei retain totipotency during germ-cell differentiation.

The spermatozoon has an unspecialized nucleus, a generally differentiated middle piece and tail, and a highly specialized distal acrosome.

An egg is a cell specialized with regard to size, its primary function being to yield a large number of small cells through successive divisions.

An egg is packaged with reserve material for growth and typically programmed to follow a special developmental course.

The inequality of division characteristic of the maturation stages of the egg is a device for retaining the maximum size and store of nutrients already attained.

Oocytal growth alone can attain only a certain critical size, and larger sizes depend on assistance from extracellular sources.

READINGS

AUSTIN, C. R., 1965. "Fertilization," Prentice-Hall.

——, 1961. "The Mammalian Egg," Oxford.

BEERMANN, W., 1966. Differentiation at the Chromosomal Level, in W. Beermann (ed.), "Differentiation," North-Holland Publishing Company, Amsterdam.

CLEMENT, A. C., 1935. The Formation of Giant Polar Bodies in Centrifuged Eggs of *Ilyanassa*, *Biol. Bull.*, **69**:403–418.

DAVIDSON, E., 1969. "Gene Activity in Early Development," Academic.

GALL, J. G., 1963. Chromosomes and Cytodifferentiation, in M. Locke (ed.), "Cytodifferentiation and Macromolecular Synthesis," Academic.

GRANT, P., 1965. Informational Molecules and Embryonic Development, in R. Weber (ed.), "Biochemistry of Animal Development," vol. I, pp. 483–594, Academic.

KING, P. E., J. H. BAILEY, and P. C. BABBAGE, 1969. Vitellogenesis and Formation of the Egg Chain in *Spirorbis borealis* (Serpulidae), *J. Marine Biol. Assoc., U.K.*, **49**:141–150.

POLLACK, S. B., and W. H. TELFER, 1969. RNA in Cecropia Moth Ovaries: Sites of Synthesis, Transport, and Storage, *J. Exp. Zool.*, **170**:1–24.

SMITH, L. D., and R. E. ECKER, 1969. The Role of the Oocyte Nucleus in Physiological Maturation in *Rana pipiens*, *Develop. Biol.*, **19**:281–309.

TELFER, W. H., 1960. The Selective Accumulation of Blood Proteins by the Oocytes of Saturniid Moths, *Biol. Bull.*, **118**:338–351.

TYLER, A., 1955. Gametogenesis, Fertilization and Parthenogenesis, in B. H. Willier, P. Weiss, and V. Hamburger (eds.), "Analysis of Development," Saunders.

WILLIAMS, J., 1965. Chemical Constitution and Metabolic Activities of Animal Egg, in R. Weber (ed.), "Biochemistry of Animal Development," vol. I, pp. 14–72, Academic.

WILSON, E. B., 1925. "The Cell in Development and Heredity," 3d ed., Macmillan.

WISCHNITZER, S., 1967. The Ultrastructure of the Nucleus of the Developing Amphibian Egg, *Advance. Morphogenesis*, **6**:173–198.

——, 1966. The Ultrastructure of the Cytoplasm of the Developing Amphibian Egg, *Advance. Morphogenesis*, **5**:131–181.

CHAPTER TEN

FERTILIZATION

Fertilization is a complex process involving the fusion of a male and female gamete, which, in animals, are the spermatozoon and the ovum, respectively. The process has been studied mostly in marine invertebrates, particularly sea urchins and various wormlike creatures (e.g., polychaetes, echiuroids, and balanoglossids) whose eggs are typically small, translucent, and without much yolk, the events of fertilization therefore being more readily followed, and certain vertebrates whose development is of special interest, particularly amphibians and mammals. Attention has focused on amphibians because they are available and suitable for analytical and experimental procedures, and on mammals because man himself is a mammal and the need to know what happens is great.

CONTENTS

The dual function of fertilization
The meeting of gametes
Capacitation
Sperm and egg fusion: the acrosome reaction
Activation of the egg
Experimental activation
Migration of pronuclei
The centriole
Blocks to polyspermy
Concepts
Readings

THE DUAL FUNCTION OF FERTILIZATION

Fertilization fundamentally has a dual function: (1) to cause the egg to start developing, and (2) to inject a male haploid nucleus into the egg cytoplasm. As already emphasized, the introduction of the male pronucleus is not essential for development and the single set of chromosomes of the female pronucleus suffices for virtually normal development. Fusion of the male and female pronuclei, however, restores the diploid state, regulates the normal alternation of haploid and diploid states of the life cycle, and introduces genetic variation. Fusion, or conjugation, of the pronuclei is the final event in the fertilization process; by the time it occurs the development of the egg has already been well and truly launched.

The importance given to fertilization, therefore, depends on the point of view. According to C. R. Austin, fertilization finally emerges as part of a generalized system of gene dispersal, the full influence of which is as yet but vaguely understood. In contrast, Albert Tyler defines fertilization as the series of processes by which the spermatozoon initiates and participates in the development of the egg. As such it includes all steps from the approach of the spermatozoon to the fusion of the pronuclei within the egg. James Gray emphasizes the two physiological aspects, stating that fertilization effects two changes in the egg: (1) by a modification which starts at the cortex, the metabolism of the egg is increased to its full level: the egg is, in fact, "activated"; (2) by penetration into the egg, the sperm nucleus with its attendant aster enables the egg to form the requisite machinery for normal cleavage.

As a mechanism of mitotic stimulation, fertilization has long been regarded as a biphasic process, i.e., the sperm initiates in a general way a series of changes in the egg cortex and then in the cytoplasm, as a result of which the egg becomes activated. At the same time the spermatozoon normally introduces its own centriole

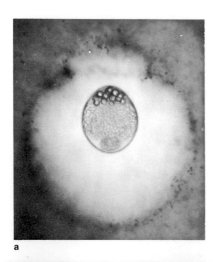

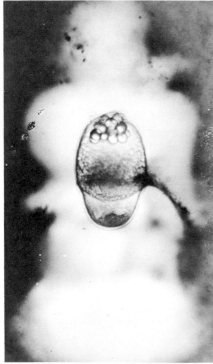

FIGURE 10.1 (a) Centrifuged egg of polychaete *Nereis* photographed in india ink suspension. (b) Inklined sperm funnel through jelly is seen on right, where particles are drawn in behind path of spermatozoon during fertilization. [*Courtesy of D. P. Costello.*]

into the egg cytoplasm. The nature of the stimulus to mitosis, however, remains obscure, although it is probable that the several mechanisms underlying mitosis are waiting in readiness in the unfertilized egg and require only the activating stimulus to go into action.

THE MEETING OF GAMETES

All gametes are single cells, no matter how large an egg may be. Male gametes are always microscopic and are minute even in comparison with the smallest egg. Eggs are immobile, while motile sperm cells, animal or plant, must cope with the viscosity of the surrounding medium because of the very high ratio of surface to volume associated with such minute and attenuated bodies. Consequently a major problem in sexual reproduction is the bringing together of male and female gametes. It is important that ripe eggs and sperm be brought together in the same general locality and that individual sperm reach the surface of the ova.

In the early days of experimental developmental biology a substance named *fertilizin* was believed to be present in the cortical regions of the egg cytoplasm, at least in the eggs of sea urchins, which have always been the most intensively studied. Fertilizin was considered to diffuse into the surrounding medium and so to set up a diffusion gradient, thereby giving orienting or directional information to nearby spermatozoa, i.e., sperm cells found their way by means of chemotaxis. The chemotaxis hypothesis accounted for the fact that repeated washing of the eggs tended to make them unfertilizable, as did time itself, merely by allowing all the fertilizin to diffuse away. Good evidence for chemotaxis, however, has so far been found only in ferns, liverworts, and mosses. In these plants the egg, or oosphere, lies at the base of a flask-shaped archegonium, the neck of which is filled with mucilage. Some of this mucilage escapes and releases a chemical into the water lying on the surface, which causes sperm in the vicinity to change direction toward regions of higher concentration. A protein is the effective substance in a liverwort, a sugar in a moss, and L-mallic acid in a fern. Chemotaxis has not yet been adequately demonstrated for animal spermatozoa. On the other hand the motility of spermatozoa does increase under the influence of material emanating from the eggs of certain sea urchins, and some small molecular sperm-activating agent may exist.

In general, animals accomplish the close approximation of eggs and spermatozoa through special devices or particular forms of behavior. The primary consideration is to deliver large quantities of spermatozoa close to numbers of ripe eggs at the right time. The variety of means employed is very great indeed. The most primitive situation is probably that of many marine invertebrates, again exemplified by sea urchins, where communities of sexually mature adults shed eggs and sperm freely into the surrounding water. In such cases, where developmental hazards are greatest, astronomical numbers of eggs as well as sperm are spawned during any one spawning period. Timing is vitally important. The adult members of a species usually become sexually mature depending on the environmental temperatures of the preceding weeks or months, and as a rule, one or more ripe females spawn at dawn or dusk, stimulated by the rapid change in light intensity. Substances liberated with the oviductal fluid at the time the eggs are shed in turn stimulate other ripe females and also the ripe males in the vicinity. Consequently clouds of eggs and sperm form in the water at the same time, and mass fertilization occurs. Even so, spermatozoa may have to travel distances which are enormous when compared with their own dimensions, and the amount of sperm produced to ensure high probability of egg and sperm unions is remarkable.

The importance of timing relates to two factors: (1) the extent of diffusion of the spermatozoa, increasing with the distance they have to travel; and (2) the limited life, with regard to fertilizability, of the reproductive cells, especially the sperm. In the sea, where the complex salt-water solution itself is essentially a primeval physiological medium, small, freely shed eggs are generally fertilizable for less than a day, although they may survive as healthy cells for much longer, while spermatozoa expend all their potential energy for motility in a few hours at the most. In freshwater the timing becomes more crucial, for spermatozoa remain active in freshwater usually for minutes rather than hours. Hence the common mating procedures of fish, amphibians, and freshwater invertebrates, where sperm is delivered directly to the eggs of an individual female at the moment of laying, if not before.

Above all, on land, mature spermatozoa are commonly stored in a physiological medium capable of maintaining their life and potential activity for days, weeks, or months, either in moisture-conserving capsules or in compartments of the male or female body, to be picked up, transferred, or utilized in one way or another. In many forms, particularly where eggs are completely enclosed in impermeable envelopes before being laid, or where they are retained within the maternal body throughout development, spermatozoa must be delivered internally. Accordingly copulation, involving the insertion of a sexual appendage, is common among terrestrial animals and is typical of mammals and insects. Yet even in mammals, human beings included, millions of spermatozoa must be ejaculated in order that a few thousand shall reach the upper end of a fallopian tube where one or more ripe eggs may be descending; with smaller numbers of sperm at first and last, an egg is unlikely to be fertilized. Together the biology and physiology of reproduction constitute an immense field of interest and investigation which relates mainly to the phenomena of adaptation and evolution, rather than to developmental problems.

CAPACITATION

There is evidence that at least in some kinds of animal the spermatozoa undergo a process called *capacitation* before they are fully capable of fertilizing an egg of the same species. In most forms, homologous egg water, i.e., water containing large numbers of unfertilized eggs, has an agglutinating effect on spermatozoa from the respective species of echinoderms, polychaetes, mollusks, tunicates, and vertebrates. In strong egg water the reaction is usually visible within a few seconds, and spermatozoa are seen to clump together head to head or, less commonly, tail to tail, a reaction similar to the agglutination reaction produced by an immune serum. These were the basic observations underlying the "fertilizin" theory formulated by F. R. Lillie more than a half-century ago, a theory whose greatest value lies in the wealth of work on the fertilization problem that it has stimulated. In sea urchins, in dilute egg water, such a reaction reverses after a few seconds or a few minutes, but the spermatozoa have by then lost their fertilizing capacity, although they are still fully motile. This phenomenon relates to the specificity of fertilization, namely, the capacity of spermatozoa to fertilize eggs of the same species but not of other species, and has been interpreted in terms of the mutual multivalence theory of antigen-antibody reactions.

The reacting substances of egg and sperm have been called *fertilizin* and *antifertilizin,* respectively. The fertilizin is thought to be produced by the ripe unfertilized egg (sea urchin) and accumulated in the external gelatinous coat; spermatozoa fertilize such coated eggs more readily than they do eggs from which the coat has been removed. According to the theory, fertilizin (which is a gel-forming glycoprotein located in the egg plasma membrane) acts as a receptor for antifertilizin, a protein located in the surface of spermatozoa, and it is the union of the two substances which is supposed to be the functional event in fertilization and to be responsible for the various fertilization phenomena. These substances have been extracted from the gametes of various invertebrate species, and their interaction is of the same order of specificity as that of fertilization. Either of these substances can be used, in solution, to block fertilization, i.e., spermatozoa can be rendered nonfertilizing by treatments with a solution of fertilizin, and eggs can be made nonfertilizable with a solution of antifertilizin. In either case the respective receptors for attachment would be blocked.

SPERM AND EGG FUSION: THE ACROSOME REACTION

All animal eggs are enveloped by one or more membranous or gelatinous layers, external to the plasmalemma. These layers constitute barriers to penetration by spermatozoa, and they serve in preventing fertilization by more than one spermatozoon or by sperm of alien species. The acrosome is the component of the spermatozoon primarily concerned with the process of penetration.

The entry of the spermatozoon into the egg was first described in 1883 in the nematode *Ascaris* and in 1854 in the frog; the explanation of penetration by means of a solvent action on egg coverings by sperm enzymes has been widely accepted. Extracts of the sperm of the keyhole limpet (*Crepidula*) readily dissolved the enclosing membrane of the egg. The lysin responsible was shown to be typically enzymatic; and similar lysins have been obtained from spermatozoa of the mussel (*Mytilus*), abalone (*Haliotis*), and serpulid polychaete worms (*Hydroides* and *Pomatoceros*). In *Hydroides* and *Mytilus*, actual holes have been seen in the vitelline membrane where spermatozoa have passed through. Nevertheless in sea urchins spermatozoa appear to make their way through the jelly coat and membrane without the aid of jelly-splitting enzymes. In mammals the enzyme hyaluronidase, associated with the sperm acrosome, liquefies the matrix of the follicular layers surrounding ovulated eggs.

In many forms that have been studied [notably the river lamprey and marine invertebrates such as sea urchins, starfish, holothurians, certain bivalve mollusks, and the balanoglossid worm (enteropneust) *Saccoglossus*], a filamentous process forms from the tip of the acrosome when spermatozoa are exposed to egg water. During fertilization the filament (or, as in some polychaete worms, a number of short processes) forms either upon contact of the sperm head with the jelly coat of the egg (e.g., in starfish) or upon contact with the vitelline membrane investing the egg surface proper. Lytic enzymes may well be involved in making possible such penetration by a protoplasmic filament.

In the past, fertilization has been seen as a process in which the spermatozoon penetrates the cortex of an egg, as a rule leaving the sperm tail at the surface. The

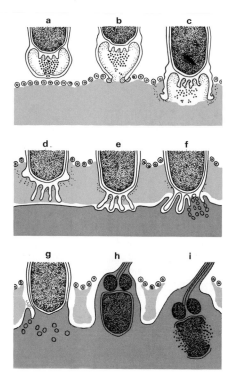

FIGURE 10.2 Stages of sperm-egg association (of polychaete *Hydroides*). (**a**) Unactivated spermatozoon at about the time of initial contact with the egg envelope. (**b**) Activated spermatozoon soon after initial contact: dehiscence of the apex has occurred. (**c,d**) Successive stages of eversion of the wall of the acrosomal vesicle as the sperm apex moves through the egg envelope. (**e**) Eversion has been completed. The acrosomal tubules have made their initial contact with the egg plasma membrane, and interdigitation of the two gamete plasma membranes has started. (**f**) Later stages of interdigitation at the stage of an early fertilization cone and a stage in which continuity between the sperm and egg plasma membranes has been established. (**g**) A very early zygote: the mosaic composition of the zygote plasma membrane is evident; the plasma membranes of the two former gametes are in continuity. (**h**) Young zygote with the profile of both gametes, since the sperm parts still protrude into the egg envelope and beyond. (**i**) Young zygote in which the internal sperm organelles lie further within the egg cytoplasm than in (**h**). [*After Colwin and Colwin, 1967.*]

STAGE OF EGG MATURATION AT WHICH SPERM PENETRATION OCCURS IN VARIOUS ANIMALS*

Young Primary Oocyte	Fully Grown Primary Oocyte	First Metaphase	Second Metaphase	Female Pronucleus
Brachycoelium	*Ascaris*	*Aphryotrocha*	*Amphioxus*	Coelenterates
Dinophilus	*Dicyema*	*Cerebratulus*	Most mammals	Echinoids
Histriobdella	Dog and fox	*Chaetopterus*	*Siredon*	
Otomesostoma	*Grantia*	*Dentalium*		
Peripatopsis	*Myzostoma*	Many insects		
Saccocirrus	*Nereis*	*Pectinaria*		
	Spisula	Ascidians		
	Thalassema			

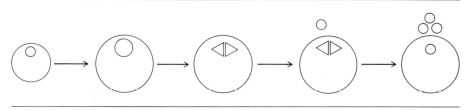

* From C. R. Austin, "Fertilization," © 1965. Reprinted by permission of Prentice-Hall, Inc. Englewood Cliffs, N. J.

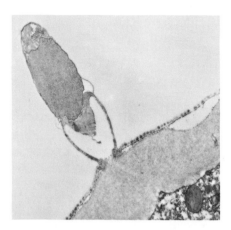

FIGURE 10.3 Electron micrograph of stage of initial contact of sperm and egg of *Hydroides*. Sperm plasma membrane meets egg envelope and ruptures, permitting the interior of acrosomal vesicle to open to outside. Egg plasma membrane is still separated from sperm cell by bulk of egg envelope. [*Courtesy of A. L. Colwin and L. H. Colwin.*]

process was thought to be aided by a fertilization cone of cortical egg cytoplasm which more or less engulfs the sperm head at the time of contact. In this view the spermatozoon is the active member, or stimulator, and the egg responds by forming the fertilization cone and by undergoing the various surface and internal changes collectively known as *activation*. Recent analyses by means of electron microscopy have produced a somewhat different concept, namely, that the fertilization event is a mutual affair in which the spermatozoon is as much influenced by the egg as the egg is by the spermatozoon. Quite apart from the evidence, this view is plausible if we regard the fertilization process as an evolutionary descendant of protistan conjugation, where fusion occurs between equal partners, as in most flagellates and ciliates.

The fertilization of an animal egg has been best described for the hemichordate *Saccoglossus* and the polychaete *Hydroides*, by the Colwins. The process as seen here appears to be essentially common to many other animals, although differing in detail according to minor differences in spermatozoon structure and egg membranes, at least in those forms where eggs are fertilized by motile spermatozoa in a watery medium.

The living spermatozoon of *Saccoglossus* shows the generally primitive form of the type of cell which is widespread among marine invertebrates. It has a simple flagellum and a nearly spherical nucleus. A flat middle piece, or mitochondrial region, is closely applied to the nucleus, and at the opposite or anterior end of the nucleus the acrosome is present as a small projection. One continuous plasma membrane encloses all four regions. The nucleus has its own nuclear membrane; and the acrosome consists of an acrosomal membrane forming the wall of the acrosomal vesicle, the vesicle being nearly filled by a large dense acrosomal granule. So-called periacrosomal material surrounds all but the apical part of the acrosome. The plasma membrane, the acrosomal membrane, and the two membranes which compose the nuclear envelope are typical unit membranes of the kind discussed in Chapter 2. The plasma membrane of the ripe, liberated *Saccoglossus* egg is surrounded by two barrier envelopes, whose combined thickness is about twice the length of the arriving sperm head, which must be penetrated.

Soon after attachment, the sperm head extends a long slender apical projection through the two egg envelopes, originally named the *acrosome filament* but now more explicitly called the *acrosomal tubule*. The changes which the acrosome undergoes during the formation of this tubule are very important and need to be followed carefully. They are:

1 The apex of the acrosome undergoes dehiscence or rupture, so that the sperm and acrosomal membranes open apically and consequently expose the interior of the acrosomal vesicle to the outside. The two membranes join around the margin of the opening.

2 The acrosomal granule is released, comes into contact with the egg envelope, and shortly after disintegrates and disappears. There is evidence that the granule

contains lytic enzymes responsible for clearing a pathway through the egg envelope material.

3   The shallow depression of the acrosomal membrane close to the nucleus now deepens and soon lengthens into a long slender tubule (in *Saccoglossus*) or into a number of short tubules (in *Hydroides*). In the former, within a few seconds, the tubule becomes twice as long as the sperm nucleus.

4   The rest of the acrosomal membrane everts and is added to the tubule at its base. This is simply an unfolding of the already continuous membrane.

An acrosomal tubule which is pointed directly toward the egg surface and extends through both envelopes is then able to reach the egg plasma membrane, fuse with it, and activate the egg. Both cells thus become activated. The egg activation is indicated by release of cortical granules or the elevation of a fertilization membrane, discussed later. The sperm activation results from dehiscence of the acrosomal region or the elongation of the acrosomal tubule.

Summarizing the events observed in both *Saccoglossus* and *Hydroides*, two forms representing widely separated animal phyla, as stated by A. L. and L. H. Colwin, *a structurally specialized apical region responds to the activation stimulus in much the same way: by dehiscence. After dehiscence the sperm plasma membrane is a mosaic, and the inserted piece is the former acrosomal membrane. The released granule rapidly disappears. The basic phenomenon of elongation occurs in both species. Elongation occurs very rapidly, requiring less than 9 seconds. Increase in length greatly exceeds increase in width. Not the entire acrosomal membrane, but only a very limited region of the adnuclear part of this membrane, undergoes the initial great increase in tubular length.*

The continuation of the reaction is described as follows: The extended acrosomal tubule, which is now the apical portion of the sperm plasma membrane, spans the egg envelopes, and its apex meets the egg plasma membrane. Over this limited area of contact the two membranes fuse to form a single continuous mosaic membrane; they thus become adjacent segments of the plasma membrane of a single cell, namely, the zygote. Then the egg part of this zygote undergoes its first fertilization reaction: the cortical granules release their contents into the perivitelline space, the space enlarges, and two layers of cortical material join the inner envelope or "vitelline membrane," transforming it into the "fertilization membrane"; beneath the region where the gamete plasma membranes have fused, a large fertilization cone is elevated above the general egg surface. The sperm nucleus and other sperm structures pass into the cone, the latter recedes, and then, at last, the zygote loses its appearance in profile of being two gametes and achieves the semblance of the single cell which it is.

Clearly the spermatozoon does not enter the egg cytoplasm intact, nor is it swallowed, so to speak, as was once thought. Its acrosome goes into action *before* meeting the egg cytoplasm, and probably none of the contents of the acrosome actually enters the egg cytoplasm. The acrosome is primarily a delivery mechanism whereby the rest of the sperm cell is brought to the egg plasma membrane; in the act of

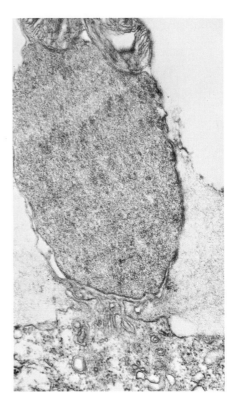

FIGURE 10.4 Electron micrograph of contact stage of sperm and egg of *Hydroides*, showing acrosomal tubules indenting egg surface, and the beginning of the rise of the fertilization cone. The two gamete plasma membranes closely confront each other in the region of interdigitation. [*Courtesy of A. L. Colwin and L. H. Colwin.*]

delivery, the acrosome is expended. When fusion occurs between the egg plasma membrane and the tubule or tubules of the acrosome to form a funnel-like connection, the egg cytoplasm becomes continuous with the periacrosomal material, not the acrosomal material, and it is the periacrosomal material that may be responsible for egg activation. In any case, whatever role the acrosomal region may play in egg activation must be mediated through the tip of the acrosomal tube.

It has long been known that spermatozoa can unite with eggs, either unfertilized or fertilized, whose enclosing envelopes have been removed. Egg envelopes, therefore, may regulate the process in some way but are not essential to it. Further studies have been made with *Saccoglossus* concerning the fusion of spermatozoa with denuded 8- to 16-cell embryos. Even cells of later stages can be "refertilized" after removal of the egg envelope. Fusion accordingly can take place whether the egg cell or its descendant is haploid or diploid and whether egg activation has already occurred or not.

The purpose of such experiments has been to eliminate circumstances relating to the normal envelope-enclosed egg; for example, when the egg envelope is removed, all parts of the *surface of the sperm cell* have an opportunity of meeting egg plasma membrane already in process of cleavage. In every case, however, a spermatozoon first becomes activated and forms an acrosomal tube, and fusion is always accomplished by this organelle. Spermatozoa that remain unactivated, i.e., that fail to form an acrosomal tubule, never fuse with the denuded cells; neither is fusion accomplished by any other part of the spermatozoon plasma membrane except the apical region. Moreover, since the plasma membrane constituting the fusion area of the tubule originally derives from the acrosomal membrane, rather than from the sperm-cell plasma membrane proper, it has a different developmental history and may well have acquired a special fusion-mediating property.

We may conclude, therefore, that membrane fusion involving the acrosomal tubule of an activated spermatozoon is the only means by which gametic union can be established in species with the same general reproductive pattern as that of *Saccoglossus* and *Hydroides*. On the other hand the egg cortex itself plays an essential role, for when spermatozoa are injected into the interior of unfertilized sea urchin eggs, the injected spermatozoa do not activate the egg but remain intact and even motile within the endoplasm.

ACTIVATION OF THE EGG

It is customary to speak of the activation of the egg by a spermatozoon as though the egg is stimulated into action by some property of the spermatozoa. This is not untrue, but it overlooks the phenomenon of mutual activation just described. In a real sense the spermatozoon and the egg are one from the moment of contact, and the combined cell or zygote is activated throughout. In each component, however,

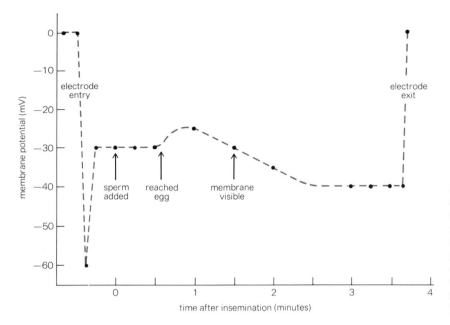

FIGURE 10.5 Membrane potential of the egg of *Asterias forbesii* and its changes following fertilization. Entry of the electrode into the egg causes a sudden appearance of a −60-mV membrane potential, which rapidly decreases to −30 mV. The previously steady potential decreases by 5 mV, as soon as the sperm contacts the egg. Subsequently, the potential again increases and remains steady at −40 mV, until the electrode is removed. [*After Tyler et al.*, Biol. Bull., **110**:*184(1956)*.]

the activation phenomenon is a procedural event, commencing with almost instant changes in the cell cortex and apparently ending with the fusion of the pronuclei and the preparation for the first cleavage of the fertilized egg.

So much of the investigation of the activating process, especially with regard to the cortical reaction in the egg, has been carrried out in sea urchin and starfish eggs during the whole history of developmental biology that we need to look at the events occurring in this particular kind of egg in detail, in spite of certain outstanding but exceptional features, before attempting to generalize about the nature and significance of the cortical reaction in a wider context.

The first visible sign is the appearance of the fertilization cone which, as we have already seen, is a response to contact with the acrosomal tube. From this site a wavelike change, visible as interference color change when observed with dark-field illumination, travels rapidly around the egg cortex and is shortly followed by the elevation of a membrane, the fertilization membrane, from the egg surface. Electron micrographs show that the cortex of the unfertilized egg is bounded by two membranes, the outer one being the vitelline membrane, about 30 Å thick, and the inner one being the plasma membrane, about 60 Å thick. Beneath the plasma membrane is the layer of cortical granules. The cortical reaction following cone elevation consists of two important phenomena:

1   The outer, vitelline, membrane becomes separated from the plasma membrane, undergoes an expansion, and gives rise to the outer layer of the fertilization membrane.

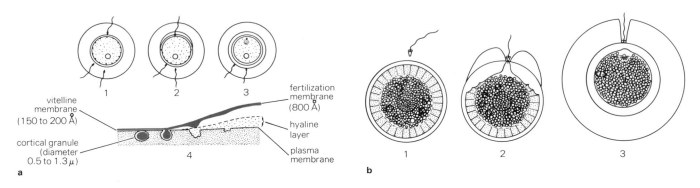

FIGURE 10.6   (a) Drawings 1 to 3 show the course of events in cortical-granule breakdown and fertilization-membrane elevation in the sea urchin egg. Drawing 4 illustrates the mechanism inferred to underlie these events. (b) Reaction of the *Nereis* egg to sperm penetration. The cortex of the egg (drawing 1) contains many alveoli filled with fine granular material. At sperm penetration (drawings 2 and 3) the alveoli break down, and the granular material, apparently becoming strongly hydrated, swells to form a thick jelly coat that eventually surrounds the egg. [*After C. R. Austin, "Fertilization," © 1965. Used by permission of Prentice-Hall, Inc., Englewood Cliffs, N.J.*]

2 The cortical granules explode and release three components: dark bodies, or the folded, laminar parts of the granules, which unfold and then fuse with the inner side of the already elevating membrane; globules which fuse together and build up a new surface of the egg, forming the hyaline layer; and some liquid material which contributes to the vitelline fluid lying between the new egg surface and the now-completed and elevated fertilization membrane.

The presence of the fertilization membrane is the most obvious sign that an egg has been fertilized. With the separation of the fertilization membrane and other constituents from the original egg cortex, the new egg plasma membrane is entirely different from that of the unfertilized egg, and, altogether, the changes involved imply a complete rearrangement of the molecular organization of the cortical layer. This reorganization is associated with changes in permeability. At the same time, in sea urchin eggs at least, a large increase in oxygen consumption occurs. This last relates to evidence that a cytochrome-oxidase inhibitor is present in the unfertilized egg which is removed during activation. Protein metabolism also seems to be involved, for activated eggs are well able to incorporate amino acids into proteins, whereas the unfertilized egg cannot do so, even though the necessary enzymes are present. It is suggested that the bar to protein synthesis in the unactivated egg results from suppression of the functional state of messenger RNA or of the ribosomes. Activation results in polyribosome formation, which results in protein synthesis.

Comparable changes take place during the activation of the eggs of other species. In amphibians, granules at the surface of the mature egg contain mucopolysaccharides; following fertilization these granules break up and mucopolysaccharides simultaneously appear in the perivitelline space. Similarly in the mammalian egg, cortical granules are present at first but disappear following fertilization. In fishes, cortical vacuoles, probably equivalent to cortical granules, disintegrate following fertilization, the disintegration starting at the animal pole, where the sperm enters by way of a micropyle through the egg membrane, and proceeding toward the vegetal pole.

In invertebrate eggs other than those of echinoderms, structural changes in the egg cortex are difficult to discern, although comparable metabolic changes occur in the egg. Changes in the rate of oxygen consumption at fertilization by the different species of eggs differ both in direction and magnitude, the very high increase (six- to sevenfold) in sea urchin eggs being unusual if not exceptional. Whatever the nature of the change, the effect of activation is to make the rate of oxygen consumption about the same for all, a rate which is much the same as that of normally growing somatic cells. In general, the emerging biochemical picture is that of a release from inhibition of both anabolic and catabolic systems. The fact that maturation of the egg reaches completion in only a few animals before fertilization suggests that the inhibiting factor reaches a threshold concentration at different stages in different animals—during maturation in some or even before maturation begins in others.

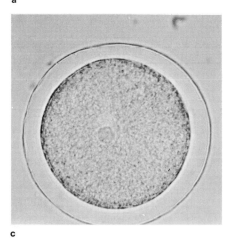

a

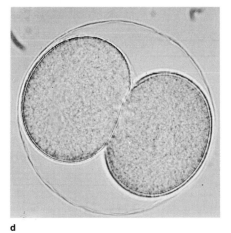

b

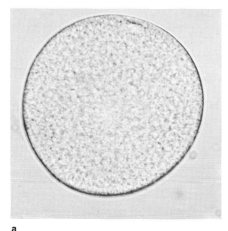

c

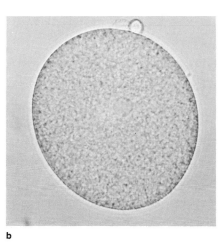

d

FIGURE 10.7 Maturation and fertilization in egg of sea urchin. (a) Egg showing metaphase of first maturation division. (b) Maturation complete, polar body evident, and pronucleus at center. (c) Fertilization membrane lifted off. (d) First cleavage. [*Courtesy of T. Gustafson.*]

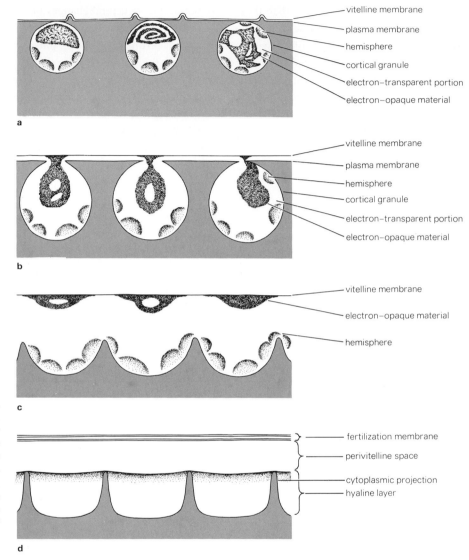

FIGURE 10.8 Changes of the egg cortex of the sea urchin *Clypeaster japonicus* following fertilization. (a) Unfertilized egg. (b) The explosion of the cortical granules. (c) Adhesion of electron-opaque material to the vitelline membrane now lifted up; complete fusion of this material with the membrane will give rise to the fertilization membrane. Meanwhile, the hemispheric bodies remain at the surface of the egg and give rise to the hyaline layer. (d) The egg surface upon completion of these changes. [*After Endo*, Exp. Cell Res., **25**:*383* *(1961)*.]

According to Monroy, "The main point is that, at least as far as the sea urchin is concerned, the block involves the energy-yielding systems; hence all the reactions requiring a large amount of energy are also inhibited. The accumulation of inhibitory substances during maturation seems to be the most likely explanation and, indeed, one such inhibitor has been identified. However, how such inhibitions are produced during the final phase of maturation and how they are inactivated or disposed of by the egg following fertilization still remains to be discovered. . . . Repressor substances manufactured by the egg in the later phases of maturation would inhibit both the metabolic activities residing in the cytoplasm and the genetic activities of the nucleus. The key reaction of fertilization would thus be the removal of the repressor(s), releasing at the same time the cytoplasmic metabolic activities and the activity of the nuclear genetic system." [1]

EXPERIMENTAL ACTIVATION

Eggs of many species can be activated by chemical or physical treatment without participation by spermatozoa. Jacques Loeb long ago discovered that the eggs of sea urchins and other species could be activated by short exposure to a dilute solution of butyric acid in seawater and that normal cleavage and development could be obtained if the treatment was followed by exposure to a hypertonic solution. He concluded that activation resulted from a cytolytic and corrective factor, but a later demonstration that a hypertonic solution alone can be effective took away much of the validity of this concept. Nevertheless, Loeb's procedure for inducing activation and development parthenogenetically remains extremely effective.

Recent studies show that after treatment with butyric acid alone the eggs undergo cortical changes often indistinguishable from the events in fertilized eggs, including amino acid incorporation. However, even after the release of protein synthesis, at least one more step is required for complete activation, as shown by the failure of eggs exposed to butyric acid to undergo normal cleavage and development. Subsequent exposure to hypertonic seawater, Loeb's corrective factor, enables the eggs to complete activation, but since this subsidiary treatment is effective whether applied after or before the butyric acid treatment, it is evident that the events induced by the two treatments must be somewhat independent of each other.

Several later interpretations of the activation reaction have been proposed. In the bimolecular theory of R. S. Lillie, based mainly on his work on starfish eggs, it is assumed that an activating substance (A) is formed from the union of two substances (B and S) already present in low concentration in the cortical cytoplasm of the egg. Substance B is assumed to result from hydrolytic processes in the egg which can be stimulated by agents, such as acid and heat, that can act under anaerobic conditions. Substance S is supposed to be formed only in aerobic conditions, from synthetic processes stimulated by agents such as hypertonicity. Increase in either B or S substance may result in the attainment of the threshold-activating level of substance A. The theory accounts for a large mass of data, particularly for the additive and sensitizing effects of agents such as acids, heat, hypertonicity, and anaerobiosis when employed in various combinations.

Other hypotheses, more or less compatible with the bimolecular theory, relate to the role of calcium. It should be remembered that seawater, body fluids, and

[1] A. Monroy, Biochemical Aspects of Fertilization, in R. Weber (ed.), "The Biochemistry of Animal Development," vol. I, p. 128, Academic, 1965.

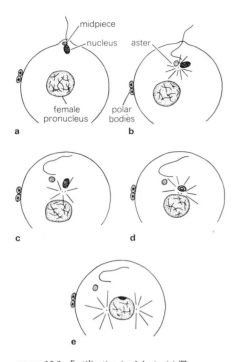

FIGURE 10.9 Fertilization in *Arbacia*. (**a**) The sperm head has entered the cytoplasm. (**b–d**) It turns through 180°. The sperm aster becomes evident, the midpiece is detached, and the aster seemingly leads the sperm nucleus toward the point of union with the female pronucleus. (**e**) Nuclear union appears to be complete and involves a male pronucleus that has not enlarged appreciably. The sperm aster divides to become the amphiaster for the first cleavage division. [*After C. R. Austin, "Fertilization," © 1965. Used by permission of Prentice-Hall, Inc., Englewood Cliffs, N.J.*]

intercellular fluids are extremely complex solutions of mineral salts, with chlorides and carbonates of sodium, potassium, calcium, and, in seawater, magnesium present in various ionic and molecular forms. The balance between the monovalent and the bivalent cations is vitally important to the maintenance of cell structure, particularly the cell cortex. Experiments show that solutions of calcium chloride isotonic with or hypertonic to seawater are effective activators in many species of animals, that activation by other agents is dependent on the presence of calcium, and that depriving eggs of calcium sensitizes them to the subsequent action of calcium. Calcium is undoubtedly intimately involved in the activation process, and its role may well relate to its action on various enzymes.

MIGRATION OF PRONUCLEI

The activation of the egg so far discussed relates mainly to the restoration of normal metabolic activity in the egg. There is merit in restricting the use of the term to this phase of fertilization, for the activities set in motion by the initial activation continue without pause and lead directly to development.

What happens following fusion of the acrosomal tube with the plasma membrane of the egg depends on the stage of arrest of the unfertilized egg, i.e., whether the maturation process is already completed, is incomplete, or has not even begun. Completion of the maturation process, if not already performed, proceeds immediately following activation. At one extreme, in some annelids and flatworms, the sperm nucleus enters the primary oocyte early in its ovarian existence and remains unchanged while the oocyte completes its growth and maturation; at the other, in coelenterates and echinoderms, where maturation is complete, the female pronucleus may be already formed at the time of gamete fusion. The male pronucleus forms from the compact spermatozoon nucleus introduced through the acrosomal tube, by a process of hydration. Both pronuclei assume the general form of an interphase nucleus and begin to move through the egg cytoplasm to a more or less central meeting place.

As the spermatozoon nucleus moves inward from the site of the fertilization cone, it soon rotates through 180°, so that the mitochondria and centrosome of the associated midpiece assume the leading position. At the same time the sperm aster forms in the egg cytoplasm around the centrosome. As the male pronucleus develops and migrates, the sperm aster seems to lead it through the cytoplasm to the site where union with the female pronucleus will take place. During this time the female pronucleus also begins to move and reaches the place of union. Some interaction between the male and female pronucleus is clearly involved in this common assignment, for in polyspermic salamander eggs, which have more than one male pronucleus, progress of extra pronuclei ceases as soon as one male pronucleus meets the female. Explanations in terms of an attractive force operating between the two pronuclei, or a pulling together by astral rays, cannot account for the fact that experimental removal of either the male or the female pronucleus fails to stop the remaining nucleus from continuing on its normal path. Moreover in artificially activated eggs the egg nucleus can migrate in the same manner as if it were fertilized, while in eggs that have been enucleated and subsequently fertilized, the sperm nucleus attains the normal position for subsequent cell cleavage.

THE CENTRIOLE

Following the completion of the maturation divisions, the egg centriole disappears or at least ceases to function under prevailing conditions. The sperm must introduce its own centriole if the egg is to form a mitotic spindle and divide in the normal way. In experimental activation a wide variety of simple physical and chemical agents induce the various cortical changes characteristic of normal fertilization, and the egg nucleus subsequently undergoes cyclical changes like those of mitosis. In step with this cycle, which may be repeated a number of times, a monaster appears and disappears, but since no amphiaster is formed, there is no true mitosis and no cleavage. Under these circumstances the egg centriole persists but is unable to divide and give rise to a mitotic spindle, which function is normally the function of the sperm centriole. Certain modifications of the activating treatment can, as we have seen, render the egg centriole capable of division and so lead to normal cleavage. Therefore, if an egg is to divide and develop, activation by means of a sperm or an artificial agent must include either the importation of a sperm centriole or the rendering of the egg centriole capable of division. Why the egg centriole usually loses this capacity is unknown, although the loss appears to be associated with the general inhibition typical of advanced stages of egg maturation.

In experimentally activated sea urchin eggs the amphiaster responsible for cleavage is evidently derived from cytasters induced by the treatment. In certain polychaetes and mollusks the amphiaster arises as a result of suppression of the polar body or maturation divisions, with the polar-body mitotic spindle serving as a cleavage spindle. However, any general hypothesis concerning the origination of centriolar bodies under various circumstances must consider that nonnucleate fragments of unfertilized sea urchin eggs can be induced to form cytasters capable of multiplication, with consequent cleavages reported, suggesting the possibility of *de novo* formation of self-duplicating centrioles in the cytoplasm.

Frog eggs can be activated readily by artificial means, such as osmotic pressure, electric shock, and even puncture with a glass or metallic needle. These serve as a "first factor" which stimulates limited development involving obvious structural changes, such as the completion of maturation and elevation of the vitelline membrane, but development ceases after a few abortive cleavages. A second factor, of biological origin, must be introduced into the egg cytoplasm for the eggs to undergo normal

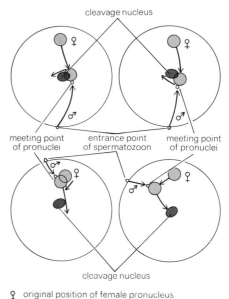

♀ original position of female pronucleus

♂ track of male pronucleus

FIGURE 10.10 Tracks of the pronuclei of the sea urchin in their movements toward each other and to the site of the cleavage nucleus, with sperm entering at opposite side of egg from site of egg nucleus and from same side, respectively. [*After E. B. Wilson, 1925.*]

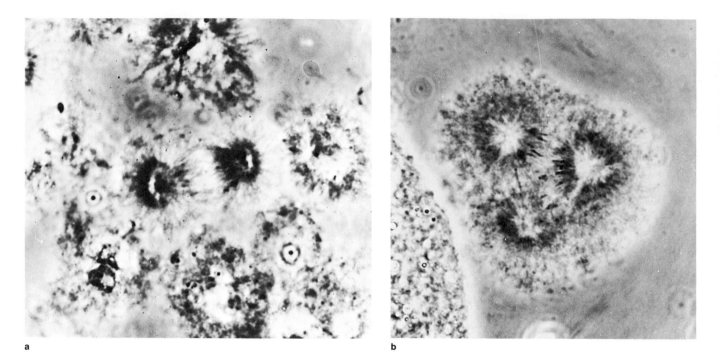

FIGURE 10.11 **(a)** Double and **(b)** triple asters of artificially activated sea urchin egg, after isolation by partial digestion of the egg. [*Courtesy of E. R. Dirksen.*]

cleavage and development. This factor is associated with the particulate fraction from a variety of tissues such as blood from mouse and chicken tissues and from both adult and embryonic frog tissues. Whatever the cleavage-initiating substance ("second factor") may be, it is apparently produced in artificially activated eggs, and although it does not initiate cleavage in the eggs that produce it, it is capable of producing cleavage in newly activated eggs. Since it is produced in enucleated eggs, the zygote nucleus evidently is not involved. A number of questions arise. What is the nature of the particle? Is it a cytoplasmic particle capable of self-replication, comparable to or perhaps antecedent to a centriole? Is it synthesized during oogenesis and stored, and if so by what means? There is some evidence that it is centriole-related.

BLOCKS TO POLYSPERMY

Limitation of fertilization to union of a single spermatozoon and an egg is usually mandatory if subsequent development is to be normal. If more than one sperm centriole and nucleus are introduced into the egg cytoplasm and become active, the outcome is generally lethal. The situation arises when two or more sperm fuse with an egg, a condition known as *polyspermy*. Various abnormal unions are possible, although it is usual and apparently normal for many spermatozoa to enter moderately yolk-laden eggs such as those of urodele amphibians (though not the eggs of anurans, which are almost as large) and some insects, and particularly the very large, yolky eggs of birds, reptiles, and elasmobranch fish, where the process is known as *physiological polyspermy* and in which only one sperm centriole and nucleus actively participate in the union of pronuclei and the initiation of cleavage.

Polyspermy in its pathological form is incompatible with normal development. If two sperm centrioles enter an egg and are either partly or wholly active, a triaster or a tetraster forms in place of the normal amphiaster, with three or four spindles present, respectively, so that the first cleavage gives rise to three or four cells. If this were all, it might not matter, but in such circumstances the chromosomes of the united pronuclei are distributed in a most irregular manner and no daughter cell has a normal complement. This irregularity is passed on to subsequent cell generations, and development soon comes to an end. Most, if not all, species, therefore, have some sort of structural or physiological mechanism whereby fertilization is restricted to the active participation of a single spermatozoon.

One such mechanism is the elevation of the fertilization membrane, common in sea urchins, starfish, polychaetes, amphioxus, frogs, and others. By itself it is a simple device which increases the distance between the egg surface and the attaching spermatozoa beyond the capacity of the acrosomal tubes to traverse it. This, however, seems to be an imperfect protection even in the case of a small egg such as the sea urchin. The cortical change leading to visible elevation starts at a point on the surface and proceeds as a wave traveling around the egg. At 18°C the wave completes its circuit in about 20 seconds. Sperm travel at about 200 μ per second, and at sperm densities of 10^6 and 10^7 the number of collisions per second should be 1.6 and 16, respectively. Since these densities do not give rise to polyspermy, it is concluded that the block to polyspermy must be established much more rapidly, or else that only a small percentage of sperm collisions result in fertilization. There is evidence, however, that the cortical change is diphasic, with a fast component covering the surface in 1 or 2 seconds and conferring partial protection, which is probably the loss from the vitelline surface of the capacity to form attachment with spermatozoa, although no visible change has as yet been linked with it. The slow component, the elevation of the fertilization membrane, is more effective once it is completed,

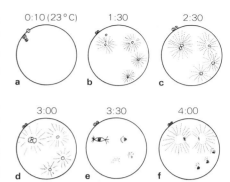

FIGURE 10.12 Polyspermy in the urodele *Triton*. (a) Entry of spermatozoa. The egg is in the metaphase of the second meiotic division. (Time, 10 minutes after semination; temperature of medium, 23°C.) (b–d) For the first 2 hours and 20 minutes all pronuclei develop similarly; then, with one male pronucleus involved in syngamy with the female, the supernumerary male pronuclei begin to regress, the nearest one showing the change first. (e,f) As the first cleavage mitosis proceeds, 3 hours and 20 minutes to 3 hours and 50 minutes after semination, suppression of all supernumerary male pronuclei is completed. [*After G. Fankhauser, 1948.*]

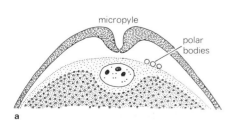

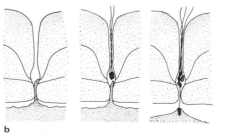

FIGURE 10.13 (a) The egg of the cephalopod *Argonauta*. The micropyle lies in a deep pit. [*After M. Ussow*, Arch. Biol., **2** (*1881*).] (b) Entry of spermatozoa into the egg of the sturgeon. The micropyle traverses the three layers of chorion, and into its lower reaches projects a tongue of vitelline cytoplasm. Contact with and attachment to the cytoplasm appear to be necessary for the spermatozoon to complete its journey through the micropyle. [*After A. S. Ginsburg*, Cytologia, **1**:510 (*1959*).]

the fast block being transient, since removal of the fertilization membrane permits fusion with a second spermatozoon.

Not all kinds of eggs produce a fertilization membrane, and other mechanisms blocking polyspermy have been evolved. In teleost eggs and in the sturgeon, the egg is surrounded by an egg membrane too thick and tough for spermatozoa to penetrate, and sperm reach an egg only through a minute hole, the micropyle, situated at one end. This opening is so narrow that only one sperm at a time can pass through, and when the first has entered the egg cytoplasm, a gelatinous substance flows up and plugs the canal; subsequently a wave of contraction passes over the egg, resulting in an impervious hyaline layer on the surface. In mammals a change of some kind occurs in the zona pellucida surrounding the egg that renders it impermeable to spermatozoa, a change known as the *zona reaction.* Even in species where physiological polyspermy is normal, only one male pronucleus ordinarily unites with the female pronucleus, and all the extra male pronuclei appear to be overwhelmed by an inhibitory influence which originates close to the conjugating pronuclei and spreads outward.

CONCEPTS

Fertilization is a dual event consisting of cortical activation of the egg and fusion of male and female pronuclei.

Attachment of sperm to egg surface is a species-specific reaction.

Junction of sperm with egg is primarily membrane fusion of the two cells.

Activation is mainly a reorganization of the egg cortex which restores the egg to a normal metabolic state.

The migratory behavior of the pronuclei is dependent on the establishment of the activated state.

Cleavage asters may be produced by the sperm centriole, by a reactivated egg centriole, or possibly by newly induced or introduced centriolar precursors.

The activation of the egg at fertilization initiates the patterns of DNA replication, spindle formation, and energy coupling associated with mitosis and cell division.

Functional protein-forming systems and an energy source sufficient to carry the egg through cleavage are preformed in the egg.

READINGS

AUSTIN, C. R., 1965. "Fertilization," Prentice-Hall.
BISHOP, D. W., and A. TYLER, 1956. Fertilization of Mammalian Eggs, *J. Exp. Zool.*, **132**:575–595.

COLWIN, A. L., and L. H. COLWIN, 1967. Membrane Fusion and Sperm-egg Association, in C. B. Metz and A. Monroy (eds.), "Fertilization," vol. 1, Academic.

—— and ——, 1964. Role of the Gamete Membranes in Fertilization, in "Cellular Membranes in Development," Academic.

COSTELLO, D. P., 1958. The Cortical Response of the Ovum to Activation after Centrifuging, *Physiol. Zool.*, **311**:181–188.

DIRKSEN, E. R., 1964. The Isolation Characterization of Asters from Artificially Activated Sea Urchin Egg, *Exp. Cell Res.*, **36**:256–269.

FANKHAUSER, G., 1948. The Organization of the Amphibian Egg during Fertilization and Cleavage, *Ann. N.Y. Acad. Sci.*, **49**:684–708.

FRANZEN, A., 1958. On Sperm Morphology and Acrosome Filament Formation in Some Annelids, Echiuroidea, and Tunicata, *Zool. bidrag Uppsala*, **33**:1–28.

GRANT, P., 1965. Informational Molecules and Embryonic Development, in R. Weber (ed.), "The Biochemistry of Animal Development," vol. I, Academic.

HINEGARDNER, R. T., 1967. Echinoderms, in F. H. Wilt and N. K. Wessells (eds.), "Methods in Developmental Biology," Crowell.

LILLIE, F. R., 1913. The Mechanism of Fertilization, in B. H. Willier and J. M. Oppenheimer (eds.), "Foundations of Experimental Embryology," Prentice-Hall.

LOEB, J., 1913. "Artificial Parthenogenesis and Fertilization," University of Chicago Press.

MONROY, A., 1965. "Chemistry and Physiology of Fertilization," Holt, Rinehart and Winston.

——, 1965. Biochemical Aspects of Fertilization, in R. Weber (ed.), "The Biochemistry of Animal Development," vol. I, Academic.

PINCUS, G., 1965. "The Control of Fertilization," Academic.

ROTHSCHILD, LORD, 1958. Fertilization of Fish and Lamprey Eggs, *Biol. Rev.*, **33**:372–392.

RUNNSTROM, J., 1966. The Vitelline Membrane and Cortical Particles in Sea Urchin Eggs and Their Function in Maturation and Fertilization, *Advance. Morphogenesis*, **4**:222–325.

TYLER, A., 1955. Gametogenesis, Fertilization and Parthenogenesis, in B. H. Willier, P. Weiss, and V. Hamburger (eds.), "Analysis of Development," Saunders.

——, 1941. Artificial Parthenogenesis, *Biol. Rev.*, **16**:291–336.

WILSON, E. B., 1925. "The Cell in Development and Heredity," 3d ed., Macmillan.

CHAPTER ELEVEN

CLEAVAGE, POLARITY, DETERMINATION, AND GRADIENTS

A ripe egg is a relatively very large cell usually packed with reserve materials and in a depressed metabolic state. Activation restores metabolic normality and, after completing whatever remains of the maturation process, the egg begins a series of cleavages which divide its substance into an increasing number of cells of progressively decreasing size. The divisions continue until the average cell sizes are those characteristic of the differentiated cells of the parental organism. The number of successive divisions accordingly depends on the volume of the egg compared with the volume of typical somatic cells, and on the availability of intracellular or extracellular nutritive reserves, such as yolk.

CONTENTS

Cleavage and development
Polarity in animal eggs
Establishment of the polar axis
Displacement of cell inclusions
Inversion of the polar axis
Cleavage pattern
Potencies
Cleavage and differentiation
Gradient theory
  Combination results
Concepts
Readings

CLEAVAGE AND DEVELOPMENT

Cleavage, from its onset following activation to cessation when minimum cell sizes and exhaustion of reserves have been reached, is clearly a process with an initial momentum and a final state of rest, a process of gradual attainment of a state of equilibrium represented by standard cell size. The cleavage rate, i.e., how fast one division follows upon another, follows a regular sigmoid curve generally characteristic of processes that attain a steady state. In other words, during cleavage, there is an early phase when one division follows closely on the one before, with the interphase nucleus existing only for a matter of minutes between the end of one mitosis and the preparation for the next; a later phase when, after a number of such divisions, the intermitotic interval noticeably lengthens; and a terminal phase when, towards the end of cleavage, the intervals become very long indeed, until finally cell division ceases.

Such is the overall event during which the subdividing material of the original egg accomplishes its cytodifferentiation and during which the various morphogenetic activities are performed by the mass of increasingly numerous and increasingly small cells produced by the divisions. Whatever the way these several processes may be geared together, the cleavage process sets the total available time, and the progressive retardation of the cleavage process has its own profound effect. The pace, or rhythm, of cleavage is determined by the cytoplasm rather than by the nucleus, a fact long known but confirmed by the observations that haploid, diploid, and triploid frog embryos all undergo cleavage with the same rhythm. There is also strong circumstantial evidence that nuclei are synthetically active only during the interphase state and,

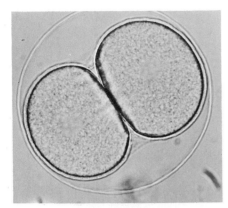

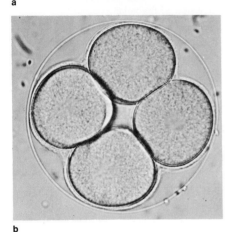

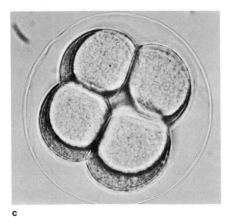

FIGURE 11.1 First three cleavages of sea urchin egg, within fertilization membrane, showing cortical hyaloplasm, flattening of interface between adjoining cells, and curved free surfaces. [*Courtesy of T. Gustafson.*]

conversely, are synthetically inactive during most of mitosis, when the chromatin is more or less condensed.

Opportunity for nuclear activity, apart from chromatin replication, accordingly increases as cleavage progresses and interphase periods of the nuclei become longer. On this basis alone we can look on the whole developmental period as consisting of three phases:

1   A cleavage phase proper, when the dividing egg gives rise to a blastula
2   A morphogenetic phase, when the blastula undergoes form changes and tissue rearrangement
3   A cytodifferentiation phase, when the constituent cells mature as specific types, depending on their cytoplasmic inheritance and their position in the system

Obviously each phase, and subphase, must be prepared for in some way during the phase preceding it, and the change from one condition to another may occur at different times in different parts of an embryo.

In speaking of rates of developmental processes we are concerned with relative or biological time rather than clock time. Except at the extremes of the temperature range compatible with normal development, differences in temperature affect the duration of development as a whole, together with its constituent phases, but do not as a rule affect the rate of development of one phase or one part relative to another. Importantly, temperature change affects the rate of respiratory or oxidative metabolism, which appears to be the common base for all other cell activities. Respiration varies with the temperature, and so does cleavage rate. In the absence of oxygen or the presence of excess of carbon dioxide, cleavage is inhibited.

Nevertheless, the general metabolic background of a developing egg, such as the sea urchin, is shown by the S-shaped respiratory curve, similar to that already mentioned for cleavage except that a second high rise appears after the cleavage rate has leveled off. The shape of the curve indicates that the respiratory rate of the developing embryo as a whole follows closely the number of cells produced. Since the total mass of living material is not increasing but the aggregate cell surface of the multiplying and ever smaller cells is rapidly increasing, the increase in total oxygen consumption may well reflect this increase in relative area of surface available for gaseous exchange. Alternatively or concomitantly the comparable increase in aggregate surface of interphase nuclear membrane is suggested as the significant feature by evidence that a respiratory rhythm accompanies cell division, with increasing rates at the stages where nuclei reappear and persist. In the sea urchin at least, both cleavage and respiratory increase level off together at about the tenth cleavage and resume again at the late blastula stage following hatching. Whatever the significance of the particular shape of the curve, the two processes of cleavage and respiratory metabolism are geared together.

In summary, the following processes occur together:

1   During early cleavage, at least, the blastomeres have presumably normal metabolic maintenance requirements but little opportunity for intake except for oxygen.

2 During early cleavage, mitosis follows mitosis, with time between only for preparation of the next division, i.e., with little or no meaningful interphase.

3 During cleavage, new protein must continually be made available to form new plasma membrane and mitotic fibers.

4 Chromatin replication must keep pace with cell and nuclear multiplication, a pressing need during early cleavage.

5 Some chemical preparation must precede visible morphogenesis.

6 Some cytoplasmic regional specialization must occur, sooner or later, leading to cytodifferentiation.

Therefore the energy and material reserves necessary to support the above must accumulate preparatory to the onset of development. Much of what is accumulated in the growing oocyte and released at activation is necessarily consumed during the phase of rapid early cleavage. In other words a head start is necessary. This is seen in one form as the change in nucleocytoplasmic volume ratio during cleavage. In the salamander *Triturus palmatus*, for instance, the progressive and inevitable decrease in the cytoplasmic volume of blastomeres during cleavage is not at first accompanied by a corresponding decrease in nuclear volume. Only at the blastula stage is the normal nucleocytoplasmic relationship established, after which the nucleocytoplasmic volume ratio remains constant, i.e., they decrease at the same rate.

POLARITY IN ANIMAL EGGS

Development proceeds from the general to the particular, as in any sort of progressive construction. The major axes of an organism are established at the start, and these serve as guidelines for regional differentiation.

The primary organizational feature of an egg is its polar axis. All animal eggs and most plant eggs have such an axis already built in; in animals it may be discernible even in the primary oocyte and is very evident in the secondary oocyte, where it is commonly expressed as a gradient in ratio of cytoplasm and yolk, extending from the animal to the vegetal pole.

Animal eggs, especially telolecithal eggs, show evidence of a polar axis before fertilization, even if this evidence is no more than an eccentric position of the nucleus and an accumulation of granules at one side of the cell. Polar bodies are formed at the animal pole, while yolk inclusions tend to accumulate at the vegetal pole of the axis, i.e., at opposite poles of the animal-vegetal axis.

The primary polarity leads to a visible polarity of differentiation and to other features such as the direction of cleavage. In the starfish egg, for example, the cleavage planes of the first and the second division pass through the poles, although at right angles to each other, while the third is equatorial. Subsequent cleavages cause the eight-cell stage thus formed to transform into the *blastula*, which is a hollow ball consisting of a single spherical layer of smaller cells. The polar effect is seen again

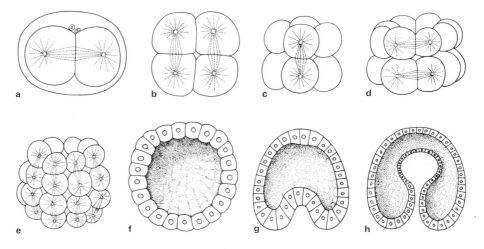

FIGURE 11.2 Development of comparatively un-specialized type of egg (e.g., of echinoderms such as *Synapta* and *Asterias*), showing **(a–c)** early cleavage, **(d–f)** conversion of morula to blastula, **(g)** invagination of blastula to form a gastrula, and **(h)** evagination from the invaginated enteron to form a mesodermal or coelomic pouch, thereby establishing the three primary germ layers: mesoderm, endoderm, and ectoderm.

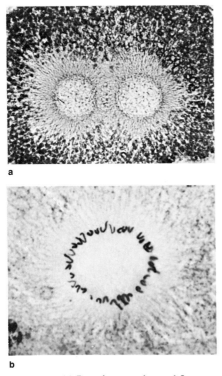

FIGURE 11.3 **(a)** First cleavage of egg of flatworm *Polychoerus carmelensis* in metaphase, showing "polar sun" stage of growth of astral region. **(b)** Polar view of central spindle of metaphase of first cleavage, showing 34 chromosomes. [*Courtesy of D. P. Costello.*]

in the transformation of the blastula into the *gastrula,* brought about by an inpushing, or invagination, from the center of the vegetal pole into the internal cavity, or blastocoel. The distal blind end of this invagination then gives off outpushings, or evaginations, which pinch off as mesodermal pouches, the remaining part of the original invagination persisting as the endoderm, or gut rudiment, and the noninvaginated tissue as the ectoderm.

The primary polarity of the animal egg is established at a very early stage and is recognizable even in the oocyte. Immediately following maturation, however, when the viscosity of the egg cytoplasm has greatly decreased, striking changes are visible, at least in certain kinds of eggs.

According to Joseph Spek (1934), "In a study of the bipolar differentiation of the eggs of the polychaete *Nereis dumerilii,* it could be shown that the cell substances of these eggs were being distributed in such a way that substances of high pH migrated toward the animal pole, substances of a low pH towards the other pole. In the egg cells of *Nereis dumerilii* there is a homogeneously dissolved yellow pigment which is a natural indicator changing color at about pH 5.4 from yellow to violet from the alkaline towards the acid side. During the bipolar differentiation the pigment is accumulated in the vegetative half of the egg cells, while the animal half becomes colorless. The first four blastomeres separated by meridional cleavage furrows show essentially the same organization. The bipolar differentiation of these blastomeres is still proceeding. When the accumulation of the substances on the vegetative pole reaches a certain amount, the color of the natural pigment changes from yellow to violet, indicating an increase of acidity, i.e., a pH of at least 5.4 in the vegetative part of each of the four blastomeres. This happens just before the third cleavage. Vital staining experiments with indicators also show an *increased acid reaction of this part of the blastomeres* from this time on. Moreover, these dyes also stain the other

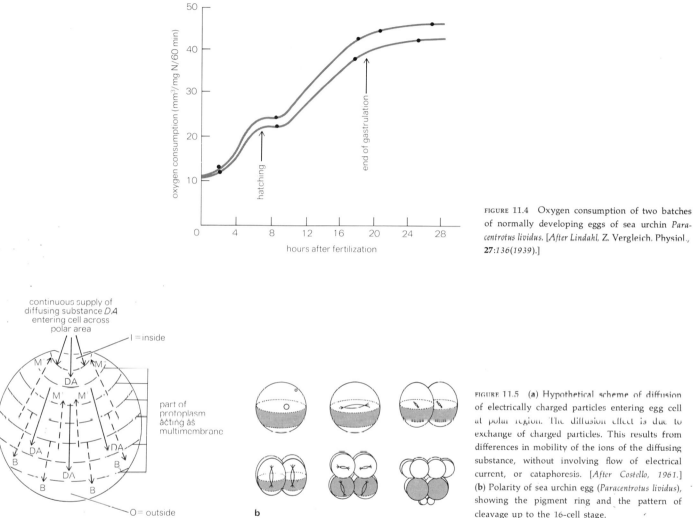

FIGURE 11.4 Oxygen consumption of two batches of normally developing eggs of sea urchin *Paracentrotus lividus*. [*After Lindahl*, Z. Vergleich. Physiol., **27**:*136(1939)*.]

FIGURE 11.5 (**a**) Hypothetical scheme of diffusion of electrically charged particles entering egg cell at polar region. The diffusion effect is due to exchange of charged particles. This results from differences in mobility of the ions of the diffusing substance, without involving flow of electrical current, or cataphoresis. [*After Costello, 1961.*] (**b**) Polarity of sea urchin egg (*Paracentrotus lividus*), showing the pigment ring and the pattern of cleavage up to the 16-cell stage.

(animal) half of the blastomeres and here they indicate an *alkaline color* more pronounced than before. Each of the four blastomeres now consists of an alkaline part and an acid part. Between them the indicators have an intermediate color.

"The preliminary explanation of these observations is that the bipolar differentiation of the egg cells is a *cataphoretic phenomenon*, that somehow there is established in the living cell an electrical field and that this causes a migration of the microscopical and ultramicroscopical particles according to their electrical charge."[1]

[1] Joseph Spek, Uber die bipolare Differenzierung der Eizellen von *Nereis limbata* und *Chaetopterus pergamentaceous*, Protoplasma, **21**:394–405 (1934).

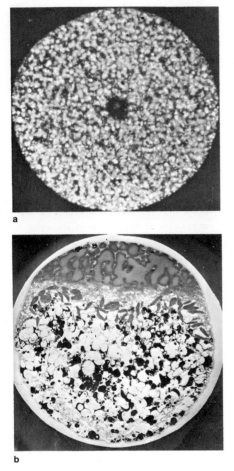

In certain plants the eggs are shed earlier and are more amenable to experimentation; most of what is known concerning the factors underlying the polarity of the egg cell comes from study of these types. These are the eggs of certain brown algae, particularly the intertidal seaweeds *Fucus* and *Pelvetia*. The eggs are haploid, naked, spherical, about 70 μ in diameter, contain numerous small chloroplasts, and are fertilized, after being shed into the sea, by a minute sperm cell. At this time no polar axis has been formed. The egg is densely stocked with carbohydrate reserves, sinks quickly, and at once secretes a glue that fastens it to any available substratum. Such is the first adaptation to the environmental circumstances of development. The second is that the egg divides into two cells in such a way that one becomes the rhizoid cell, which gives rise to the branching holdfast, and the other becomes the thallus cell, which gives rise to the fronds. For successful attachment and growth, the orientation of the polar axis, indicated by the elongation of the egg cell and the plane of division, is crucial and is subject to environmental influence.

ESTABLISHMENT OF THE POLAR AXIS

The first sign of visible differentiation in *Fucus* occurs about 12 hours after fertilization and consists of a bulging at one pole. The egg thus becomes pear-shaped and then divides unequally, so that the protrusion remains as part of the smaller, rhizoid cell, the other end becoming the thallus cell. In the absence of external factors the point of sperm entry determines the future rhizoid or basal pole, although even this localizing influence is not essential, since experimentally (parthenogenetically) activated eggs can develop normally.

Whatever the innate determination of the polar axis may be, however, a variety of environmental agents are able to override it. This has been known since 1889, and over the years the following agents have been found to be effective: direct electrical current by E. J. Lund (1923) and others, mainly D. M. Whitaker (1931–1944); unilateral light, especially at the short end of the spectrum; the presence of neighboring eggs; a pH gradient; a temperature gradient; and, of a different category, centrifugation. Susceptibility to agents, particularly to light, is at a maximum about 8 hours after fertilization. All such responses are first evident as elongation of the cell. In fact when recently fertilized eggs of *Fucus* are gently sucked into a small pipette while the cell wall is hardening and are then blown out into seawater, the elongated shape is retained and the long axis thus produced becomes the polar axis, or axis of differentiation.

More recently the mechanism by which localization, or pattern formation, occurs in a single cell, as represented by the egg of *Fucus*, has been investigated by L. F. Jaffe, following up an old hypothesis of Joseph Spek that such patterns are established by self-generated cataphoresis. He suggests that localized changes in plasma membrane potential are an obvious source of the driving force underlying the polarizing process.

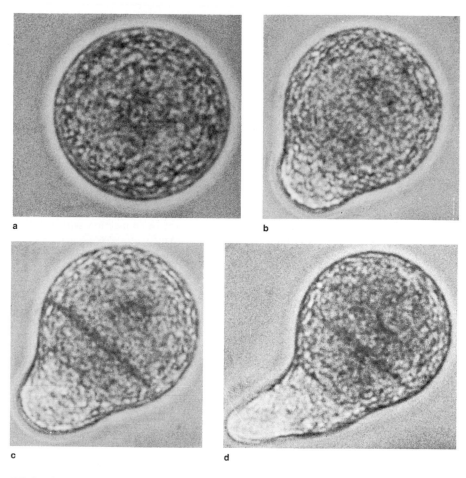

a

b

c

d

FIGURE 11.7 Development of *Fucus* egg 4, 16, 18, and 26 hours after fertilization, respectively, seen by phase contrast, showing establishment of primary polarity, formation of rhizoid, and first cleavage. [*Courtesy of C. B. Bouck.*]

"If the developing system were a single cell, then such inhomogeneities in membrane potential would drive current in loops going through the cytoplasm and back through the medium. Measurements of the extracellular currents, then, would reveal the internal ones. . . . Efforts to measure the electrical potentials across individual developing *Fucus* eggs so far have failed, but if a hundred or so eggs are placed in a long loose-fitting capillary tube and are so illuminated as to form rhizoids towards one end of the tube, in series, then as all undergo development, this end becomes measurably electronegative. Details show that these tube potentials are produced by the return past each egg of currents which each drives *through itself* from rhizoidal to thallus end. Such measurements indicate that at least 60 picoamps (6×10^{-11} amps) begins to flow *through* the normally developing *Fucus* egg when it begins to elongate, or 'germinate.' This current, considered as a flow of positive ions, enters the egg at its basal or elongation pole. It continues to flow as the embryo elongates, at a rate which seems to be proportional to the elongation. Calculation indicates that such a current may generate a sufficient potential gradient across the cytoplasm to concen-

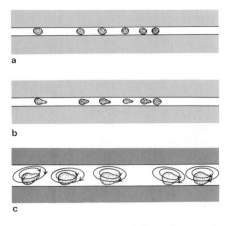

FIGURE 11.8 Measurement of electrical potentials in fertilized *Fucus* eggs. (a) Eggs drawn into glass capillary tube of diameter slightly less than that of the eggs. (b) A day later, all eggs have germinated in the same direction, dividing into a rhizoid and a thallus cell. (c) Suggested scheme of inferred current pattern in a tube. [*After Jaffe, 1966.*]

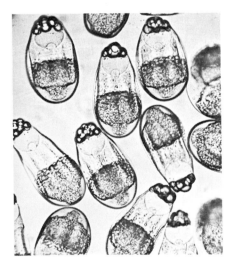

FIGURE 11.9 Centrifuged unfertilized eggs of *Nereis*, showing zones of stratification, with oil drops at top, hyaline protoplasmic layer containing the nucleus, a zone of yolk spheres, a second hyaline zone, and a zone of jelly-precursor granules. [*Courtesy of D. P. Costello.*]

trate significantly some relatively large and negatively charged particles at the expansion pole, and also may directly produce substantial ion gradients through the embryo."[1]

In other words, "current (considered as a flow of positive ions) moves through the egg from its rhizoidal to its thallus pole and returns through the medium. It starts to flow at about the time of axis formation. Since the imposition of comparable currents may determine the egg's polarity it is suggested that the self-generated current is a cause as well as a consequence of localization."[2] There is evidence that the various environmental agents capable of influencing the originally apolar egg leave traces in the cytoplasm which become progressively amplified by the flow of current set up, and that these traces lie in or near the plasma membrane. In spite of our meager knowledge of processes of this sort, there is reason to think that they, together with other processes already discussed, are vitally important in connection with cellular differentiation and multicellular development. Thus self-cataphoresis may be added to self-assembly and macrocrystallinity as three principles operating in developmental phenomena.

DISPLACEMENT OF CELL INCLUSIONS

Most eggs, as we have seen, need to be activated in order to complete the maturation processes. In many the oocyte nucleus remains intact until activation, when the nuclear membrane disappears and the large amount of nuclear fluid that has accumulated during the growth phase of the oocyte mingles with the surrounding cytoplasm. In consequence, the viscosity of the cytoplasm is greatly decreased and various streaming processes occur. Eggs centrifuged before the disruption of the oocyte nucleus (germinal vesicle) show little effect, presumably because of the high viscosity or gelation of the cytoplasm. Eggs centrifuged subsequently readily undergo stratification of contents according to relative densities, typically with fatty and lipoidal material concentrating at the centripetal pole, yolk granules and pigment granules toward the centrifugal pole, and nucleus and more or less "pure" translucent cytoplasm in between. The visible inclusions thus displaced have some tendency to return to their original positions if centrifugation has not been too extreme or prolonged, suggesting to earlier workers that the egg cell contained an invisible elastic cytoskeleton which could be distorted under centrifugal force but not destroyed, anticipating the later discovery of cytoplasmic reticular structure.

More important was the discovery that displacement of visible particulate substances in the egg, once thought to have "organ-forming" or determinative properties, has no significant influence on the course of development. This and other evidence

---

[1] L. F. Jaffe, Electrical Currents through the Developing *Fucus* Egg, *Proc. Nat. Acad. Sci.*, **56**:1103 (1966).
[2] *Ibid.*

led to the concept that the controlling agency for developmental or cleavage patterning lay in the cortical region, or ectoplasm, of the egg and that this region, assumed to be more physically structured, remained unaffected by centrifugation, whereas the endoplasm permitted diffusion or passage of various components and was itself subject to flow.

In a few species the original polarity of the animal egg has been experimentally inverted, by exploiting gravitational force. The eggs of the nematode worm *Ascaris*, long used in experimental and analytical studies, are particularly suitable in this respect because the cell arrangement and differentiation established during the first few cleavages of the egg are so distinctive: there is no need to await late stages in development to see if polarity change has been effected.

The course of cleavage in *Ascaris* is highly distinctive, in that the first cleavage plane is horizontal to the vertical polar axis of the egg; the second cleavage plane is vertical in the animal-pole cell, i.e., in the plane of the polar axis, but is horizontal in the vegetal-pole cell, thus giving rise to a unique configuration known as the T-shaped embryo, although it consists of only four cells. Any experimental inversion of the polar axis would, therefore, become almost immediately discernible. Boveri himself, however, was unable to invert the polarity of *Ascaris* eggs by centrifuging, although he did succeed in showing that chromosomal diminution took place, depending on the cytoplasmic content of the cell.

The mitotic apparatus in *Ascaris* first appears at right angles to the animal-vegetal axis, which is usual for most kinds of eggs and would in that position divide the egg vertically, the cleavage plane passing through the poles. In this case the still-growing mitotic apparatus rotates at once for 90°, to become parallel with the polar axis, and consequently divides the egg cell through the equatorial plane. Before the onset of this division, differential distribution of ectoplasm and endoplasm has already been established, with ectoplasm increasing in concentration toward the animal pole and endoplasm increasing toward the vegetal pole. The first cleavage, as in *Fucus*, therefore separates two halves of the egg that are already to some extent different in composition.

FIGURE 11.10 Polarity and inversion of polarity in the egg of *Ascaris*. (a) Egg at metaphase of first cleavage, with polar body at animal pole. (b) Two-cell stage, with chromosome elimination occurring in animal cell (S_1AB) but not in vegetal cell (P_1). (c) Four-cell stage: the animal-pole cell divides vertically, the vegetal cell horizontally, so that a T-shaped embryo results from the second cleavage. (d) Centrifuged eggs showing the five zones referred to in text. (e) When the egg is centrifuged so that the contents are inverted relative to the polar axis, an inverted T-shaped embryo results. (See also Figures 12.6 and 12.7.) [*After Tadano, 1962.*]

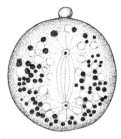

a

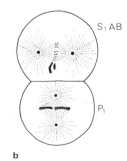

b

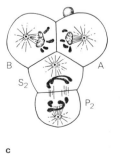

c

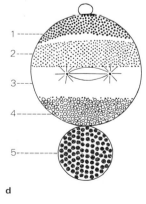

d

e

Thus the *Ascaris* egg divides into a primary somatic cell in the direction of the animal pole and into a primary propagative cell toward the vegetal pole. The S_1 blastomere then divides in the plane of the polar axis into the A and B blastomeres, in which chromatin diminution takes place; while the P_1 blastomere divides in the plane at right angles to the polar axis into S_2 and P_2 blastomeres, making the T-shaped configuration. Then, because this is physically an unstable arrangement, in the same sense that a T-shaped arrangement of four soap bubbles would be, the P_2 blastomere slides along the surface of S_2 toward blastomere B, so that the four cells assume a rhombic arrangement, in which the free surface of the whole is reduced to a minimum. Diminution does not occur in the P_2, or primordial germ line stem cell.

INVERSION OF THE POLAR AXIS

Inversion of the egg axis of *Ascaris* has now been effected by Tadano in Japan. When the egg is weakly centrifuged, after fusion of the pronuclei but before the first cleavage, its material stratifies into five layers; when strongly centrifuged, the first of the five layers at the heavy, or centrifugal, side separates into the mitochondrial and the heavy hyaloplasmic layers. After the cessation of centrifugation, the various stratified materials regain their normal distribution. Inversion of the polar axis depends on whether or not the eggs are free to rotate in the centrifuge medium. If they are not fixed in place, the eggs rotate so that the lighter, animal pole points toward the centrifugal end of the tube and the centrifugal force operates merely to augment the eggs' normal gradient of ectoplasm and endoplasm. However, if the eggs are first held in place so that the heavier, vegetal pole is centrifugal, the effect of centrifuging is to cause a large amount of ectoplasmic material to shift and reform at the vegetal pole, thereby inverting the gradient polarity of the egg.

The new animal pole in all cases forms where the ectoplasm is most concentrated. Moreover, eggs with an inverted polar axis develop normally, and it is evident that the differentiation of the early blastomeres as primary somatic or propagative cells depends on the relative amounts of ectoplasmic and endoplasmic material assigned to them. According to Tadano, the polarity is essentially represented by a nonspecific gradient differential in metabolic activity associated with the gradation in the distribution of ectoplasm in the fertilized but uncleaved egg.

CLEAVAGE PATTERN

The pattern of cleavage in *Ascaris* is unique and is an example of an extreme type of eggs and their development, namely, eggs of the so-called mosaic or determinate category, which are in contrast to the regulative or determinate kinds. The eggs of many species do not fit into either category, being intermediate in character. In fact

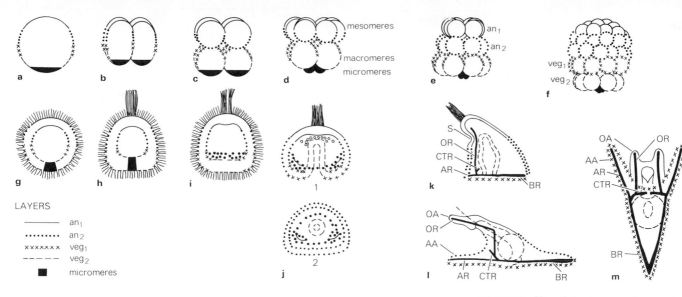

a

b

c

d
mesomeres
macromeres
micromeres

e
an₁
an₂

f
veg₁
veg₂

g

h

i

1

2

j

LAYERS

————	an₁
··········	an₂
xxxxxxx	veg₁
— — — —	veg₂
▬	micromeres

k
S
OR
CTR
AR
BR

l
OA
OR
AA
AR CTR BR

m
OA
OR
AA
AR
CTR
BR

no egg is completely and exclusively one or the other. Although expedient and still currently used, the terms *determinate* and *indeterminate,* or *mosaic* and *regulative,* have never been fully accepted, and they are employed here only in a provisional way. In general, however, the development of eggs of mollusks, annelids, flatworms, nematodes, ctenophores, and ascidians is more or less determinate, whereas that of echinoderms, balanoglossids, coelenterates, and amphibians is regulative, or indeterminate.

FIGURE 11.11 Normal development of sea urchin *Paracentrotus lividus,* (a) to (f) showing differentiation of layers (see text), (g) and (h) showing later development. (a) Uncleaved egg. (b) 4-cell stage. (c) 8-cell stage. (d) 16-cell stage. (e) 32-cell stage. (f) 64-cell stage. (g) Young blastula. (h) Later blastula, with apical organ, before formation of the primary mesenchyme. (i) Blastula after formation of the primary mesenchyme. (j) Gastrula: secondary mesenchyme after the two triradiate spicules have formed (drawing 1); transverse optical section of the same gastrula (drawing 2), bilateral symmetry established. (k) The so-called prism stage; stomodaeum invaginating. (l) Pluteus larva from the left side; broken line indicates position of egg axis. (m) Pluteus from the anal side. *AA,* anal arm; *AR,* anal rod; *BR,* body rod; *CTR,* central transverse rod; *OA,* oral arm; *OR,* oral rod; *S,* stomodaeum. [*After Hörstadius, 1939.*]

CELL NUMBERS AT END OF CLEAVAGE OF EGG*

Phylum or Class	Species	Egg Diameter, mm	Number of Cells Formed (Approximate)
Coelenterata	*Aurelia* sp.	0.15–0.20	2,600
Polycladida	*Thyzanozoon*	0.12	950
Polychaeta	*Serpula* sp.	0.07	920
	Lumbrinereis nasuta	0.16	4,100
Echinoderma	*Toxipneustes variegatus*	0.11	2,100
Tunicata	*Ascidiella aspersa*	0.16	3,400
	Ecteinascidia turbinata	0.72	132,000
Cephalochordata	*Amphioxus*	0.12	9,000
Cyclostomata	*Entosphenus (Petromyzon) wilderi*	1.00	300,000
Amphibia	*Bufo americanus*	1.30	450,000
	Rana sylvatica	2.10	700,000
	Triturus viridescens	2.60	860,000

*From Berrill, 1935.

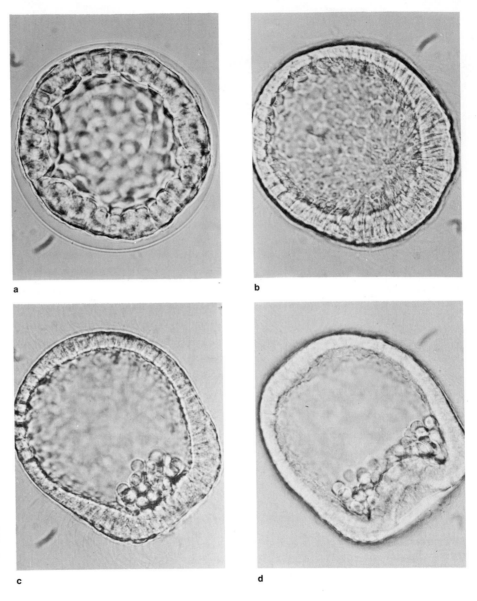

FIGURE 11.12 (a) Early blastula, (b) late blastula, (c) migration of primary mesenchyme cells, and (d) early invagination of enteron of sea urchin. [*Courtesy of T. Gustafson.*]

There is variation even within each group; e.g., sea urchin development is more determinate than that of starfish, and frogs are more determinate than salamanders. Thus the first two cleavages of the starfish egg pass through the poles and the third is equatorial, as in the sea urchin; the following (fourth) cleavage of the vegetal quartet of cells does not give rise to macromeres and micromeres, as in the sea urchin, but is more or less equal. This is in keeping with the fact that starfish larvae have no spicular skeleton, have no related precocious migration of primary mesenchyme cells, and have a relatively weak oxidation-reduction gradient as demonstrated by Janus green experiments.

Cleavage pattern itself has been classified as determinate or indeterminate, according to whether or not the cells formed during the first few cleavages of the egg segregate potentially very different histogenetic regions. The first two cleavages of the *Ascaris* egg do so; the first two cleavages of the sea urchin egg do not. Yet when sea urchin eggs develop in hypotonic seawater, the first two cleavages may yield a T-shaped stage similar to that of *Ascaris*, with the cells having much the same prospective fate. Both with regard to the basic polar gradients and to the commonly very distinctive patterns of cleavage exhibited by many types of eggs, the egg cortex appears to be deeply involved. Analysis of development accordingly becomes more and more a study of cortical change, nuclear activity, and the influence of the one on the other.

POTENCIES

Although polarity is already established in animal eggs by the time the oocytes are mature, little else may be determined. In many species the activated egg is an open

FIGURE 11.13 **(a)** Late gastrula (prism larva) and **(b,c)** pluteus larvae, showing extension and subdivision of invaginated digestive tube, its union with the stomodaeal invagination, and the relation of tips of larval skeleton to location of outgrowth of pluteus arms. [*Courtesy of T. Gustafson.*]

a

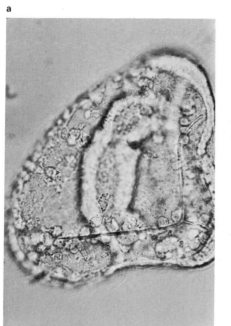

b

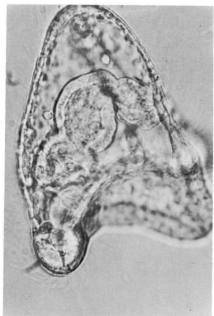

c

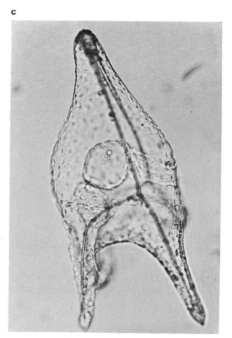

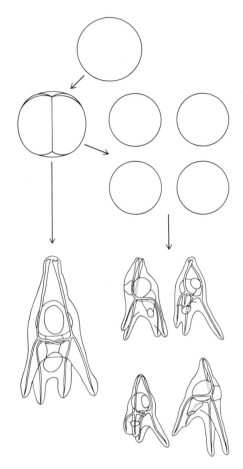

FIGURE 11.14 Development of isolated blastomeres of four-cell stage. Each develops into a pluteus larva of normal proportions but reduced size.

system, capable of developing in various ways according to circumstances, although by no means necessarily to form a viable organism. So-called normal development is the path taken in "normal" or standard conditions, to which the developing system has progressively adapted during the course of evolution. Seawater, for instance, is a mineral salt solution of great stability, antiquity, and complexity. Truly marine species usually can develop normally only in unmodified seawater. Artificial seawater from which calcium has been omitted fails to support the normal adhesive properties of the cell surface of cleaving eggs.

Sea urchin eggs in process of cleavage give rise to two daughter cells, or blastomeres, which are virtually spherical at the moment of completion of the cleavage furrow and are in mutual contact at a very small area of the newly formed, additional, cell surface. Almost at once, however, the area of contact extends rapidly until the two cells are closely apposed as a flat interface representing about one-third of the total surface of each cell. Yet if eggs about to cleave are placed in calcium-free seawater, division is completed but each cell remains virtually spherical, with minimal contact with the other. Mild shaking then causes the blastomeres to fall apart. Beginning with the pioneering experiments of Herbst and Driesch nearly a century ago, this procedure has been employed for the separation of blastomeres, resulting in subsequent independent development when they are returned to normal seawater.

Blastomeres of the two-cell stage of the sea urchin, each of which normally gives rise to one-half of the embryo and larva, behave differently when separated from each other, and each gives rise to a whole larva, though of somewhat reduced size. Blastomeres isolated at the four-cell stage show the same capacity. Together they form one whole; separately each becomes a whole. Similarly in several species of newts, each of the first two blastomeres can, under certain circumstances, form half-sized but structurally normal embryos. Fusion of two newt eggs at the two-cell stage may result in development of a large but normally constructed embryo.

Driesch described the cleaving egg as a harmonious, equipotential system, in which every part has potentially the properties of the whole. This still remains a good description of the prevailing state, for it highlights the phenomenon and also our general ignorance concerning its physical basis. The only advance in our understanding has come recently, from further experimental analysis involving new techniques and, most importantly, from Loewenstein's hypothesis of junctional communication between adjoining cells, described in Chapter 5. Electrochemical communication exists between the first two blastomeres, or among the first four blastomeres of the dividing egg, for example, but is disrupted when the blastomeres are isolated from one another. In other words, adjoining blastomeres in a normally cleaving egg constitute a single closed-circuit system acting as a unit, each cell subject to and contributing to the whole. When isolated, each cell reverts to the original state if electrical cataphoresis has not already brought about irreversible regional distribution of content.

CLEAVAGE AND DIFFERENTIATION

The course of cleavage of the egg of most small, totally dividing eggs follows a seemingly general sequence in which the first cleavage divides the egg in a plane passing through the poles of the egg axis, the second cleavage plane is also meridional and passes through the poles but is at right angles to the first, and the third cleavage plane is horizontal and often more or less equatorial. To produce this sequence the mitotic apparatus must be in a plane at right angles to the polar axis during the first two cleavages, and parallel to it in the third. At one time it was considered that the mitotic apparatus grew in the direction of greatest cytoplasmic space, which is a fair description of the events.

Such is the course in the cleavage of the sea urchin egg. Yet the fourth cleavage of the sea urchin egg shows that more is involved, for whereas each of the cells at the animal pole divides as before (in the plane of the polar axis, together forming a tier of eight), the vegetal four cells each divide horizontally, at right angles to the axis. Further, the division of these four vegetal cells is extremely unequal, forming four very small cells, the "micromeres," at the vegetal pole, and four large cells, "macromeres," as a layer between the micromeres and the eight cells, "mesomeres," at the animal pole. The question arises whether a particular succession or pattern of cleavage planes subdivides the egg in a causally differentiating way or whether the succession is itself determined by progressive differentiation. In *Fucus*, differentiation clearly precedes the first cleavage.

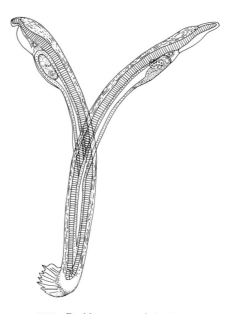

FIGURE 11.15 Double monster of *Amphioxus*, produced by mechanical disarrangement and partial separation of the blastomeres of the two-cell stage. [*After Conklin*, J. Exp. Zool., **64**:*373(1933)*.]

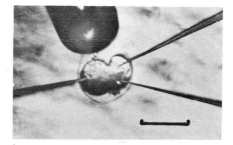

a   b

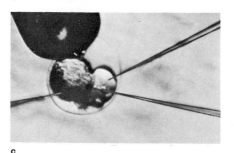
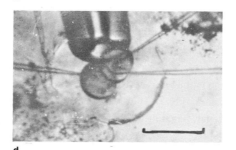

c   d

FIGURE 11.16 Intercellular electrical coupling at a forming membrane junction in a dividing egg of the starfish *Asterias forbesii*. (a) Fertilized egg impaled by two microelectrodes. (b–d) Various stages of cleavage with three microelectrodes in intracellular position (calibration = 0.1 mm). Ion communication between the two blastomeres diminishes progressively during division as a cell-membrane junction forms at the plane of cleavage. [*Courtesy of W. R. Loewenstein.*]

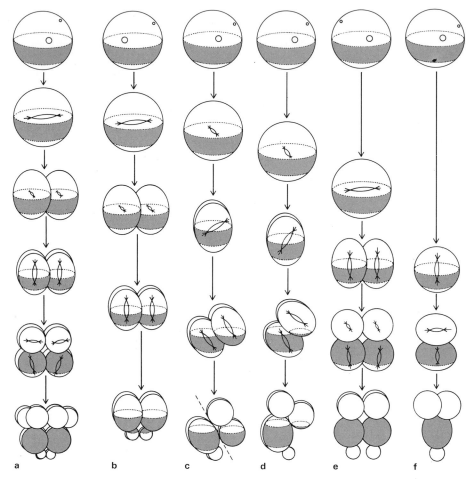

FIGURE 11.17 Normal and retarded cleavage in sea urchin. (a) Normal cleavage. (b–f) Cleavage delayed in relation to the process of determination, affecting the position of the cleavage spindles and causing the activation of the micromere substance. The delays were induced by hypotonic seawater. [*After Hörstadius, 1939.*]

a　　b　　c　　d　　e　　f

A simple experiment on the sea urchin egg shows that the differentiation process is primary and that cleavage pattern reflects but does not cause it. When fertilized sea urchin eggs are placed in hypotonic seawater, the process of cleavage becomes somewhat delayed, so that by the time eggs in normal seawater have divided four times, those in the hypotonic medium have done so only three or two times. This retardation in the rate of cleavage, which probably derives from retardation of either the rate of growth of the division apparatus or replication of centrioles, is not accompanied by a corresponding retardation in the rate of change in the orientation-control agency, whatever that may be. Comparing only the extremes, by the time the egg developing in normal seawater attains the 16-cell stage, described in the previous paragraph, the most retarded case has divided twice, the first time horizontally, separating the animal half from the vegetal half of the egg, and the second time so that the animal cell divides vertically and the vegetal cell again horizontally to form a T-shaped four-cell stage.

Evidently a progressive process occurs in the fertilized egg which controls the orientation of the mitotic apparatus growing at different times and is unaffected by hypotonicity in the environmental medium. In sea urchin development this process reaches a critical phase after several hours, when, in normal seawater, several divisions of the egg have already occurred. The relative slowing of the cleavage rate, experimentally, reduces the number of cleavages that can take place before the critical orientation phase is reached. In *Ascaris* a similar mitosis-orienting phase is reached before the first cleavage occurs, not because of cleavage retardation but presumably because the determination process leading to the particular orienting situation proceeds relatively very much faster. The same sort of differentiation-cleavage process goes on in each, but the gearing ratio is different.

GRADIENT THEORY

Since the time of Driesch's discovery that sea urchin blastomeres of the two-cell or four-cell stage can develop into a whole larva when isolated, the sea urchin egg has been subject to an immense number of investigations, partly because of the challenging nature of the early discoveries and partly because the eggs are readily obtained in large quantities, are easily fertilized artificially, and are easy to maintain through the process of gastrulation to the pluteus larva stage. Various operative and chemical procedures have been employed in attempts to analyze the developmental processes leading to or involved in the formation of the blastula, the gastrula, and the actively swimming larva.

To begin with, separation of the animal- and vegetal-cell layers following the third or initial equatorial cleavage showed that some segregation has already occurred by this time. The four blastomeres of the animal half develop to form a hollow blastula but develop no further, whereas the four vegetal blastomeres together develop into a blastula which gastrulates and may become a whole larva. Cutting the fertilized but uncleaved egg equatorially into animal and vegetal halves, however, showed that this difference in potential already exists shortly before cleavage but is not established closer to the time of fertilization. The implication is that something in the newly fertilized egg becomes concentrated in the vegetal half within a short period.

Most of the operative experiments with developing sea urchin eggs have been made with embryos at the 32- and 64-cell stages, i.e., one and two further cleavages beyond the 16-cell stage already described. The 16-cell stage consisted of a ring of eight animal blastomeres, four large blastomeres or macromeres, and four small micromeres at the vegetal pole. At the next division the animal half consists of two rings of eight cells each, called an_1 and an_2 for identification. At the 64-cell stage the macromeres have also divided into two rings of eight cells, called veg_1 and veg_2. For convenience, therefore, the developing egg may be said to consist of five layers: an_1,

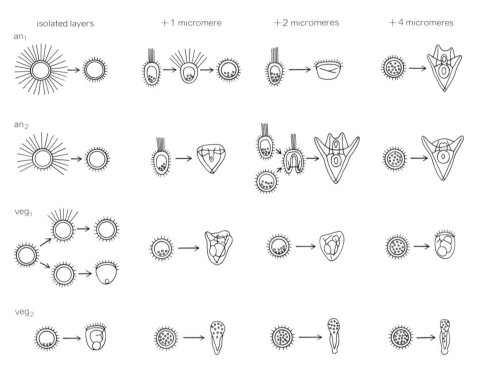

FIGURE 11.18 Development of the layers an_1, an_2, veg_1, and veg_2 isolated (*left column*) and with one, two, and four micromeres implanted. [*After Hörstadius, 1939.*]

an_2, veg_1, veg_2, and the micromeres. The development of these several layers, representing strata along the animal-vegetal axis, has been studied separately and in various combinations. By projection the strata may be traced back to equivalent zones of the undivided egg.

As development continues, a typical blastula stage forms which consists of a single layer of cells enclosing a central cavity, the blastocoel. The blastula soon acquires cilia at its external surface, relatively long, stiff cilia appearing at the animal pole. At the vegetal pole the cells derived from the original micromeres migrate into the blastocoel, where, as the primary mesenchyme, they will later lay down the spicules forming the larval skeleton. Following this migration the cell layer remaining at the vegetal pole flattens and begins to push inward as the invaginating archenteron, to form the gastrula. Vital-staining experiments show that the migrating primary mesenchyme cells are exclusively derived from the micromere material, and that all of the subsequently invaginating archenteron is derived from the veg_2 layer. An_1, an_2, and veg_1 together give rise to the larval ectoderm.

With further development the tip of the archenteron buds off (evaginates) the secondary mesenchyme and coelom; following this an ingrowth from the oral field area of the outer wall, or ectoderm, forms the stomodaeum, which fuses with the tip of the archenteron to form a continuous digestive tube through the interior of the larva. Bilateral symmetry is thus manifest, the axis of symmetry coinciding with the meridion of the animal-vegetal axis which passes through the site of oral invagination.

Following the foundations laid by Driesch and Herbst for the analysis of sea urchin development, the prototype for animal development as a whole, voluminous work has been carried out by Runnström, Lindahl, von Ubitch, C. M. Child, and especially by Hörstadius, who has performed spectacular micromanipulative recombination experiments with the several layers of cells just described. The results are interpreted in terms of gradients along the polar axis.

Isolation results are essentially as follows:

an_1 Develops to form a blastula entirely covered with long stiff cilia, but it develops no further; i.e., it is unable to invaginate to form the gastrula.

an_2 Develops like an_1 except that long cilia do not cover all the blastula surface.

veg_1 May become a ciliated blastula only or may develop a small archenteron by invagination.

veg_2 Gives rise to an ovoid larva with an archenteron and one or two spicules, i.e., veg_2, which normally gives rise only to endoderm and mesoderm, now produces ectoderm as well.

micromeres Continue to divide to form a specific number of small isolated cells representing the primary mesenchyme cells they would normally have become.

Combination Results

If four micromeres are added to an animal half or to an entire presumptive ectoderm, i.e., an_1 + an_2 + veg_1, gastrulation occurs and a normal larva develops.

If one or more micromeres are added to veg_2 (veg_2 being self-sufficient), an exogastrula forms, i.e., the archenteric layer is evaginated instead of invaginated.

Four micromeres together with veg_1 give rise to a larva similar to that produced by veg_2 alone.

Accordingly the most normal development, culminating in a pluteus larva, results from veg_2 alone, veg_1 plus one micromere, an_2 plus two micromeres, and an_1 plus four micromeres. In fact, two normal larvae can be obtained from a single cleaving egg by two horizontal separations if the middle layer is left intact and the micromeres are joined to the most animal layer. Considerable regulation is required to accomplish such development.

According to the Runnström-Hörstadius theory, a double gradient system is assumed to control morphogenesis and differentiation in the early development of sea urchins, an assumption which they state has been a useful conceptual tool in

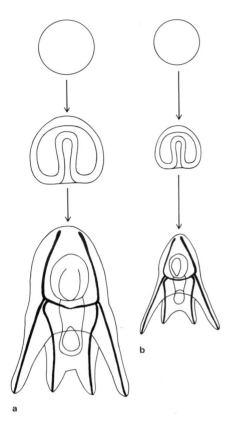

FIGURE 11.19 Development of (**a**) normal and (**b**) reduced eggs of sea urchin, the reduced eggs having had 25 to 50 percent of the contents withdrawn by micropipette. Apart from the reduced size of the gastrula and larva, development is normal.

experimental studies. At the animal pole (the "most animal" region of the egg), the power to form structures characteristic of that region is most intense; at the vegetative pole (the "most vegetative" region), the power to invaginate and form structures normally arising there (archenteron, mesenchyme) is at a maximum. Owing to the interception of the gradients, characteristic levels arise in the egg, and at each level the state is determined by the relation between animalizing and vegetalizing agents.

Each gradient is considered to extend to the opposite pole, so that gradients overlap through the whole egg, and to be of qualitatively different nature, based on differences in metabolism. One suggestion was that the animal gradient might be cortical and the vegetal gradient might be in the interior. This was shown not to be so by experiments involving the withdrawal of much of the egg cytoplasm by means of a micropipette. Even when more than 50 percent of the cytoplasm has been removed, and eggs subsequently have been fertilized, no deviation from normal development and normal larvae can be detected. The typical development may thus be a consequence of unaltered conditions in the cortex although this has yet to be proved.

To sum up, assuming that some sort of gradient hypothesis is valid: a factor initially somewhat diffused along the animal-vegetal axis becomes progressively concentrated toward the vegetal pole, this concentration being determinative with regard to the process of gastrulation and to the specialization of cells as spicule-forming mesenchyme. It is notable that the direction of gastrulation may be changed by an excess of this factor and that the proportion of ectoderm to endoderm may also be radically altered.

A chemical approach to the analysis of development of the sea urchin egg has gone hand in hand with the operative approach right from the start. As early as 1892 Herbst discovered that lithium ions added to seawater caused an increase in the amount of endoderm and led to exogastrulation. Exogastrulation, in which the archenteron everts instead of inverts, may occur in the absence of any change in proportion of endoderm to ectoderm; and exogastrulation is not itself a sign of so-called vegetalization. Exogastrulation results when there is lack of invagination, but with constitutional normality. Vegetalization leads to an enlargement of the endoderm at the expense of ectoderm and is a different phenomenon. It is now customary to speak of vegetalizing agents as those which have such an effect, and animalizing agents as those which increase ectoderm at the expense of endoderm. These are clearly related to the problem of gradients. The question of invagination, whether relating to gastrulation or to some other event, is primarily one of tissue mechanics and is left for later discussion. The gradient concept was introduced in 1901 by Boveri, who considered that the vegetal region of the sea urchin egg exerted a dominant control over more animal regions, the influence decreasing toward the animal pole.

Vegetalization by means of lithium has been commonly employed, while NaCNS and trypsin have been generally used to cause animalization, although a variety of

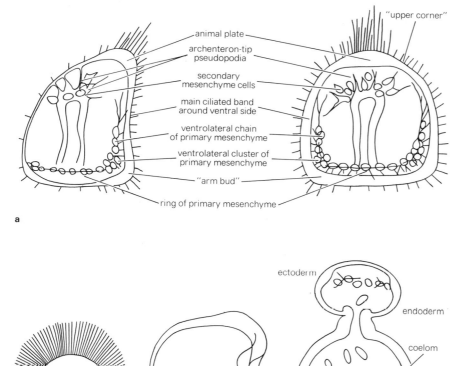

"upper corner"

animal plate

archenteron-tip
pseudopodia

secondary
mesenchyme cells

main ciliated band
around ventral side

ventrolateral chain
of primary mesenchyme

ventrolateral cluster of
primary mesenchyme

"arm bud"

ring of primary mesenchyme

a

ectoderm

endoderm

coelom

esophagus

ANIMALIZED

VEGETALIZED

b

FIGURE 11.20 Animalization and vegetalization of sea urchin development. (a) Normal advanced gastrula showing distribution of primary mesenchyme cells. (b) Abnormal larvae induced by various chemical agents. (*Left and center*) Animalized larvae with long cilia and mesenchyme cells only or with very reduced endoderm which forms a small invagination; (*right*) vegetalized exogastrula larva with only small ectoderm and a corresponding increase in the size of the endoderm. [*After Gustafson, 1965.*]

other substances have been found to have comparable effects. The value of these two agents is not so much that they throw light on the chemical nature of gradients as that they are useful as tools for modifying the gradient or gradients in various ways. Complete animalization of the vegetal half of the egg can be effected by trypsin so that it develops to form a typical animal-half type of blastula, complete with long apical cilia; while lithium can vegetalize isolated animal halves so that they gastrulate and become pluteus larvae. Thus lithium and implanted micromeres have a similar effect. The basic question is: What is the nature of the gradient systems which are affected by these agents?

One possibility is that gradients in the rate of oxidative metabolism are involved. Using the rate of reduction of vital dyes, particularly Janus green, as an indicator of the rate of oxidative metabolism, C. M. Child showed that such a reduction gradient is present in the oocytes, cleavage stages, and blastulas of sea urchin and starfish. This

FIGURE 11.21 Sea urchin larvae of about the same age but varying in the animal and the vegetal tendencies. (e) Normal gastrula. (a–d) Larvae animalized to a varying extent. (f–g) Larvae vegetalized to a different extent. In (g) the ectoderm is reduced to a small knob. The hypothetical double-gradient system corresponding to each larva is indicated. Animalization can be brought about by a reduction of the amount of vegetal material, by treatment with various chemical agents such as SCN before fertilization, or by treatment with o-iodosobenzoic acid or 2,5-thiomethylcytosine during cleavage stages. Vegetalization can be brought about by lithium or chloramphenicol treatment during cleavage. [After Gustafson, 1965.]

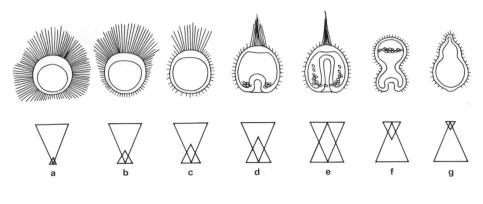

FIGURE 11.22 A proposed model for the interactions between cytoplasmic factors and groups of genes A, V, Z′ in early development of sea urchin. The first gene groups to become active (A and V) are controlled by the animalizing (an) and vegetalizing (veg) agents. The open circles indicate derepressive effects; filled circles, repressive effects. Genetic replenishment of animalizing and vegetalizing agents is indicated by arrows. Combined gene products from A and V evoke the derepression of the genes of groups Z. These stablize intermediate zones and exert a control on the formation or activity of the animalizing and vegetalizing agents. The products of the Z genes are also responsible for the ready interactions between ectoderm and the underlying tissues. [After Runnström, 1967.]

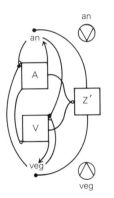

was in support of his contention that a single physiological gradient controls morphogenesis. Certain amino acids, such as glutamine and lysine, have an animalizing effect, while others have a vegetalizing effect, suggesting that both difference in type and differential rates of protein synthesis may underlie the phenomena described. This is supported by the recent discovery by Hörstadius and Josefsson that specific animalizing and vegetalizing substances, namely, one containing a tryptophan and the other a nucleotide, respectively, are present in mature unfertilized eggs.

The general indication is that the animal factor promotes synthesis of ribonucleic acid and proteins, while the vegetal factor seems to inhibit the onset of these processes for a while, so that the time pattern of early protein synthesis will be different in the two regions. In a review, Lallier states the current point of view as follows: "The introduction of the concept of inhibition in the analysis of the development of the sea urchin egg suggests an interpretation of differentiation and the mode of action of the animalizing and vegetalizing agents. At the very beginning of development the nuclei would emit information of animal or vegetal type. This information would specifically orientate protein synthesis. The activation of the synthesis of animal proteins would inhibit the synthesis of the vegetal pole proteins and vice versa. At the beginning of development, the synthesis of animal proteins would be favored at the animal pole, while synthesis of vegetal proteins would be favored at the vegetal pole. The local advantage favoring a special type of protein synthesis at each pole might be connected to some structural peculiarity at the pole of the eggs, perhaps in relation to the cortex. . . . According to the general scheme that we have just indicated, animalizing and vegetalizing agents would influence morphogenesis by acting directly as inhibitors, or indirectly by favouring the formation of inhibitors of the synthesis of specific proteins."[1]

[1] R. Lallier, Biochemical Aspects of Animalization and Vegetalization in the Seaurchin Embryo, Advance. Morphogenesis, 3:191(1964).

CONCEPTS

The rate of cleavage is under cytoplasmic rather than nuclear control.

Cleavage proceeds until normal somatic cell sizes and nucleocytoplasmic ratios are attained.

Development proceeds from the general to the particular.

Polarized differentiation of the egg is established by means of self-cataphoresis.

The polar axis is associated with ectoplasmic, or cortical, substance and is reversible with difficulty.

Nuclear chromatin responds differentially to regional differences in cytoplasmic constitution.

The cleaving egg is a harmonious equipotential system, based on electrochemical junctional communication.

The relative rates of cleavage and determination process vary according to species and according to environmental factors.

Double gradient systems relating to differential qualitative and quantitative control of protein syntheses may be responsible for progressive differentiation and morphogenesis.

READINGS

BERG, E., 1967. Some Experimental Techniques for Eggs and Embryos of Marine Invertebrates, in F. Wilt and N. K. Wessels (eds.), "Methods in Developmental Biology," Crowell.

BERRILL, N. J., 1935. Cell Division and Differentiation in Asexual and Sexual Development, *J. Morphol.*, **57:**353–427.

BOVERI, TH., 1902. On Multiple Mitosis as a Means of Analysis of the Cell Nucleus, in B. H. Willier and J. M. Oppenheimer (eds.), "Foundations of Experimental Embryology," 1964, Prentice-Hall.

CZIHAK, G. and S. HÖRSTADIUS, 1970. Transplantation of RNA-labeled Micromeres into Animal Halves of Sea Urchin Embryos. A Contribution to the Problem of Embryonic Induction, *Develop. Biol.* **22:**15–30.

CHILD, C. M., 1940. Lithium and Echinoderm Exogastrulation: With a Review of the Physio logical-gradient Concept, *Physiol. Zool.*, **13:**4–42.

COSTELLO, D. P., 1961. The Orientation of Centrioles in Dividing Cells and Its Significance, *Biol. Bull.*, **121:**368–387.

————, M. E. DAVIDSON, A. EGGARS, M. H. FOX, and C. HENLEY, 1957. "Methods of Obtaining and Handling Marine Eggs and Embryos," Marine Biological Laboratory, Woods Hole.

DRIESCH, H., 1892. The Potency of the First Two Cleavage Cells in the Development of Echinoderms, in B. H. Willier and J. M. Oppenheimer (eds.), "Foundations of Experimental Embryology," 1964, Prentice-Hall.

GUSTAFSON, T., 1965. Morphogenetic Significance of Biochemical Patterns in Sea Urchin Embryos, in R. Weber (ed.), "Biochemistry of Animal Development," vol. I, Academic.

HARVEY, E. M., 1956. The American *Arbacia* and Other Sea Urchins, Princeton University Press.

HINEGARDNER, R. T., 1967. Echinoderms, in F. Wilt and N. K. Wessels (eds.), "Methods in Developmental Biology," Crowell.

HÖRSTADIUS, S., 1955. Reduction Gradients in Animalized and Vegetalized Seaurchin Eggs, *J. Exp. Zool.*, **129**:249–256.

———, 1939. The Mechanics of Sea Urchin Development, Studied by Operative Methods, *Biol. Rev.*, **14**:132–179.

JAFFE, L., 1968. Localization in the Developing Egg and the General Role of Localizing Currents, *Advance. Morphogenesis*, **7**:295–328.

———, 1966. Electrical Currents through the Developing *Fucus* Egg, *Proc. Nat. Acad. Sci.*, **56**:1102–1109.

——— and W. NEUSCHLER, 1969. On the Mutual Polarization of Nearby Pairs of Fucaceous Eggs, *Develop. Biol.*, **19**:549–565.

JOSEFSSON, L., and S. HÖRSTADIUS, 1969. Morphogenetic Substances from Seaurchin Eggs: Isolation of Animalizing and Vegetalizing Substances from Unfertilized Eggs of *Paracentrotus lividus*, *Develop. Biol.*, **20**:481–500.

LALLIER, R., 1964. Biochemical Aspects of Animalization and Vegetalization in the Seaurchin Embryo, *Advance. Morphogenesis*, **3**:147–196.

LUND, E. J., 1923. Electrical Control of Organic Polarity in the Egg of *Fucus*, *Bot. Gaz.*, **76**:288–301.

MONROY, A., and P. R. GROSS, 1967. The Control of Gene Action during Echinoderm Embryogenesis, in "Morphological and Biochemical Aspects of Cytodifferentiation," Karger.

NEEDHAM, J., 1942. "Biochemistry and Morphogenesis," Cambridge University Press.

ROUX, W., 1888. Contributions to the Developmental Mechanics of the Embryo, in B. H. Willier and J. M. Oppenheimer (eds.), "Foundations of Experimental Embryology," 1964, Prentice-Hall.

RUGH, R., 1962. "Experimental Embryology," Burgess.

RUNNSTRÖM, J., 1967. The Mechanism of Control of Differentiation in Early Development of the Seaurchin, in "Morphological and Biochemical Aspects of Cytodifferentiation," Karger.

SPEK, J., 1934. Über die bipolare Differenzierung der Eizellen von *Nereis limbata* und *Chaetopterus pergamentaceous*, *Protoplasma*, **21**:394–405.

TADANO, M., 1962. Artificial Inversion of the Primary Polarity in *Ascaris* Eggs, *Jap. J. Zool.*, **13**:329–355.

WHITAKER, D. M., 1938. The Effect of Hydrogen Ion Concentration upon the Induction of Polarity in *Fucus* Eggs, III: Gradient in Hydrogen Ion Concentration, *J. Gen. Physiol.*, **21**:833–845; also in R. A. Flickinger (ed.), "Developmental Biology," Wm. C. Brown.

WILSON, E. B., 1925. "The Cell in Development and Inheritance," 3d ed., Macmillan.

CHAPTER TWELVE

CLEAVAGE
AND
NUCLEAR ACTIVITY

During cleavage of an egg the successive divisions occur at first relatively rapidly one after the other and then more and more slowly as cell sizes are approached which are characteristic of mature somatic tissues. As we have seen, rate of cleavage and rate of respiratory metabolism follow each other closely, and change in one affects the other.

CLEAVAGE COURSE AND CELL CYCLES

In holoblastic eggs especially, mitosis follows mitosis as quickly as is possible, with a minimum G_1 phase in each cell cycle. During later cleavage, particularly during late blastula stages, the intercleavage intervals become progressively greater, although the mitotic process itself is accomplished in about the same time as before. In fact, in spite of the progressive retardation in cleavage rate as a whole, the S-G_2-M portion of the cell cycle, i.e., DNA replication, mitotic protein synthesis, and the mitotic event, apparently remains as a unit that is not significantly affected. The progressive increase in the duration of the interphase is due mainly to increase in the G_1 phase, i.e., in the interval between the completion of one division and the onset of DNA synthesis signaling the next. It is during this phase that cells prepare for differentiation leading to courses other than continuing mitoses.

In terms of the cell cycle alone, therefore, we may expect to find the developing system expressing the two somewhat incompatible processes of cell division and cell (and, consequently, tissue) differentiation more or less at different developmental phases. In holoblastic eggs, both small and large, the intercleavage interval as a whole is so short—about 30 to 60 minutes, depending on size (at normal temperatures)—that the G_1 phase can hardly be said to exist. Cell reserves piled up in one form or another in the mature unfertilized egg carry it through the cleavage stages with little need or opportunity for syntheses not related to the mitotic process. During later development the intercleavage interval becomes greatly drawn out and is typically from 12 to 24 hours, as in cultures of somatic cells. It should be realized, however, that cells in this state are generally of standard somatic-cell sizes and must grow to double their minimal volume during each cell cycle, i.e., extensive cytoplasmic growth is necessary, in addition to DNA synthesis, between one cell division and the next.

CONTENTS
Cleavage course and cell cycles
First proteins and masked messengers
Nucleic acid synthesis
Equivalence of nuclei
The role of the nucleus
Transplantation and potency of nuclei
Differential gene action
Nuclear activity at gastrulation
Concepts
Readings

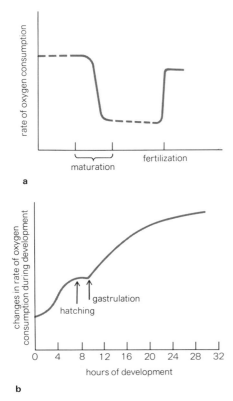

FIGURE 12.1 Oxygen consumption during development of sea urchin egg. (a) Decrease and restoration of rate of cell respiration during maturation and at fertilization. (b) Change in rate of respiration during development, the arrows representing the time of hatching and the onset of gastrulation respectively. [*After Lindahl*, Z. Vergleich. Physiol., *27:136(1939)*.]

FIRST PROTEINS AND MASKED MESSENGERS

The unfertilized egg typically has an enormous supply of ribosomes packaged into its cytoplasm, mainly in the form of monoribosomes. They are commonly associated with annulate lamellae, derived from the nuclear membrane at specific times during oogenesis, which remain in stacked array in the cytoplasm until they dissociate upon commencement of protein synthesis. Similarly the messenger RNA which controls protein synthesis after fertilization is *presumably* already synthesized in the unfertilized egg and enclosed in vesicles, derived from the lamellae, to prevent its destruction prior to fertilization. In addition an amino acid pool is present at the time of fertilization and onset of cleavage, a pool which must be renewed, however, at an early stage; this in turn implies that the necessary enzymes must be available from the start, presumably in the oocyte.

Protein synthesis either is greatly stimulated or is actually switched on at fertilization, with conversion of ribosomes to polyribosomes. Evidence accumulates for the concept that templates (mRNA) preexist in the unfertilized egg and that the early acceleration of protein synthesis following fertilization cannot depend on newly synthesized messenger RNA (mRNA). In other words, the first proteins are made by the embryo independently of immediate genomic control, from templates present in the unfertilized egg. Moreover, many or all of these templates bear information about the proteins needed for cell division but not about differentiation beyond the early blastula. Synthesis of new messenger RNA is needed, however, for normal development beyond the blastula stage. The demonstration of postfertilization protein synthesis in the absence of RNA synthesis, presumably resulting from stored, i.e., specifically protected, messenger RNA held in the unfertilized egg for future use, has yielded the "masked-messenger" hypothesis.

Yet the relatively impermeable mature unfertilized egg *is* metabolically active although not as active in protein synthesis as the fertilized egg. The crucial evidence, then, becomes whether any of the proteins made after fertilization are different from those made before fertilization. If not, the masked-messenger hypothesis is weakened, since the alternative explanation that some messages may be translated at one rate before fertilization and at a faster rate after fertilization then appears to be equally valid. More precisely, a translational-level regulation of long-lived mRNA activity is as plausible as the concept of an unavailable form of mRNA which becomes "unmasked" at fertilization. In any case, the ribosomes seem to be the controlling element preventing protein synthesis before fertilization. Ribosomes obtained from unfertilized eggs do not support protein synthesis in vitro, whereas those obtained from fertilized eggs do so. In terms of the masked-messenger concept, which suggests that at fertilization protein synthesis is initiated by the release of a protein that protects a template-ribosome complex present in the fertilized egg, once this protein is removed, ribosomes engage with messenger to form polyribosomes and protein synthesis starts.

With regard to protein synthesis in relation to cleavage and differentiation in the sea urchin, Monroy sums up the situation as follows:

1 "The period of the first 5 or 6 cleavages, which is the phase during which the rate of oxygen consumption increases rather slowly, is fairly active in protein synthesis; on the other hand, during the following period of rapid increase in the rate of oxygen consumption, the rate of incorporation of labelled amino-acids into protein undergoes a slight decrease on the whole.

2 "The proteins synthesized at an early stage, or a fraction thereof, are of importance for the differentiation of the primary mesenchyme and perhaps of the whole vegetative territory of the embryo. It may be appropriate to remember here that the early cleavage (but particularly the one coinciding with the rapid increase of oxygen consumption) is the phase of highest lithium-sensitivity.

3 "During the first wave of protein synthesis there is no evidence of synthesis of *new* proteins (i.e., proteins not already present in the unfertilized egg), but it seems likely that the egg only resumes the synthetic processes that were going on during oogenesis and which had been arrested at a certain stage during the course of maturation. This may explain why no new antigens—i.e., those different from those present in the unfertilized egg—have been detected until the gastrula or blastula stage.

4 "Each cleavage involves the formation of new cell surface. Whatever the mechanism of its production, there is no doubt that this must involve some kind of protein synthesis. Therefore a large part of the protein synthesis taking place during cleavage may possibly be accounted for by the production of new cell surfaces."[1]

A number of peaks of newly synthesized proteins appear during early development, as indicated by sensitivity to actinomycin D, while during blastulation new protein species begin to be synthesized, with a rate increasing to that of gastrulation. Some part of the heterogeneous RNA synthesized during early development may be a functional messenger for this. Yet the inhibition of gastrulation and the later development by actinomycin D can be reversed by returning embryos to normal seawater as late as 18 hours after the beginning of development. The embryos gastrulate and become normal pluteus larvae. This reversibility of gene action implies that no stringent timing of specific protein synthesis is necessary for later normal development.

A critical change in the course of development in the sea urchin begins at the time of migration of the cells of the primary mesenchyme, i.e., the descendants of the micromeres, into the blastocoel. During this phase a new increase in oxygen consumption occurs, together with a new increase in the rate of protein synthesis. At the same time the adult-type proteins make their first appearance, at least as far as serological evidence shows.

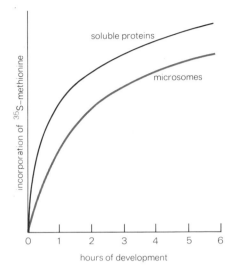

FIGURE 12.2 The first proteins, as indicated by the incorporation of ^{35}S-methionine in the soluble proteins and in microsomes. [*After Monroy, 1960.*]

[1] A. Monroy and R. Maggio, Biochemical Studies on the Early Development of the Seaurchin, *Advance. Morphogenesis,* **3:**110 (1964).

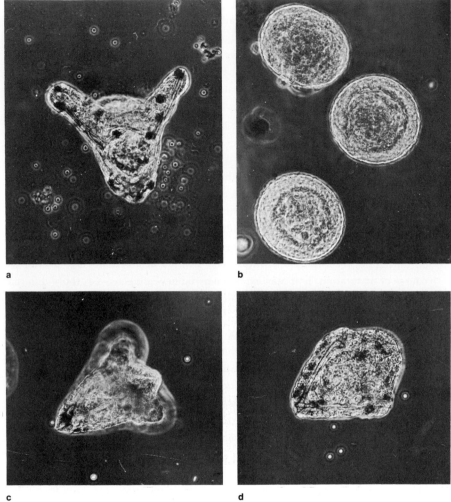

FIGURE 12.3   Effect of actinomycin D on sea urchin development. (a) Normal *Arbacia punctulata* pluteus larva. (b) Arrested *Arbacia* blastulas after continuous treatment with actinomycin D during the first 26 hours of development. (c) Larva of *Arbacia* 24 hours after removal from continuous exposure to actinomycin D at the eighteenth hour of development. (d) Larva 20 hours after continuous exposure to actinomycin D for the first 24 hours of development. Gastrulation has begun, but no sign of skeletal spicule formation that is normally present at this stage. [*Courtesy of C. H. Ellis.*]

NUCLEIC ACID SYNTHESIS

Synthesis of DNA begins immediately following fertilization. In ascidian and sea urchin eggs the pronuclei incorporate thymidine even before they fuse, which at least indicates a rapid turnover. Some DNA reserve, however, may generally exist, since the sea urchin egg does contain a small amount, sufficient to carry it through the first four or five cleavages, although from fertilization to the early blastula stage DNA synthesis is the most conspicuous phenomenon.

According to Ellis, "the majority of *early* differences in protein synthesis are under the influence of one or more mechanisms of control at the level of translation, or even further beyond the gene. But by the time of gastrulation and the beginnings

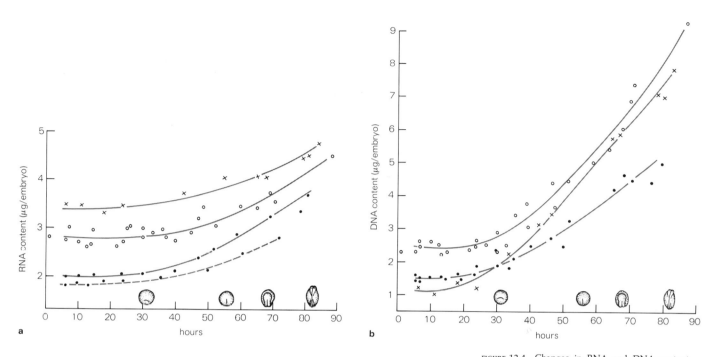

FIGURE 12.4 Changes in RNA and DNA content during early urodele development, for three species of *Triturus*. [*After Chen, 1967.*]

of visible differentiation toward the pluteus larva, a shift in levels of control has occurred such that the bulk of the embryo's protein is synthesized using newly synthesized messenger RNA as template, rather than utilizing the stable messenger RNA as in early development."[1]

A remarkable separation between the cleavage and morphogenesis phases is seen in the development of the brine shrimp *Artemia*. The mature artemia lays two types of eggs, a thin-shelled summer egg that hatches almost immediately after being laid, and a thick-shelled winter egg (an encysted dry egg) which can survive 10 or more years under desiccated conditions but renews its development when soaked in appropriate salt water. The eggs, which are centrolecithal, are fertilized as they enter the brood pouch of the female and at once undergo numerous divisions to form a virtually solid gastrula stage. Summer eggs continue to develop to naupliuslike larvae; winter eggs arrest at the gastrula stage.

At the time of arrest the number of cells constituting the blastula is approximately 4,000. No organization or embryonic structure is discernible at this stage except that cells with large nuclei and cells with small nuclei are both present. When suspended in 2 percent NaCl solution (at 27°C) the shells crack open, and embryos emerge after 12 to 24 hours. A few hours later they have become free-swimming nauplius larvae. No further cell division occurs throughout this period of morphogen-

[1] C. H. Ellis, The Genetic Control of Seaurchin Development: A Chromatic Study of Protein Synthesis in *Arbacia punctulata* Embryos, *J. Exp. Zool.*, **163**:15 (1966).

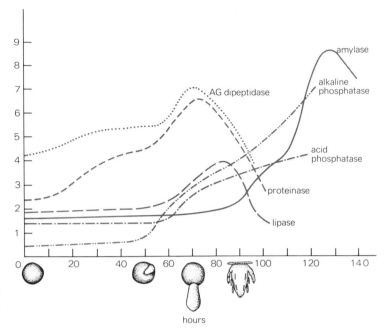

FIGURE 12.5 Enzyme activity, expressed in arbitrary units, during the embryonic development of *Artemia salina*. Development stages of particular significance are illustrated. [*After Urbani, 1962.*]

esis. In other words, the spherical mass of cells assemble in such a way that a highly organized, specifically structured larva is formed. The total cell number remains at about 4,000 until after the free-swimming stage is produced, after which cell proliferation and larval growth proceed. Accordingly, the *Artemia* embryo provides a system in which the relationship between protein synthesis and differentiation can be examined without the added complexity of concomitant cell division.

Dormant encysted eggs are packed with ribosomes, all of which are monosomes and are unattached to endoplasmic membranes. Protein synthesis, however, begins as soon as the eggs are hydrated, and polyribosome formation occurs rapidly at the expense of monosomes after dormancy ends in the encysted embryo; small polyribosomes are evident within a few minutes of reactivation, with heavy polyribosomes appearing as development proceeds. The sequential attachment of ribosomes to messenger RNA is suggested. This is clear evidence for the existence of stored messenger RNA.

EQUIVALENCE OF NUCLEI

The regional specialization of cells in a developing system has been a problem to developmental biologists from the beginning. Weismann proposed that such cell diversification arises in development through differential, or qualitative, division of nuclei during the course of cleavage. The early development of the *Ascaris* egg (already described, page 245) supported this view. In them the nuclei of prospective somatic cells undergo chromosomal diminution, whereas those of prospective reproductive cells do not. The concept was soon abandoned when the work of cytologists made the exquisite, equal division of the chromosomes between two daughter cells overwhelmingly clear. This led to the dogma that nuclei descended from a zygote nucleus during development are equivalent in every way.

Driesch demonstrated the equivalence of nuclei, at least up to the 16-cell stage of the sea urchin, by subjecting a fertilized egg to pressure between two glass plates so that the cleavage of the flattened egg resulted in a flat, disc-shaped aggregate of blastomeres instead of the normal sphere arrangement. In this experiment some of the nuclei which would normally have been located in endodermal cells in the vegetal region come to lie in ectodermal cells, and vice versa. A rounding up of the cell mass follows release from pressure, with subsequent normal development. In a later, comparable experiment consisting of ligaturing a fertilized but uncleaved salamander egg so that the zygote nucleus becomes confined to one half, cleavage followed only in the part containing the nucleus. When, as the result of the successive divisions, the cells near the ligature became sufficiently small, the nucleus from one slipped into the nondividing half, which then underwent cleavage and development. On the other hand the discovery that in centrifuged eggs of *Ascaris,* chromosome diminution or the absence of diminution depends on the particular cytoplasmic environment in which a nucleus is thrown, shows that a nucleus is responsive to cytoplasmic conditions.

THE ROLE OF THE NUCLEUS

The role of the nucleus in development, particularly during early stages, is a subject of debate and continuing investigation. Cleavage can proceed to some extent in the absence of the nucleus, as shown by strongly centrifuging unfertilized sea urchin eggs so that they pull into two parts, only one of which contains a nucleus; subsequent experimental activation of the nonnucleate part causes cytasters to appear, and a number of divisions occur. Frog eggs fertilized by sperm previously exposed to a heavy dose of x-rays (which inactivates the nucleus), followed by removal of the egg nucleus, divide to form an early blastula stage by virtue of the sperm centriole. Cleavage is possible, therefore, without the presence of nuclei, but it fails to continue beyond a few divisions. Without a nucleus there is no truly significant development.

The egg nucleus at the beginning of development moreover may be haploid, diploid, or polyploid, all conditions being able to support more or less normal development. Whatever the number of chromosome sets, it is vital that each cell formed during cleavage receive the normal complement. Boveri long ago showed that abnormal assortment of chromosomes in the sea urchin egg, resulting from its penetration by

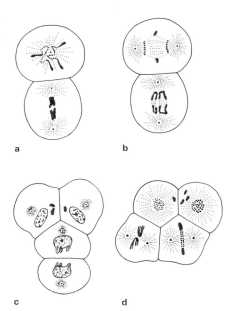

a   b

c   d

FIGURE 12.6 *Ascaris.* Development of an egg from the two-cell stage to the end of the four-cell stage. (a) Two-cell stage, upper cell undergoing chromosome diminution, lower cell showing normal chromosome behavior. (b) Two-cell stage, slightly later than in (a), upper cell in side view. The ends of the diminished chromosomes remain in the center of the cell between the separating chromosome fragments. (c) Four-cell stage, resting nuclei. In the upper two cells, degenerating chromosome ends lie outside the nuclei. In the lower two cells, the undiminished chromosomes are recognizable by the irregular shape of the nuclei. (d) Four-cell stage, nuclei dividing. Upper cells show the earlier diminished chromosomes; lower right cell undergoes diminution; lower cells show normal chromosome behavior. The lower left cell corresponds to the lowermost cell of (c), which has changed its position. [*After Boveri, 1899.*]

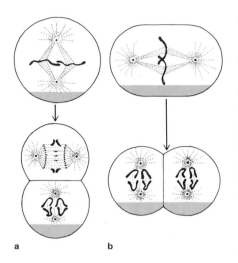

FIGURE 12.7 *Ascaris.* (a) Normal development, first and second divisions. There is diminution in upper cell. (b) Abnormal development in a centrifuged egg, first and second divisions. There is no diminution in either cell. The horizontal lines indicate regional differentiation of the cytoplasm. [*Diagrammatic, after Boveri, 1910.*]

two sperm and the consequent formation of three or more astral centers and spindles in the egg before cleavage, gives rise merely to a mass of cells. He concluded that a single lagging chromosome could throw the developing system out of kilter.

The degree of ploidy also has an effect, although associated with more or less normal development. The degree of ploidy is conveniently indicated, at least in salamanders and some other animals, by the number of nucleoli per cell. (Each set of chromosomes has a single nucleolus.) Among many invertebrates both haploid and diploid eggs develop naturally from unfertilized and fertilized eggs to form male and female individuals, respectively, as in *Daphnia* and other cladoceran crustaceans. In salamanders a haploid egg may also develop to form a normal larva, although more commonly it becomes a larva showing stunted growth, microcephaly, and some swelling. This is equivalent to a homozygous state with all faults expressed.

None of the several explanations offered for such abnormality of development has been accepted, and the problem remains. The development of polyploid eggs with chromosome sets ranging from three to seven has been studied mainly by Fankhauser. All eggs yield larvae of the same size and normal shape (although the higher polyploids show depressed growth rate and viability). This is surprising since the size of the cells constituting the various tissues and structures of the developed larvae varies according to their nuclear constitution.

Gastrulation, neurulation, and organ formation of haploids and polyploids all proceed normally in time and character. This means that the critical events resulting in change of form, rearrangement of tissues, construction of complex organs, etc., are cytoplasmically determined and take place at predetermined times following fertilization (environmental conditions being constant), irrespective of the rate of cleavage. But cell size varies directly with the degree of ploidy, and tissues and structures of the same size are formed from progressively larger and fewer cells as the degree of ploidy increases. For example, cross sections of kidney tubules (pronephric) of haploid, diploid, and pentaploid larvae show eleven, six, and three cells, respectively, forming tubules of the same diameter and wall thickness. Accordingly, neither the number of nuclei nor the number of cells present at a particular time or point in the differentiation or formative process significantly affects the course of development, except that, as we shall see elsewhere, cell number at any stage of development is important in a permissive sense, i.e., an insufficiency of cells may prevent a particular tissue movement from taking place. What governs the progression and nature of the formative events is one question, and what governs cell size is another.

TRANSPLANTATION AND POTENCY OF NUCLEI

The present approach to the question is accordingly based on an acceptance of the general concept that the nuclei of a multicellular individual organism are basically equivalent but may be subtly modified by their cytoplasmic surroundings. Most recent work is based on a nuclear transplantation technique, evolved by Briggs and King, employing frog eggs and embryos (*Rana pipiens*). The nucleus of an unfertilized egg, which is conveniently located just beneath the animal pole, is effectively removed with a needle or inactivated by ultraviolet light. A substitute nucleus is obtained by using a micropipette with a diameter less than that of the particular donor cell, so that the cell is ruptured and the isolated nucleus is obtained. The donor cell may be in the early or late blastula stages, the gastrula stage, the neurula stage, or even from mature tissues of the functioning tadpole or frog. The last and most difficult step in the procedure involves the insertion of the donor-cell nucleus into the uncleaved enucleated egg. This technique has been extended to serial nuclear transplantation, i.e., donor nuclei, instead of being taken from cells of an embryo developed from a fertilized egg, are taken from an embryo resulting from a nuclear transplantation. The nuclei of such cells have an identical set of genes. A further refinement of the procedure involves the use of donor nuclei which can be distinguished from host nuclei, so that there can be no confusion concerning results. One of the most useful nuclear markers is a feature of the nuclei of a mutant strain of the clawed frog *Xenopus laevis*, which contain a single nucleolus instead of the two nucleoli typical of most frogs.

In the experiments of Briggs and King, nuclei were extracted from embryos, all of *Rana pipiens*, in various developmental stages, and injected into the enucleated egg of the same species. Nuclei taken from blastula cells (from the animal hemisphere) supported development which resulted in normal tadpoles in 80 percent of the transplants. Nuclei derived from gastrula cells were less supportive, while nuclei derived from endoderm cells of the tail-bud stage of the donor embryo yielded normal development to the tadpole in only 7 percent of the transplants. Many such experiments, including serial transfers, gave much the same results, i.e., the percentage of transplants showing normal development declines as the age of the donor nuclei

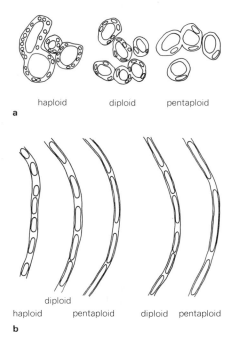

haploid diploid pentaploid

a

diploid

haploid pentaploid diploid pentaploid

b

FIGURE 12.8 Maintenance of normal structure in heteroploid tissues. (**a**) Cross sections of pronephric tubules from a haploid larva (35 days old), from a diploid, and from a pentaploid. Size of tubules and diameter of wall remain approximately the same in spite of differences in cell size, through changes in cell shape. (**b**) (*Left*) Small portions of lens epithelium of the same haploid, diploid, and pentaploid larvae. Thickness of epithelium remains nearly constant. (*Right*) Corresponding portions of lens epithelium of an older diploid larva and of a pentaploid of the same age and stage. Cells and nuclei of the pentaploid again flattened to about same diameter as in the diploid. [*After Fankhauser*, Quart. Rev. Biol., **20**:*20* (*1945*).]

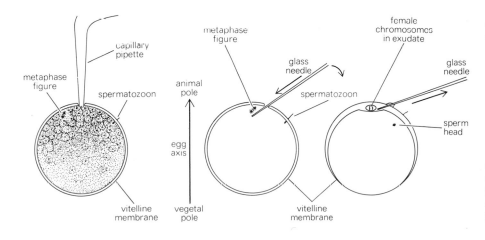

metaphase figure

capillary pipette

spermatozoon

metaphase figure

animal pole

glass needle

spermatozoon

female chromosomes in exudate

glass needle

sperm head

egg axis

vitelline membrane

vegetal pole

vitelline membrane

FIGURE 12.9 Enucleation of frog egg by means of capillary pipette or glass needle. Another method is to kill the egg nucleus by exposure to ultraviolet radiation. [*After Rugh*, "*Experimental Embryology*," *p. 180, Burgess, 1962.*]

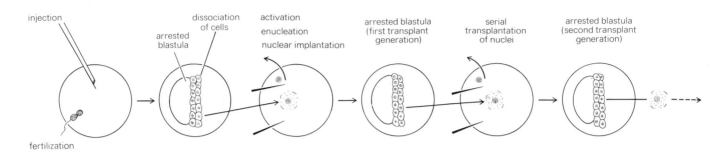

FIGURE 12.10   Serial transplantation of nuclei, from an embryo with nucleus derived from an egg arrested at the blastula stage. In each generation the development of the egg is arrested at the blastula stage, indicating that a permanent change has occurred during the cell-differentiation process, possibly the loss of certain genes unnecessary for at least certain specific types of cell differentiation but necessary to sustain the morphogenetic processes commencing with gastrulation. [*After Markert, "The Nature of Biological Diversity," p. 110, McGraw-Hill, 1963.*]

increases. These investigators concluded that nuclei undergo change in character as the blastula transforms into the gastrula. Other kinds of evidence certainly show that nuclear activity changes at or near the onset of gastrulation.

The permanence or semipermanence of such change is doubtful, however, partly because of similar experiments carried out by Fischberg and by Gurdon, on the African clawed frog *Xenopus laevis*. Any exposure of a naked nucleus to an alien medium is most harmful, and as long as *some* nuclei from any particular embryonic or tissue source support normal development, it is at least possible that those that do not do so have merely been damaged during the transfer, and that the degree of damage is propagated in the serial transfer procedure. Another interpretation of the relatively small percentage of nuclear-transplant development is an initial incompatibility between the very slow rate of division of differentiating cells (once in every two or more days) and the rapid rate of division of the egg (roughly once in every hour); if the injected nucleus cannot adjust to the increase in replication rate, it will suffer great harm in the division process.

In *Xenopus,* not only do nuclei from gastrula cells support normal development in a significant percentage, but nuclei derived from fully differentiated epithelial cells of the tadpole intestine have also been shown to do so. Both male and female adult clawed frogs, fertile and completely normal, have been produced by eggs containing only a transplanted intestine nucleus; the resultant adults carry the nucleolar marker in all their cells. In serial transfer experiments starting with a single intestine nucleus, 70 percent of the transplant embryos developed at least as far as the muscular-response stage, in which both muscle and nerve cells are necessarily differentiated and functional. Results such as these show that normal differentiation of cells, at least in epithelial cells of the intestine and, of course, in sperm cells themselves, does not involve permanent inactivation of genes.

More surprisingly, nuclei from frog kidney tumor can support normal development, whether the nuclei are diploid or triploid. Triploid nuclei serve as nuclear markers, although in both cases normal tadpoles are produced, which undergo metamorphosis. Triploid tadpoles are obtained by subjecting newly fertilized eggs to hydrostatic pressure sufficient to suppress the formation of the second polar body.

PREPARATION OF
DONOR CELLS

INJECTION OF DONOR NUCLEUS
INTO RECIPIENT EGG

PREPARATION OF
RECIPIENT EGG

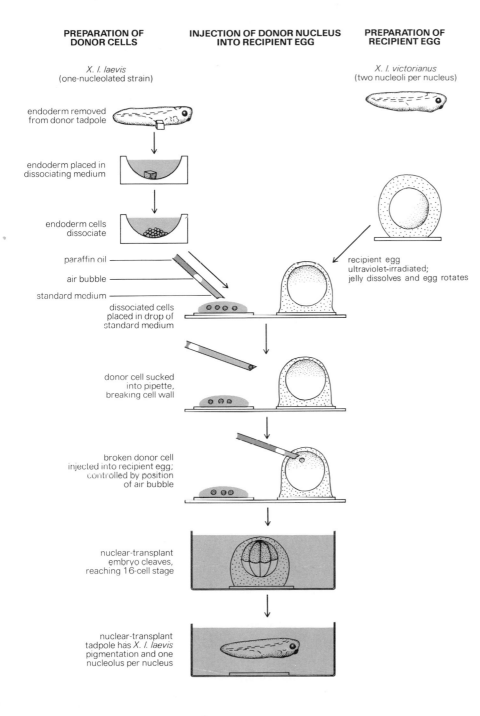

X. l. laevis
(one-nucleolated strain)

endoderm removed
from donor tadpole

endoderm placed in
dissociating medium

endoderm cells
dissociate

paraffin oil
air bubble
standard medium

dissociated cells
placed in drop of
standard medium

donor cell sucked
into pipette,
breaking cell wall

broken donor cell
injected into recipient egg;
controlled by position
of air bubble

nuclear-transplant
embryo cleaves,
reaching 16-cell stage

nuclear-transplant
tadpole has *X. l. laevis*
pigmentation and one
nucleolus per nucleus

X. l. victorianus
(two nucleoli per nucleus)

recipient egg
ultraviolet-irradiated;
jelly dissolves and egg rotates

FIGURE 12.11 The principal stages involved in transplanting nuclei in *Xenopus laevis*. Donor nuclei are taken from a strain of the subspecies *Xenopus laevis laevis*, which has only one nucleolus in each nucleus. The young tadpoles of this subspecies have many pigment cells on their body. Recipient eggs are taken from the subspecies *Xenopus laevis victorianus*, in which nuclei of most diploid cells have two nucleoli and in which the young tadpoles have no body pigment. The nuclear-transplant tadpoles have the characteristics of the nuclear parent, not of the cytoplasmic one, showing that their nuclei are derived from the transplanted nucleus. [*After Gurdon*, Endeavour, **25**:96(*1966*).]

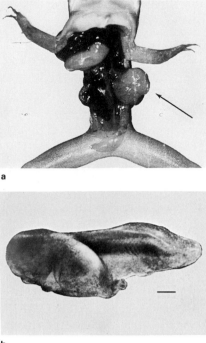

a

b

FIGURE 12.12 (a) Left renal tumor (arrow) of recently metamorphosed triploid frog. (b) Tadpole
developed from egg with transplanted triploid
tumor nucleus, with well-formed head, body, and
tail (scale = 1 mm). [*Courtesy of R. G. McKinnell.*]

Triploid tumors are induced in such tadpoles by injecting them with extract of frog
renal adenocarcinoma. Nuclei are then obtained from dissociated cells of this tissue,
and these when transplanted to enucleated eggs can support full development. Following metamorphosis the majority of both diploid and triploid frogs obtained by these
means develop massive renal adenocarcinoma. The tumor cells accordingly retain
a multipotent genome capable of replacing the zygote nucleus of the egg. The tumor-
inducing virus is inevitably transmitted during nuclear transplantation.

Nuclei transferred to a new cytoplasmic environment change their function
within an hour or two to the new circumstances by assuming the condition and activity
characteristic of the nucleus that has been replaced:

1   This is seen in normal fertilization when, after penetration, the compact sperm
    nucleus swells and assumes the appearance of the female pronucleus.
2   Nuclei from cells of the mid-blastula stage, which normally synthesize DNA but
    little RNA, when injected into the unfertilized egg (mature oocyte) synthesize RNA
    but not DNA. When these nuclei are injected into an activated egg, DNA is
    synthesized.
3   Neurula nuclei which synthesize all kinds of RNA stop doing so when first placed
    in cytoplasm of an uncleaved egg, but again synthesize transfer ribosomal and
    heterogeneous RNA when transplant embryos are reared through the gastrula and
    neurula stages.
4   Nuclei from frog brain cells, which normally synthesize RNA and never divide,
    switch from RNA to DNA synthesis when injected into activated eggs; the same
    nuclei injected into maturing oocytes cease nucleic acid synthesis altogether and
    are rapidly converted into groups of chromosomes and spindles.

It is very evident that the state or nature of the cytoplasm controls the particular
state of the nucleus as long as the cytoplasmic condition persists.

In all cases, whether an injected nucleus is of the type to support development
or not, the immediate visible change is a pronounced swelling. The sperm nucleus
increases about fiftyfold in volume in half an hour, to become 20 μ in diameter,
while the egg pronucleus swells to the same size at the same time. Blastula nuclei,
still comparatively large, increase in volume about threefold when injected into an
unfertilized egg. Nuclei from tadpole intestinal epithelium swell about fortyfold, while
brain cells swell about sixtyfold. How such swelling occurs is not known. During
enlargement, the chromatin becomes dispersed and some cytoplasmic protein passes
into the swelling nucleus. The primary effect of egg cytoplasm is to promote a single
reaction (the polymerization of deoxynucleotide triphosphates into DNA by DNA
polymerase), the product of which is easily recognized biochemically or autoradio-
graphically by incorporation of tritiated thymidine.

Gurdon's hypothesis is that nuclear swelling and chromatin dispersion facilitate
the association of cytoplasmic regulatory molecules with chromosomal genes, thereby
leading to a change in gene activity of a kind determined by the nature of the molecules

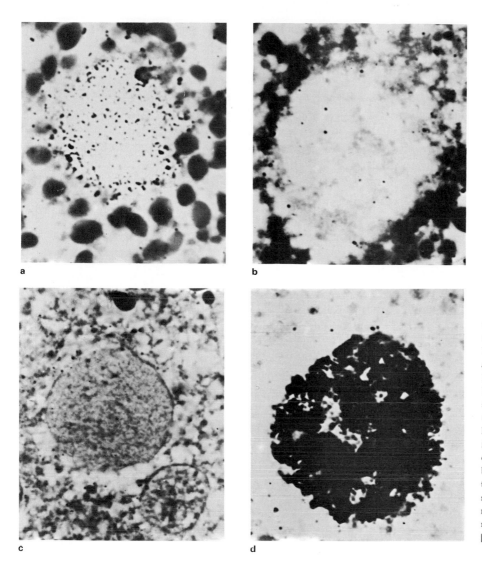

FIGURE 12.13 RNA and DNA of transplanted nuclei. Mid-blastula nuclei, which synthesize DNA but not RNA, are injected into (a) an oocyte and (b) an egg. Oocytes normally synthesize RNA but not DNA; eggs normally synthesize DNA but not RNA. Nuclei injected into the oocyte synthesize RNA, but similar nuclei injected into the egg do not. Black granules are RNA precursors labeled with a radioactive isotope. (c,d) Results following injection of brain nuclei into oocyte and egg, respectively. Frog brain nuclei synthesize RNA but rarely DNA. In the oocyte where RNA synthesis is progressing, brain nuclei do not synthesize DNA, but in the egg the injected brain nuclei switch from RNA to DNA synthesis in response to conditions in the host-cell cytoplasm. [Courtesy of J. B. Gurdon.]

that enter the nucleus. Nuclear genes are not necessarily lost or permanently inactivated in the course of cell differentiation, and major changes in chromosome function as well as different kinds of gene activity can be experimentally induced by normal constituents of living cell cytoplasm. The identity of such cytoplasmic components and their mode of action are now under investigation. The changes induced experimentally in transplanted nuclei greatly resemble those seen in nuclei normally reconstituted following mitosis, and the two processes may well be equivalent. It is possible that in normal development, during the telophase of mitosis, chromosomes are repro-

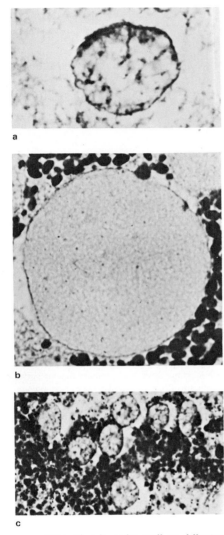

FIGURE 12.14 Blastula nuclei swell up following their injection into (a) an egg or (b) an oocyte. The nucleus in the oocyte enlarges 200 times. (c) Blastula nuclei of normal dimensions. [*Courtesy of J. B. Gurdon.*]

grammed with regard to potential gene activity by association with cytoplasmic proteins, for interphase nuclei show no such change except when enlarging in a new cytoplasmic environment.

To what extent chromosomal change takes place in varying cytoplasmic environments, apart from the gross diminution in somatic cells of *Ascaris,* is known only in dipteran insects whose salivary glands have nuclei with giant banded chromosomes that exhibit localized areas of enlargement, or puffs. Only in *Drosophila* are the chromosomes well enough known to give any prospect of identifying chemical changes at particular genetic sites. The distribution of puffs along the chromosomes exhibits distinctive patterns for the cell type and for the stage of its differentiation. The dependence of nuclear activity on the cytoplasmic environment is shown by transplanting nuclei from drosophila salivary gland cells into younger cytoplasm obtained from embryos of different ages. In each case the puffing patterns change and are characteristic for each cytoplasm.

DIFFERENTIAL GENE ACTION

From the foregoing it is clear that nuclear transplantation is a stringent test for nuclear differentiation during development. Yet the examples given concerning the equivalence or totipotency of nuclei in a developing system prove not that the genetic material of differentiated cells is unchanged, but rather that it has not been irreversibly altered. Moreover, since we cannot yet with certainty write off the high percentage of failure of transplanted nuclei to support normal development, the choice between two hypotheses concerning the nature of gene action remains. The preponderance of evidence favors "differential gene action," which postulates that all types of cells within a single organism contain the same genes in equal number, and that cells of one tissue differ from those of another tissue according to which group of genes is expressed. The alternative is "differential gene alteration," which proposes that one cell type differs from another as a result of a true modification of the genes themselves, such as a modification of DNA bases. Although evidence in support of the former view continues to grow, it is conceivable that both processes occur and that differential gene action need not be an entirely exclusive means of control.

Histones, the basic nuclear proteins of animal and plant cells, seem to be involved in the regulation of gene activity; however, the nonbasic proteins may play a more important role in this respect. Histones inhibit DNA-primed RNA synthesis in such unrelated forms as bacteria, pea, and mammal, and generally diminish the activity of DNA polymerase. It has been suggested that structural genes are "turned off" before gastrulation because they are associated with histones, and are later "turned on" by changes in the DNA-histone association. However, the experimental inhibitory effect of histones during the period from cleavage to the blastula stage suggests that genes are turned on to a large extent and are unselectively inhibited by histone treatment. In other words, histones generally have an essentially negative influence, and some other source of control is called for. Nonhistone proteins occurring on chromosomes have been suggested for this role, and the model proposed for gene regulation is that histones serve as repressors, nonhistones as derepressors, and in early embryogenesis these derepressions are localized in the cytoplasm.

There is need to be clear about what exactly is meant by differential gene action, as well as when a gene is considered to be turned on or turned off. Both the synthesis and activity of a gene are controlled at many different levels, from chromosomal to cytoplasmic, and in the analysis of development it becomes important to distinguish between the time of synthesis of enzymes (which may or may not be immediately active) and the time of activity of such enzymes and the synthesis of specific proteins.

According to one hypothesis, the chromomeres are the important units of activity. As seen in the lampbrush chromosomes of amphibian oocytes, the coiled, compact chromomere granule unwinds into a loop on which the DNA-dependent RNA activity (i.e., gene activity) occurs. Whatever inhibits loop formation, or chromomere unwinding, inhibits the activity of the locus involved, actinomycin D being an example of such an inhibitor. Schultz sums up a discussion of this concept as follows: "What are the lessons from this analysis? They are, essentially, that the control mechanisms for the functioning of individual loci are highly integrated, as expected from the product of a long evolutionary history. Disturb them, and the immediate neighborhood of a locus determines some new activity, the normal conditions having been disturbed. The coordination itself is dependent, not only on what happens in the immediate neighborhood, but on the balance of gene activity which the optimum function requires. The primary molecular phenomenon concerned with these genetic activities is most probably the compaction and extension of the template. . . . When the template is compacted, replication is inhibited and a loss of information ensues. Extend it and, according to the geometry, disproportionate replication may take place or transcription be facilitated, according to the type of activity of the genome."[1]

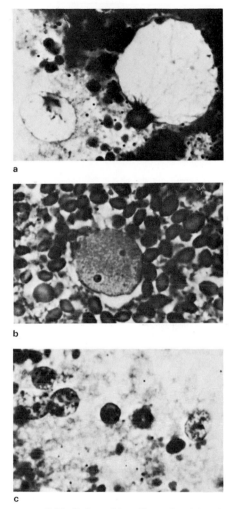

FIGURE 12.15 Brain nuclei swell up when injected into (a) an egg or (b) an oocyte, though less so than do injected blastula nuclei (see Figure 12.14). (c) Brain nuclei of normal size. [*Courtesy of J. B. Gurdon.*]

NUCLEAR ACTIVITY AT GASTRULATION

As a rule the first signs of morphogenesis in a developing egg are the tissue movements leading to gastrulation. Gastrulation may be invaginative, as in small, sparsely yolky eggs, or it may be accomplished in some other way, depending on the nature of the egg. Essentially it is associated with differentiation to form the primary germ layers, namely, the inner endomesoderm and the outer ectoderm. Whatever the form

[1] J. Schultz, Genes, Differentiation, and Animal Development, Brookhaven National Laboratory, Symposium in Biology, no. 18, p. 143, 1965.

of gastrulation, the onset of the event is marked by change in nuclear activity, particularly the first synthesis of ribosomal RNA, which has been shown in such diverse animals as brine shrimp, insect, and frog. In the sea urchin new ribosomal RNA is first detectable in the cytoplasm at the mid-gastrula. A variety of old and new experiments, the more recent studies having been made mostly on developing *Xenopus* (clawed frog) and sea urchin eggs, clearly indicate that newly synthesized gene products are essential for normal gastrulation. In outline, the sequence of RNA synthesis in *Xenopus* is as follows:

1   No nuclear RNA synthesis is detectable during the first 10 cleavages (up to the early blastula stage consisting of about 1,000 cells).

2   At the mid-blastula stage, large RNA molecules, probably including messenger RNA but no ribosomal RNA, are synthesized.

3   In the late blastula stage (consisting of about 30,000 cells), transfer RNA is first detectable.

4   Ribosomal RNA first appears at the start of gastrulation.

The conclusion is that without continuous nuclear guidance the late blastula cells cannot proceed with morphogenesis. Nuclear activity increases dramatically between the blastula and gastrula stages, when the per-nucleus rate of high-molecular-weight, nonribosomal RNA increases more than twentyfold within 1 hour in presumptive endoderm and mesoderm immediately preceding gastrulation, although the presumptive ectoderm becomes activated somewhat later. That gastrulation is a critical phase of development has been known for a long time. At this stage, many lethal mutations are known to kill, hybridization in amphibians often leads to developmental arrest, and actinomycin arrests development in frog, sea urchin, and ascidian, although it is not known at what point in the biosynthetic chain the block occurs. Apparently many types of newly synthesized proteins appear during this period, representing activation of a *program* of structural genes, not of just a few. The process of gastrulation in fact is not only critical in development but will remain a focus of analytical investigation for a long time to come.

CONCEPTS

In early development of the egg the first proteins are made from templates present in the unfertilized egg, coincidently with the conversion of free ribosomes to polyribosomes.

The first wave of protein synthesis does not involve formation of new kinds of proteins but relates to protein necessary for new cell surface and for mitotic spindles.

Adult-type proteins first appear shortly before the first indication of tissue specialization and formative movements; they are preceded by intense DNA synthesis.

Nuclei remain totipotent, and are equivalent, throughout cleavage stages and may either persist in this state or revert to it.

The state of the nucleus with regard to general or local nuclear activity is determined by the state of the surrounding cytoplasm.

Genes are turned on or off according to the degree of puffing or swelling of the chromatin.

Following cleavage, cells cannot proceed with morphogenesis without nuclear guidance.

READINGS

BOVERI, T., 1902. On Multipolar Mitosis as a Means of Analysis of the Cell Nucleus, in B. Willier and J. Oppenheimer (eds.), "Foundations of Experimental Embryology," 1964, Prentice-Hall.

BRIGGS, R. W., and T. J. KING, 1952. Transplantation of Living Nuclei from Blastula Cells into Enucleated Frog Eggs, *Proc. Nat. Acad. Sci.,* **38**:455–463.

CHEN, P. S., 1967. Biochemistry of Nucleo-cytoplasmic Interactions in Morphogenesis, in R. Weber (ed.), "The Biochemistry of Animal Development," Academic.

DAVIDSON, E. H., 1968. "Gene Activity in Early Development," Academic.

DENIS, H., 1968. Role of Messenger Ribonucleic Acid in Embryonic Development, *Advance. Morphogenesis,* **7**:115–152.

ELLIS, C. II., 1966. The Genetic Control of Seaurchin Development: A Chromatic Study of Protein Synthesis in *Arbacia punctulata* Embryos, *J. Exp. Zool.,* **163**:1–22.

FANKHAUSER, G., 1955. The Role of Nucleus and Cytoplasm in Development, in B. Willier, P. Weiss, and V. Hamburger (eds.), "Analysis of Development," Saunders.

GOLUB, A., and J. S. CLEGG, 1968. Protein Synthesis in *Artemia salina* Embryos, I: Studies on Polyribosomes, *Develop. Biol.,* **17**:644–656.

GOULD, M. C., 1969. RNA and Protein Synthesis in the Unfertilized and Fertilized Eggs of *Urechis caupo, Develop. Biol.,* **19**:460–481.

GRAHAM, C. F., and R. W. MORGAN, 1966. Changes in the Cell Cycle during Early Amphibian Development, *Develop. Biol.,* **14**:439–460.

GROSS, P. R., L. I. MALKIN, and W. A. MOYER, 1964. Templates for the First Proteins of Embryonic Development, *Proc. Nat. Acad. Sci.,* **51**:407–414.

GURDON, J. B., 1968. Transplanted Nuclei and Cell Differentiation, *Sci. Amer.,* Dec.

—— and H. R. WOODLAND, 1968. The Cytoplasmic Control of Nuclear Activity in Animal Development, *Biol. Rev.,* **43**:233–267.

HARRIS, H., 1968. "Nucleus and Cytoplasm," Oxford University Press.

HULTIN, T., and J. E. MORRIS, 1968. The Ribosomes of Encysted Embryos of *Artemia salina* during Cryptobiosis and Resumption of Development, *Develop. Biol.,* **17**:143–164.

KING, R. L., and H. W. BEAMS, 1938. An Experimental Study of Chromatin Diminution in *Ascaris, J. Exp. Zool.,* **77**:425–443.

KING, T. J., and R. W. BRIGGS, 1956. Serial Transplantation of Embryonic Nuclei, in R. A. Flickinger (ed.), "Developmental Biology," 1966, Wm. C. Brown.

KROEGER, H., 1960. The Induction of New Puffing Patterns by Transplantation of Salivary Gland Nuclei into Egg Cytoplasm of Drosophila, Chromosoma, 11:129–145.

MANO, Y., 1970. Cytoplasmic Regulation and Cyclic Variation in Protein Synthesis in the Early Cleavage Stage of the Sea Urchin Embryo, Develop. Biol., 22:433–460.

MCKINNELL, R. G., B. A. DEGGINS, and D. D. LABAT, 1969. Transplantation of Pluripotential Nuclei from Triploid Frog Tumors, Science, 165:394–395.

MONROY, A., 1960. Incorporation of S^{35} Methionine in the Microsomes of Soluble Proteins during the Early Development of the Sea Urchin Egg, Experientia, 16:114–117.

—— and P. R. GROSS, 1967. The Control of Gene Action during Echinoderm Embryogenesis, in R. Weber (ed.), "Morphological and Biochemical Aspects of Cytodifferentiation," Karger.

—— and R. MAGGIO, 1964. Biochemical Studies on the Early Development of the Seaurchin, Advance. Morphogenesis, 3:95–146.

SCHULTZ, J., 1965. Genes, Differentiation, and Animal Development, Brookhaven National Laboratory, Symposium in Biology, no. 18, pp. 116–147.

SMITH, L. D., 1966. The Role of a "Germinal Plasm" in the Formation of Primordial Germ Cells in Rana pipiens, Develop. Biol., 14:330–347.

SPIEGEL, M., E. S. SPIEGEL, and P. S. MELTZER, 1970. Qualitative Changes in the Basic Protein Fraction of Developing Embryos, Develop. Biol., 21:73–86.

THALER, M. M., M. C. L. COX, and C. VILLEE, 1969. Actinomycin D: Uptake by Seaurchin Eggs and Embryos, Science, 164:832–834.

TYLER, A., and B. S. TYLER, 1966. Physiology of Fertilization and Early Development, in R. A. Boolootian (ed.), "Physiology of Echinodermata," Wiley.

URBANI, E., 1962. Comparative Biochemical Studies on Amphibian and Invertebrate Development, Advance. Morphogenesis, 2:61–108.

WEISZ, P., 1946. The Space-time Pattern of Segment Forming in Artemia salina, Biol. Bull., 91:119–140.

WHITELEY, H. R., B. J. MCCARTHY, and A. H. WHITELEY, 1970. Conservatism of Base Sequences in RNA for Early Development of Echinoderms, Develop. Biol., 21:216–242.

CHAPTER THIRTEEN

THE EGG CORTEX
AND DEVELOPMENT

Most eggs studied so far (particularly holoblastic eggs such as various echinoderm, mollusk, annelid, and amphibian eggs) have two principal regions in common: (1) a cortex, which is relatively stable to centrifugation, and (2) an endoplasm, which is displaceable, containing yolk and other inclusions that exhibit a graded distribution along an axis. This description, however, holds for many other kinds of cells, notably *Amoeba*, and merely affords a point of departure in this discussion.

THE CORTEX

In view of the lability of so many kinds of activated but uncleaved eggs, perhaps it should not be surprising that so little tangible organization of either the cortical or endoplasmic region of the egg is discernible, at least so far as it relates to development. Yet many developmental biologists believe that the cortex of the egg is possibly the most important agent in primary development. Both polarity and dorsoventral organization, as we have seen, have been related to gradients or fields depending on cortical structure. Also, there are many indications that the localization of nuclei and spindles is dependent on properties of the cortex.

The main argument for this dependency is that although no structure internal to the cortex can resist severe centrifugation, perfect morphogenesis still ensues. For instance, sea urchin and clam eggs centrifuged at several thousand times gravity for 5 minutes show, at least when examined with the electron microscope, that the only structure not affected is the plasma membrane. Much depends, therefore, on what the cortex consists of. In neither sea urchin nor clam egg can any material underlying the plasma membrane be seen in electron micrographs that is different from the inner cytoplasm.

Yet strong evidence has been found through microdissection of sea urchin eggs by various workers, including Robert Chambers, who first developed and exploited microdissection apparatus, that a gel-like layer is present, either in direct contact with the plasma membrane or in actual continuity with it, and that this layer and the membrane act together as a unit, between 1.6 and 5 μ thick. This plasmagel layer has obvious elastic properties, visible through manipulation under the light microscope, although it is structurally transparent to the electron microscope.

When eggs (whether of sea urchin, clam, or amphibian) divide, movements or displacements of the plasma membrane always affect the underlying cytoplasm, which is carried into the cleavage furrow by the stretching or growing membrane. Therefore, with some assurance in spite of failure to detect cortical ultrastructure

CONTENTS

The cortex

Bilateral symmetrization

Patterns of differentiation

Determination and cleavage

Gastrulation and topographical reorganization

Synthesis in regulative and determinative development

Duality of development

Reassembly of a multicellular organism

Concepts

Readings

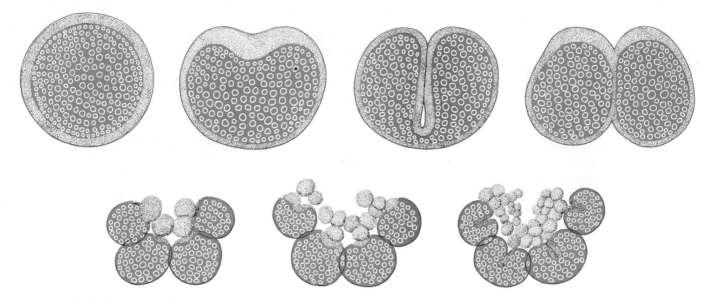

FIGURE 13.1   Redistribution of cortical (micromere-forming) cytoplasm during cleavage of the cteno-phore *Beroë ovata*, as seen in dark field. The ani-mal pole is uppermost in each stage. [*After Spek, Arch. Entwickl.-mech.,* **107**:54 (1926).]

electron-microscopically, we may define the egg cortex as the plasma membrane plus some underlying gel-like or highly viscous cytoplasm. Chemical studies suggest that this cortical layer as a whole consists of a colloid complex in which proteins, phosphatides, and probably ribonucleic acid take part, with calcium ions being essential for stability and physiological integrity.

Cortical localizations, in fact, are extremely evident in the very early developmental stages of living eggs of certain animals, particularly with regard to polar plasm and bilateral symmetrization. This was shown by Spek's studies of the living egg and cleavage stages of the ctenophore *Beroë ovata.* Conklin's earlier observations of the yellow crescent of cortical material in the fertilized egg and cleavage stages of the ascidian *Styela partita,* marking the axis of embryonic symmetry, are comparable. The primary question is whether the pattern of cleavage determines or is determined by the regional cytoplasmic differences, or whether both are perhaps determined by some other agency. Events in the cortical region, however, give definite indication of an active transitional zone between plasma membrane and the egg endoplasm.

This cortical zone appears brilliant green in *Beroë* under dark-field illumination and forms a well-defined, uniform layer between membrane and endoplasm. (See Needham, "Biochemistry and Morphogenesis" for color reproductions.)

Cleavage furrows begin near the animal pole, associated with the eccentric location of the nucleus and spindle in the animal hemisphere. As the furrow cuts down through the egg, the green material is swept towards the vegetal pole, only to re-form uniformly. It is dislocated by the cleavage process but returns to its original state. The same procedure follows during the second cleavage. In the third cleavage, also vertical, the green material remains at the animal pole, and the nucleus and spindle move further towards that pole, becoming aligned with the polar axis; the fourth cleavage cuts off the green caps to form the first octets of animal-pole micromeres. Thus, in this cortical fourth cleavage, regional segregation of cortical material precedes the division process, and, if a choice has to be made, we must say that the cortical segregation process controls the location and orientation of the division apparatus.

BILATERAL SYMMETRIZATION

Bilateral symmetry is characteristic of most kinds of animals. In chordate development bilateral symmetry develops early or late depending on the type of egg. Chordate eggs fall into three categories: **(1)** the very small, translucent eggs of the protochordate ascidians and amphioxus, of a size comparable to the freely shed eggs of many invertebrates; **(2)** the much larger but otherwise very similar eggs of several groups of vertebrates, namely, the lamprey, sturgeon, lobe-finned fish, and lungfish, and amphibians, all of which lay a distinctive type of egg that we may call the primitive freshwater vertebrate egg; and **(3)** the very yolky eggs of most other vertebrates. The first two types are holoblastic, i.e., cleavages are complete, and such yolk as they possess is consequently distributed among the individual cells. The third type is meroblastic, with cleavage restricted to a polar cap of cytoplasm, so that a sheet of dividing cells forms at the surface of a mass of undivided yolk. The development of meroblastic eggs generally, which includes such invertebrate kinds as those of the cephalopod mollusks and the insects, is so different from that of holoblastic eggs that it calls for separate treatment. In brief, however, holoblastic chordate eggs establish the bilaterality of the future organism essentially before cleavage of the egg begins, whereas bilaterality is established in the development of meroblastic vertebrate eggs only after cleavage has virtually ended.

The eggs of ascidians and of amphioxus are notable in that the first cleavage plane coincides with the plane of bilateral symmetry of the organism-to-be, i.e., with the future embryonic axis. According to Conklin, whose early studies on the visible changes occurring in the living, developing egg of the ascidians *Styela* and *Ciona* are classical: "In the ascidian egg, which is one of the most highly differentiated types known, the different cytoplasmic substances before maturation are nearly concentric in position; there is a granular peripheral layer which later goes into the mesoderm and may, therefore, be called mesoplasm, a hyaline layer thickest at the animal pole, especially after the escape of the contents of the germinal vesicle, which may be called the ectoplasm, since it gives rise chiefly to ectoderm, and an inner yolk-rich layer the endoderm, which largely goes into the endoderm. After fertilization and before the first cleavage these concentric layers rapidly take the characteristic positions which they occupy in the embryo and larva.

"The method of segregation and localization of visibly different substances in the polar or bilateral axes or in the organ-forming areas can be followed especially

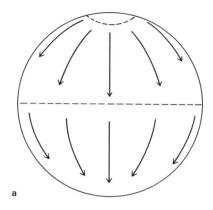

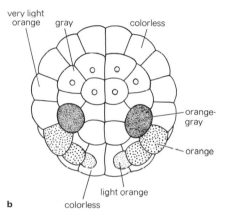

FIGURE 13.2 **(a)** Flow of peripheral cytoplasm from animal pole to equator and from equator to vegetal pole at time of fertilization of ascidian egg. **(b)** Vegetal aspect of cleaving egg of ascidian *Boltenia echinata*, just before onset of gastrulation, showing segregation of different ooplasms within distinct cell groups, with colors and patterns as seen in living embryo. [*After Berrill, 1929.*]

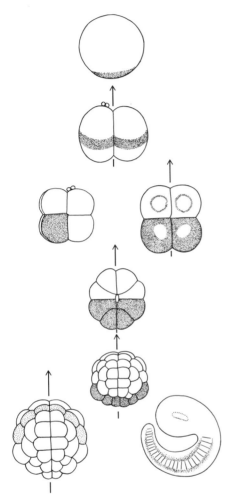

FIGURE 13.3 Distribution of mesodermal crescent material (represented by naturally colored mitochondria in some species, and by vitamin C) in cleaving the ascidian egg, and the relationship between the embryonic axis and the symmetry of cleavage. Lowest two figures show distribution of crescent material in tail muscle of embryo (*right*), and (*left*) neural crescent cells. Arrows show embryonic axis.

well in the living eggs of the ascidian *Styela partita*. When the germinal vesicle breaks down in the prophase of the first maturation division there is liberated into the cell body near the animal pole a very large amount of transparent nuclear sap, linin and oxychromatin; the surface of the egg is at this time covered by a layer of hyaloplasm in which are imbedded yellow spherules of a lipoid substance and many mitochondrial granules; the remainder of the egg contains gray yolk spherules imbedded in a hyaloplasmic matrix.

"At the moment when the spermatozoon enters the egg near the vegetative pole, the superficial layer containing the yellow spherules flows rapidly to the point of entrance where it forms a yellow cap on the surface, and at the same time most of the transparent cytoplasm collects in a zone immediately above the yellow cap, leaving most of the yolk in the animal half of the egg. The sperm nucleus and aster then move up to the equator on the posterior side and the yellow material follows them, and as the cleavage amphiaster is formed, this yellow material spreads out on the surface in the form of a crescent immediately over the amphiaster. At the same time the zone of clear cytoplasm moves to the equator where it forms a clear crescent just above the yellow crescent, while most of the yolk is rotated into the anterior half of the egg. During the first cleavage this rotation of the clear cytoplasm and yolk continues until, at its close, they occupy their definitive positions, the clear cytoplasm in the upper half of the egg, opposite the yellow crescent, which corresponds in position and in potency to the gray crescent of the frog's egg."[1]

These segregations and localizations are the results of movements within the egg which are apparently induced by the entrance of the spermatozoon, its movement through the egg, and the formation of the cleavage amphiaster. But, on the other hand, there is good evidence that the region of entrance of the sperm, its path within the egg, and the location of the cleavage spindle posterior to the middle of the egg and at right angles to its chief axis are determined by the organization of the cytoplasm. The polar differentiation is accentuated by the flow of mesoplasm to the vegetative pole. The mesoplasm then forms a crescent around the posterior side of the egg, parallel with the first-cleavage spindle, and the chordaneuroplasm forms a crescent around the anterior side; the endoplasm lies on the vegetative side of these crescents, and the ectoplasm on the animal pole. Thus bilaterality and the definitive pattern of localization are established.

Specialized regions of the egg cytoplasm (ooplasm) accordingly assume a bilateral organization, which therefore defines the axis of bilateral symmetry of the individual. The first cleavage divides the egg into prospective left and right halves. Isolated blastomeres of the two-cell stage give rise to left or right half-larvae, respectively; isolated anterior or posterior pairs of four-cell-stage blastomeres develop into anterior or posterior half-larvae, respectively. Dalcq has shown, by cutting eggs in two in various planes and at various times between the onset of maturation and the first

[1] E. G. Conklin, Problems of Development, *Amer. Natur.*, **63**:17–18 (1929).

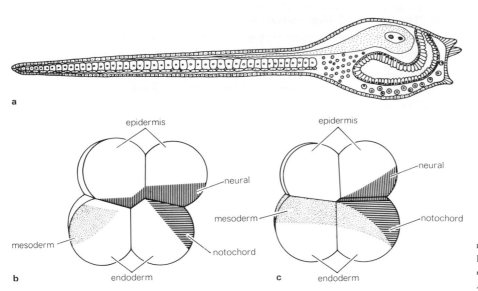

a

b

epidermis

neural

mesoderm

notochord

endoderm

c

epidermis

neural

notochord

mesoderm

endoderm

FIGURE 13.4 (a) Fully developed ascidian larva. (b) Fate map of eight-cell stage of ascidian embryo. (c) Fate map of eight-cell stage of *Amphioxus*.

cleavage, that the bilateral organization of the crescents is established before fertilization in the ascidian *Ascidiella* in essentially the same way that it is established in *Styela* and *Ciona* after fertilization.

In *Amphioxus* the pattern of localization is essentially like that of the ascidian egg, namely, there is a mesodermal crescent around the posterior side, together with a chordaneural crescent around the anterior side. Also, as in ascidians, the first cleavage is always in the plane of bilateral symmetry, dividing the crescents into exactly equivalent halves. In both the ascidian and *Amphioxus* all the poles and axes are irreversibly fixed as early as the first cleavage, while in at least one ascidian species, and probably in others, the poles and axes are evidently established even before fertilization. The common feature in *Styela* and *Ascidiella* is that the bilateral ooplasmic organization appears immediately following the escape of a large quantity of nuclear sap from the ruptured germinal vesicle. In *Styela* this occurs after fertilization; in *Ascidiella* it may occur before, depending on the prefertilization stage of arrest of the oocyte.

The primitive freshwater vertebrate type of egg ranges in size from about 1 to 3 mm in diameter, representing a volume increase relative to the protochordate egg of from 1,000 to about 10,000 times. Yet the general organization is remarkably similar. Almost all experimental studies, however, have been made on the eggs of amphibians, those of the lamprey, sturgeon, and primitive bony fishes being relatively difficult to obtain and maintain. The eggs of frogs have been mostly employed. The unfertilized frog egg outwardly consists of two regions, a dark brown animal region which extends to below the equator, and a whitish vegetal region, the animal region having a black layer of pigment under the vitelline membrane, absent in the vegetal

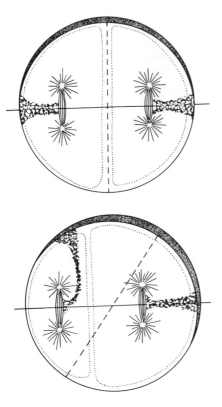

FIGURE 13.5 Equatorial section through dispermic frog egg showing that the gray crescent (position indicated by thin outline opposite thickest part of outline) forms at the midpoint between the two points of sperm entry. Broken line indicates plane of symmetry. [*After Herlant,* Arch. Biol., **26** (*1911*).]

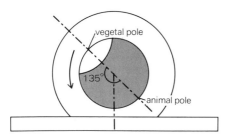

FIGURE 13.6 Rotation experiment to determine the median plane of the frog's egg. An unfertilized egg in its jelly membrane is mounted on a glass plate; the egg axis forms at an angle of 135° with the vertical. [*After Ancel and Vintemberger,* Bull. Biol., *suppl.* **31**:*377* (*1948*).]

region. The pigmented layer extends to the same level below the equator all the way round. The plasma membrane, together with a peripheral layer of cytoplasm of uncertain depth, constitutes the egg cortex and gradually becomes thinner as it nears the vegetal pole. Internally a cup-shaped mass of white yolk granules occupies the vegetal region, with the central concavity opening toward the animal pole. The cavity of the mass is occupied by a somewhat pigmented mass of cytoplasm, while the endoplasmic region of the animal hemisphere is occupied by less pigmented cytoplasm. Yolk granules are largest near the vegetal pole, smaller and less densely packed towards the animal pole. There is, accordingly, a polar axis with egg constituents arranged radially around it and exhibiting a concentration gradient from pole to pole.

Following fertilization, various changes occur in the uncleaved egg comparable to those seen in the ascidian egg: (1) a temporary shifting of the pigments resulting from a raising and concentration of the cortical layer toward the animal pole, the components soon returning to their original place; (2) a displacement of the egg contents brought about by a rotation of the cortical layer around the underlying layers, with relative concentration of the pigmented layer towards a particular meridian. This is accompanied by a corresponding thinning of the cortex on the opposite side of the egg, thereby producing a crescent-shaped surface known as the *gray crescent,* which resembles the yellow crescent of the *Styela* egg. The plane passing through the center of the crescent and the animal pole defines the plane of bilateral symmetry in each case, i.e., it coincides with the embryonic axis, and is the only plane which separates the egg into two equivalent parts, each containing half the crescent material. In *Styela* the crescent is made visible by its contained yellow, mitochondrial granules. In the frog the crescent represents material no longer covered by the pigmented cortical layer, and it is gray because it lacks reflective yolk platelets.

According to Pasteels, who has worked intensively in this field, two periods should be recognized between fertilization and the first cleavage: (1) the labile period between fusion of the pronuclei and the appearance of the gray crescent and (2) the time when bilateral symmetry is irreversibly fixed.

As in the *Fucus* egg, where an innate tendency to establish a polar axis is evident but is readily overcome and redetermined by a variety of agents, so it is in the establishment of the bilateral axis of the frog egg. Eggs activated by a nonlocalized agent such as heat form a gray crescent of normal shape and location, which corresponds to the axis of symmetry of the future embryo. Such bilateral symmetry seems to be due to a slight shifting of the polar axis relative to the position of the vegetal yolk. This developing organization of the egg may itself influence or determine the point of sperm entry or the direction of rotation, although this is not known. It is known that the point of sperm entry, occurring normally or directed experimentally, defines the position of the crescent in a high percentage of cases, although the point of sperm entry and the center of the crescent are on opposite sides of the egg. The other factor of great importance is the effect of free rotation of the egg within its membrane, during which the inner, heavy yolk material of the egg may rotate faster than the cortex. Whatever it may be, during the labile period, the last-acting or the most powerfully acting agent fixes the axis of bilateral symmetry.

This axis becomes irreversibly fixed as the result of the movement of the cortex around the inner mass. The movement has been called the *rotation of symmetrization*, a movement more easily described than explained.

The cortical pigment of the frog egg not only makes visible the shift of the cortex relative to the underlying layers but permits determination of the path taken by the sperm head within the egg cytoplasm. As the sperm nucleus and centriole move inward from the surface, a train of pigment granules is drawn along. This is the *sperm track*, and it serves as a record of both point of penetration and path taken. Early observations that the sperm track is always to be found in the plane of bilateral symmetry have been well confirmed. We can now definitely state that in normal conditions of fertilization and also in experimentally controlled fertilization sites, the point of entrance of the spermatozoon determines the future position of the gray crescent and therefore of the axis of bilateral symmetry. This is accomplished by the sperm's causing a descent of pigment on one side, the side on which it enters, and an ascent of pigment on the other side, the future dorsal side, of the egg.

The symmetrizing effect of the spermatozoon can be counteracted, however, by making an egg rotate within its membranes, after its connections with the chorion have been broken. This rotation can be decisive not only with fertilized eggs but also with experimentally activated eggs. The process is effective as long as the gray crescent has not been actually formed; the greater the degree of rotation, the more effective the process is. After the gray crescent appears, as stated earlier, it is fixed. Before this event, if two or more rotations are made, in different directions, the last one is always determinative.

The determination of bilateral symmetry in other eggs of the same type appears to follow a generally similar course. Eggs of urodeles such as *Triturus* and *Ambystoma* show a clear crescent corresponding to that of anurans but smaller and less well defined. Study of the process is hampered by the mobility of the eggs within their membranes and by the fact that polyspermy is normal in the fertilization of urodele eggs, making it impossible to identify the point of entrance of the effective spermatozoon, if one there be; more than one sperm may contribute to activation of the urodele egg, although only one male pronucleus and its centriole remain active in the cytoplasm and enter into pronuclei fusion and the cleavage of the egg. In the sturgeon, which has eggs very similar to the frog egg though somewhat ovoid and more yolky, a micropyle exactly above the animal pole directs sperm penetration to take place at that level. The egg then becomes fixed to its substrate by its envelope, becomes round, and rotates within the envelope so that the animal pole becomes uppermost, following which, at the onset of cleavage, a clear crescent appears and bilateral symmetry is established. Since the development of the lamprey egg and that of the primitive bony fishes *Protopterus* and *Ceratodus* are so similar, it is inferred

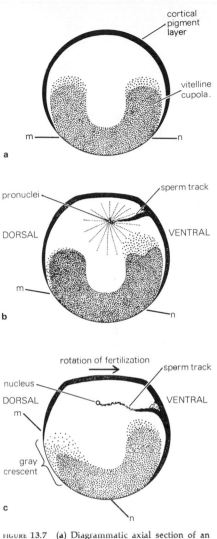

FIGURE 13.7 (a) Diagrammatic axial section of an unfertilized egg of *Rana fusca*, m–n representing the interior limit of the cortical pigment layer. (b) The reaction of the egg to the spermatozoon when it enters. Note in particular the 10° inclination of m–n, and that the vitelline horn has moved closer to the future dorsal side. [*After Ancel and Vintemberger*, Bull. Biol., *suppl.* 31:377 (1948).] (c) An axial section through the center of the gray crescent of *R. fusca*, showing the modifications which have occurred in the egg after the rotation of fertilization (*rotation of symmetrization*). The arrow indicates the direction of the rotation of fertilization. The gray crescent, which marks the dorsal side, is formed. Within this dorsal area, only slight pigmentation remains after the pigmented cortical layer has receded. The vitelline horn on the dorsal side has moved nearer the cortex, and the line m–n has inclined through 30°.

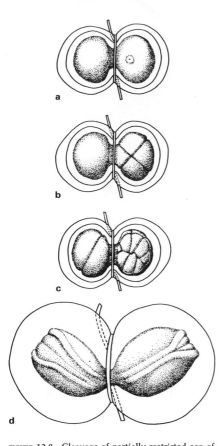

FIGURE 13.8 Cleavage of partially restricted egg of *Triturus viridescens* illustrating "delayed nucleation" of left-hand part. (a) Immediately after construction. (b) Second cleavage; one of the four cells connects with a bridge, allowing its nucleus to enter left part. (c) Fourth cleavage, with first division of left part. (d) Neurula stage, showing normal twin embryos but with difference in time of beginning of cleavage of the two halves still clearly seen. [*Figures 13.8 and 13.9 Courtesy* Annals of the New York Academy of Sciences, *vol. 49, art. 5, G. Fankhauser.* © *The New York Academy of Sciences, 1948. Reprinted by permission.*]

that the symmetrization process is also the same. Accordingly we may conclude that there is but a single type of freshwater vertebrate egg, representing three classes of vertebrates (cyclostomes, fishes, and amphibians), and that it has the same fundamental organization and symmetrizing features as the much smaller eggs of the marine protochordates. In eggs which contain yolk so dense that cleavage is meroblastic, or discoidal, particularly teleosts, selachians, reptiles, and birds, the situation is very different and symmetrization is retarded.

Inversion and ligature of the amphibian egg before or during the first cleavage have yielded interesting phenomena. If frog eggs are maintained in an inverted position following the appearance of the gray crescent (vitally stained for the purpose of the experiment), the heavy vegetal yolk flows down through the egg from the now upper vegetal pole towards the lower animal pole, displacing pigment as it reaches the animal cortex and disrupting the gray crescent, although leaving a reduced white, yolky area at the vegetal pole. In both places a junction usually forms between the yolk margin and material derived from the central part of the original crescent. Each junction then serves as a determinant of bilateral symmetry, and two embryos develop accordingly. An interaction between central crescent material and the yolk margin may therefore be more significant in inducing an embryonic axis than the crescent material by itself. In urodeles, ligaturing the egg before the first cleavage yields different results depending on the relation of the plane of ligature to the position of the crescent. If the ligature is vertical and cuts through the center of the crescent, twin embryos develop which are complete except for some deficiency along the side of the ligature. If the ligature divides the crescent into a larger and smaller part, a normal and a microcephalic embryo (larva) develop, respectively. If the ligature, still vertical, leaves the whole crescent in one part, that part gives rise to a normal larva, while the other merely continues cleavage for a while and dies. A similar result follows separation of animal and vegetal halves: only the animal half, which contains the crescent, develops into an embryo. Salamander twin larvae resulting from ligature of the egg shortly after it is fertilized are notable in that one is a mirror image of the other with regard to the asymmetrical arrangement of the viscera. The larva derived from the original prospective left side is normal, the other exhibits *situs inversus* of the heart and viscera.

PATTERNS OF DIFFERENTIATION

Although eggs may be conveniently designated as determinate (mosaic) or indeterminate (regulative), and apart from the fact that some kinds fall between the two extremes, no egg is rigidly determinate from the start or entirely indeterminate. The same kinds of processes and events occur in each but at different rates relative to the rate of cleavage. However, the remarkable patterns of cleavage seen in the early

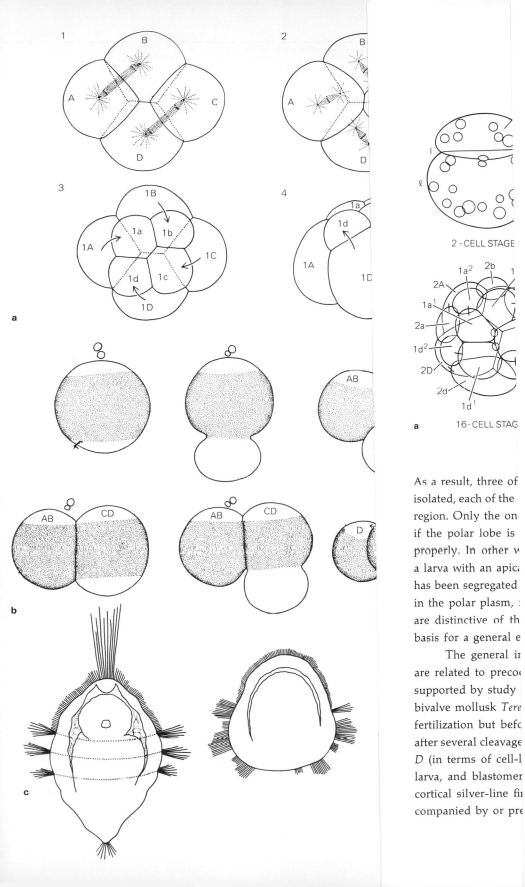

development of the so-called determinate eggs, notably those of spiralia and ascidians, have long attracted attention.

1. The early studies, known as *cell-lineage* studies, consisted of painstaking tracing of the sequence of cleavages in each developmental type until cells became too small and numerous to follow any further. These early investigations seemed for a while to be dead-end studies, but they appear now to be coming into their own as the basis for new experimental analysis.

2. Other, related studies combined a study of cell lineage with extirpation, or defect, experiments in which one or more blastomeres were picked off or destroyed in order to see what might consequently be missing from the embryo subsequently produced. The pioneering experiment was that of Chabry, in 1887, who devised the simple but ingenious procedure of drawing fertilized eggs (of the ascidian *Ascidiella*) into a capillary tube of slightly smaller diameter and pricking off one or another of the early-cleavage blastomeres by means of a glass needle inserted through one end of the tube, all the while observing under the microscope. This was the first of all microdissection devices.

3. Isolation experiments have included those in which the development of isolated blastomeres was followed to determine their respective potentials. More recent work has extended this general inquiry by employing vital staining; small groups of cells of the early embryo are stained *in situ* with dyes such as neutral red, Nile blue, or Janus green, and their developmental destiny is followed. Analysis of this sort has not led to insight into the nature of developmental processes but does serve as an essential observational record of the events with which we are concerned and which we wish to explain.

The course and significance of the strikingly ordered pattern of cleavage seen in the more determinate types of eggs are of great interest. The unequal cleavage that gives rise to the micromeres and macromeres in sea urchin development is a case in point, although the sea urchin egg does not fall in either of the extreme categories; the pattern of cleavage of the *Ascaris* egg, also briefly discussed before, is classically determinate. In such cases and many others, the pattern of cleavage was originally thought to establish the pattern of regional differentiation or specialization, and it was this concept that supplied the motive for many of the earlier studies. However this may be, a correlation exists between pattern of cleavage and regional specialization of cells or tissue. What causal relationship exists between the two is the question. Primarily the patterns relate to the polar axis of the egg, to the axis of bilateral symmetry, if present, and to asymmetry of development.

Development is so remarkably similar in various annelids and mollusks, particularly the marine species with relatively small eggs, that it may be treated as a single type.

Cleavage exhibits a spiral pattern, for example, in the polychaete *Nereis* and the gastropod *Patella*. Although the first two cleavages are vertical, passing through

FIGURE 13.9 (a) Two-headed larva developed from partially constricted egg, shown in neurula stage in Figure 13.8. (b) Normal control larva. (c) Twins produced by partial constriction of unsegmented egg in median plane, the connection between the two halves of the egg having broken after gastrulation; the eye, balancer, and forelimb bud are smaller on the right side of the left-hand twin and on the left side of the right-hand twin. [*After Fankhauser, 1948.*]

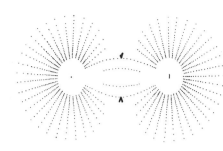

FIGURE 13.11 Resting first-cleavage metaphase spindle of *Polychoerus*, showing centriolar orientation. The two centrioles are at right angles to each other, and each gives rise to a daughter centriole oriented at right angles to itself. In subsequent cleavages, daughter centrioles form somewhat obliquely to the parent centriole, in keeping with an oblique (spiral) cleavage pattern. [*After Costello, 1961.*]

Centrifugation e
properties in fertilizec
stances such as yolk ɲ
that when the eggs of
ugation into pieces of
dark, inclusion-contaiɪ
of subsequent cleavag
size and content, exhiɩ
normal eggs.

More recently, C
lobe, which can be pr
in normal proportions,
vegetal fragments proc
differentiate lobe-depe
of development has a
to subsequent cleavag
a whole but also, more
this primary mechaniɛ
of resolution of the e
of something that maɣ

To sum up, the

1   A cortical region oɪ
    brane.
2   The cortex is labile
    change.
3   The pattern or couɪ
4   Early cortical speciɩ
5   The combination oɪ
    cortical cytoplasmiɩ
6   Cells undergoing ɲ
    for mitosis, and coɪ
    ing of comparativeɪ

DETERMINATION ANI

Further support for the
ment of ascidian eggs
tablishment of bilateraɪ
As already stated, the c
echinata eggs, present

confers bilaterality on the egg embryo. In the present connection the significant fact is that the first cleavage always divides the egg in the plane of the embryonic axis. As successive cleavages proceed, the material of the crescent becomes segregated in two of the first eight blastomeres, and these continue to divide to form a total of only 36 cells, the future muscle cells of the tadpole larva. Similarly the material in the anterior crescent becomes segregated as the 38 to 40 large cells which differentiate as notochord cells. Together these two groups of cells give rise to the basic structure of the chordate tail of the larva. They do so precociously, i.e., not only is the presumptive material discernible in the uncleaved egg, but the tail tissues derived from it differentiate and become functional long before other tissues or organs in the body do so. In correlation with this precocity, the regions of the egg cytoplasm giving rise to these tissues undergo comparatively very few cell divisions (about five divisions in each case), so that the derivative tissues consist of relatively very large and very few cells.

It is clear from the foregoing and from more detailed evidence that the presumptive muscle and notochordal regions of the fertilized egg are mainly cortical (endoplasm, for instance, can be withdrawn by micropipette without disturbing the normal development, as in the sea urchin). Cleavage takes place in conformity with regional symmetries and cytoplasmic differences already established, so that cellular compartmentalization of diverse differentiating regions is effected. And as a consequence of some general incompatibility between rapid structural cell differentiation and continuing mitoses, a precocious or accelerated cytodifferentiation results in premature cessation of cell division of the material involved. In other words, the precise patterning of cleavages in so-called determinate, or mosaic, eggs reflects, and at least to some extent is caused by, differentiating processes associated with the egg cortex. Although the first cleavage plane of an ascidian egg always divides the egg into presumptive right and left halves of the embryo-to-be, and in the frog does so in about 40 percent of cases, this is not important in the long run. After an egg has subdivided into a large number of cells, it makes little difference whether or not early divisions coincide with this or that embryonic axis. Nevertheless, cellular segregations and associated phenomena do have consequences with regard to developmental mechanics and possibilities.

Two questions, differing in kind, accordingly arise. One concerns the nature of the egg substances or properties responsible for the precocious differentiation of certain cytoplasmic territories. The other concerns the developmental need, and also the consequences, of such special development.

Since the yellow crescent material becomes mainly confined to the precociously functioning muscle tissue of the larval tail, considerable effort has been made to identify the responsible component or process, beginning with centrifugation experiments of Conklin and continuing to the present day. The yellow material, or myoplasm, is of lipid material, but its displacement by centrifuging does not disturb normal muscle development, nor do displacements of other egg inclusions, such as yolk;

so far we have no certain knowledge of the chemical nature of the special ooplasmic zones appearing at or shortly after fertilization in the ascidian or any other kind of determinate egg.

Whatever the basis of the yellow crescent and other specialized ooplasmic regions may be, the presence of such identifiable ooplasms is an indication that certain vital developmental processes have commenced. And there is no doubt that at the two-cell stage the egg definitely is a "mosaic" egg, while at the eight-cell stage most of the blastomeres already have an irrevocable destiny: the two anterior animal blastomeres will produce head epidermis, attachment papillae, and the brain with its two pigmented sense organs; the two posterior animal blastomeres give rise only to epidermis; the two anterior vegetal blastomeres contain the territory for spinal cord, notochord, and a part of the intestine; while the two posterior vegetal blastomeres give rise to part of the intestine, to the mesenchyme, and to muscle tissue. When these four pairs of blastomeres are isolated, respectively, the results for three of the pairs are the same as their prospective fate; but this is not the case for the anterior animal pair. The fates of the vegetal pairs are fixed for intestine, notochord, muscle, and mesenchyme, all derived from the vegetal cells; but isolated anterior animal blastomeres fail to produce neural structure of any kind. They do so only when they are in contact with anterior vegetal blastomeres.

Apart from this stimulator-response effect, the developmental course of the ascidian is set at the two-cell stage and is rigidly set by the time the third cleavage has occurred. This is remarkably early. Yet there is good evidence that whatever the patterning or localizing process may be, it is at least in a labile state in the unfertilized egg. Such eggs, when subjected to high-speed centrifugation, separate into "dark" and "hyaline" fragments. In *Ciona* the dark fragments which contain yolk granules, mitochondria, and pigment, when fertilized develop in an entirely normal manner, with typical cleavage pattern, gastrulation, 38- to 40-cell notochord, etc., differing only in having cells of subnormal size. The hyaline fragment, devoid of mitochondria, never undergoes cleavage. On the other hand, in *Ascidia*, mitochondria occur in both kinds of fragments, and each is capable of developing to form a swimming tadpole. Mitochondria are clearly involved in some way, both in muscle formation and in development as a whole, though there is no indication that they have anything more than a general metabolic function.

The precision of the differentiation process relative to cleavage in ascidian development is not limited to histogenetic differentiation but applies to the onset and completion of gastrulation, and later events, as well. The presumptive territories of all the various tissues of the future tadpole larva are clearly defined at the 64-cell stage. This stage represents the sixth cleavage of the egg. Gastrulation by invagination occurs following the sixth cleavage and is completed by the time of the seventh cleavage, i.e., between the 64-cell and 128-cell stage (at the mid-gastrula stage there are approximately 76 cells). The large ventral opening of the gastrula closes, and the embryo begins to lengthen along the embryonic axis indicated by the plane of the

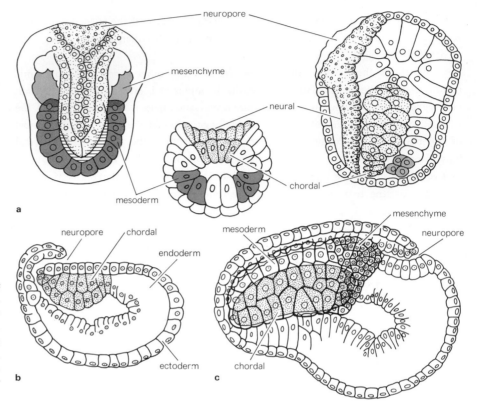

FIGURE 13.13 Neurula stage of development of ascidian. (a) Dorsal view (*left*) and side view (*right*) of neurula of *Styela*, and cross section of open-neural-plate stage of younger embryo. [*After E. G. Conklin, 1905.*] (b,c) Early and late neurula of *Clavelina* from side, showing neuropore, neural tube, notochord (stippled), and superimposed tail mesoderm and mesenchyme. [*After Van Benedin and Julin, Arch. Biol., 2:317 (1884).*]

first cleavage. The upper surface, or *neural plate*, of the elongating gastrula then invaginates along the midline to form a groove which subsequently becomes a tube, the *neural tube*, and the stage during which this process occurs is the *neurula*. With further development the neurula transforms into a minute tadpole-shaped larva complete with a typically chordate tail (i.e., with central notochord, dorsal neural tube, and lateral muscle bands) and a trunk containing a sensory vesicle equipped with gravity- and light-sensitive organs and a neural ganglion.

In this connection a comparison of three of the lower chordates, namely, *Amphioxus* (Branchiostoma), *Styela* (ascidian), and *Oikopleura* (a larvacean tunicate), is illuminating. All three have translucent eggs of roughly the same size (about 120 μ in diameter). All develop into the basic chordate organism, with notochord, dorsal tubular nerve cord, lateral tail musculature, and gill slits. In *Amphioxus* there are further growth and elaboration without any significant change in character. In *Styela* and other ascidians, apart from the gill slits, the chordate structures are more precociously developed as a transient larval organization persisting in functional form for only a few hours, when a destructive metamorphosis ensues. In *Oikopleura* development closely resembles that of *Styela*, although it is yet more precocious; but neither metamorphosis nor marked postlarval growth occurs.

In these three types, at or before fertilization, a flowing cytoplasm establishes specialized crescental zones, imposing bilaterality on the egg. Crescentic zoning, bilateral cleavage pattern, the association of the anterior crescent with chordaneural tissue and of the posterior crescent with mesoderm (muscle) indicate a fundamental relationship between the three. The crescents must include in their substance histogenetic agents in some form.

GASTRULATION AND TOPOGRAPHICAL REORGANIZATION

While the crescental plasms represent histogenetically the essential chordate-type tissues, they are not in proper mutual relationship. This is established only as the result of gastrulation. *Only as the gastrulation process nears completion, and as a direct result of this process do the anterior and posterior crescents come to be in juxtaposition and make possible the orderly development of the chordate swimming tail and trunk.*

In each of the three forms under discussion, cell division within the crescental zones ceases as histological differentiation becomes visible. In all three there is a close correlation of the rate of differentiation of notochordal and mesodermal tissues, and a correlation between this and the time of onset (in terms of cleavage) of gastrulation. In *Amphioxus*, gastrulation occurs between the ninth and tenth cleavages; in *Styela* between the sixth and seventh; and in *Oikopleura* between the fifth and sixth. Corre-

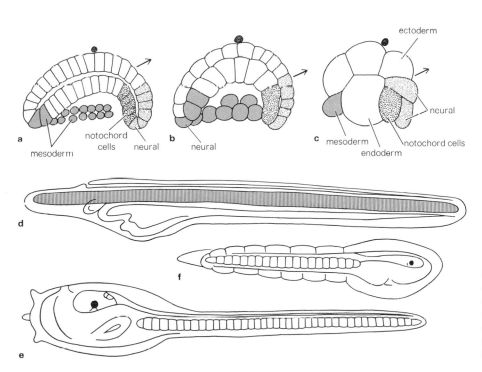

FIGURE 13.14 Gastrulas and larvae of *Amphioxus*, an ascidian, and a larvacean tunicate, showing effect of varying ratios of differentiation and cleavage rates. (**a,d**) Gastrula and larva of *Amphioxus*. (**b,e**) Gastrula and larva of ascidian *Ciona*. (**c,f**) Gastrula and larva of *Appendicularia*. [*After Berrill, 1941.*]

spondingly, the number of notochordal and muscle cells in the *Amphioxus* embryo is 300 to 400 in each case, representing from eight to nine divisions of the initial territory; in *Styela* there are about 40 cells in each, representing from five to six divisions; and in *Oikopleura* the number of each kind of cell is 20, or from four to five divisions. In each form the size of the cells produced is inversely proportional to their number. Differentiation relative to cleavage rate is accelerated in ascidians compared with *Amphioxus,* and in *Oikopleura* compared with ascidians.

The conclusion drawn from these facts is that the precocious histogenetic processes and the process culminating in the visible movements of gastrulation represent a single coordinated mechanism initiated in these eggs at or before fertilization, and responsible primarily and possibly exclusively for the development of the essentials of chordate structure and organization. The rate of action of this mechanism varies as a whole, and differences may be due to differences in the initial concentrations of reacting substances or in some related catalytic agent.

It is unlikely that the rearrangement of originally widely separated ooplasms, or "histozones," established before gastrulation is a primary function of the gastrulation process, for it is by no means a universal phenomenon, being most obvious in the most condensed and probably the most specialized types of development.

Two interpretations of the pregastrular localizations are possible. (1) Histogenetic localizations may have originally been postgastrular, in coordinated relationship to one another with regard to collective structure and function, and may have been pushed back gradually through gastrulation into earlier and earlier cleavage stages, even to the unsegmented egg. (2) Alternatively, mutations affecting the ripening egg may have resulted in the creation of local histogenetic patterns in the egg which make sense only after gastrulative rearrangement, any placement errors being eliminated by natural selection. This possibility seems less likely.

SYNTHESIS IN REGULATIVE AND DETERMINATIVE DEVELOPMENT

The first comparison to be made at the molecular level of the differences between determinative and regulative development, by Markert and Cowden, emphasizes the difference in timing of crucial events. The conceptual framework of this comparison is that the developmental autonomy of the cytoplasm of invertebrate eggs varies greatly in different regions of the egg and from one species to another; this variation, itself, suggests that information must be stored in the cytoplasm. Isolated blastomeres from developing determinative eggs develop into partial embryos in accord with their original fate. The cytoplasm of such blastomeres may be regarded as having considerable fixed informational content at the outset of development, with correspondingly low dependence on immediate supplies of information from the nucleus. At the opposite end of the spectrum of developmental behavior are the highly regulative eggs in which isolated blastomeres give rise to complete embryos.

Some indication of the molecular basis for determinative and regulative development has been obtained from an investigation of protein and nucleic metabolism during the early development of organisms representing these two extremes in development, namely, ascidian and sea urchin, employing actinomycin, puromycin, and antimycin as the principal analytic agents, i.e., as inhibitive agents of nuclear RNA translation, ribosomal transcription, and mitochondrial electron transport, respectively. Although the synthesis or release of messenger RNA may be required for the initiation of new programs in development, the synthesis of ribosomal RNA and transfer RNA is not immediately necessary as long as sufficient concentrations remain from oogenesis to sustain synthetic activity.

It is significant, therefore, that nucleoli and increasing levels of cytoplasmic RNA first appear in ascidian development *after* the swimming-tadpole-larva stage has been attained and during the process of metamorphosis. In the sea urchin, RNA synthesis begins at least as early as the eight-cell stage. Actinomycin D, inhibiting DNA-dependent RNA synthesis, produces marked effects at very different stages of development of ascidian and urchin, respectively, development being arrested in the sea urchin either before hatching or at the swimming-blastula stage, whereas in the ascidian development proceeds through the swimming-larval stage as far as metamorphosis before it is blocked. Ascidian development is blocked at the tail-bud stage, however, by antimycin, which blocks mitochondrial function, which is in keeping with the observations that the mitochondria-rich crescent material of the fertilized but uncleaved egg is destined to give rise to the differentiating muscle tissue of the tail.

In the sea urchin the principal synthesis of protein occurs after gastrulation and during the period of larval skeleton formation, i.e., when the primary mesenchyme cells, descendants of the micromeres, have ceased dividing and are secreting the skeletal spicules. All amino acid analogues, which inhibit protein synthesis, so far tested, produce retarded and abnormal development at this stage. In the ascidian, amino acid analogues and other antimetabolites allow development to proceed to the swimming-tadpole stage but disturb or arrest the process of metamorphosis. Further, very little incorporation of carbon 14 into amino acids occurs during development of the tadpole larva, but extensive incorporation occurs in all amino acids at the onset of metamorphosis.

According to Markert and Cowden, "Some protein synthesis occurs throughout the development of *Ascidia nigra,* but with the onset of metamorphosis extensive degradation of proteins to amino acids occurs, followed by accelerated synthesis of many new proteins. This period of transition is especially vulnerable to inhibition by analogs of amino acids and by other antimetabolites affecting protein synthesis. . . . The cumulative evidence from biochemical, cytochemical, and autoradiographic studies on this species implies that the principal synthesis of ribosomes and of protein occurs at the onset of metamorphosis. By inference, the RNA synthesized prior to gastrulation is probably messenger RNA. Since, however, determination is complete in mosaic ascidian embryos by the four- to eight-cell stage, nucleus-directed synthetic events are not likely to influence determination. <u>Precocious determination</u> <u>is more probably mediated by selective segregative cytoplasmic constituents.</u>"[1] They suggest that the differences between mosaic (determinative) and regulative eggs may well reside in the extent of the program that is unblocked at fertilization. This is comparable to the conclusion already made (page 296) in comparing cleavage/determination rates in larvacean, ascidian, and amphioxus, namely, that the same determinative pattern is set in motion at fertilization but that the rate of reaction or the concentration of the determinants varies among these three types.

DUALITY OF DEVELOPMENT

In making comparisons among the several protochordates or between the development of the determinative ascidian egg and the regulative sea urchin egg, one must remember that all are determinative in some degree and that all are regulative to some extent to begin with. The tail of the tadpole larva and the skeleton of the pluteus larva are each temporary larval structures, each precociously differentiated, although with different degrees of precocity. And in ascidian and sea urchin alike, the special larval structure undergoes self-destruction after a more or less short period of functioning, and metamorphosis into respective miniature adult-type organisms proceeds.

We are confronted, in fact, by two forms of development: (1) the development of the egg into a specific type of temporary larval organism, highly specialized in structure and function, and (2) a development of the permanent or adult-type organism of a very different character. Metamorphosis represents the destruction of the one and the consequent acceleration of the other.

The difference in the developmental cycle of ascidian and sea urchin relates in part to difference in egg size, a matter of economy. The urchin egg is typically small, the *Arbacia* egg having a diameter of about 70 μ. It develops into a correspondingly small pluteus larva. The pluteus larva is equipped to feed on the smallest forms of unicellular phytoplankton and, by means of its ciliated skeleton-supported arms, to sustain a planktonic existence. Only after considerable growth is enough tissue present to permit transformation into an adult type of organism, however small. Ascidian eggs of the kind that are typically shed into the water to be fertilized and develop, as in sea urchins, are larger by several times in volume. Even the smallest tadpole larvae do not feed, and they contain enough tissue mass to transform directly into the adult form. There is no growth during the larval period to confuse the picture. The two phases of development are more sharply presented.

Accordingly, in the ascidian, the egg at the time of fertilization is essentially preprogrammed to develop into a tadpole larva. All the special properties of the

---

[1] C. L. Markert and R. R. Cowden, Comparative Responses of Regulative and Determinate Embryos to Metabolic Inhibitors, *J. Exp. Zool.,* **160**:44 (1965).

egg appear to be related to this production. Not all of the egg is directed to this process, for some features of the permanent organism slowly develop during the rapid formation of the tadpole structure. When the period of tadpole larval activity comes to an end, however, and larval tissue is resorbed, the residual tissue takes a great leap forward, so to speak, being released from the dominating presence of tadpole organization and at the same time receiving a new nutrient supply in the form of autolyzed tadpole-tail tissue. To a very great extent, a second and permanent organization arises from the ashes of the old.

The most striking demonstration of developmental duality is seen in the many ascidians and related tunicates that are capable of both sexual and asexual reproduction and development. A single example must suffice, the common, worldwide, seacoast compound ascidian *Botryllus*. Briefly, *Botryllus* produces a comparatively large egg, which cleaves and invaginates to form a blastula, gastrula, and neurula in succession. It becomes a fairly typical tadpole larva, which metamorphoses into a sessile, adult-type ascidian after a short free-swimming period. The ascidiozooid thus produced, however, forms minute buds on each side, which in turn produce buds in the course of their own growth, and so on, until extensive flat sheets of small ascidiozooids unite in a common colonial system. Each bud is minute at its inception compared with the ascidian egg. The bud develops in a direct and seemingly monumentally simple manner, by expansion (growth) of epithelia together with local infoldings and outfoldings and some cell segregation, to form a more or less sexual mature adult individual. While the initial state of the bud may be compared with a simple gastrula in that it consists of an outer and an inner layer of cells, separate from one another, nothing remotely comparable to neurulation and the development of a chordate embryo and swimming tadpole larva occurs during its development. Yet it segregates cells which grow into mature eggs and sperm. Everything that is known concerning the early development of the ascidian egg relates to the precocious and particular development of the special chordate tissues and their organization.

The inevitable conclusion, therefore, is that somatic cells (in the initial form of a small disc of unspecialized cells) possess the potential of developing the adult organization in full multicellularity, sexuality, and structure. The fertilized egg can do the same, but in addition it can undergo precocious organization, at or before the time of fertilization, which relates only to the development of the chordate larval structure. In other words the developing egg represents a dual system: the egg as an egg develops the precocious and special larval structure, of transient nature; and the egg as a cell develops the sexually mature adult structure. It is unlikely that this duality in egg development is peculiar to the ascidian. Probably it is only more obvious in this group, and accordingly any egg should be regarded as a special developmental device which may incorporate secondary forms of development or special reinforcements for primary developmental procedures inherent in the egg as a cell. What is certain is that the egg as an egg is a highly evolved and differentiated cell that is adapted to various environmental circumstances on the one hand and,

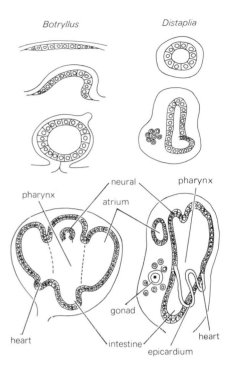

FIGURE 13.15 Direct development in the buds of two types of ascidians, showing development of double-layered disc of *Botryllus* bud into stage with six primary organizational divisions by invaginative and evaginative foldings, and comparable stages of bud of *Distaplia*. [*After Berrill, 1955.*]

on the other, to the biological engineering problems inherent in any continuously growing construction.

REASSEMBLY OF A MULTICELLULAR ORGANISM

Reassembly, or self-assembly, of enzymes, viruses, individual cells, and tissues has already been described in earlier chapters. Reconstitution of a sponge from dissociated differentiated cells may be regarded as the reassembly of an organism, although the organization of a sponge is more comparable to tissue organization than to that of metazoan organisms. An impressive reassembly process, however, has been demonstrated in the ascidian *Amaroecium* by Scott (1959). The egg of *Amaroecium* develops into a chordate tadpole larva, which metamorphoses into a young functional ascidian. Later, by a process of abdominal strobilation, the individual gives rise to a number of buds and becomes a colonial organism.

Working with fully formed but nonactive individuals, each derived from a single metamorphosed tadpole and each enclosed in its protective coat of tunicin, Scott fragmented the individual zooid within the cavity of the tunic, by gentle pressure, until it was a mass of floating bits of tissue. Reconstitution of a new individual from the tissue fragments occurs within 24 hours.

If the epidermis is only slightly damaged during the experimental procedure, it spreads until it becomes whole again. In all such cases the fragmented tissues floating within the epidermal envelope reorganize into an essentially single digestive canal consisting of a branchial basket and a typically differentiated U-shaped intestine. Fragments of all organs reassemble into their organ systems and establish themselves in the proper linear axis.

In contrast, when experimental fragmentation ruptures the epidermis too widely for this tissue to reestablish its continuity as a whole, the epidermis forms two or three separate envelopes of smaller size. All masses contained within each of these envelopes then reorganize within each into a complete system of digestive tract, heart, etc., so that twins or triplets result. Apparently, therefore, the property of wholeness and some of the information relating to the ordering of the contained reconstituting cell mass reside in the epidermis.

CONCEPTS

The egg cortex, consisting of plasma membrane together with subjacent cytoplasm, is the site of morphogenetic patterning relating to polar, bilateral, and general organization of the developing egg.

The material basis for cortical patterning is below the limits of resolution of the electron microscope.

The capacity to exhibit the specific patterning of the whole, as expressed in cleavage patterns, is inherent in small egg fragments before fertilization.

Cortical organization is at first labile but becomes regionally set and irreversible.

Specific cleavage patterns are determined by the progressively changing cortical organization, and do not cause it.

Cleavage tends to segregate cytoplasmic territories having different (cortical) histogenetic properties.

Early local cortical specialization leads to precocious regional cytodifferentiation.

Precocious cytodifferentiation inhibits cell division, so that precociously differentiating territories yield small numbers of large cells, in contrast to other regions that continue division to become large numbers of small cells.

The rate of action of the determination complex released at fertilization varies as a whole, relative to cleavage rate, among the eggs of different but related animals.

Eggs represent a dual system of development: if it is mechanically feasible, the egg as a cell develops directly the adult-type organization; the egg as an egg in addition develops a secondary organization which may divert, reinforce, or temporally inhibit the development of the final organization.

READINGS

BERRILL, N. J., 1961. "Growth, Development and Pattern," Freeman.

——, 1955. "The Origin of Vertebrates," Oxford University Press.

——, 1950. Size and Organization in the Development of Ascidians, in P. Medawar (ed.), "Essays on Growth and Form," Oxford University Press.

——, 1941. Spatial and Temporal Growth Patterns in Colonial Organisms, in "Growth," 3d Growth Symposium, vol. 5, pp. 89–111; also in Bobbs-Merrill Reprints in The Life Sciences, B-13.

CLAVERT, J., 1962. Symmetrization of the Egg of Vertebrates, *Advance. Morphogenesis*, **2**:27–60.

CLEMENT, A. C., 1968. Development of the Vegetal Half of *Ilyanassa* Egg after Removal of Most of the Yolk by Centrifugal Force, Compared with the Development of Animal Halves of Similar Visible Composition, *Develop. Biol.*, **17**:165–186.

CONKLIN, E. G., 1933. The Development of Isolated and Partially Separated Blastomeres of *Amphioxus*, *J. Exp. Zool.*, **64**:303–375.

——, 1932. The Embryology of *Amphioxus*, *J. Morphol.*, **54**:69–119.

——, 1931. The Development of Centrifuged Eggs of Ascidians, *J. Exp. Zool.*, **60**:1–80.

——, 1905. The Organization and Cell Lineage of the Ascidian Egg, *J. Acad. Nat. Sci.* (Philadelphia), **8**:1–119.

COSTELLO, D. P. 1961. The Orientation of Centrioles in Dividing Cells and Its Significance, *Biol. Bull.*, **120**:285–312.

———, 1945. Segregation of Oöplasmic Constituents, *J. Elisha Mitchell Sci. Soc.*, **61**:277–289.

———, M. E. DAVIDSON, A. EGGERS, M. H. FOX, and C. HENLEY, 1957. "Methods for Obtaining and Handling Marine Eggs and Embryos," Marine Biological Laboratory, Woods Hole.

CURTIS, A. S. G., 1965. Cortical Inheritance in the Amphibian *Xenopus laevis:* Preliminary Results, *Arch. Biol.*, **76**:523–546.

FANKHAUSER, G., 1948. The Organization of the Amphibian Egg during Fertilization and Cleavage, *Ann. N.Y. Acad. Sci.*, **49**:684–708.

HARRISON, R. G., 1945. Relations of Symmetry in the Developing Embryo, *Trans. Conn. Acad. Arts Sci.*, **36**:277–330.

HOLTFRETER, J., and V. HAMBURGER, 1955. Progressive Differentiation. Amphibians, in B. Willier, P. Weiss, and V. Hamburger (eds.), "Analysis of Development," Saunders.

MARKERT, C. L., and R. R. COWDEN, 1965. Comparative Responses of Regulative and Determinate Embryos to Metabolic Inhibitors, *J. Exp. Zool.*, **160**:37–45.

NEEDHAM, J., 1942. "Biochemistry and Morphogenesis," Macmillan.

PASTEELS, J. J., 1964. The Morphogenetic Role of the Cortex of the Amphibian Egg, *Advance. Morphogenesis*, **3**:363–388.

RAVEN, C. P., 1964. Mechanisms of Determination in the Development of Gastropods, *Advance. Morphogenesis*, **3**:1–32.

REVERBERI, G., 1961. The Embryology of Ascidians, *Advance. Morphogenesis*, **1**:1–54.

SCOTT, F. M., 1959. Tissue Affinity in *Amaroecium*, I: Aggregation of Dissociated Fragments and Their Integration into One Organism, *Acta Embryol. Morphol. Exp.*, **2**:209–226.

TUNG, T. C., S. C. WU, and Y. Y. F. TUNG, 1962. The Presumptive Areas of the Egg of *Amphioxus*, *Sci. Sinica*, **11**:629–644.

———, ———, and ———, 1960. The Developmental Potencies of the Blastomere Layers in *Amphioxus* Eggs at the 32-cell Stage, *Sci. Sinica*, **9**:119–141.

WILSON, E. B., 1929. The Development of Egg-fragments in Annelids, *Arch. Entwickl.-mech.*, **117**:180–210.

———, 1925. "The Cell in Development and Heredity," Macmillan.

WILT, F. H., and N. K. WESSELS, 1967. "Methods in Developmental Biology," Crowell.

WINESDORFER, J. E., 1967. Marine Annelids: *Sabellaria*, in F. H. Wilt and N. K. Wessells (eds.), "Methods in Developmental Biology," Crowell.

CHAPTER FOURTEEN

DEVELOPMENTAL MECHANICS OR FORMATIVE MOVEMENTS

The animal egg is typically a relatively large and dense, more or less spherical mass of protoplasm. No matter what processes of chemical and ultrastructural differentiation occur during development, the initial mass must be converted into a large number of small cohering cells in order for even the simplest metazoan form of life to evolve. Cleavage begins shortly after fertilization and continues until a large number of small cells have been produced. Cell numbers and cell movements play vital roles throughout development, in conjunction with cytoplasmic differentiation. Cell affinities and cell incompatibilities underlie much of the visible events of development; in a physical sense, development is the continually changing behavior of a continually increasing population of progressively diversifying component cells. Although the movements of individual cells are important in development, most of the movements that mold the organism consist of the spreading and folding of sheets of cells, particularly as invaginations and evaginations, involution, and lateral and axial extensions.

CONTENTS

Cleavage and blastula formation
Invaginations and morphogenesis
Organization of the amphibian blastula
Blastulation in meroblastic vertebrates
Integration of gastrulation
Blastodermic gastrulation
Multiple induction
   Fish
   Bird (or reptile)
   Mammal
Neurulation in amphibians
Concepts
Readings

CLEAVAGE AND BLASTULA FORMATION

Unless modified by precocious determinative processes, the successive cleavage stages of an egg assume configurations resembling clusters of soap bubbles. Blastomeres and bubbles, both with viscous surfaces, are subject to two forces: (1) the tendency of any mass with fluid or semifluid properties to assume a spherical shape, whereby free surface energy is reduced to a minimum, and (2) the tendency to adhere to any adjacent surface that has the same physical and chemical properties as its own. If surface tension is high, a more or less spherical form is maintained; if it is low, a cell tends to spread out on a surface to which it has become attached. Thus cells tend to round up during mitosis, when surface tension becomes high, and to flatten out to form an interface with a neighboring cell when division is complete.

As in other studies, echinoderm eggs have been mostly used for analysis of the physical processes involved in cleavage and blastula formation. Cleavage, as elsewhere, results in the production of many blastomeres, the number increasing and the individual size decreasing with every division. At an early stage when they are comparatively few and large, the blastomeres adhere together as a more or less solid ball—the morula. A hyaline plasma layer associated with the overall outer surface appears to bind them more firmly together. If this layer is removed or its formation prevented by treatment with calcium-free seawater, blastomeres simply pile up. In this circumstance the blastomeres are seen to be attached to one another in clusters by protoplasmic strands.

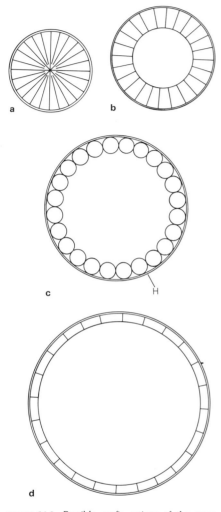

FIGURE 14.1 Possible configurations of the cross section of a tube comprising 24 cells in a single layer attached to a supporting membrane (*H*), with no change in cell volume. (**a**) Considerable contact between cells and no cavity. (**b**) Moderate contact, and a cavity is present. (**c**) Point contacts between the rounded-up cells. (**d**) Considerable contact with the basement membrane. [*After Gustafson and Wolpert, 1967.*]

In normal development, as cell number mounts and individual cell size decreases, the morula transforms into the blastula, consisting of a spherical envelope of cells, one cell thick, enclosing a central fluid-containing cavity, the blastocoel. The problems are the mechanism of volume increase of the developing blastula and how a hollow sphere of cells is formed rather than a compact cluster.

The hyaline layer seems to be necessary for blastula formation in echinoderms, but since no such layer is discernible in some other forms (e.g., certain coelenterate eggs), other factors such as protoplasmic connections and general adhesiveness must serve. Nevertheless the role of the hyaline layer has received much attention, and its importance in the development of the sea urchin and starfish is undeniable. The hyaline layer first appears in the echinoderm shortly after activation through the release of material at the egg surface by the rupturing cortical granules described in an earlier chapter. It completely invests the egg, and its continuity is retained as a continuous externally investing layer during later stages of development.

In the first place, the general increase in size of the blastula and blastocoel takes place in steps correlated with the synchronously occurring cell divisions characteristic of early development. The cells are known to adhere strongly to the hyaline layer and even to be anchored to it by cell processes; in fact, the layer itself undergoes change, being at first diffuse and fibrous but later becoming two layers complexly formed by proteins and polysaccharides. One explanation, put forward by Dan, is that blastula formation results from adherence of cells to the hyaline layer and osmotic accumulation of fluid in the blastocoel, the latter exerting positive pressure as though a balloon were being inflated. Alternatively, and more likely, blastula formation in echinoderms results from properties of the cell membrane, the fact that cleavage planes are radial, and, again, firm adherence to the hyaline layer. During division, individual cells round up and temporarily lose contact with adjoining cells but retain attachment to the hyaline layer.

The hyaline layer accordingly serves as a surface coat, comparable to the surface coat described for developing amphibian eggs. Since such coats have not been seen in electron micrographs, some workers have denied their existence. This invisibility merely indicates that the coat material is not electron-dense and probably contains little protein, being mainly mucopolysaccharide. The coats can be seen under the light microscope and can be mechanically manipulated. The surface coat is not the only intercellular or supracellular structure formed in the sea urchin blastula; a fibrous layer is also secreted at the inner surface of the blastula wall, as a thin basement membrane or inner surface coat, while desmosomes link cell to cell at their outer edges.

Gustafson and Wolpert summarize the whole process succinctly as follows: "One may thus visualize blastula formation in the following terms. The cells are firmly attached to the hyaline membrane. At each cell division the cells round up owing to increased tension in their own membrane and their contact with each other is reduced. Since the plane of cleavage is radial, the elongating and constricting cells exert a tangential force upon each other leading to both an increase in blastula and blastocoel volume. After cleavage, there is a reduction in tension, and contact in cells is increased, leading to a reduction in volume and blastocoel. There remains, however, an increase in volume which arises from a new packing of an increased number of cells in a single layer."[1] *The original surface of the egg then remains as the outer surface of the blastula,* the new smoother surface being formed in relation to cell division.

The importance of radial cleavage is that cells remain side by side, whereas if the division planes were tangential, multiple layers of cells would result. Increasing the number of cells in the single-layer wall, however, does not necessarily cause expansion. If divisions merely broke up cells into narrower units of the same depth as before, the blastula wall would not expand. In actuality after each division the daughter cells assume the same cuboidal shape as before although on a smaller scale. Inevitably, therefore, the cell depth, or wall thickness, diminishes while the aggregate cell surface greatly enlarges with each successive cleavage. As long as the cells thus formed manage to stay in place, blastula formation and enlargement are inevitable. Since the hyaline layer is not universally present in blastulas of different groups of animals, it may be best regarded as a reinforcement of the innate developmental process.

INVAGINATIONS AND MORPHOGENESIS

The formation of the gastrula in small holoblastic eggs results from the activity of a simple epithelium, i.e., of a blastula wall, one cell thick. Normally the process is an invagination, although, as we have seen, chemical or cell recombination procedures can reverse the relative tissue movements to produce evagination, or exogastrulation. During the development of an egg, gastrulation is prepared for by processes occurring during the blastula stage. It is itself a process whereby the basic organization of the organism is established as a tube within a tube, or as endomesoderm within an envelope of ectoderm. Primarily the event is an infolding of the cells or tissue comprising the vegetal hemisphere within the expanding layer representing the animal hemisphere. We are concerned here mainly with the cell and tissue mechanics responsible for the actual performance, particularly with the initial flattening of the vegetal region of the blastula to form the vegetal plate, and the in-pushing of the center of the plate as an invagination.

Many attempts have been made throughout the present century to explain how the invaginative movement is accomplished, the early interpretations being based on the assumption that the forces which bring about gastrulation result from the pressure of dividing cells acting in a plane of the polar axis, so that the cells of one hemisphere are turned in. This concept was soon tested by Moore in a series of

[1] T. Gustafson and L. Wolpert, Cellular Movement and Contact in Seaurchin Morphogenesis, *Biol. Rev.,* **42**:461 (1967).

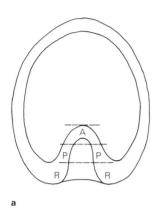

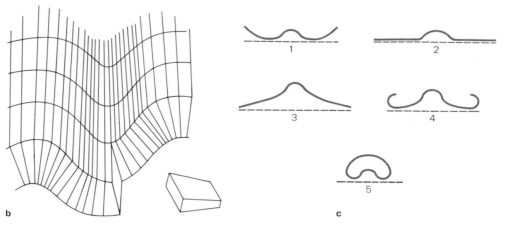

FIGURE 14.2 (a) Outline of normally invaginating gastrula. *A*, the originally active zone derived from the central spot of the gastral plate; *P*, the active zone derived from the plastic zone of the plate, which formed the periphery of the plate surrounding *A*; *R*, the new active ring. (b) Changes in cell shape associated with folding of columnar epithelium. (c) Horizontal view of the positions assumed successively by the gastral plate of starfish *Patiria miniata* during the hour following excision. [*After Moore, 1941.*]

direct operational experiments on starfish embryos, principally by cutting away parts of the embryo above the vegetal plate.

The surprising result of excising the vegetal plate is the continuation of the process of invagination in the isolated plate. The in-pushing deepens, and the rim of the isolated plate rolls up and closes over the central invagination to form a comparatively small gastrula. Experiments such as these show that the forces producing gastrulation act in the plane of the vegetal plate to produce an inward movement of the center, rather than through the vertical, or median, plane. If a radial cut is made passing from the periphery to the center of the isolated plate, the cut edges spring apart, suggesting that the whole plate is normally under tension. It is also noteworthy that in such small, reconstituted gastrulas the periphery of the plate, which normally would become invaginated as endoderm, is changed into functional ectoderm. Moore concluded, "The forces having to do with invagination exist in the endodermal plate and, in intensity and direction, are disposed from the center outward in the form of a radial gradient. It is suggested that differential cohesion of cells may be an important factor in producing pocketing."[1]

This tension hypothesis is supported in various ways. One is that the hyaline layer of the late blastula exerts a tensile force which causes the vegetal cells to buckle in. In starfish embryos particularly, this layer is about 2 μ thick and is present at all stages; yet when it is removed at an early stage by trypsin digestion, gastrulation still takes place, even though it is not entirely normal. The conclusion is that the only proved function of the hyaline layer remains that of holding the blastomeres together and that it is an agent in forming a normal blastula. Other experiments with trypsin indicate that in general an essential part of the forces which determine form is intrinsic rather than extrinsic to the cells themselves.

[1]A. R. Moore and A. S. Burt, On the Locus and Nature of the Forces Causing Gastrulation in Embryos of *Dendraster excentricus*, *J. Exp. Zool.*, **82**:167 (1939).

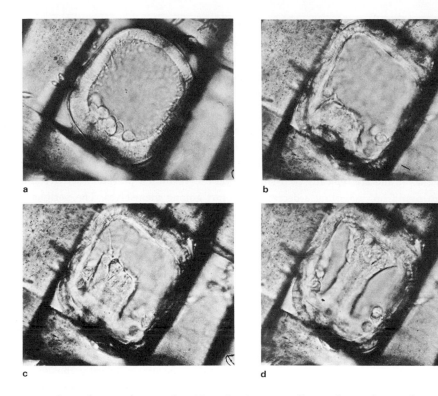

a

b

c

d

FIGURE 14.3   Invagination in the sea urchin showing (a) immigration of primary mesenchyme cells
from vegetal-plate region of embryo, (b) onset
of invagination, (c) cells at anterior tip of invagination extending pseudopodia toward far side of
blastocoel, and (d) completed invagination with
anterior end making broad contact with ectodermal
wall. [*Courtesy of T. Gustafson.*]

Gastrulation of sea urchin blastulas is essentially similar to that in the starfish.
Differences relate mainly to the formation of micromeres in sea urchin development
and to their subsequent role as precociously in-pushing cells destined to construct
the larval skeleton, a structure absent in starfish larvae. Yet in both forms the forces
are confined to the vegetal plate. The most intensive investigations have concerned
the relatively complex gastrulation process seen in sea urchin development, and it
is unfortunate that the simpler process characteristic of starfish and many other
comparable cases of gastrulation, such as that of *Amphioxus*, has not received as much
attention. However, the first phase of gastrulation in all is a bending inward of the
vegetal plate to form a more or less hemispherical structure. It is of interest that
early treatment with trypsin, which is hydrolytic in its action and can break down
a protein, thereby eliminating the action of the protein in development, reduces the
extent of invagination or completely inhibits it.

We have already seen that in sea urchin development the micromeres form
at the vegetal pole. Micromeres transplanted into the wall of the animal half of an
early blastula produce an invagination at that site, in addition to the normal invagination of the host embryo. In addition to their role as cells specialized for skeletal
spicule secretion, they undoubtedly contain an agent actively involved in the gastrulation process. Normally the four initial micromeres continue to divide until they have
formed from 40 to 60 small cells, depending on the species. When the blastula as

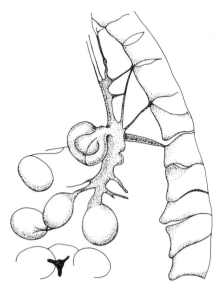

FIGURE 14.4 The pseudopodal attachments of the primary mesenchyme cells to the ectoderm wall. The wall has been stretched, and the pseudopodal contacts at the cell junctions are clear. [*After Gustafson, 1967.*]

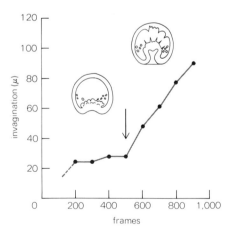

FIGURE 14.5 The time course of invagination of the gut during gastrulation. Ordinate, extent of invagination, in microns; abscissa, the time scale, in film frames. The time between frames is 20 seconds. The arrow indicates the time when pseudopodia were formed. [*After Gustafson, 1967.*]

a whole has become about 1,000 cells, these 40 or so micromere descendants exhibit a pulsatory activity, loosen contact with one another and with the hyaline layer, and move into the blastocoel, before the vegetal plate as a whole shows signs of invagination. Once inside they throw out numerous thin, long, temporary pseudopodia and move about, eventually taking up a characteristic pattern relating to the form of the skeleton finally produced. This phenomenon is itself a matter of great though special interest but is outside the present discussion of invaginative gastrulation.

Following the migration of the primary mesenchyme cells, i.e., the micromere descendants, the cells remaining in the vegetal plate do not lose contact with the hyaline layer and are typically pear-shaped; yet they exhibit a pulsatory activity and tend to round up, indicating a decreased contact among themselves. The periphery of this region consists of cells with full contact. Gustafson has suggested that for geometrical and mechanical reasons this combination of a fixed ring of cells surrounding an area of cells with decreased mutual contact brings about an invagination. This is a debatable point, for it is uncertain to what extent the changes in cell contact primarily depend on (1) decreases in adhesion, (2) the effect of increase of tension at the surface, produced by rounding-up of cells, and (3) active pulsatory movements of cells. However this may be, there is an interesting difference between invaginative gastrulation of *Amphioxus* and of the sea urchin. In *Amphioxus* the invaginating area is relatively very large. Virtually the entire vegetal hemisphere of the blastula is involved, and the invagination is so extensive as to occupy practically the whole blastocoel. The phase of invagination just described for the sea urchin appears to hold for the whole invaginative process, but in the sea urchin itself, where the initial invagination has to be extended into a narrow tube, only the first phase of gastrulation is thus accounted for and a different, second phase completes the process.

The first phase of gastrulation takes the tip of the invaginating tube about one-third of the way through the blastocoel. Then there is a pause, followed by vigorous pseudopodial formation in place of the pulsatory activity. These pseudopodia attach to the outer wall of the gastrula, and their subsequent contraction appears to pull the tip of the archenteric tube forward. The extending cells belong to the sides of the tube, and they in effect pull out two sacs which form the right and left coelomic sacs of the larva. As pseudopodia reach for the gastrula wall ahead, similar pseudopodia reach toward them from the gastrula wall (ectoderm) and at least aid in drawing in the mouth invagination, an in-pocketing similar to the first phase of gastrulation, which meets and fuses with the tip of the archenteric tube, thus forming a continuous tube, the intestine. It is significant that the archenteric pseudopodia extend toward a particular region of the gastrula wall where the mouth invagination is destined to appear, and at the same time pseudopodia from cells of the future mouth region, followed by actual invagination, extend toward them. There is an effect clearly seen at a distance, but it is not clear whether one of the two active regions is stimulating the other or whether there is a mutual reaction between two innately active regions. It is uncertain whether the contractile force of the pseudo-

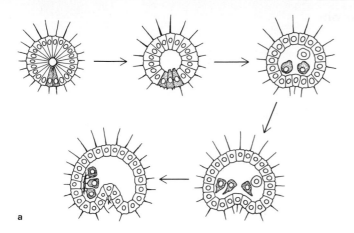

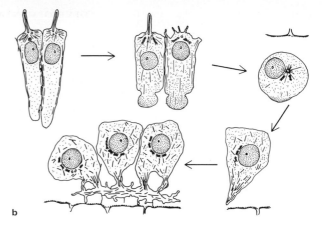

a

b

pods is sufficient to draw the invaginations toward each other, or whether increase in cell number or cell stretching in the invaginating tubes is responsible.

The eggs of many echinoderms, brachiopods, enteropneusts, and *Amphioxus*, all relatively indeterminate, regulative eggs, undergo form changes almost entirely by expansion and foldings of a simple epithelium. The development of *Amphioxus* illustrates this even more strikingly than it is seen in echinoderms. The blastula invaginates to form a gastrula between the ninth and tenth cleavages, when there are about 760 cells present; the invaginating archenteron involves the entire vegetal hemisphere of the blastula and soon obliterates the blastocoel, first as a hemispherical two-layered cup and then as an elongating gastrula with a progressively narrowing blastopore. Subsequently the elongating archenteron evaginates in the median dorsal line to form the presumptive notochord, and as a succession of sacs along each dorsolateral line to form the series of mesodermal pouches. Simultaneously the dorsal ectoderm flattens as the neural plate and invaginates along the median line to form the neural groove and tube. The effect of the various foldings, whether invaginations or evaginations, is to segregate the several regions of the blastula wall corresponding to special ooplasmic territories which are recognizable in the fertilized egg and which are comparable to those seen in the eggs of some ascidians.

FIGURE 14.6 (**a**) Cell-shape changes (shown in black) that each primary mesenchyme cell undergoes during its formation and differentiation. Stage 1, early blastula; stage 2, late blastula; stage 3, newly formed primary mesenchyme cells; stage 4, exploratory stage of the primary mesenchyme cells; stage 5, formation of the cable syncytium. (**b**) Distribution of microtubules at each stage in the sequence. Note that these elements parallel the long axis of the asymmetry of the cell. When the cell is spherical (stage 3) they radiate radially from a central spot. [*After Gibbons et al., 1968.*]

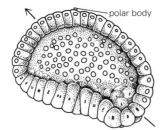

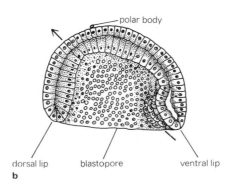

polar body

dorsal lip   blastopore   ventral lip

a

b

FIGURE 14.7 Gastrulation in *Amphioxus*. (**a**) Flattened vegetal-plate stage. (**b**) Complete invagination, with obliteration of blastocoel. Arrow indicates anteroposterior axis of embryo. Polar body marks original animal pole. [*After Conklin, 1932.*]

NORMAL STAGES OF DEVELOPMENT OF SALAMANDER (HARRISON STAGES)

Harrison Stages of *Ambystoma* at 18°C	Age		Developmental Stage
1			Newly laid
2	15½	hours	First cleavage incomplete
	20	hours	Second cleavage incomplete
3	21¾	hours	Four-cell stage
	23½	hours	Third cleavage incomplete
4	25	hours	Eight-cell stage
	26	hours	Fourth cleavage beginning
5	27	hours	Sixteen-cell stage
6	31	hours	Thirty-two cells
9	2	days	Middle blastula
10	4	days	Early gastrula, dorsal lip blastopore
11⁺	4½	days	Middle gastrula, blastopore large
12	5	days	Late gastrula, yolk plug
13	6	days	Blastopore a slit, neural groove
15	7	days	Early neural plate with folds
16⁻	7–7½	days	Neural folds closing, gill plate
17	7½	days	Late neural folds
19	8	days	Neural folds fusing
20	8½	days	Neural folds fused, five or six somites
27	10	days	Optic vesicle, otocyst, gill plate
31	13	days	Olfactory pit, optic cups, gill plate
35	15	days	Balancer, gill plate, rhombocoel
37	17	days	Forelimb bud, external gills, balancer
40⁺	3	weeks	Balancer, two digits on forelimb
46⁻	6	weeks	Stomodaeum, hind-limb bud, coiled gut

ORGANIZATION OF THE AMPHIBIAN BLASTULA

Large holoblastic eggs, all vertebrate, are of another order of size compared with the holoblastic eggs of invertebrates; typically they are about 10 times greater in diameter, or about 1,000 times greater in volume. It is probably safe to regard this kind of egg as the primitive vertebrate egg which has been retained by many groups that continue to shed eggs in freshwater. The development of the eggs of the lamprey, the most primitive of all living vertebrates, is known from one extensive experimental study to be remarkably similar to the development of amphibian eggs up to the formation of a larva. Although difficult to obtain, further experimental analysis is greatly to be desired.

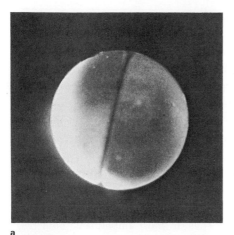
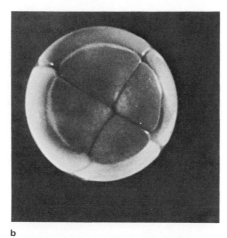
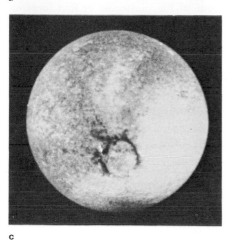
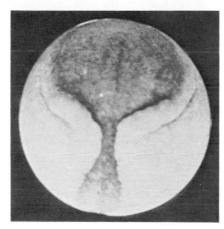

FIGURE 14.8 Stages in development of amphibian egg: (a) two-cell; (b) eight-cell; (c) end of gastrulation and onset of neurulation; and (d) open-neural-plate stage of neurula. [N. J. Berrill.]

The amphibian egg not only is larger than the invertebrate kinds just discussed but contains proportionately more yolk. If a frog egg is centrifuged, for example, so that all the yolk is compacted toward one side, the yolk is seen to occupy about one-half the egg interior. Normally it is distributed as a gradient, so that the ratio of yolk to cytoplasm is least toward the animal pole and greatest toward the vegetal pole. As a result, the egg nucleus resides considerably above the equator. The relative concentration of yolk in the vegetal half of the egg and the location of the nucleus in the animal half have important consequences with regard to cleavage, blastula formation, and gastrulation.

The first two cleavages are both in the plane of the polar axis, at right angles to each other. The mitotic apparatus, however, forms in conjunction with the eccentrically placed nucleus, so that the cleavage furrows first appear near the animal pole and progressively extend toward the opposite pole. The third, or horizontal, cleavage is also influenced by the position of the nucleus, in each of the first four blastomeres, and divides each cell into a relatively small animal cell, or micromere, containing relatively little yolk, and a relatively large, yolky, vegetal macromere. By so doing, it sets up a differential. From this time on, the micromeres divide relatively rapidly, being free from the burden of cleaving through a yolk-dense region, while the macromeres divide very slowly, since the cytoplasm is now loaded with yolk throughout. Yolk granules or platelets have the double effect of conferring physical inertia on the dividing mass and diluting the active cytoplasm with metabolically inactive inclusions.

As cleavage proceeds, a stage is reached which consists of an upper, animal region of numerous small, rapidly and synchronously dividing cells covering a blastocoel, and a lower vegetal region, amounting to at least half the whole embryo and consisting of relatively large, slowly dividing cells. In most amphibians, both frogs and salamanders, the cells constituting the roof of the blastocoel at first form an epithelium one cell thick but soon establish a layer two cells thick, presumably because their rate of multiplication exceeds the capacity of the blastula to expand. In *Xenopus*, however, the blastula roof remains as a single layer. In either case it is possible to distinguish three main regions in the blastula—an upper part around the animal pole, a lower part around the vegetal pole, and an intermediate, or marginal, zone between the others which extends around the equator. In many amphibian species the animal region is most deeply pigmented, while the marginal zone, also pigmented, is typically grayish, in contrast with the commonly white vegetal region. The marginal zone includes the gray crescent, which has already been discussed in connection with the establishment of bilateral symmetry in the fertilized egg. The three regions roughly represent the future three primary germ layers.

The animal region constitutes the presumptive outer germ layer (or ectoderm), the vegetal region constitutes the future inner layer (or endoderm), and the marginal zone becomes the middle layer (or chordamesoderm). Because of the large number of cells constituting the blastula as a whole, on the order of 50,000 to 100,000, the

FIGURE 14.9 Mapping of movement of various areas of salamander egg and embryo by means of vital-dye staining. (a) The method of staining devised by Vogt [Arch. Entwickl.-mech., **106** (*1925*)]. The egg is held against pieces of agar stained with Nile blue or neutral red. (b) Gastrulation movements as seen from the side.

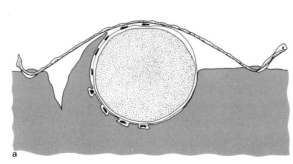
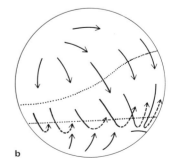

differentials in yolk content and cell numbers between the two hemispheres, and other factors, a full comprehension of the gastrulation movements has been difficult to obtain. Much of what is known comes from Vogt's vital-dye method of analysis, which consists of pressing small pieces of agar impregnated with various kinds of vital dye, particularly neutral red and Nile blue sulfate, against the surface of the blastula and following subsequent shape changes and migrations of the stained patches of the blastula. Vogt was able to construct a fate map showing the future ectoderm, mesoderm, and endoderm—the ectoderm comprising the presumptive central nervous system and epidermis, the mesoderm including the material for the notochord, with myotome material on each side, and the endoderm comprising the primitive gut and its derivatives.

BLASTULATION IN MEROBLASTIC VERTEBRATES

The animal-vegetal cleavage differential seen in the amphibian egg is carried to an extreme in the eggs of bony fishes and of the truly terrestrial vertebrates, two groups which are assumed to have evolved from primitive freshwater fish with amphibian-type eggs and embryos. In both cases, irrespective of egg size, yolk is so densely packed and so fills the interior of the egg that only a cap of polar cytoplasm, or blastodisc, containing the nucleus, is capable of division. The first cleavages consequently produce blastomeres which are separate from one another where they adjoin but are continuous with the underlying mass of yolk. Subsequent mitoses segregate them entirely, as a cellular cap, the blastoderm, from the undivided yolk region of the egg. Early cleavage is therefore partial, and development is meroblastic. The blastoderm which develops on top of the yolk increases by continued cell division and gradually envelops the yolk, although the embryo forms from only a part of it, the so-called embryonic shield. Gastrulation in these forms is, therefore, at least in large part, the process of overgrowth, or epiboly, whereby the undivided yolky region of the egg is engulfed by the spreading margin of the blastoderm.

In meroblastic eggs generally, the cleavage and blastula phase is a period during which a cell size characteristic of the species is being established in the blastoderm, which thus acquires the properties of a stable epithelium. The blastoderm, consisting of the layer, or mound, of cells now completely separated from the underlying yolk, is anchored to the yolk at its margin by the periblast, which is a peripheral continuation of the epithelial cellular layer on to the yolk syncytium, where partial division of the cell cortex continues. In the teleost the rapid spreading of the blastoderm is combined with thickening and invagination of material at the peripheral edge, or germ ring, where the blastodermal cells roll inward above the periblast.

Both the blastoderm and the periblast have an intrinsic capacity to spread. Normally blastoderm and the periblast spread over the yolky sphere toward the vegetal

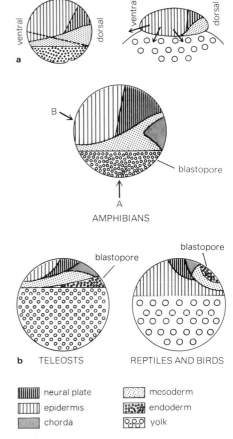

FIGURE 14.10 (a) (*Left*) The basic vertebrate map, with a line of folding ("the primitive edge") indicated by a broken line. (*Right*) The blastula squashed down on to a mass of yolk, having folded along the primitive edge. If such a deformation is to produce the bird map, there will also have to be an expansion of the epidermal area. (b) Maps of the presumptive areas in blastulas, seen from the left side. In the upper figure, *A* indicates the place into which the yolk mass would have to be inserted to produce the teleost condition, while *B* shows the place of insertion of the yolk mass to produce the amniote arrangement. [*After Waddington, 1952.*]

FIGURE 14.11 (**a–c**) Behavior of a chick blastoderm explanted with normal orientation onto a vitelline membrane. In (**b**) adhesive undersurface attaches to the vitelline membrane. In (**c**), normal expansion follows, as shown by diagram of whole preparation. (**d–f**) Behavior of a blastoderm inverted on a vitelline membrane. In (**e**), the blastoderm edge curls under to bring adhesive surface against the vitelline membrane. In (**f**), expansion is reversed in direction and results in formation of a hollow vesicle. Shading in (**a**), (**b**), (**d**), and (**e**) indicates adhesive surface. Thick line in (**c**) and (**f**) denotes ectoderm; broken line denotes endoderm. [*After New, 1959.*]

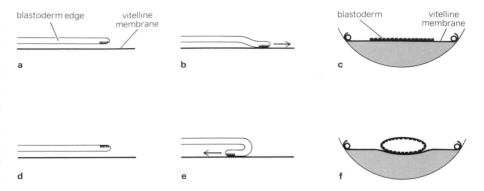

pole, the periblast extending ahead of the blastoderm. In fact the periblast can complete epiboly in the total absence of the blastoderm. The blastoderm, however, ordinarily completes its epiboly by spreading over the periblast and catching up with the periblast margin, until finally the whole of the yolk is enclosed by the two layers. The periblast substratum in fact appears to be essential for blastoderm spreading. Much the same picture seems to hold for the bird and reptile egg, although the processes here are less readily observed and may be more complex.

Again, however, the need for a proper substratum seems to be important in the expansion of the cell sheets. Chick blastoderm, for example, does not expand when cultured on agar or plasma substrates, but when supplied with the normal substratum, i.e., the inner surface of its vitelline membrane, it adheres and expands over it at the normal rate, even in culture. Only the extreme edge of the blastoderm adheres strongly to the vitelline membrane, the remainder being more or less unattached, and the only cells that adhere and spread are those normally at the margin, suggesting that the marginal cells are significantly different from the rest. Mechanical tension imposed by the spreading margin is thought to be necessary for blastoderm expansion.

As Trinkaus has said, "Although analysis of morphogenetic movements is as yet in a rudimentary state, it has already revealed certain modes of behavior that may apply generally. Epithelial layers expand primarily at the margin in a number of systems. How the marginal cells do this and how the central cells respond to the tension thus created are urgent problems. It now seems possible that organisms may achieve cellular rearrangements unrivaled for their complexity without recourse to highly specific interactions, delicately oriented substrata, long-range taxes and the like. Conceivably they might manage all this by means of quantitative differences in a few properties, such as cellular adhesiveness and motility, expressed in a simple, but defined topography."[1]

[1]J. P. Trinkaus, Mechanisms of Morphogenetic Movements, in R. L. DeHaan and H. Ursprung (eds.), "Organogenesis," p. 97, Holt, Rinehart and Winston, 1965.

INTEGRATION OF GASTRULATION

The first indication of the onset of gastrulation in amphibian development is an infolding starting on the borderline of the vegetal region and exactly below the center of the gray crescent area, i.e., in the marginal zone at a point lying on the axis of bilateral symmetry, so as to form a narrow groove or slit. The fold above the groove represents the upper, or dorsal, lip of the developing blastopore.

As the inward movement of tissue continues, the two ends of the groove extend horizontally around the embryo. Eventually they meet on the side opposite the starting point and consequently complete the circle of the blastopore. Although this in-rolling, or involution, of tissue is generally called *invagination,* it is not strictly so, since it is not a substantial infolding of a sheet of tissue, as in the forms previously described, but is a flowing or streaming movement of the cells toward and around the blastoporal lip, as shown by the elongation of dye spots that were initially round in shape.

Meanwhile the upper, ectodermal layer of cells, in-rolling along its edge, appears to extend progressively over the lower, vegetal region. This is particularly apparent in frog embryos, where animal-pole cells are deeply pigmented and vegetal cells are white. The ectodermal sheet does in fact expand, as the cells continue to divide and to spread out, and eventually the originally sickle-shaped blastopore becomes first an open and then a closed circle of steadily diminishing size. Finally a small yolk plug surrounded by dark ectoderm is all that remains to be seen of the mass of vegetal cells, and at last only a narrow vertical slit denotes the blastopore and the termination of gastrulation. At this time the ectoderm alone covers the surface of the embryo. However, some involution (rolling-under) continues in the lips of the blastopore.

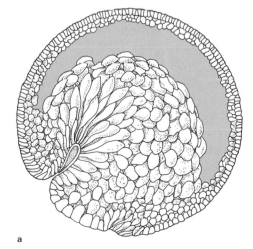

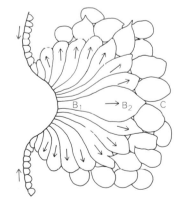

a   b

FIGURE 14.12 Invagination in amphibian egg. **(a)** Slightly schematized section through an advanced gastrula. **(b)** The mechanics of gastrulation, showing active extension of cells. [*After Holtfreter, 1943.*]

The apparent epibolic growth of an expanding pigmented ectoderm over a seemingly passive unpigmented vegetal region is to some extent an illusion; the vegetal region is actually as dynamically involved as the animal region (even though the constituent cells are comparatively large, few, and yolky), for simultaneously with the flowing-downward and -inward stream of presumptive notochord and mesoderm around the extending rim of the blastopore, the presumptive endoderm streams upwards and inwards in a similar but grosser manner, the upper and lower streams rolling in together in a marvelously integrated motion. The full impact and beauty of this event can be conveyed only by time-lapse cinemaphotography, for in nature the process is too slow to be visually appreciated and neither words nor still pictures can adequately portray the reality. Such a film is one on the development of the salamander made by L. S. Stone some time ago but not likely to be surpassed.

Internally the gastrulation event includes the movement of the chordamesoderm and the endoderm into their final positions. As the first shallow groove of the blastopore deepens, the archenteron, or primitive gut, begins to form, and during the continuing process of invagination the archenteron lengthens, somewhat in the manner of the lengthening of the archenteric tube in the sea urchin gastrula. It does so until the blind end reaches the inner surface of the ectoderm opposite the blastopore, where later a small invagination of the ectoderm fuses with the archenteron. The ectodermal invagination forms the mouth cavity, or stomodaeum; the adjacent endoderm becomes the pharynx.

As the archenteron begins to elongate, its anterior roof is formed by material of the prechordal plate which rolls over the dorsal lip of the blastopore, advancing in front of the anterior end of the notochordal material. Its floor and sides consist of endoderm. Most of the mesoderm invaginates by rolling over the lateral and ventral lips of the blastopore. The notochord and the mesodermal material at this stage form one continuous sheet, the chordamesodermal mantle, lying between the ectoderm and the endoderm, the lateral parts of which grow downwards between the ectoderm and the free margin of the endoderm. The three germ layers are then in their definitive positions.

Various theories have been proposed to account for the orderly integrated movement of cells responsible for amphibian gastrulation. Principally the debate is whether some supercellular agency acts to produce the overall movement, with individual cells caught up and passively carried along, or whether the whole action can be explained as the sum of the parts, i.e., as stated by Holtfreter in his classical account of this process: "The directed movements can be traced back to basic faculties of the single cells and to their specific response to changes of environment. The unitarian character of their combined effort is mainly the result of the predisposed arrangement of cells with a locally different kinetic behavior. The controlling supercellular forces can be localized and defined in physico-chemical terms."[1] Much of this interpretation rests on the properties of a surface coat of the egg and embryo

[1] J. Holtfreter, A Study of the Mechanics of Gastrulation, *J. Exp. Zool.*, **94**:262 (1943).

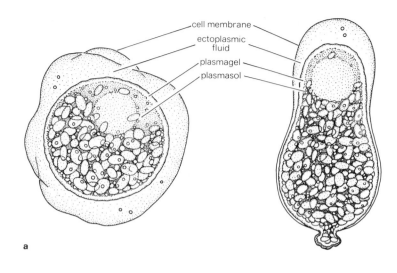

cell membrane
ectoplasmic fluid
plasmagel
plasmasol

a

b

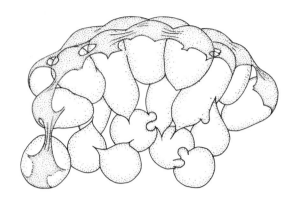

FIGURE 14.13 (**a**) Protoplasmic structure of a globular and an elongate cell from an amphibian gastrula. (**b**) Fragment from early blastula showing dark surface coat and filiform and knoblike processes interconnecting the blastomeres. [*After Holtfreter, 1943.*]

comparable to the hyaline layer of the echinoderm blastula. The interpretation remains controversial because the very existence of the coat has been challenged by workers who find no sign of it in electron micrographs.

Concerning the hyaline layer of the sea urchin egg, this contention may mean no more than that the coat consists of material which is not electron-dense, and the negative evidence cannot weigh too heavily against the circumstantial description of coat properties given by Holtfreter. For instance, "The significance of the coat for mass movements derives from its mechanical properties. The coating substance possesses plastic elasticity. Isolated parts of it can be stretched considerably between two glass needles, after which they contract into small lumps. The tension at the surface of an amphibian egg is not an expression of the surface tension of the naked cytoplasm but is due to the elastic properties of this film which is solid and non-sticky at the surface, but sticky and more fluid at a deeper cortex, where it is closely amalgamated with the liquid cytoplasm."[1] It is, in other words, an integral part of the egg cortex, and it persists as the outer surface of cells, at least through cleavage and blastula formation. Positive evidence for the existence of such a coat comes from focused ultrasonic treatment, which has been used to isolate a coat from young frog embryos. Later, it cannot be isolated without attached cells.

Summing up, Holtfreter describes the roles of cells and surface coat as follows: "(**1**) The shape of individual cells reflects the direction of their movement. (**2**) Invagination is everywhere associated with a temporary transformation of the constituent cells into flask-shapes. (**3**) The boundaries of the inner cells are not continuous like the wax in a honeycomb, but each cell is a discrete unit, interconnected with its neighbors only by slender processes. (**4**) All peripheral cells are firmly held together by a common surface film of plastic elasticity.

[1] *Ibid.,* p. 277.

"When the cells of an amphibian gastrula enter into the phase of morphogenetic movements, it is in the first place the investing coat which integrates the amoeboid activity of the individual cells into the coordinated and synchronized movements of entire germ layers. By virtue of this superficial elastic sheet which cements the peripheral cells together in a continuous layer, gliding movements of the cells in any region of the gastrula are transmitted to adjacent regions. Similar effects of a tangential pull can be observed in a layer of ectoderm which is proceeding to close in on a wound inflicted to this layer. It should be emphasized, however, that the motive forces for the gastrulation movements of spreading and invagination cannot be ascribed to the properties of the coat, but originate within the living cells proper. Embryonic cells which are not held together by a syncytial coat are perfectly capable of spreading over an organic or inorganic substratum, and they will slip, singly or in groups, into the depth of a layer of endoderm with which they have been brought into contact. The coat merely regiments these amoeboid activities so that they become collective events. The extraordinary faculty of epidermal wound healing demonstrates that the spreading potency is by far greater than is normally manifested. It resides in every isolated ectodermal cell where, in the presence of a proper substratum, it is displayed in the form of centrifugally expanding fans and membranes radiating substructures."[1]

During invagination every cell and every part of the embryo participate; the total event consists of three simultaneous movements. The future ectoderm expands and covers the whole surface of the gastrula stage, partly by continuing cell division but especially by thinning or spreading of the epithelial layer. This spreading tendency is inherent in the cells of both the superficial and the inner ectoderm. A grafted piece spreads in all directions over a surface, and the expansion of the presumptive ectoderm in normal development is readily explained as a spreading which merely follows the path of least resistance, replacing the mesoderm and endoderm as they move inward.

One layer thus imposes direction of movement on another. The endoderm also has an innate tendency to spread, although its spreading capacity is limited, by the spreading ectodermal sheet, to the amount required for normal gastrulation. The general picture of amphibian gastrulation that emerges may be summarized as follows:

1   The superficial endodermal cells of the marginal zone sink into the deep endoderm as bottle cells. Since their outer ends cohere tightly together, the surface begins to pocket and the archenteron begins to form.

2   Prospective mesoderm, which already has an inherent tendency to stretch in the anteroposterior direction, follows and spreads on the highly adhesive lower surface of the ectoderm.

3   At the same time the spreading ectoderm cells expand over the invaginating mesoderm and replace it as it disappears from the surface.

[1] J. Holtfreter, Significance of the Cell Membrane in Embryonic Processes, *J. N.Y. Acad. Sci.*, **49**:740 (1948).

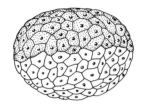
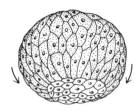

a

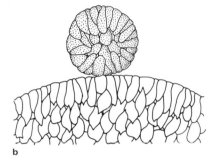
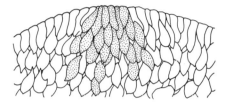

b

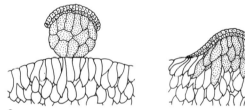

c

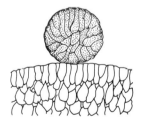

SECTIONS

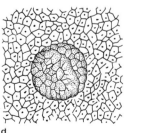
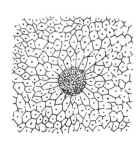

d

SURFACE VIEWS

FIGURE 14.14 Spreading and segregation of tissues in reaggregated cell masses of amphibian embryos. (a) Coated endodermal cells spread over uncoated area. (b) Aggregate of uncoated endoderm becomes incorporated into endodermal substratum. (c) Endodermal graft covered by ectoderm invaginates into endoderm; ectoderm spreads at first and then becomes isolated. [*After Townes and Holtfreter, 1955.*] (d) A graft of blastoporal cells partly covered by the coat sinks into an endodermal substratum and forms a blastoporal groove. [*After Holtfreter, 1943.*]

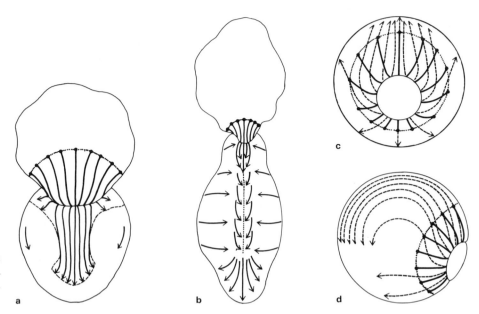

FIGURE 14.15 Exogastrulation in amphibians (*Axolotl*). **(a,b)** The mass movements are compared with **(c,d)** normal gastrulation. [*After Holtfreter, 1933.*]

The superficial endoderm, which is similar to the ectoderm in having nonadhesive outer surfaces and therefore low average cohesiveness, spreads extensively to line and consequently enlarge the archenteron. The greater spreading tendency of the ectoderm confines the spreading of the endoderm to the lining of the archenteron and thus prevents exogastrulation. The above account, however, is still hypothetical, and questions remain as to whether the quantitative differences in cell-surface adhesiveness actually exist and, if so, how cells adhere to one another during these formative movements, etc.

Studies on aging, unfertilized frog eggs have shown some remarkable similarities with gastrular movements of material. Unfertilized eggs kept in good condition exhibit streaming movements very similar to those seen in fertilized eggs, and the pigmented surface layer covering the animal hemisphere of the egg proceeds to expand over the unpigmented lower hemisphere until the whole egg is engulfed by it. Moreover, in most eggs a gray crescent appears, separating the upper pigmented zone from the more or less obliterated white zone; at this stage it is difficult to distinguish such eggs from the normal late-stage blastula. Again, diffuse streaks of darker pigment radiate in some cases from the animal zone across the dorsal gray crescent towards an imaginary blastopore; in other cases the pattern of surface pigment resembles the arrangement of marginal cells in a gastrula. In other words, both the epiboly of the prospective ectoderm and the cellulation and convergent stretching of the dorsal mesoderm are simulated by the aging eggs. These observations led Holtfreter to suggest that the ectodermal spreading movement in normal development is the expression of a maturation process of the protein compounds involved, comparable to the unfold-ing of reversibly denatured protein molecules in monolayers spreading on the surface of a substratum with appropriate adsorption properties.

BLASTODERMIC GASTRULATION

The process of gastrulation in meroblastic vertebrate eggs is in part included in the process of epiboly already described, inasmuch as epiboly results in the enclosure of the yolk and periblast by the blastoderm. The future embryo, on the other hand, forms within a localized area of the dorsal part of the blastoderm.

In teleosts a local thickening appears at the edge of the germ ring and extends centrally for some distance toward the animal pole in the middle of the blastoderm,

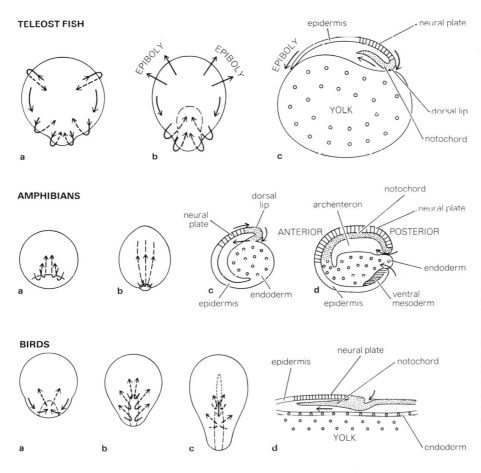

FIGURE 14.16 Different types of gastrulation in vertebrates. *Teleost fish:* (a) surface view of embryonic disc at early gastrula stage; (b) same, at late gastrula stage; (c) sagittal section of mid-gastrula stage. *Amphibians:* (a) external view of early gastrula; (b) external view of late gastrula; (c) sagittal section of early gastrula; (d) sagittal section of late gastrula. *Birds* (surface views of embryonic disc): (a) early gastrula stage; (b) mid-gastrula (definitive streak) stage; (c) late gastrula (head process) stage. Full arrows indicate surface tissue movements; dotted arrows, movements of cells below surface. (d) Sagittal section of stage (c). [*After Deuchar, 1965.*]

forming the primitive shield; with further growth the adjacent edges of the germ ring converge and fuse. During this whole process, which shows clear resemblance to amphibian blastoporal gastrulation, the embryonic axis results from a combination of outward migration and axial convergence of deep cell layers of the blastoderm, according to Ballard (1966).

The role of the posterior embryonic shield has been analyzed by Brummett (1968) by grafting and deletion experiments on *Fundulus* embryos. The posterior shield, removed from early gastrulas and inserted in the pericardial cavity of slightly older host embryos, exhibits considerable growth and differentiates typical trunk and tail structures involving all germ layers. On the other hand, the donor embryos develop normal heads and pectoral regions but lack almost all trunk structures. These and many collateral experiments show that the posterior embryonic shield supplies the bulk of the material for the formation of the trunk and tail, as might be expected.

MULTIPLE INDUCTION

Fish

In fish, conditions at the time of induction are labile, contrary to expectations derived from experience with amphibians. If, for instance, halves of two blastoderms from either side of the incipient embryonic shield of a fish are removed and fused together, a new organization center forms, and a normal embryo is produced. On the other hand, if a blastoderm sector containing an already defined embryonic axis is removed, two embryos arise, with their axes not at right angles to the blastodermal edge but actually lying along it, facing each other across the gap. The two territories thus involved may well correspond to the lateral regions of tissue converging toward the original organization center. If so, regional territories possess the total potentiality of the whole.

Twinning, which is the production of two or more embryos from a single blastoderm, can be induced by nonoperative treatments, however. Stockard, half a century ago, found that subjection of *Fundulus* eggs to cold or to reduced oxygen tension, for a period shortly before embryonic shield formation begins, results in a percentage of double monsters, either partial or complete twins, of various degrees of separation. Later, Hinrichs (1938) found that exposure to ultraviolet radiation results in complete double embryos. In all such cases, since only the embryonic axes are initially duplicated, attachment is to one and the same yolk sac, and embryos end by being conjoined abdomen to abdomen, like Siamese twins.

The chick blastoderm appears to be comparable in many ways to that of fish, although in detail the cell and tissue movements may be markedly different. Multiple embryos may be induced in essentially the same ways and at much the same critical period as in teleosts, but, mainly for practical reasons, much more experimental investigation has been carried out on fish.

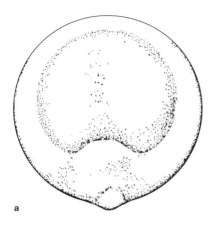

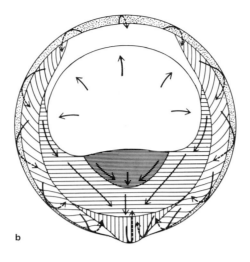

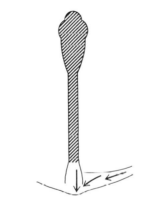

FIGURE 14.17 Gastrulation movements in the trout. (a) Early stage, showing the whole blastodisc. (b) Movements taking place in the surface layer (solid arrows) and those occurring below the surface (dotted arrows). (c,d) Stages in the formation of the embryonic axis. [*After Pasteels*, Arch. Biol., **51**:335 (1940).]

FIGURE 14.18 (a) Twins and (b) partial triplet in teleost.

Bird (or Reptile)

In the chick the two aspects of gastrulation, namely, the advance of the blastodermal margin and the establishment of a visible embryonic axis, are regionally separate processes. The growth of the blastoderm margin is exclusively concerned with enclosure of the yolk within a yolk sac, while a primitive streak, corresponding to the primitive shield of the teleost, first appears at some distance from the edge of the blastoderm and extends toward the center. The thickening of the blastoderm constituting the primitive streak represents the dorsolateral margins of the amphibian blastopore, here fused as a single structure; while the thickened anterior end of the streak, known as the *primitive knot,* or *Hensen's node,* represents the dorsal lip; the primitive ridges correspond to the lateral lips.

Though the processes of cell migration and proliferation involved in the establishment of the primitive shield and primitive streak, respectively, have long been the subject of intensive investigation they are still imperfectly understood. Marking experiments by Spratt (1946), made on blastoderms, just before onset of streak formation, the blastoderms then being grown on plasma clots so that movements could be more readily followed, showed that surface cells move as a sheet, at first from the posterior margin and later from the lateral blastoderm as a whole, to congregate as the narrow, elongate streak which thus defines the future embryonic axis. Many of the cells then sink in the middle line to a deep position to take part in the formation of the endoderm. Later, cells push forward from the region of Hensen's node to form the notochord. The principal movements of blastoderm tissues are shown in the accompanying diagrams.

In general, tissue sinks in along the center line of the streak and then moves laterally and forward beneath overlying ectoderm and above subjacent endoderm, to form the chordamesoderm, which is equivalent to the chordamesodermal sheet

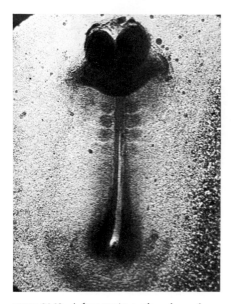

FIGURE 14.19 A four-somite turtle embryo, showing head fold and developing brain, neural tube open posteriorly as a neural groove, four pairs of mesodermal somites with lateral mesodermal plate extending posteriorly on each side of the neural groove, and primitive node. [N. J. Berrill.]

FIGURE 14.20 General relationship of the regressing node center to the elongating axial parts of the chick embryo. The heavy lines and dots illustrate the changing positions of carbon-marked epiblast (ectoderm) cells. Time increases from left to right in the diagrams. Note the more rapid and extensive regression of the primitive streak relative to marked areas lying lateral to it. [*After Spratt, 1946.*]

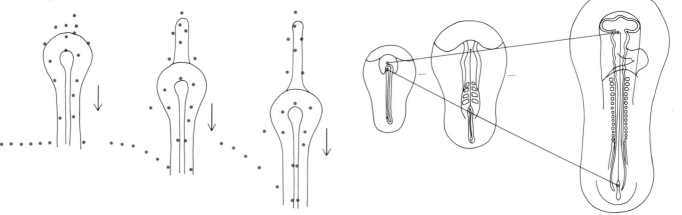

of the amphibian embryo. The presumptive notochordal tissue underlies the midline of the embryonic axis, with lateral mesoderm at each side. As this process continues, with such tissue moving forward to the limits of the embryo-to-be, the primitive-streak tissue undergoes a corresponding attrition at its anterior end, so that in effect the streak becomes progressively shorter and seems to shift posteriorly.

The potentiality and lability of the chick blastoderm preceding the establishment of the primitive streak have been demonstrated, as in fish, both by environmental conditioning and by operative procedures. Because development of a bird egg normally proceeds at an incubation temperature of 40°C, the system is comparatively sensitive to temperature change. Cooling the developing egg of a chick at the critical preprimitive-streak stage retards or arrests the developmental process. Such cooling may be produced by a simple physical change such as cytoplasmic gelation; in any case it is reversible. Following the return to normal incubation temperature, two or three primitive streaks commonly appear on the blastoderm in place of the one which would have developed normally, and consequently two or three well-formed embryos develop. However, since they are associated with the same yolk sac although each has its own amnion and allantois, they become abdominally conjoined at hatching, as in fish.

Operative procedures in chicks offer a wider range of experiments:

1 If the early blastoderm is cut into two pieces while it is still on top of intact yolk, twins develop.

2 If the blastoderm is cut into four pieces by two cross-cuts and grown on a culture medium, each piece develops a well-formed embryo with its head toward the original center of the blastoderm.

3 Conversely, if three blastoderms, each having had its posterolateral parts cut away, are joined so that the three prospective organization centers unite in the middle of the fused disc, three embryos form, each with its head toward the middle, i.e., the original blastodermal polarity is reversed.

4 Similar results are obtained in giant blastoderms (four times normal size) made up by fusion of up to eight parts of eight individual blastoderms.

In all such cases, embryos are formed from cell populations that otherwise would have contributed either to the formation of extraembryonic tissues or to entirely different parts of the embryo. The blastoderm is evidently a two-layered disc of tissue that is essentially undetermined with regard to specific embryonic axes until after considerable growth or expansion has occurred. The number of cells present in the blastoderm at the period critical for irreversible axis determination is very great. In fact, even by the time an egg has been laid, after having been fertilized as it entered the upper end of the oviduct, some 60,000 small cells are already present, constituting the initial blastoderm, about 2 mm across and already consisting of an upper and a lower layer of cells. However, probably only about 500 of these cells, and their descendants, participate in forming the embryo body.

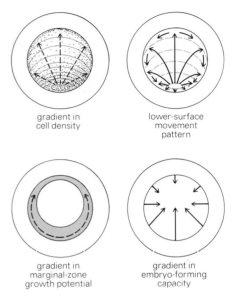

gradient in cell density

lower-surface movement pattern

gradient in marginal-zone growth potential

gradient in embryo-forming capacity

FIGURE 14.21 Important properties of the unincubated blastoderm. [*After Spratt and Haas, 1960.*]

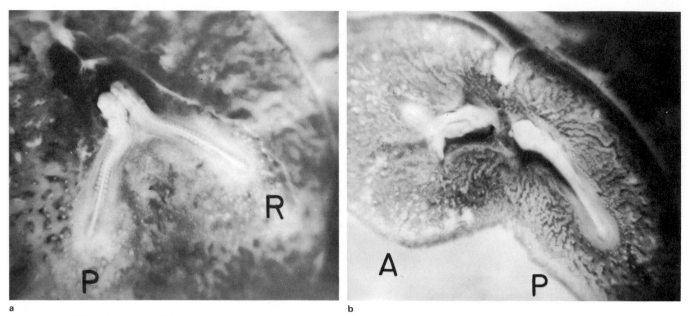

a

FIGURE 14.22 Chick twins produced by cutting unincubated blastoderms. [*Courtesy of Nelson Spratt, Jr.*]

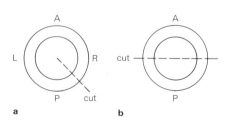

FIGURE 14.23 The angle and extent of cuts yielding the results shown in the photographs of Figure 14.22.

Upon incubation of the egg outside the body, differential growth occurs in the blastoderm, which exhibits radial and polarized patterns of differential rates of cell multiplication in various regions of the cell population. The outcome of this differential growth is a ring-shaped, lower-level *cell-population-density gradient* on which is *superimposed a bilaterally symmetrical cell-population-density gradient* with maximal density on one side where the primitive streak is normally initiated.

According to Spratt, there is a gradient in the embryonic-axis-forming capacity of isolated pieces which is congruent with the bilaterally symmetrical gradient in cell-population density and growth potentiality. He considers that the important property of the blastoderm or isolated part of a blastoderm is the presence of a group of rapidly dividing cells located eccentrically. Here the embryo body is initiated. The cells of this growth center tend to spread out radially as a lower cell layer (hypoblast) along the underside of the radially symmetrical and immobile upper layer (epiblast). The spreading lower cells form a coherent sheet which moves like a viscous fluid away from its source until it reaches the marginal zone of the blastoderm, when the cells at the edge move circumferentially, to the right and to the left. The course, or pattern of movement, of the lower sheet of cells is accordingly like a fountain, and it appears to be entirely a consequence of the geometry of the blastoderm.

In normal development the axis of symmetry of the lower-layer movement pattern and the position of the head-tail axis of the future embryo coincide. If the main axis of the movement pattern is experimentally shifted, the head-tail axis is correspondingly shifted. If the lower-layer movement is blocked, no embryo develops.

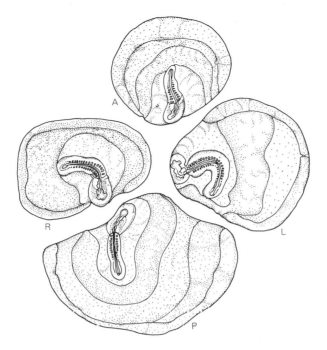

FIGURE 14.24 Potentiality and polarity in chick blastoderm. Four individuals resulting from division of blastoderm into four segments. [*After Spratt and Haas, 1960.*]

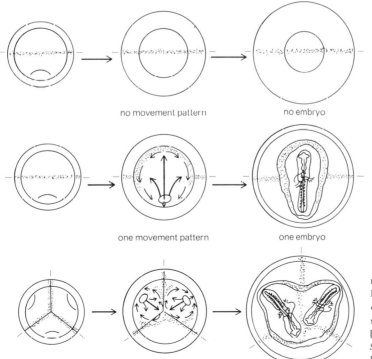

no movement pattern

no embryo

one movement pattern

one embryo

two movement patterns

two embryos

FIGURE 14.25 Bilateralizing role of the lower-layer movement pattern (as shown by movements of applied carmine particles), which gives rise to single or twin individuals as the case may be. [*From "Introduction to Cell Differentiation" by N. T. Spratt, copyright © 1964 by Litton Educational Publishing, Inc., by permission of Van Nostrand Reinhold Publishing, Inc.*]

If early blastoderms are fused to form a giant blastoderm, several independent move-ment patterns appear in the lower layer, each establishing its own embryo axis. The establishment of an axis or axes, therefore, appears to be a consequence of a dynamic flow or spreading of the lower-level cell sheet, the sheet representing a more or less equipotential system as long as movement persists, and each portion, if large enough, possessing the capacity to form a whole when isolated from the rest.

Mammal

In the true mammals, where yolk and a confining shell have been lost, cleavage results in the formation of a cap, or disc, of cells at the animal pole and a cellular envelope enclosing the space where the yolk would have been. The embryo proper develops from the polar disc. In fish, reptile, bird, and monotreme mammal, the disc expands as the blastoderm, and marginal growth eventually encloses the yolk mass within a yolk sac, the wall of the yolk sac consisting of ectoderm, mesoderm, and endoderm. In the placental and the marsupial mammal, the outermost layer of the sac persists as the *trophoblast,* separated from the endodermal innermost layer which retains the name of yolk sac.

Early mammalian development The mature mouse ovum has a diameter of about 95 μ, and it has five to ten times as much DNA, some of which is mitochondrial, as

DEVELOPMENTAL TIMETABLE OF MOUSE

Days of Gestation	Stage
1	One- to two-cell, in uppermost part of oviduct
2	Two- to sixteen-cell, in transit to uterus
3	Morula, in upper part of uterus
4	Free blastocyst in uterus, shedding of zona pellucida
4½	Beginning of implantation
5	Inner cell mass elongating, primitive streak evident, and proamniotic cavity
6	Implantation complete, extraembryonic parts developing
7	Ectoplacental cone, amniotic folds, primitive streak, heart and pericardium forming, head process evident
7¼	Early neurula, neural plate, chorioamniotic stalk, allantoic stalk beginning, inner cell mass with three cavities, somites beginning to differentiate, foregut present
19–20	Birth

normal diploid somatic cells. It is surrounded by a noncellular follicle-cell secretion known as the *zona pellucida,* which plays an important role in the fertilization process. Cleavage is total, as might be expected in such a small and almost yolkless egg. The timetable of the early developmental events, up to the establishment of the primary embryonic structures, is shown in the accompanying table.

In size and general appearance the mammalian egg closely resembles the small eggs of many marine invertebrates and the protochordates. Nevertheless its development, except for the return to holoblastic cleavage, differs in important ways. To begin with, eggs do not all cleave at the same time intervals, nor are all the cleavages of a single egg complete, so that in many cases there may be uneven numbers of cells. By the 16-cell stage nucleoli have become prominently active centers of RNA synthesis. The morula becomes a blastocyst in the 32- to 64-cell stage by acquiring an eccentrically placed, slitlike cavity filled with fluid. The blastocyst, however, differs from the blastula seen in the sea urchin or in *Amphioxus* and shows kinship with meroblastic development, for a single layer of lining cells, the trophoblast, encloses the cavity of the blastocyst except at one side, where a knob of cells forms the *inner cell mass.* The inner cell mass represents the embryo-to-be, while the trophoblast represents the extraembryonic blastoderm that spreads over the yolk mass of the meroblastic egg.

The blastula shortly becomes the free blastocyst, escaping, or "hatching," from the enclosing zona pellucida, a process which includes both an enlargement and a rhythmic undulating movement. However, if the zona is digested off with pronase, normal hatching takes place. The functions of the zona includes maintenance of the normal cleavage pattern and prevention of egg fusion. Implantation begins shortly after, during which the trophoblast cells in contact with the uterus transform to giant cells, while the inner cell mass changes in form and size. Similar changes in the rabbit are thought to be initiated and controlled by a protein, named *blastokinin,* which is secreted by the uterus into the uterine cavity. The inner cell mass, during implantation, undergoes gastrulation, inasmuch as the part nearest to the blastocoel splits off to form the endoderm, while the remaining cells form the ectoderm. Accordingly we see a situation resembling that in the chick blastodisc, where there are upper- and lower-level sheets of cells. Mesoderm forms, however, from cells moving in between ectoderm and endoderm, while the primitive streak becomes evident as a thickening of the embryonic ectoderm. Usually a single embryonic axis (primitive streak) forms, but twinning is frequent among mammals generally; in one species of armadillo the production of quadruplets from a single blastodisc is the rule.

The question is: What goes on during the cleavage phase, that is, during the process of blastulation, that gives rise to the fully formed blastoderm of the fish and bird, preceding the first sign of embryonic shield or primitive streak, and to the mammalian blastocyst at the time of implantation, when it consists of trophoblast and disclike inner cell mass at one pole?

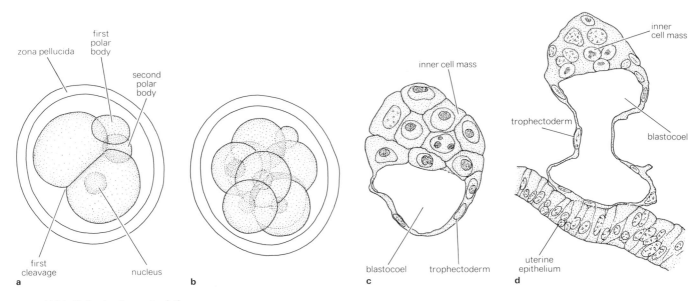

FIGURE 14.26 Early development of the mouse, showing **(a)** first cleavage (with polar bodies), **(b)** the eight-cell stage, **(c)** the blastocyst with inner cell mass and trophectoderm, and **(d)** the implantation stage. [*After Rugh, "The Mouse: Its Reproduction and Development," Burgess, 1968.*]

Two related questions arise: **(1)** When does the developing egg become activated with regard to germ-layer and embryo formation? **(2)** What is responsible for the frequent arrest of development prior to implantation such as is typically seen in lactating mice?

With regard to the first question, conventional chemical methods, sensitive bioassays, as well as application of labeled ions, show that a far-reaching change in the metabolic picture takes place during the period of high growth accompanying implantation. Concerning the second question, implantation, i.e., continuous development, does occur in the absence of ovarian hormones in extrauterine sites. Weitland and Greenwald (1969) suggest that the absence of estrogen is responsible for delayed implantation in pregnant, lactating mice and that the ability of blastocysts to implant in extrauterine sites without hormonal influence is due to the effect of progesterone in the absence of estrogen. However, since blastocysts do not begin protein synthesis, or implantation, in the absence of progesterone, the progesterone cannot be the inhibitor.

NEURULATION IN AMPHIBIANS

At the close of gastrulation of the amphibian embryo, when the yolk plug finally disappears and the blastopore closes to a dorsoventral slit, the embryo is still almost spherical and evenly pigmented. Shortly afterwards a slight longitudinal furrow appears on the dorsal side, anterior to the blastopore slit, around which forms a flattened pear-shaped area, the neural, or medullary, plate. The cells of this area become

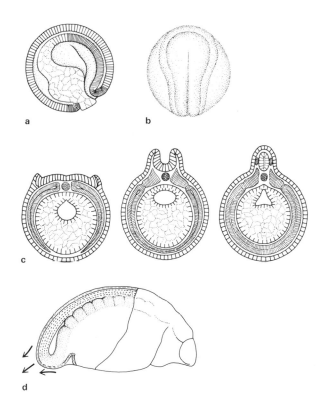

FIGURE 14.27 Neurulation in urodele amphibian embryo. (a) Sagittal section through late gastrula showing chordomesoderm (archenteron roof) underlying ectoderm. (b) Surface view of open-neural-plate stage. (c) Cross sections of three stages in neurulation. (d) Side view of post-neurula showing somite formation and outgrowth of tail bud.

elongated and constitute a columnar epithelium, so that the plate thickens, whereas the ectodermal cells elsewhere remain more or less flattened.

During this process the embryo lengthens along its anteroposterior axis, and at the same time the edges of the neural plate raise up as ridges. Continuing the process, the neural plate narrows and the neural folds become higher, so that the neural area contained between them forms a wide neural groove. The neural folds then meet each other in the dorsal middle line and fuse. In this way the neural tube is formed. Finally the lateral ectoderm (epidermis) of each side also meets and fuses at the mid-dorsal line, while the neural tube segregates and comes to lie below the completed epidermal layer. The neurula is the stage from the beginning to the end of the above process.

The neural tube, however, is not prospectively neural throughout. The broad anterior end gives rise to the brain and its cavities, and the middle part to the spinal cord and canal, but the most posterior part undergoes involution and forms notochord and somite tissue of the future tail, although the somites of the rest of the body arise from the mesodermal mantle.

As the neural folds fuse in the mid-dorsal line, and the neural tube thus formed separates from the covering epidermis, a flattened mass of cells, the neural crest, runs like a peripheral band around the whole neural-plate area. It is derived from a strip of cells, identifiable even in the early gastrula stage. When the neural folds meet each other dorsally, the neural material in one fold fuses with the neural material of the other, while the epidermal material in the two folds also fuses. The neural-crest material separates from both folds and comes to lie between epidermis and neural tube. Following this, the *neural-crest cells migrate* laterally and ventrally on each side and stream along the inner side of the epidermis, external to the lateral mesoderm, penetrating between other tissues and eventually making contact with one another at the midventral line of the embryo. The stream of cells has been followed by means of tritiated thymidine, a cell-specific, nontoxic cell marker. Many experiments, involving grafts between embryos of different genera which differ in cell size, color, etc., extirpations of parts of the neural crest, and changes in surroundings, show that neural-crest material gives rise to an assortment of nonneural tissue of general mesodermal character. Pigment cells, parts of the skeleton, and the sympathetic system and its associated tissues form from the neural crest.

Neurulation, though specifically the process of neural-tube formation, is the phase of development during which mesoderm and endoderm also undergo primary differentiation into particular structural entities, as already noted in *Amphioxus.* During neural-plate formation, the mesoderm, which has already delaminated from the endoderm of the archenteric roof, splits into five strips of tissue, representing the median dorsal notochord and, on each side, the somites and lateral plates. In *Amphioxus,* segregation of the equivalent mesodermal derivatives occurs by means of evaginations of an epithelium.

In the Amphibia the rudiment of the notochord separates from the mesodermal sheet as a narrow rod of cells, seen in transverse section as a longitudinal cell column but with cells not visibly different from those of the adjoining mesoderm; shortly after, fluid-containing vacuoles appear in the notochord cells and press the nucleus and cytoplasm against the cell borders, the whole rod becoming round and turgid, and lengthening. The lengthening occurs because the notochord cells swell within a confining collagenous sheath, which allows them to slide and interdigitate in the long axis of the sheath but not in other planes.

Simultaneously the strip of mesoderm at each side of the notochord thickens and subdivides, beginning at the anterior end, into a series of cell masses, or somites. Each is separate from its neighbors but remains joined to the lateral mesodermal plate by a strand of cells. This strand, known as the *nephrotome,* is the source of future kidney tissue. Subsequently, the part of each somite next to the notochord separates and becomes the sclerotome, or skeleton-forming tissue around the notochord, while the remaining, major part of the somite becomes the myotome, the cells of which differentiate into the striated muscle fibers of the somatic muscles.

Lastly, the lateral mesodermal plate of each side splits into two layers, an outer, or parietal, layer next to the ectoderm and an inner, visceral, layer next to the endoderm. The slitlike cavity between the two layers is the coelomic cavity. The inner layer gives rise to the smooth muscle of the intestine and to the blood and blood

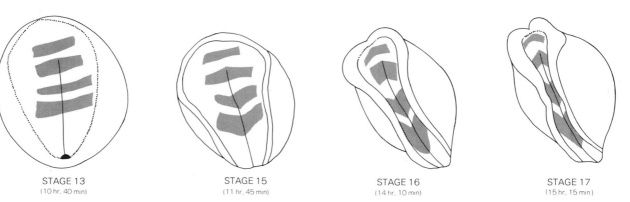

STAGE 13
(10 hr, 40 min)

STAGE 15
(11 hr, 45 min)

STAGE 16
(14 hr, 10 min)

STAGE 17
(15 hr, 15 min)

FIGURE 14.28 Changes in the size and position of the color marks applied to the neural plate of a salamander at stage 13. Development is followed up until stage 17. [After Jacobson, 1962.]

vessels. The endoderm also undergoes further development during the neurulation phase. At the end of gastrulation it is still an open trough. As neurulation begins, the free margins of the endoderm unite in the dorsal middle line beneath the notochord, to complete the formation of the definitive gut, or enteron; the floor of the enteron persists as a thick layer of large yolk-filled cells. Anteriorly the enteron makes contact with the anterior ectoderm, where, later on, the mouth invagination forms and fuses with the blind end of the enteron to form a continuous gut cavity. Later again, the lungs, liver, and pancreas develop from evaginations from the gut.

Such are the major events during neurulation, consisting of infolding or invaginations, tissue thickenings, tissue splitting, and cell and tissue segregation, all proceeding in an ordered manner to produce a particular organization of differentiating structure. In the context of this chapter we are concerned mainly with the changes in form and configurations of the cells and tissues involved in these morphogenetic events, and especially with the processes resulting in the formation of the neural tube.

Attempts to analyze the process of neurulation, as in the study of most problems of development, go back to the nineteenth century, in particular to Roux (1885), who maintained that the course of events was the result of certain changes in the neural plate itself, rather than the effect of outside forces such as expansion, and therefore pushing, by the surrounding ectoderm. Since the cells of the neural plate, commencing with those of the neural ridges, become taller and then flask-shaped, their basal and free surfaces necessarily become smaller. And since the free surface becomes more constricted than the basal surface, the layer of cells tends to curve. These two changes in cell shape have been held to be responsible for the neurulation process. Another suggestion is that differential rates of cell division in neural plate and lateral ectoderm cause neurulation; this might well be so except for evidence that no such differences exist.

The lateral ectoderm does move upwards toward the plate, and this movement has been regarded as an important factor in initiating the contractile forces. Moreover,

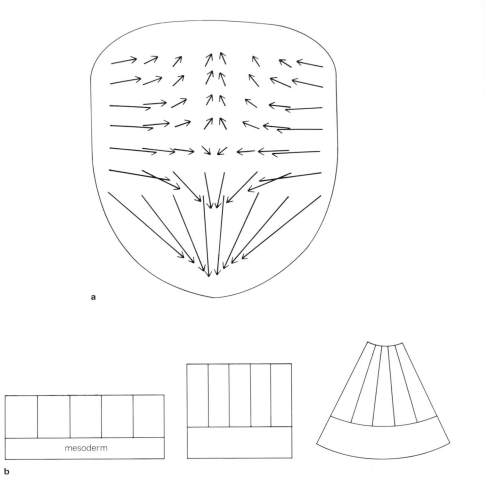

FIGURE 14.29   (a) Movements in the neural material during early neurula stages. The cells move in the direction of the arrows. The length of the arrows corresponds to the distance traveled by the cells. (b) Cells from the neural plate as seen in transverse section, and overlying archenteron roof, indicating increased adhesion between the cells and marked adhesion to the underlying material, which contracts at the same time; the cells become columnar. Increased intercellular adhesion and strong adhesion to the underlying material cause the area of the base to become greater than that of the apex, so that the whole structure consequently becomes curved. [*After Jacobson, 1962.*]

as shown by local vital staining, cellular mass movements resulting in elongation of the plate are known to occur. These have recently been followed in detail by means of vital staining combined with time-lapse photography. In this way it is shown that the plate lengthens by about two-thirds, the lengthening affecting only the prospective spinal cord, the length of the prospective brain remaining unchanged.

On the basis of these observations some enlightening experiments have been performed on the neurula of the *Axolotl* by C-O. Jacobson, who has removed the neural plate inside the neural ridge, either alone or together with various adjoining tissue. His observations are summarized here:

1   Excision of the entire neural plate, alone, leaving the ridge intact, merely retarded the neurulation process, which was otherwise normal.
2   Excision of the entire neural plate within the neural ridge, including the underlying archenteron roof, resulted in failure of neurulation.

3 Partial excision of the neural ridge was followed by near-normal neurulation.

4 Partial excision of the notochord caused failure of the neural tube to close in the area operated upon and also, significantly, inhibited longitudinal extension.

The surprising conclusion from these experiments is that the neural plate, which normally gives rise to the brain and spinal cord, is not at all necessary for the movement and the dorsal fusion of the lateral ridges. Theories placing the responsibility for the whole neurulation process within the plate itself, therefore, seem at least to be exaggerated, even though pieces of isolated neural plate show a strong tendency to roll up to form a tube. In order to form a tube *in situ*, however, the presence of an intact archenteron roof is required, i.e., the chordamesodermal plate at this stage. There is a strong adhesion between the two layers of tissue, and it is especially strong between the plate and the prospective notochord material. Time-lapse films show that movements similar to those already seen in the neural plate also occur in the underlying chordamesoderm, or archenteron roof. Neurulation fails only when the drawing power of the archenteric roof has been eliminated by removing the neural plate and archenteron roof, or the archenteron tissue alone, or by making a slit through both.

The operating factor generally thought to be the cause of the movement of material towards the middle line is the transformation of cuboidal neural-plate cells to cylindrical form during early neurulation, particularly their acquisition of a truncated pyramidal form. Evidence suggests that this is a secondary phenomenon; another explanation of the truncated pyramidal shape of the cells of the neural plate and the formation of columnar epithelium could be local increase in intercellular adhesive power.

C-O. Jacobson explains neurulation initiation as follows: "Lengthening of the notochord results in movements inwards involving both mesodermal and neural material. The upward (dorsal) movement of the lateral ectoderm raises the ridges, while drawing powers in the neural plate and underlying material give the result that the ridges approach each other. Thus in order for neurulation to take place at the normal site, both the pulling forces of the mesoderm and the autonomous tendency of the neural plate to roll are essential requirements."[1]

The primary problem which remains concerns the basis of the invaginating process, whether of gastrulation, of neural-tube formation, or of the various in-pocketings and out-pocketings characteristic of so much of morphogenesis. Few discussions of the general process, as distinct from specific cases, exist. A common basis for directed contractility and extension of tissues probably exists, and this common agency should therefore underlie, for instance, gastrulation processes of echinoderm and amphibian alike, even though they are manifestly different in detail. One general interpretation was presented by Lewis in terms of an increase or decrease of a superfi-

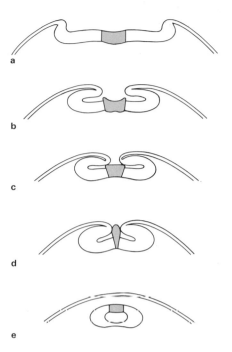

FIGURE 14.30 Transverse sections through the central nervous system, primordia having been subject to operations as follows: (a) A median strip of the neural plate has been replaced by a lateral strip from another larva; the extension of the operation is shown. (b–d) The graft is expelled by the intact tissue, an event observed from stage 18. (d,e) The graft then unites with the neural-ridge tissue. [*After Jacobson, 1962.*]

[1] C-O. Jacobson, Cell Migration in the Neural Plate and the Process of Neurulation in the *Axolotl* Larva, *Zool. bidrag Uppsala*, **35**:444 (1962).

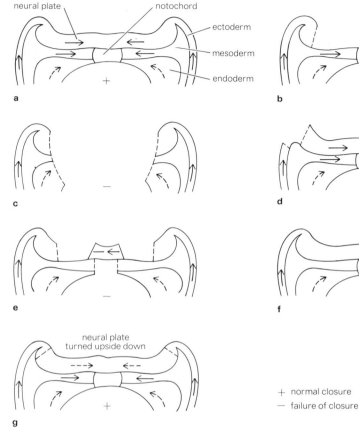

FIGURE 14.31 Transverse sections through a dorsal part of an axolotl embryo at stage 15. Migration (pulling or pushing forces) is indicated by arrows. (a) Intact embryo. (b) Neural plate extirpated within the ridges (operation I). (c) Neural plate plus archenteron roof extirpated (operation II). (d) Parts of both neural ridges removed (operation III). (e) Notochord material removed (operation IV). (f) Incision through the neural plate and archenteron roof (operation V). (g) Inversion of the neural plate (operation VI). [*After Jacobson, 1962.*]

cial gel layer on one side of a group of adherent epithelial cells that offer some resistance to distortion. Invagination would then occur on the side of greater tension. The peculiar characteristics of each invaginating organ are determined by a great variety of modifying factors. The superficial gel layer, in this view, is the surface layer of a cell. It is firm, tough, pliable, and more viscous on its outer aspect than on its deep aspect, where it merges into endoplasm. However, the factors responsible "for local changes of its contractile tension that result in invagination and inward migration of cells from the invaginating surface are unknown." The weakness in this model has been failure to identify the presence of such a gel layer with the necessary properties.

This interpretation has since received some support from the discovery of microfilaments in a wide variety of cells in embryos and adult tissues. At least in some cases, e.g., the filaments in the ectoplasm of the slime mold *Physarum*, such filaments are contractile. Cortical contractile microfibrils may accordingly take the place of the more abstract postulated gel layer. During the metamorphosis of the ascidian tadpole larva, the epidermis of the tail contracts toward the base; Cloney has shown this contraction to be due to the alignment and contraction of microfilaments throughout the tail epidermal epithelium, close to the plasma membrane.

Microtubules, in addition to microfibrils, also have been seen in the amphibian embryo in the cells of the neural groove and neural folds, and in elongated blastopore cells; the suggestion has been made that they may play an important role in the neurulation process insofar as change in cell shape is responsible for the neurulation process. However, actively elongating microtubules alone do not seem sufficient to explain the formation of the neural groove and tube, and reduction of the apical cell surfaces appears to be necessary for the invagination process. The apices of neural-plate cells contain a thin band of filaments aligned parallel to the outer surface of the plate. The cell apices narrow as the neural groove narrows and deepens, and the apical band becomes broader and crowded with oriented filaments, while the cell body assumes a long columnar shape, with cell apices joined to one another. The question is whether the cells are changing in shape because of changes in microfibril orientation and possibly contraction, or whether the fibril changes are responding to changing cell shape and merely serve as supportive structures.

CONCEPTS

Cleavage, or cellulation, of the egg mass is a necessary enabling condition for most formative movements.

Morphogenetic movements are instigated at various times during the progressive cleavage of holoblastic eggs.

The original surface layer, or cortex, of the egg persists as the outer surface of the subdivided egg, or blastula.

The primary events of early development are the demarcation and segregation of the three prospective germ-layer tissues, namely, ectoderm, endoderm, and mesoderm.

Folding processes, as invaginations or evaginations, are the effective segregating agencies when single epithelial layers are initially involved. Delamination segregates multiple layers.

Yolk concentration commonly displaces cytoplasm and nucleus to eccentric locations, and indirectly results in various degrees of unequal division and consequent modifications of blastulation and gastrulation.

Tissues spread primarily as epithelial sheets over suitable substrata, typically with an active leading edge.

Supracellular junctional continuity appears to be an integrating or controlling agent of tissue movement.

Changes in the shape of constituent cells within a tissue appear to be primarily responsible for change in tissue conformities and foldings.

One layer of tissue can impose directional movement on another.

Different combinations of unlike tissues establish different mutual topographic relationships.

Microfibrils and possibly microtubules are mainly responsible for change in cell shape and form of tissue.

READINGS

ABERCROMBIE, M., 1965. Cellular Interactions in Development, in J. Moore (ed.), "Ideas in Modern Biology," Natural History Press.

BAKER, P. C., and T. E. SCHROEDER, 1967. Cytoplasmic Filaments and Morphogenetic Movement in the Amphibian Neural Tube, *Develop. Biol.*, **15**:432–450.

BALLARD, W. W., 1966. Origin of the Hypoblast in Salmo, II: Outward Movement of Deep Central Cells, *J. Exp. Zool.*, **161**:201–219.

BARTH, L. G., 1964. The Developmental Physiology of Amphibia, in "Physiology of the Amphibia," Academic.

BRUMMETT, A. R., 1968. Deletion-transplantation Experiments on Embryos of *Fundulus heteroclitus*, *J. Exp. Zool.*, **169**:315–334.

BURNSIDE, M. B., and A. G. JACOBSON, 1968. Analysis of Morphogenetic Movements in the Neural Plate of the Newt, *Taricha torosa*, *Develop. Biol.*, **18**:537–552.

CLONEY, R. A., 1961. Observations on the Mechanism of Tail Resorption in Ascidians, *Amer. Zool.*, **1**:67–88.

CONKLIN, E. G., 1932. The Embryology of *Amphioxus*, *J. Morphol.*, **54**:69–118.

COSTELLO, D. P., 1955. Cleavage, Blastulation and Gastrulation, in B. Willier, P. Weiss, and V. Hamburger (eds.), "Analysis of Development," Saunders.

CURTIS, A. S. G., 1967. "The Cell Surface: Its Molecular Role in Morphogenesis," Academic.

DAN, K., 1960. Cytoembryology of Echinoderms and Amphibia, *Int. Rev. Cytol.*, **9**:321–367.

———, 1952. Cyto-embryological Studies of Seaurchins, II: Blastula State, *Biol. Bull.*, **102**:74–83.

DeHAAN, R. L., 1964. Cell Interactions and Oriented Movements during Development, *J. Exp. Zool.*, **157**:127–138.

DEUCHAR, E. M., 1965. Biochemical Patterns in Early Developmental Stages of Vertebrates, in R. Weber (ed.), "The Biochemistry of Animal Development," Academic.

DEVILLERS, C., 1961. Structural and Dynamic Aspects of the Development of the Teleostean Egg, *Advance. Morphogenesis*, **1**:379–424.

GIBBONS, I. R., L. G. TILNEY, and K. R. PORTER, 1968. Microtubules in the Formation and Development of the Primary Mesenchyme of *Arbacia punctulata, Develop. Biol.,* **18**:523–539.

GILLETTE, R., 1944. Cell Number and Cell Size in the Ectoderm during Neuralization (*Ambystoma maculatum*), *J. Exp. Zool.,* **96**:201–222.

GUSTAFSON, T., 1969. Fertilization and Development, in "Chemical Zoology," vol. 3, pp. 149–206, Academic.

—— and M. TONEBY, 1970. The Role of Serotonin and Acetylcholine in Sea Urchin Morphogenesis, *Exp. Cell Res.,* **62**:102–117.

—— and L. WOLPERT, 1967. Cellular Movement and Contact in Seaurchin Morphogenesis, *Biol. Rev.,* **42**:442–498.

HOLTFRETER, J., 1943. A Study of the Mechanics of Gastrulation, *J. Exp. Zool.,* **94**:261–318; **95**:171–212.

——, 1943. Properties and Functions of the Surface Coat in Amphibian Embryos, *J. Exp. Zool.,* **93**:251–323.

JACOBSON, C-O., 1962. Cell Migration in the Neural Plate and the Process of Neurulation in the *Axolotl* Larva, *Zool. bidrag Uppsala,* **35**:433–449.

KÄLLÉN, B., 1965. Early Morphogenesis and Pattern Formation in the Central Nervous System, in R. L. DeHaan and H. Ursprung (eds.), "Organogenesis," Holt, Rinehart and Winston.

LEWIS, W. II., 1947. Mechanics of Invagination, *Anat. Rec.,* **97**:Zoology, 139–156.

MOORE, A. R., 1941. On the Mechanics of Gastrulation in *Dendraster excentricus, J. Exp. Zool.,* **87**:101–111.

—— and A. S. BURT, 1939. On the Locus and Nature of the Forces Causing Gastrulation in the Embryos of *Dendraster excentricus, J. Exp. Zool.,* **82**:159–171.

NEW, D. A. T., 1959. The Adhesive Properties and Expansion of the Chick Blastoderm, *J. Embryol. Exp. Morphol.,* **7**:146–164.

OPPENHEIMER, J., 1970. Cells and Organizers, *Amer. Zool.,* **10**:75–88.

——, 1947. Organization of the Teleost Blastoderm, *Quart. Rev. Biol.,* **22**:105–118.

RUDNICK, D., 1955. Teleosts and Birds, in B. Willier, P. Weiss, and V. Hamburger (eds.), "Analysis of Development," Saunders.

——, 1944. Early History and Mechanics of the Chick Blastoderm, *Quart. Rev. Biol.,* **19**:187–212.

RUGH, R., 1968. "The Mouse. Its Reproduction and Development," Burgess.

SPRATT, N. T., 1964. "Introduction to Cell Differentiation," Van Nostrand, Reinhold.

——, 1946. Formation of the Primitive Streak in the Explanted Chick Blastoderm Marked with Carbon Particles, *J. Exp. Zool.,* **103**:259–304.

—— and H. HAAS, 1960. Integrative Mechanisms in the Development of the Chick Embryo, I: Regulative Potentiality of Separate Parts, *J. Exp. Zool.,* **145**:97–137.

TOWNES, P. Z., and J. HOLTFRETER, 1955. Directed Movements and Selective Adhesion of Embryonic Amphibian Cells, *J. Exp. Zool.,* **123**:53–118.

TRINKAUS, J. P., 1965. Mechanisms of Morphogenetic Movements, in R. L. DeHaan and H. Ursprung (eds.), "Organogenesis," Holt, Rinehart and Winston.

——, 1951. A Study of the Mechanism of Epiboly in the Egg of *Fundulus heteroclitus, J. Exp. Zool.,* **118**:269–320.

WADDINGTON, C. H., 1952. Modes of Gastrulation in Vertebrates, *Quart. J. Microscop. Sci.*, **93**:221–228.

WEISS, P., 1950. Some Perspectives in the Field of Morphogenesis, *Quart. Rev. Biol.*, **25**:177–198.

WEITLAND, H. M., and G. S. GREENWALD, 1969. Influence of Oestrogen and Progesterone on the Incorporation of $_{35}$S-methionine by Blastocysts in Ovariectomized Mice, *J. Exp. Zool.*, **169**:463–469.

WESSELLS, N. K. et al., 1971. Microfilaments in Cellular and Developmental Processes, *Science*, **171**:135–143.

CHAPTER FIFTEEN

EMBRYONIC INDUCTION

Induction, or the evocative action of one tissue on another, namely, the responsive tissue, is a widespread phenomenon in development. Transplantation of micromeres to the animal half of a developing sea urchin egg, causing prospective ectoderm to develop proportionately into ectoderm, endoderm, and marginal mesoderm, is an example; the evocation of the mouth evagination by the blind end of the advancing archenteron is another. Inductive processes, however, have been studied mainly in developing eggs of chordates, both vertebrate and prevertebrate chordates, particularly in amphibians and ascidians, respectively. The general pattern of development is essentially alike in these two groups, in spite of the very great difference in egg size and in the number of cells present at any given stage. Yet the relatively small number of cells characteristic of the developing ascidian egg, their precocious determination and differentiation processes, and the relatively small amount of yolk present in the egg, make the study of determination and induction in ascidian development a useful prelude to an analysis of these processes in amphibian development.

CONTENTS
Induction by blastomeres
Induction in vertebrates
Gene activation
Neural induction
Primary inductor and the gray crescent
Foreign inductors
Regional specificity
Competence
Pattern and fields
Concepts
Readings

INDUCTION BY BLASTOMERES

As already described, the ascidian egg has a strongly developed polar axis, or animal-vegetal yolk gradient, and it has a posterior "mesodermal" ooplasmic crescent, and anterior "neural" and "chordal" crescents, one above the other, all more or less at the equatorial level of the egg. At the eight-cell stage there are four animal blastomeres and four vegetal blastomeres of approximately equal size; the mesodermal crescent material lies mainly in the two posterior vegetal blastomeres, the chordal and neural crescent material is mainly in the anterior vegetal blastomeres, and there is no special plasm in either anterior or posterior animal blastomeres.

In normal development the prospective fates of the different blastomeres of the eight-cell stage are as follows: the two animal anterior cells produce head epidermis, palps, and the brain with its two pigmented sensory structures; the two animal posterior cells produce only epidermis; the two vegetal anterior cells produce notochord, spinal cord, and part of the intestine; the two posterior vegetal cells produce mesenchyme, muscles, and part of the intestine. Blastomeres of the eight-cell stage when isolated and allowed to develop as pairs, however, present a more complex situation than first appeared. To quote Reverberi: "With great surprise it was observed that, while all territories (intestine, notochord, muscles, mesenchyme) have their destiny already fixed, and consequently differentiate according to their presumptive fate, the destiny of the neural system is not fixed; the two animal anterior blastomeres, if isolated, do not give rise to brain or palps or sensory spots but on the contrary, give rise

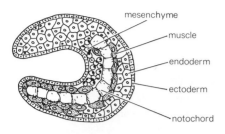

FIGURE 15.1 Determination in ascidian embryo. The vegetal hemisphere of the cleaving egg is determined; the animal hemisphere is labile and subject to induction. When the egg is treated with lithium chloride in seawater, tail-structure development proceeds, but development of brain, sensory organs, and other "head" structure is suppressed. [*After Reverberi, 1961.*]

only to epidermis. Moreover, one obtains from an isolated animal quartet an ectodermal vesicle, always lacking a brain. As a result of various other combinations, one arrives at the conclusion that the formation of the brain is 'dependent', and that the material responsible for its differentiation is situated in the vegetal anterior blastomeres. Only by combining the two animal anterior blastomeres with two vegetal anterior ones, does one obtain embryos provided with brains."[1] Are these two vegetal blastomeres "responsible" for brain formation, i.e., are they "inductors"?

The two vegetal anterior blastomeres give rise to diverse tissues, namely, endoderm, notochord, and spinal cord. At the 64-cell stage the prospective material for these three tissues has been segregated. When the four prospective spinal-cord cells of this stage are destroyed, the embryo still forms a normal brain and sensory organs. Accordingly the inductor should lie in the notochord or in the endodermal cells. But since destruction of the six prospective endodermal cells of the 64-cell stage fails to inhibit the normal development of the brain, only the chordal cells are left; as far as experimental proof is concerned, they win by default and are presumed to be inductive. This inductive capacity of chordal tissue is in keeping with what is known of amphibian development.

According to Reverberi, the destiny of the presumptive "neural" ectoderm is not fixed at the eight-cell stage, at which stage all the other territories are already "determined." The presumptive "neural" ectoderm is converted into the neural system only under the inductive action of the immediately underlying chordal (and endodermal) cells. The induction is consequent on a *contact* between correlated territories.

Similarly in *Amphioxus* the developmental potencies of the animal hemisphere are influenced to a large extent by the inductive stimulus from the material of the vegetal hemisphere, although the animal half can also exert an influence on the vegetal half of the egg to some extent. In experiments conducted by Tung, Wu, and Tung (1960), at the Institute of Oceanography of mainland China, where *Amphioxus* is so abundant as to have some commercial value, the animal hemisphere was rotated 90 to 180° on the vegetal hemisphere at the eight-cell stage, without changing the animal-vegetal axis. In spite of the shift of the original presumptive neural blastomeres to a posterior position, normal embryos developed, i.e., the blastomeres moved to the anterior position are induced to develop neural structure. In fact the whole of the prospective *Amphioxus* ectoderm appears to be undetermined during cleavage stages, since both the nervous system and tail form in normal relationship in spite of shifting the animal hemisphere blastomeres. The vegetal hemisphere, or presumptive endoderm-mesoderm, on the other hand exhibits early self-determination and has an inductive effect on the animal hemisphere.

In further experiments by the same investigators, grafting of the blastoporal lip of an early gastrula to the blastocoel of an embryo of the same stage either resulted in formation of endodermal tissue with no subsequent effect, or produced notochord

[1] G. Reverberi, The Embryology of Ascidians, *Advance. Morphogenesis*, 1:95 (1960).

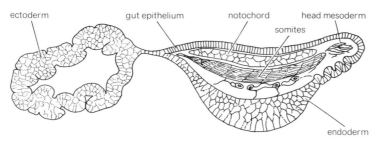

ectoderm   gut epithelium   notochord   head mesoderm
                                    somites

endoderm

FIGURE 15.2 Main differentiations in an exogastrulated embryo. [*After Holtfreter, 1933.*]

with accompanying neural tube and mesodermal somites. They concluded that chordal tissue possesses the power of neural induction, while mesodermal and endodermal tissue has little, if any, such inductive power.

INDUCTION IN VERTEBRATES

A comparable situation holds for the development of the amphibian neurula. In normal development, as described in the preceding chapter, the neural plate, destined to give rise to the brain, spinal cord, and some other tissue, forms in contact with the roof of the archenteron. In exogastrulas, which are readily produced by maintaining embryos in hypertonic saline solution, the archenteron is everted and the endoderm encloses the chordamesodermal tissues, while the ectodermal layer survives as an empty, crumpled envelope attached to the posterior endoderm only by a narrow stalk. Under these circumstances all the tissue except the ectoderm proceeds with normal differentiation, although turned inside out, but the ectoderm fails to differentiate in any way. Neural differentiation occurs, apparently, only when dorsal ectoderm is in contact with the roof of the archenteron or tissue derived from it. On the other hand this conclusion has been questioned because the exogastrulating treatment itself alters the prospective potency of the dorsal ectoderm.

Although the first analysis of induction by Spemann (1901–1912) concerned the development of the eye, specifically the induction of a lens by the optic-cup rudiment, to be discussed further, the classical experiment upon which the concepts of "embryonic induction" and "organizer" are founded came later. This was the discovery in 1924 by Spemann and Mangold that the dorsal blastoporal lip from a salamander (newt) gastrula, when implanted into a ventral or lateral position of another gastrula, turns inward, develops into a notochord and somites, and induces the host ectoderm to form a neural tube, i.e., it induces the formation of the axial organs of a more or less complete secondary embryo. This secondary embryo accordingly may have a notochord, somites, and neural tube composed partly of grafted, partly of induced host material. Various other structures may also be induced.

There is an apparent basic similarity among the developing eggs of echinoderms, protochordates, and primitive vertebrates. In all three groups the animal region of the egg appears to be undetermined as to its fate and incapable of self-differentiation under ordinary circumstances. In each group the tissue associated with the initial site of gastrular invagination, when isolated, is a self-differentiating tissue but unable to produce an organized embryo.

GENE ACTIVATION

Nuclear reaction is required for gastrulation, as experiments on enucleate embryos have shown; development of hybrid embryos lacking normal nuclear function resulting from lethal chromosome combinations is usually arrested at about the time of onset of gastrulation. Newly synthesized gene products, in fact, appear to be essential for gastrulation, inasmuch as messenger RNA synthesis per cell increases before the process begins. Widespread and sudden activation of nuclear RNA synthesis becomes evident in the presumptive endoderm and inner mesoderm during the late blastula stage. Mesodermal nuclei are continually activated as the tissue invaginates, and a high level of RNA synthesis is established by the late gastrula stage.

According to E. H. Davidson, the gene activation observed in endoderm and inner mesoderm of the late blastula, amounting to a twentyfold increase within 1 hour, is the molecular event underlying the overt differentiation which appears at gastrulation. Within 2 hours of the synthesis of messenger RNA, at least part of it is translated into differentiated cell properties, as shown by the capacity of the cells to proceed with their specific differentiations when isolated or when transplanted to other sites. This is in contrast to the ectoderm, which can still be directed to form either epidermal or neural tissue, or even mesodermal tissues under certain circumstances. In keeping with this, ectodermal nuclei undergo a transient burst of RNA synthesis only after mid-gastrulation, much later than the endomesoderm.

NEURAL INDUCTION

One of the first questions to be answered concerns the time of induction of the neural system. Pieces of the prospective neural plate, i.e., the dorsal ectoderm, of the early gastrula transplanted into the ventral region of another gastrula develop according to their new position, i.e., they become belly epidermis; pieces removed from a late gastrula and transplanted to a ventral position develop according to their original position, i.e., they become neural plate. Induction evidently occurs during the time when the material initially present in the dorsal lip moves inward and forwards to the anterior region of the embryo. The vital-staining experiments of Vogt showed that the material successively forming the dorsal blastoporal lip moves forward as

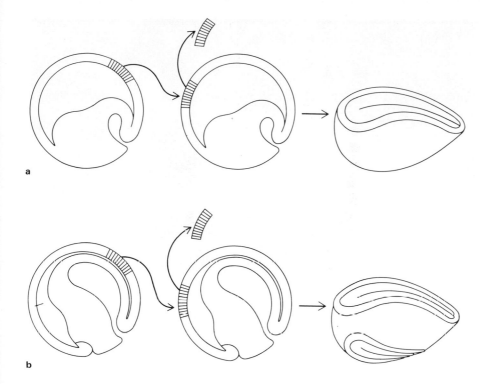

a

b

FIGURE 15.3 (a) Piece of presumptive neural plate removed from an early newt gastrula, and grafted onto the ventral side of another of the same age. No induction occurs, but the graft develops in accordance with its new surroundings. (b) A corresponding piece removed from a late gastrula develops in the new surroundings according to its original fate. [*After Saxén and Toivonen, 1962.*]

the archenteric roof. Transplants taken from this region are also able to induce a secondary embryo or the belly of a new host, i.e., the archenteron roof acts as a primary inductor in essentially the same way as does the dorsal-lip tissue proper.

The inductive capacity of the blastoporal lip, however, varies both regionally and temporally. Most of the dorsal and dorsolateral blastoporal material is necessary for a graft to induce a more or less complete secondary embryo. If only prospective endomesodermal material is grafted into another gastrula (or alternatively into an isolated vesicle of ectoderm!), it induces a partial embryo containing ectodermal head structures such as brain, sense organs, mouth structures, etc. Moreover, explants or grafts of the median portion of the dorsal lip tend to form eyes and olfactory pits, lateral parts tend to form posterior cephalic structures, and still more lateral parts of the marginal zone tend to induce spinal cord and mesenchyme of a trunk-tail.

As invagination continues, however, and the dorsal lip no longer consists of prospective head endomesoderm but progressively becomes prospective trunk mesoderm, it acts as a trunk-tail inductor. The most caudal region of the archenteron roof, in fact, specifically induces tail somites and probably other mesodermal tissue. The archenteron roof, in other words, normally induces entirely different classes of tissue: various neural and mesoectodermal tissues by its anterior region, and various mesodermal tissues by its most posterior region. This has led to the conclusion that

FIGURE 15.4 Dorsal blastopore lip of an early newt gastrula, transplanted into the blastocoel of another, induces a complete secondary embryo. [*Courtesy of L. Saxén and S. Toivonen.*]

differences in specific induction capacities exist between head and trunk levels of the archenteron roof, and are related to the regional differentiation of the neural tissue into archencephalic (including forebrain, eye, nasal pit), deuterencephalic (including hindbrain, ear vesicle), and spinocaudal components. However, the areas are not sharply defined, and they overlap, so that alternative interpretations are possible, namely, quantitative variation in a single inducing agent, or a segregation of qualitatively different specific inductors.

PRIMARY INDUCTOR AND THE GRAY CRESCENT

The dorsal-lip region of the blastopore at the onset of gastrulation can be traced back to the gray crescent of the undivided fertilized amphibian egg. The crescent has already been shown to be the initial expression of bilateral symmetry of the embryo; moreover, blastomeres of the two-cell stage when separated from each other give rise to twins if the first cleavage plane divides the egg midway through the crescent, but they give rise to one normal embryo and one blastula-type embryo if the cleavage plane divides the egg so that the crescent is entirely in one cell. The crescent material gives rise to the invaginating blastopore rim, the center of the crescent corresponding to the dorsal-lip region of the blastopore. The circumstantial evidence that the crescental material of the egg cortex initiates the gastrulation process is strong.

This hypothesis has been tested, in *Xenopus*, by grafting cortex from the gray-crescent region of the uncleaved fertilized egg to the opposite side of a second egg, so that the egg receiving the graft has two gray crescents on opposite sides. As a result the blastula undergoes two separate gastrulation movements and produces two

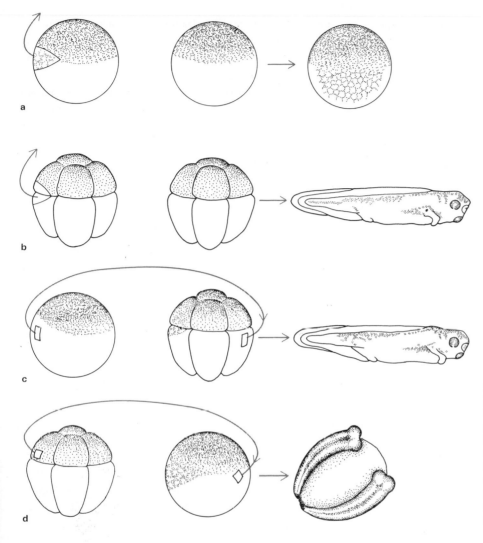

FIGURE 15.5 Experiments of Curtis in *Xenopus*.
(a) Excision of the cortical gray-crescent area at
the one-cell stage: no gastrulation. (b) Same experiment at the eight-cell stage: normal embryo. (c)
Graft of the gray-crescent cortex of the one-cell
stage to the ventral part of the eight-cell stage
does not result in the induction of a secondary
embryonic axis. (d) Graft of the gray-crescent
cortex from the eight-cell stage to the ventral
margin of the one-cell stage induces a secondary
embryonic axis. [*After Curtis, 1962.*]

separate primary nervous systems with associated somites. It becomes conjoined twins.
Conversely, if the gray-crescent cortex is excised from the fertilized egg, cell division
proceeds undisturbed but gastrulation fails to take place. The causative role of the
gray crescent appears to be established.

Similar experiments conducted on the eight-cell stage, however, show that
something has happened during the short interval represented by the first three
cleavages. Gray-crescent cortex of the eight-cell stage still retains its inductive capacity
when grafted to younger stages, but the eight-cell stage itself no longer reacts to
grafts of the crescent material. Also, excision of the gray crescent at this stage no
longer inhibits subsequent gastrulation and normal development, the missing crescent
properties being replaced from adjacent cortical regions. According to Curtis, who

made these experiments, a change in cortical organization spreads across the surface of the egg during the second and third cleavages, starting from the gray crescent; when this change is completed, interactions, probably of a biophysical nature, can take place among various parts of the cortex.

Amphibian eggs have been employed almost exclusively in these investigations because operative and biochemical analysis can be performed on them unimpeded by an overabundance of yolk. Nevertheless, the extremely yolky eggs of other vertebrates such as fish and birds, which without doubt evolved long ago from amphibian-type eggs, must inevitably be homologous in many ways. In the chick, Hensen's node has been considered, on morphological grounds, to be homologous with the dorsal lip of the blastopore of the frog egg. The living Hensen's node has been shown by Waddington and others to induce neural differentiation when it is grafted under responsive chick ectoderm. Similarly, the head process, a derivative of the node, also acts as a neuralizing factor. More recently Leikola and McCallion, by culturing, in vitro, fragments of competent chick ectoderm in intimate contact with either heat-killed or alcohol-killed Hensen's-node tissue, obtained well-differentiated neural plates or tubes. The inductive capacity persists, and accordingly the natural primary inductors of the chick embryo are in these respects comparable to those of the amphibian embryo.

FOREIGN INDUCTORS

The discovery of the primary-inductor quality of the dorsal lip of the amphibian blastopore, the "organizer" of Spemann, led to concerted attempts, by teams of embryologists and biochemists throughout the 1930s, to identify its chemical nature. Sterols, protein, nucleoprotein were all implicated, but the massive effort faded before the indigestible discoveries that whereas the only live tissue that could evoke primary embryonic structures was gray-crescent—namely, dorsal-lip—material, almost any tissue from the whole animal kingdom could do so if killed, i.e., denatured, by heat or alcohol. The concept of a universally present "masked" organizer, normally released for action only in the primary-inductor region, briefly held sway. Finally inorganic agents such as iodine and kaolin, and then even local mechanical injury, were found to be effective, so that all that remains of the early grand concept of a master-chemical embryonic organizer is Holtfreter's sublethal cytolysis, namely, the idea that reversible cell injury liberates the neural inductor, but not the mesodermal inductor. However, the main outcome of this early search for the chemical nature of the "organizer" has been the view that the answer to the problem must lie in the tissues being "organized."

The inductor becomes an evocator, i.e., it does not induce a pattern or a particular differentiation, but rather it evokes or triggers a particular response among the many that a cell or a sheet of cells may have in its repertoire. Responses are specific, whereas

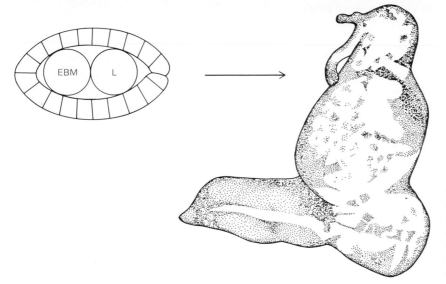

FIGURE 15.6 The effect of two heterogeneous inductors on the undetermined ectoderm of an amphibian gastrula. (*Left*) Scheme of the experiment: both inductors are wrapped in a sandwich of ectoderm. (*Right*) An embryolike complex has developed. *EBM*, extract of bone marrow; *L*, piece of guinea pig liver. [*After Toivonen and Saxén, 1955.*]

evocators of a given response may be of widely diverse character and may not be chemically related in any significant way to the event called forth, even though different responses may be made to closely related stimuli. A similar situation is seen in the cancer problem, where the malignant change can be initiated by any of a large group of chemical and physical agents, so that the problem of malignancy becomes almost exclusively part of general cell biology and is no longer a search for some factor in common among the inducing agencies. Holtfreter, who has been closely identified throughout with the spectacular discoveries concerning induction in amphibian em-

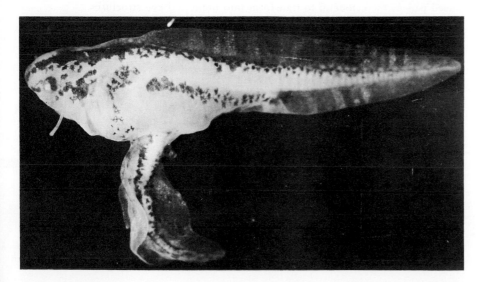

FIGURE 15.7 Implantation of a piece of alcohol-treated guinea pig kidney tissue into the blastocoel of a gastrula induces tail outgrowth with myotomes, notochord, and spinal cord. [*Courtesy of L. Saxén and S. Toivonen.*]

bryos, has recently (1968) summarized the chemical approach to the "organizer" as follows: "Despite (or because of?) the large amount of data collected during the past 36 years, the issue has remained in a sadly obfuscated state."

REGIONAL SPECIFICITY

Two groups of investigators, represented by Toivonen and Yamada, assume that two chemically distinct factors are involved in the action of the primary inductor, one a "neuralizing agent" and the other a "mesodermalizing agent." Both groups have worked mainly with denatured bone marrow and liver as the inductors. The general interpretation is as follows: The regional specificity of the embryonic axis arises from the interaction between two gradients: the neuralizing principle has its highest concentration on the dorsal side of the embryo and diminishes laterally, while the mesodermalizing principle is present as an anteroposterior gradient, with its peak in the posterior region. Anteriorly the neuralizing principle acts alone to induce forebrain structures; more posteriorly the mesodermalizing principle acts along with the neuralizing one to induce midbrain and hindbrain structures; while even more posteriorly the high concentration level of the mesodermal gradient produces spinocaudal structure.

In Holland a group of investigators, headed by Nieuwkoop, using living notochord as the inductor, postulate only one factor which first evokes ectoderm to form neural tissue and later causes ectoderm to transform into more posterior and mesodermal structures. In one experiment, consisting of combining isolated gastrula ectoderm with a piece of notochord and then removing the notochord tissue after varying lengths of time, it was found that only 5 minutes' exposure to the inductor caused part of the ectoderm to transform into brain and eye structures.

FIGURE 15.8 Oppositely directed movements of invaginating archenteron roof and epibolically extending ectoneuroderm at four successive stages of the gastrulation process (a–d). The interaction of both layers leads to a caudocraniad spreading of an activating inductive principle in the ectoderm when it overlies the prechordal portion of the archenteron roof. This primary action is followed by a caudocraniad spreading of a superimposed transforming inductive principle emanating with increasing intensity from successively more caudal regions of the archenteron roof. [*After Nieuwkoop, 1966.*]

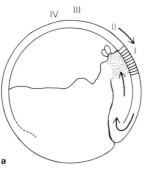
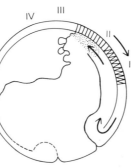
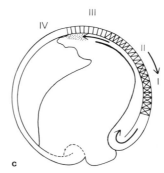
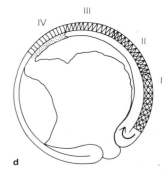

a   b   c   d

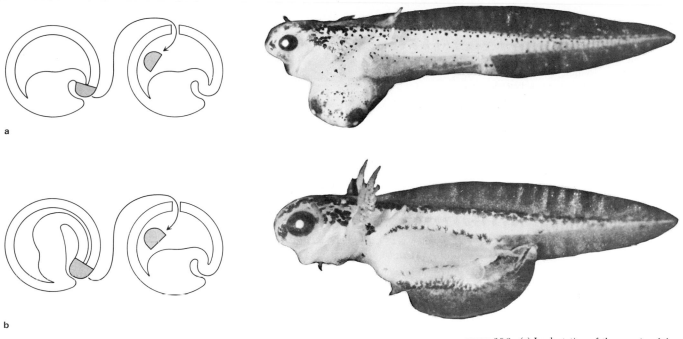

FIGURE 15.9 (a) Implantation of the margin of the blastopore lip from an early gastrula (head inductor) into another of the same age causes development of a secondary head (above). (b) The margin of the blastopore lip of a late gastrula (trunk inductor) induces a secondary trunk and tail when implanted in an early gastrula. [*Courtesy of L. Saxén and S. Toivonen*]

Saxén and Toivonen (1962) summarize the chemical analysis of inductors as follows:

1 It has been shown that most of the inductively active tissues can be separated chemically into fractions yielding different inductive actions.

2 Correspondingly, chemical analyses have shown that fractions which yield archencephalic inductions are chemically different from preparations inducing spinocaudal or mesodermal structures.

3 Some chemical or physical treatments of "combined" inductors (deuterencephalic or spinocaudal) will always destroy selectively the capacity to induce caudal structures, and following these treatments the tissues yield a pure archencephalic inductive action.

Recently (1968) L. G. and L. J. Barth, long associated with the problem of primary-inductor analysis, have clarified a concept of a different sort. In their own words, "While a wide array of unrelated chemical compounds and tissue extracts have been found to act as stimuli for induction of new cell types in the ectoderm of *Rana pipiens* gastrula, their primary site and mechanism for action are still unknown. A release mechanism as the basis for primary induction was looked for several decades ago in the form of gradients in energy-yielding systems within the early gastrula,

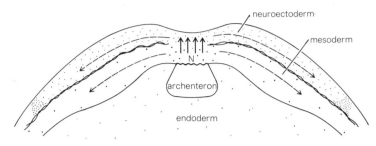

FIGURE 15.10 Mediolateral spreading of an inductive action within the mesoderm and a similar spreading of the neuralizing action in the overlying ectoderm. Parallel vertical arrows indicate the main site of the neuralizing inductive action (N) in the area of intimate adhesion of the two germ layers. [*After Nieuwkoop, 1966.*]

and competition among the various regions of the gastrula. Current thinking tends to express the problem in terms of regional differential release of factors responsible for gene activation and derepression. The search now as before is directed, as we can see it, toward understanding how simple physiological factors, such as local change in ion concentration or pH, can alter energy-yielding systems and gene function in a manner meaningful for cellular differentiation."[1]

Their own working hypothesis is that the actual process of induction may be initiated by release of ions from bound form, representing a change in ratio between bound to free ions within the cells of the early gastrula. Experimentally, induction of nerve and pigment cells in small aggregates of prospective (presumptive) epidermis of the frog gastrula has been found to be dependent on the concentration of the sodium ion, that is, in the absence of other tissue; also, normal induction of nerve and pigment cells by mesoderm in small explants from the dorsal-lip and lateral marginal zones of the early gastrula is dependent on the external concentration of sodium. Nerve and pigment cells are induced when the culture medium contains 0.088 M NaCl; at 0.044 M NaCl the mesoderm differentiates into muscle and mesenchyme but nerve and pigment cell induction does not occur. The interpretation is that normal embryonic induction depends on an endogenous source of ions and that an intracellular release of such ions occurs during late gastrulation. A fortyfold increase in intracellular sodium ions and a threefold increase in calcium ions takes place during gastrulation. The ratio of the two kinds of ions may be more significant than the actual concentrations.

The experimental approach to the problem of primary induction in amphibian embryos has been mainly concerned with the properties of inductive substances, with less attention to the target tissue, i.e., the presumptive epidermis of the gastrula. The presumptive epidermal cells have been assumed to be pluripotent, able to differentiate in whatever direction the specific action of inducing agents may determine. The constituent cells are presumed to be all alike, and neural or mesodermal differentiation is presumed to result from the action of inducing agents specific for each

[1] L. G. and L. C. Barth, The Sodium Dependence of Embryonic Induction, *Develop. Biol.*, **20**:236 (1969).

kind of induction. There is some evidence that more than one kind of cell is present in presumptive epidermis, with, for example, different electrophoretic mobilities and protein isoelectric points. This has suggested that different cell species are associated with neural and mesodermal differentiation, respectively, and that specific inductors operate by selectively killing off one cell type or the other, i.e., mesoderm-inducing substances destroy cell species destined to form neural tissue, and neuro-inducing substances destroy cells destined to form mesodermal tissue. If so, this pushes the problem further back and changes its character, but is not otherwise enlightening. Nevertheless, localized cell death during morphogenesis is a well-established fact in limb development (see Chapter 17), and the possibility of its happening at this early stage cannot be ignored. However, the following account of ectodermal, or presumptive epidermal, competence is based on the assumption that the cells of the reacting tissue are essentially alike, at least during early stages of development.

COMPETENCE

Competence is a term which has been defined as the physiological state of a tissue which permits it to react in a morphogenetically specific way to determinative stimuli, or, more in keeping with the molecular biological outlook, "a term which sums the ability of the enzyme complement of the embryonic cell to adapt to a particular ratio of metabolites." Whatever it may be, competence is always related to particular

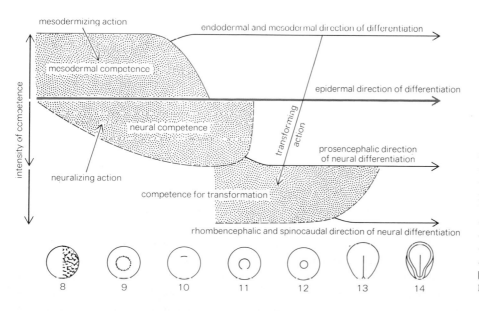

FIGURE 15.11 Diagrammatic presentation of the successive phases of competence for, respectively, endodermal and mesodermal differentiation, prosencephalic neural differentiation, and rhombencephalic and spinocaudal neural differentiation of the presumptive ectoneuroderm. In the absence of any inductive action, the ectoderm develops exclusively in an epidermal direction. On the abscissa, stages are given according to Harrison. [*After Gebhardt and Nieuwkoop*, J. Embryol. Exp. Morphol., **12** (*1964*).]

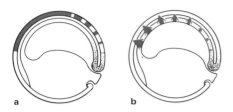

FIGURE 15.12 **(a)** The neuralization gradient in the ectoderm overlying the mesodermalization gradient of the chordamesoderm and **(b)** the neuralizing gradient of the latter. [*After Takaya*, Jap. J. Exp. Morphol., **11** (*1957*).]

stimuli and particular corresponding responses. With regard to primary induction, therefore, we may speak of neural differentiation as a primary competence of the ectoderm.

It has long been known that ectoderm transplanted from various developmental stages from blastula to early neurula gradually loses neural competence. With the aging of the ectoderm it gradually loses its capacity to respond to the inductive stimulus of the chordamesoderm by becoming neural tissue. Similarly dorsal-lip tissue transplanted under ventral ectoderm at various stages of development shows that neural competence drops sharply near the end of gastrulation and is lost by the onset of neurulation. Isolated ectoderm unexposed to neural induction and ectoderm transplanted too late differentiate into epidermis.

The "ectodermal-fold" technique of Nieuwkoop, in which ectoderm is cut out, folded between two pieces of agar and transplanted, for instance, into the dorsal midline of a young gastrula, has been useful in testing neural competence. Because of the inductive action of the host tissue, these folds of competent ectoderm differentiate into successive zones of various neural, mesectodermal, and epidermal structures. If folds are transplanted in different regions of the host, the most proximal part of the transplant always differentiates into neural structures corresponding to the regional structure of the host at the level of the fold, while the more distal parts of the folds usually show a more anterior type of differentiation, such as forebrain structures.

The decrease in neural competence with aging of the tissue has been tested by isolating fragments of gastrula ectoderm, maintaining them in isolation for various lengths of time, and then transplanting them into different locations in a neurula where various strong inductors are present. Competence to form brain structures markedly decreases at an age corresponding to the late gastrula stage of control embryos. Ectoderm corresponding to the late neurula stage is completely without neural competence; yet the two neural-crest derivatives, namely, mesenchyme and pigment cells, can still be evoked in ectoderm of the tail-bud stage. Older ectoderm, already differentiating as epidermis, is entirely without competence to do anything but proceed toward its intrinsic epidermal destiny.

At the same time, the aging of the ectoderm does not merely restrict its neural competence but it can bring about new responsiveness not present before. Late neurula epidermis, no longer convertible into neural tissue, becomes competent to respond to other inductors: under the influence of eye vesicle, hindbrain, and forebrain, respectively, it differentiates into lens, ear vesicle, and nasal pits, during post-neurula stages of development. Successive states of competence appear synchronously with the succession of primary and secondary inductors.

It is evident, therefore, that inductors play a vital role in embryonic development even though their nature and means of action are poorly understood. It is equally evident that the role of the responding tissue is greater and even less well understood. Neural induction, to take the primary example, is apparently a triggering process that sets competent ectoderm along the path of neural differentiation. The extent

INDUCTION PROCESSES IN THE HEAD

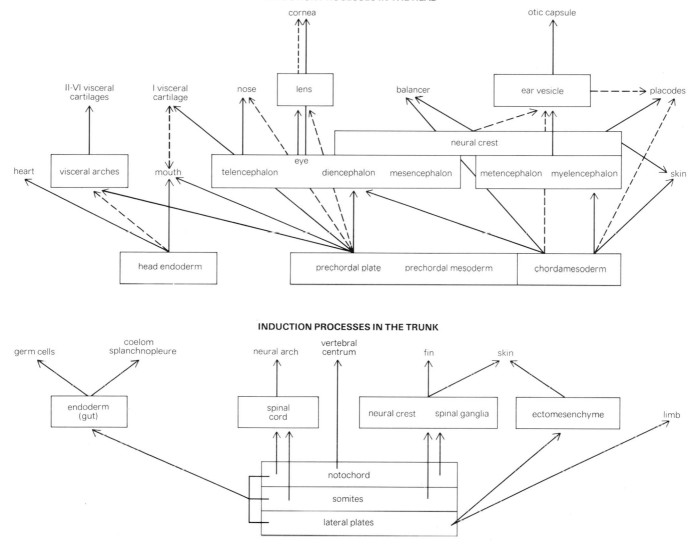

FIGURE 15.13 The induction process during early amphibian development in head and in trunk regions. The inductive action starts from the archenteron roof (inductor of the first degree); the inductors of the second, third, and following degrees effect the process by their own actions. The different strengths of the inductive action are indicated by thick, thin, and broken arrows. [*After Mangold,* Acta Genet. Med. Gemel., **10** (*1961*).]

and shape of the inductive material, either as tissue or as inductive substance diffusing from a more restricted region, determine the extent and shape of the affected ectoderm, i.e., both are wide anteriorly and narrow posteriorly. Within this territory a patterned structure develops that has no relation to any structural properties of the inductor. Yet once established, such an area of neuralized ectoderm undergoes a self-determined patterning, a development typical of so-called embryonic, or morphogenetic, fields.

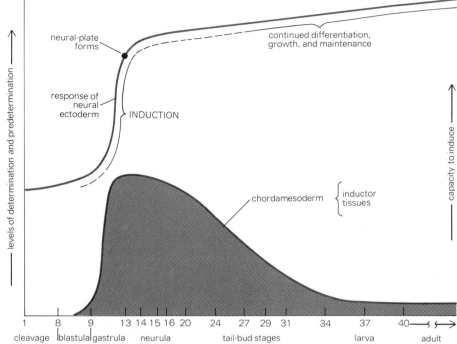

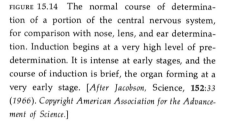

FIGURE 15.14 The normal course of determination of a portion of the central nervous system, for comparison with nose, lens, and ear determination. Induction begins at a very high level of predetermination. It is intense at early stages, and the course of induction is brief, the organ forming at a very early stage. [*After Jacobson,* Science, **152**:33 *(1966). Copyright American Association for the Advancement of Science.*]

PATTERN AND FIELDS

Field phenomena are to be seen in many forms, some of them with inductive effects. The amphibian embryo, for instance, from the early neurula stage on, has a ciliated epidermis with a strong head-tail direction of beat which has a locomotory function. Experiments show that the polarization effect is determined by an invisible structuration of the underlying mesoderm.

Transplantation of tissues between different genera or orders of amphibians (xenoplastic transplantation) demonstrates the complexity of the interacting system of inductors, fields, and, in this event, genetic competence. These experiments exploit certain morphological differences between various amphibian larvae. Most salamander (or newt) larvae possess a pair of temporary lateral head structures known as *balancers.* The axolotl, however, does not have these structures, nor do any anuran larvae. On the other hand, the anuran tadpoles possess a pair of adhesive glands on the ventral side of the head. If a piece of competent axolotl ectoderm is transplanted into the lateral head region of a salamander gastrula, no balancer is formed; but if competent salamander ectoderm is transplanted to the head region of the axolotl gastrula, a balancer is always formed. Evidently the directions, whatever they may be, for balancer

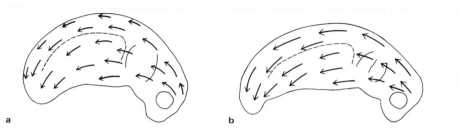

a   b

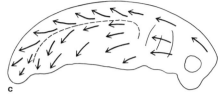

c

FIGURE 15.15 Normal ciliary action of *Triturus torosus*. The arrows indicate the pattern of ciliary action (as tested by the displacement of finely suspended carmine particles) at successive stages in embryonic development. In the earlier stages the direction of effective beat is essentially uniform over the entire surface, creating a backward sweep which follows closely the curvature of the embryonic axis. As development proceeds one may note the gradual emergence of two major currents out of the originally uniform flow. [*After Twitty and Bodenstein, J. Exp. Zool.,* **46**:*343–380 (1941).*]

development are present in the axolotl embryo but axolotl ectoderm is not competent to respond to them; however competent ectoderm from a balancer-producing species recognizes them and performs accordingly.

Even more spectacular are the results of similar experiments made between salamander and frog: competent salamander ectoderm again forms a balancer when transplanted to the anuran head region, even though so far as is known such anuran larvae have never possessed balancers. (There is little doubt that balancers once were present in the axolotl but have been suppressed.) Similarly, competent anuran ectoderm transplanted to the head region of a salamander embryo forms typical anuran adhesive glands.

Two important conclusions may be drawn from these experiments: **(1)** Transplanted competent tissue responds to a new situation according to its own genetic constitution, i.e., each species has its own specific repertoire of possible cytodifferentiations. **(2)** The guidelines (morphogenetic fields and general inductors) are essentially the same throughout the entire class of Amphibia, and possibly throughout other classes of vertebrates. Neither fields nor inductors are species-specific. Different species and orders of amphibians are endowed with a set of genetically similar or "homodynamic" modes of action and reaction, responsible for the general organ pattern of the class. Within this pattern are marked species differences, i.e., a variety of species-specific developmental processes have evolved and are added to or superimposed over the homodynamic framework.

FIGURE 15.16 **(a)** A toad embryo to which, at the beginning of gastrulation, a piece of ectoderm from a donor at the same stage of development has been grafted with inverted orientation. The donor has been previously stained with Nile blue sulfate. The cilia on the graft beat as shown by arrows, in the normal direction. **(b)** A toad embryo to which a piece of ectoderm from an early gastrula, previously stained with Nile blue sulfate, has been grafted with normal orientation. The piece of mesoderm of the embryo underneath the graft has been rotated through 180°. The operation was made when the embryo was at the neural-fold stage. The large section of cilia on the graft beats in reversed direction. **(c)** A toad embryo in which, at the circular blastopore stage, a piece of ectoderm in the flank region has been rotated through 180°. The ciliary action of the graft is opposite in direction. [*After Tung and Tung, 1940.*]

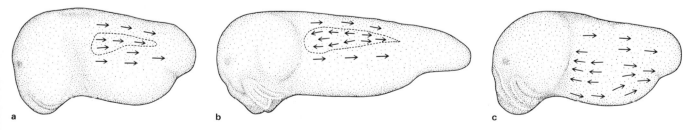

a   b   c

Nieuwkoop summarizes present understanding as follows: "After the discovery that the blastoporal lip of the early and late gastrula acts as head and trunk inductor respectively observations of several investigators have led to the conclusion that the regional organization of the axial system of the embryo depends upon two inductive actions, a mesodermalizing one and a neuralizing action. . . . For a good understanding of the phenomenon of regional organization in the central nervous system it must be emphasized that both inductive actions are exerted during the gastrulation process, viz., when the dorsal archenteron roof, formed by the invagination of the dorsal and lateral marginal zone material, moves upwards toward the animal pole, while the ectoderm simultaneously expands downwards toward the vegetal pole. . . . One inductive agent emanating from the invaginating mesoderm evokes the appearance of a neural plate in the overlying ectoderm, while a second inductive agent also produced by the archenteron roof, is responsible for the regional organization of the developing nervous system. . . . The medio-lateral organization of the mesodermal mantle is brought about by an inducing agent emanating from the notochord anlage and spreading with decrement into the surrounding mesoderm. Depending on the intensity of this factor differentiation into notochord, somites, nephrogenous tissue and blood islands occurs. This inducing action has a two-dimensional field character and leads to the characteristic spatial and structural segregation of the mesodermal mantle into a number of individual organ anlagen. Such actions are called 'morphogenetic field actions.' . . . The process of autonomous pattern formation in embryonic organ anlagen—as we have encountered it, respectively, in the endoderm prior to gastrulation, in the mesoderm in the later phases of gastrulation and at the onset of neurulation, and in the neuroectoderm during neural tube formation—is one of the most challenging problems in embryology. In contrast to the induction phenomena we know very little about the nature of this process. Morphogenetic field phenomena expressing themselves by means of morphogenetically active substances may play a role—as in the segregation of the mesodermal mantle—but biophysical phenomena may also be of great importance."[1]

[1] P. D. Nieuwkoop, Induction and Pattern Formation as Primary Mechanisms in Early Embryonic Differentiation, pp. 132, 133, 137, in W. Beermann (ed.), "Cell Differentiation and Morphogenesis," North-Holland, 1966.

CONCEPTS

Induction is the evocative action of one tissue upon another.

Tissue response is specific, whereas several chemically or physically unrelated inductors may evoke the same response.

Induction is generally consequent on contact between correlated territories.

Tissue differentiation may be dependent on inductive agents or may be independent of them, according to circumstances.

Tissue exhibiting early autodifferentiation generally acts as an inductor of adjacent undetermined tissue.

Competence, i.e., the capacity to respond to inductive influences, is characteristic of an undetermined state. Competence is progressively lost during development.

Diffusion gradients, single and double, of inductive substances or electrolytic properties may play a vital role in the induction process.

An activation-transformation hypothesis is that the induction process consists of two successive phases: a general stimulus leading to manifest differentiation of the target tissue (the activating principle), and a subsequent wave of transformation, of a quantitative nature.

In place of gradients of a quantitative nature, whether one or two, a concept of regionally specific inductors may be valid.

Because time is a factor in all formative tissue movements, such as gastrulation, the inductive stimuli, whatever they may be, exhibit a time gradient which may be crucial with regard to action and reaction events.

Most inductively active tissues can be separated chemically into fractions yielding different inductive actions.

Combined inductive agents may be selectively destroyed by certain chemical or physical treatments, thereby changing the inductive effect.

In the development of the central nervous system the initial inductive stimuli are followed by a chain of secondary tissue interactions which lead to final segregation of the tissue.

Once established, whether by self- or induced activation, a differentiating territory develops a patterned structure which has no relation to any structural properties of the inductor, and is known as an "embryonic field" or a "morphogenetic field."

Morphogenetic fields and biological-field phenomena generally, which are real though enigmatic, may have an electrochemical basis.

READINGS

BARTH, L. G., and L. J. BARTH, 1968. The Role of Sodium Chloride in Cells of the *Rana pipiens* Gastrula, *J. Embryol. Exp. Morphol.,* **19**:387–396.

—— and ——, 1967. Competence and Sequential Induction in Presumptive Epidermis of Normal and Hybrid Frog Gastrulae, *Physiol. Zool.,* **40**:97–103.

CURTIS, A. S. G., 1962. Morphogenetic Interactions before Gastrulation in the Amphibian, *Xenopus laevis,* in the Cortical Field, *J. Embryol. Exp. Morphol.,* **10**:410–422.

HOLTFRETER, J., 1968. Mesenchyme and Epithelia in Inductive and Morphogenetic Processes, in R. Fleischmajer (ed.), "Epithelial-Mesenchyme Interaction," Williams & Wilkins.

——, 1951. Some Aspects of Embryonic Induction, Symposium, *Growth,* **10**:117–152.

—— and V. HAMBURGER, 1955. Embryogenesis: Progressive Differentiation (Amphibia), in B. Willier, P. Weiss, and V. Hamburger (eds.), "Analysis of Development," Saunders.

LEIKOLA, A., and D. J. MCCALLION, 1968. Inductive Capacity of Killed Hensen's Node and Head Process in the Chick Embryo, *Can. J. Zool.,* **46**:205–207.

MANGOLD, H. O., 1931. Experimental Analysis of the Development of the Balancer in Urodeles; an Example of Interspecific Induction of Organs, in M. L. Gabriel and S. Fogel (eds.), "Great Experiments in Biology," 1955, Prentice-Hall.

NEEDHAM, J., 1942. "Biochemistry and Morphogenesis," Macmillan.

NIEUWKOOP, P. D., 1967. Problems of Embryonic Induction and Pattern Formation in Amphibians and Birds, in "Morphological and Biochemical Aspects of Cytodifferentiation," Karger.

——, 1967. The Organization Center, I: Field Phenomena, Their Origin and Significance, II: Segregation and Pattern Formation in Morphogenesis, *Acta Biotheor.,* **17**:151–177; 178–194.

——, 1966. Induction and Pattern Formation as Primary Mechanisms in Early Embryonic Differentiation, in W. Beermann (ed.), "Cell Differentiation and Morphogenesis," North-Holland Publishing Company, Amsterdam.

OPPENHEIMER, J. M., 1959. Extraembryonic Transplantation of Sections of the *Fundulus* Embryonic Shield, *J. Exp. Zool.,* **140**:247–267.

PASTEELS, J. J., 1964. The Morphogenetic Role of the Cortex in the Amphibian Egg, *Advance. Morphogenesis,* **3**:363–390.

PUCCI-MINAFRA, I., and G. ORTOLANI, 1968. Differentiation and Tissue Interaction during Muscle Development, *Develop. Biol.,* **18**:692–712.

REVERBERI, G., 1961. The Embryology of Ascidians, *Advance. Morphogenesis,* **1**:55–103.

RUDNICK, D., 1955. Embryogenesis: Teleosts and Birds, in B. Willier, P. Weiss, and V. Hamburger (eds.), "Analysis of Development," Saunders.

SAXÉN, L., and S. TOIVONEN, 1962. "Primary Embryonic Induction," Prentice-Hall.

SCHOTTÉ, O. E., and MAC V. EDDS, 1940. Xenoplastic Induction of *Rana pipiens* Adhesive Discs on Balancer Site of *Amblystoma punctatum,* *J. Exp. Zool.,* **84**:189–221.

SPEMANN, H., 1938. "Embryonic Development and Induction," Yale University Press (reissued, 1968).

—— and H. MANGOLD, 1924. Induction of Embryonic Primordia by Implantation of Organizers from a Different Species, in B. Willier and J. Oppenheimer (eds.), "Foundations of Experimental Embryology," Prentice-Hall.

SPRATT, N. T., 1964. "Introduction to Cell Differentiation," Reinhold.

——, 1958. Analysis of the Organizer Center in Early Chick Embryos, IV: Some Differential Enzyme Activities of Node Center Cells, *J. Exp. Zool.,* **138**:51–80.

TOIVONEN, S., and L. SAXÉN, 1968. Morphogenetic Interaction of Presumptive Neural and Mesodermal Cells Mixed in Different Ratios, *Science,* **159**:539–540.

TUNG, T. C., S. C. WU, and Y. Y. F. TUNG, 1962. Experimental Studies on the Neural Induction in *Amphioxus, Sci. Sinica,* **11**:805–820.

——, ——, and ——, 1960. Rotation of the Animal Blastomeres in *Amphioxus* Egg at the 8-cell Stage, *Sci. Rec.,* **4**:389–394.

—— and Y. F. Y. TUNG, 1940. Experimental Studies on the Determination of Polarity of Ciliary Action of Anuran Embryos, *Arch. Biol.,* **51**:203–218.

WOLFF, E., 1965. General Factors of Embryonic Differentiation, in W. Beermann (ed.), "Cell Differentiation and Morphogenesis," North-Holland Publishing Company, Amsterdam.

YAMADA, T., 1961. A Chemical Approach to the Problem of the Organizer, *Advance. Morphogenesis,* **1**:1–54.

PART THREE

ORGANIZATION, RECONSTITUTION, AND DIFFERENTIATION

Developmental processes do not cease with the development of the egg. They continue in many subtle and varied ways in the later stages of the organismal cycle. Development continues as growth itself, in the replacement of cells and tissues, with an ever-present opportunity for change. Regional but complete forms of development are exhibited in the development of a limb from a limb field, or an eye from an eye field, more or less independently of the surrounding tissues. Fully differentiated and functioning organisms, such as amphibian and insect larvae, undergo profound transformations into equally differentiated and functioning adult organisms of a very different kind. Organization finds its counterpart in the disorganization characteristic of malignant growth. The phenomena are endless. Nevertheless both the phenomena and the material are in many instances more amenable to experimental investigation than are developing eggs.

Developmental biology is in fact a three-phase study. It embraces the essential nature and complexity of the individual cell as the primary and irreducible unit of life. It has been somewhat obsessively concerned with the integrated developmental phenomena whereby an egg becomes a functional, differentiated, multicellular organism. And it includes the manifold phenomena associated with the growth, maintenance, and progressive reconstitution of the developed organism. On the assumption that the whole always exceeds the sum of its parts, the last phase calls for a very broad perspective, indeed, however difficult to attain.

Yet the study in depth of any segment is likely to illuminate the whole, for the basic problems are common to most forms of development. Wherever we look, we encounter the processes of differentiation within single cells, the assembly of cells to form multicellular structure, the elusiveness of the organizational agency in the multicellular state, the problem of the determination of specific patterns, the question of the role of hormones, genes, and other cellular and intercellular determinants, and, not least, the hierarchy of levels of organization which truly constitutes the organism. The subject is not only vast, it is open-ended, and this final section of the book is primarily a door to the future, leading eventually, it is to be hoped, to a broader and deeper conceptual basis for developmental biology as a living science.

Every step in development shows the cell in a double light, partly as an active participant and partly as a passive subordinate to agencies that lie outside its own competence and control. It contributes to the internal and external state of other cells, either directly from cell surface to cell surface or indirectly by way of the external microenvironment, and by the same means it is affected by other cells. Only by the extension of detail concerning this system may the gap between the dynamic living system and the underlying molecular organization be bridged. In recent years an intense concentration on isolation of differentiating cells and on developing organs and organisms has been rewarded by spectacular progress in particular channels. The accumulation of detail in fact becomes staggering, and there is an increasing danger of loss of coherence in outlook, a loss of the perspective necessary to an appreciation of the unitary quality of the developmental event wherever we find it.

At all times in a developing system we are confronted by two fundamentally distinct but interdependent phenomena, namely, the diverse differentiation of cells and the progressive establishment of form and pattern in the developing whole and in its parts. For each organism a specific variety of paths of specialization are open to the descendants of an unspecialized cell, and the knowledge of the factors, internal and external, which set a cell into this or that path is fundamental to understanding the present problem of tissue differentiation. Yet neither cells nor genes, taken by themselves, sufficiently account for the development and maintenance of pattern. This section is accordingly concerned with tissue interactions, particularly in ectodermal-mesenchymal and in lens-retinal systems; with tissue and organismal reactions to varying hormonal situations; with the problem of selective cell differentiation; and with the phenomenon of malignancy in relation to tissue organization.

CHAPTER SIXTEEN

CELL ASSEMBLY
AND INTERACTION

A developing organism, like any complex machinery, may be seen as a functioning whole or as an assembly of parts. Investigation of the inductive effects of one tissue upon another within the developing system, considered in the preceding chapter, represents analysis working from the highest level of organization to lower levels. The complementary approach is to study the capacity of completely disassociated cells of tissues, organs, and organisms to reassemble and differentiate. This phenomenon already has been discussed to some extent in Chapter 5 (pages 101 to 123), with particular regard to the nature of cell junctions and to disaggregation and reaggregation procedures and phenomena, with sponges as the experimental material. The kinds of intercellular junctions described appear to be essentially the same as those of metazoan tissues as well, and the account already given may be taken as the necessary basis for the following presentation. On the other hand, although sponge-cell aggregates do reconstitute into functional sponge organizations, cell identities are difficult to follow during the process, and a sorting-out of different histological cell types is more assumed than proved. Sorting-out of unlike cells, however, is clearly seen in mixtures of cells derived from different genera or species, and species specificity in sponges is a fact.

CONTENTS
Sorting-out
Tissue hierarchy
Tissue hybrid interaction
Development of intercellular adhesions
Intercellular communication
Maintenance of a tissue
Tissue assembly
Age and tissue differentiation
Tissue interaction
Collagen and morphogenesis
Concepts
Readings

SORTING-OUT

Disaggregation and reaggregation experiments with amphibian embryos show that the sorting-out process and mutual rearrangement of tissues depending on differences already established in the gastrula and neurula stages. According to the classical work of Townes and Holtfreter (1955), the different cell types in a composite aggregate derived from the gastrula become sorted into distinct homogeneous layers whose stratification corresponds to the normal germ-layer arrangement. (Cells were disassociated by brief treatment with a high-pH medium.) Amphibian material has been notably suited to this investigation, as it has for experimental embryology generally. Amphibian cells are relatively large, and different kinds vary greatly in degree of pigmentation. Consequently cell movements can be followed with comparative ease. Thus in an aggregate of ectoderm, endoderm, and mesoderm cells, Holtfreter was able to see the lightly pigmented mesoderm cells vanish before his eyes as they moved into the depths of the tissue mass, while nearly pigment-free endoderm cells and darkly pigmented ectoderm cells moved to the periphery to replace them.

On the basis of this experiment and many comparable ones, he concluded that the variety of tissue segregations, delaminations, dispersals, and recombinations oc-

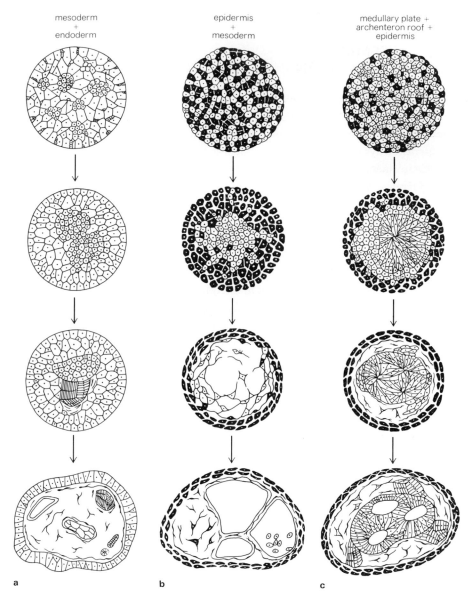

mesoderm
+
endoderm

epidermis
+
mesoderm

medullary plate +
archenteron roof +
epidermis

a

b

c

FIGURE 16.1 Combination and segregation of various combinations of dissociated tissues of amphibian embryos. (a) Mesoderm and endoderm mixtures result in centripetal migration of mesodermal cells. (b) Epidermal cells, when combined with mesodermal cells, move outward, and mesodermal cells move inward. (c) Mixtures of axial mesoderm, medullary (neural) plate, and epidermis cells sort out and re-form respective tissue types. [*After Townes and Holtfreter, 1955.*]

curring in normal embryos could, at least largely, be attributed to developmental stage-conditioned positive or negative "affinities" of the cells involved. In a culture of dispersed cells from early amphibian embryos, any of the cells may unite with any other cell, but the cells later separate and sort out according to tissue specificity. This is in line with the hypothesis of Weiss and Taylor that selective cell adhesion involves the reaction of specific molecules at the cellular interfaces in a manner analogous to that seen in an antigen-antibody complex.

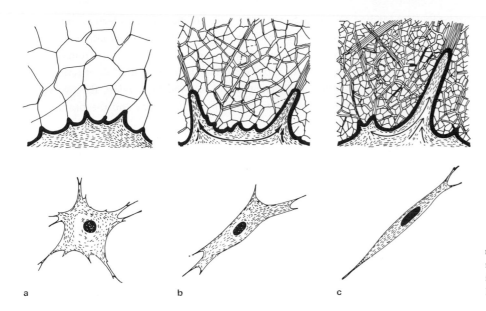

a   b   c

FIGURE 16.2 Structure of mesenchyme cells in three media of (from left to right) increasing plasma fibrous concentrations. [*After Weiss and Garber*, Proc. Nat. Acad. Sci., **38**:*270 (1952)*.]

In mixed aggregates of ectoderm and mesoderm, the ectoderm cells at first mingle with the mesoderm but are later expelled to the surface, where they form a separate epidermal layer, i.e., an initial affinity is later lost. Prospective neural-plate cells and prospective epidermal cells separate to form neural tube and epidermis, respectively. Neural and epidermal tissue can be kept from separating, however, if some mesodermal cells are present to form an intermediate layer adhesive to both.

The directed movements appear to be of two kinds: **(1)** inward movements, as manifested by the cells of the neural plate, or of the mesoderm, which have been combined with either epidermal or endodermal cells, and **(2)** outward movements and peripheral spreading, as manifested by epidermal cells. Tissue segregation becomes complete because of the emergence of a selectivity in cell adhesion: homologous cells when they meet remain permanently united to form functional tissues, whereas a cleft develops between certain nonhomologous tissues (e.g., between neural or endodermal tissues and adjacent epidermis).

Both sorting-out and reassembly processes have been more intensively studied as more sophisticated procedures have been employed, particularly by Moscona, for inducing disaggregation and reaggregation of embryonic cells of chick and mouse. The method primarily employed uses brief trypsin digestion to disrupt intercellular adhesions; this is followed by cell suspension and rotation to bring about forced reaggregations. Such a procedure is readily controlled and standardized with respect to all cellular and environmental parameters. Like mixtures of amphibian cells, mixed cell suspensions of either chick or mouse embryos first form randomly mixed aggregates and then sort out within each aggregate according to histological type. The forces initially binding cells together in aggregates are clearly nonselective and unstable; the selective adhesion to form segregated tissue within an aggregate is acquired later and is stronger. Such a sorting-out is seen when procartilage cells from a 7-day chick embryo are commingled with about ten times their concentration of retina cells from the same embryo. When coaggregated and allowed to develop, the two types of cells sort out, the retinal cells constructing retinal structures and the procartilage cells forming cartilaginous masses.

TISSUE HIERARCHY

Employing similar procedures and material, Steinberg has shown that a hierarchy exists with regard to the positioning of tissues relative to one another, a phenomenon of vital significance concerning the mutual responses of germ layers as seen in the experiments on amphibian germ layers described above, and in the construction and reconstruction of organs. Chick embryo tissues were disaggregated and caused to reaggregate as various paired combinations. Procartilage, heart, and liver cells, among others, were combined in different pairings. Typical results were as shown:

Procartilage → Heart peripheral, cartilage internal
Heart → Liver peripheral, cartilage internal
Liver → Liver peripheral, heart internal

A hierarchy of tissue types is apparent, presumably reflecting differential adhesion properites, i.e., among different motile units of a multi-unit population, adhesive tendencies of different strengths can account for sorting-out, spreading, and the internal versus external positioning, the more cohesive units going internal to the less cohesive.

Some six different tissues were dissociated and mixed in pairs. In each pair, each cell type sorted out and combined, and in each pair the tissues assumed a characteristic mutual arrangement, one exterior and one interior. The various tissue localizations follow a pattern: cells of one tissue segregated internally in all five combinations; those of a second did so in four; a third did so in three, and so on. In other words, if tissue a takes a position internal to tissue b, and b is internal to c, then a will be internal to c, thereby exhibiting a true hierarchy. Whether the specificity of sorting-out is based on qualitative chemical affinities or on quantitative differences in adhesion mechanics, or perhaps on both, remains a question.

TISSUE HYBRID INTERACTION

In the experiments performed with the red and yellow mature sponges, mixed cell suspensions from the two genera at first formed randomly mixed aggregates but later sorted out as separate aggregates, each consisting of cells of but one genus or the other. Mature tissues generally and notoriously reject each other unless from the same individual. This is not true of embryonic tissues. Racial intolerance develops but is not initially present. In fact if presumptive chondroblasts (stellate mesenchyme cells destined to become cartilage-forming cells) from the limb buds of chick and mouse embryos are cultured together for several days, both kinds contribute to the formation of typical cartilage, the cartilaginous matrix of the mouse cells merging with that of the chick cells. To quote Moscona: "Embryonic mouse and chick cells derived from homologous tissues at comparable stages of development are able to form interspecific mosaics and to participate jointly in the histogenesis of such chimeric structures. Regardless of the taxonomic distance, intermingled cells become assorted and matched selectively in accordance with their functional similarity. Kidney cells from both species join to form chimeric tubules; neural retina cells form chimeric rosettes; chondroblasts form chimeric cartilage, and so on. Needless to say cells from different tissues separate out clearly in such interspecific mixtures as they do in intraspecific combinations and, depending on conditions, form either completely separate aggregations or group in different regions of composite aggregates."[1] Tissue type dominates over genetic constitution.

DEVELOPMENT OF INTERCELLULAR ADHESIONS

During the earliest stages of cleavage of sea urchins and amphibian eggs, cytoplasmic strands may be seen connecting adjacent blastomeres, although they do not persist. The dissociated cells of the blastula and early gastrula of the frog reaggregate and also recognize themselves, i.e., a sorting-out occurs in cell mixtures of the two stages, but the aggregating capacity is low. Cells of the middle or late gastrula, however, establish strong bonds which are long-lasting. The development of the strong connections has been more closely followed in the early, primitive-streak embryo of the chick, particularly by Trelstad and Hay. This material is of particular interest since desmosomes were not discovered in either the primary mesenchyme or the primary epithelia; this raises acutely the question of the nature of the intercellular contacts. The earliest occluding junctions observed were points of fusion between plasma membranes of adjacent cells, termed *occluding* or *tight junctions*. These minute tight junctions are formed not only within the embryonic epithelia such as epiblast and hypoblast but also within migrating mesenchyme of the mesoblast layers.

In the chick these several layers arise as follows: When the egg is laid, the cellular blastoderm resulting from the meroblastic cleavage consists of two layers, (1) the hypoblast, consisting of yolk-rich cells, which will give rise to the endoderm, and (2) the epiblast, which will give rise to both the ectoderm and the mesoderm.

---

[1] A. Moscona, Analyses of Cell Recombinations in Experimental Synthesis of Tissues *in vitro, J. Cell. Comp. Physiol.,* **60**(suppl. 1):68 (1962).

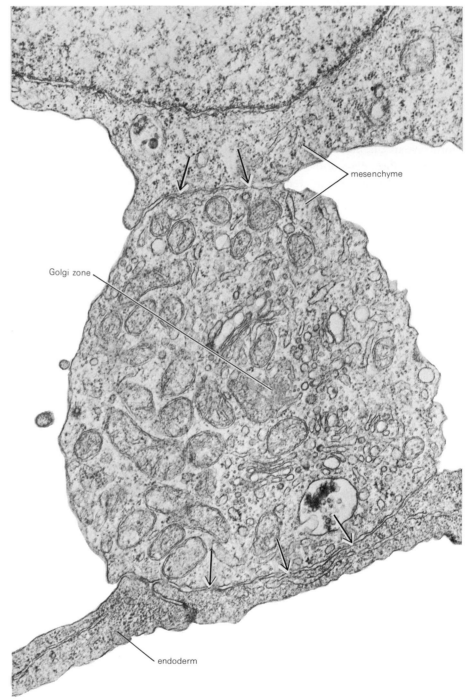

FIGURE 16.3   Close junctions (arrows) between two mesenchymal cells and between mesenchyme and presumptive endoderm, from a region lateral to the primitive streak. [*Courtesy of R. L. Trelstad, E. D. Hay, and J. P. Revel.*]

The epiblast thickens along the midline of the embryo to form the primitive streak. The mesoblast which gives rise to the mesoderm forms from cells of the primitive streak as they move laterally and anteriorly between epiblast and hypoblast. *Mesenchyme* here refers to nonepithelial cells capable of amoeboid movement; *epithelium* refers to tissue whose cells have a free surface and a basal surface. Both types are polarized cells.

At this early stage of development the intercellular contacts in the three layers of tissues consist of (1) focal tight junctions, points where no extracellular space seems to be present between adjacent plasma membranes, and (2) close junctions, where the extracellular space is of more limited width (25 to 100 Å) than elsewhere. According to Trelstad, Hay, and Revel, "Tight junctions not only contribute to cell adhesion in a general way, but also the apposed membranes are probably morphological sites of low resistance to ion flow in various embryonic and adult tissues. Because of the large surface area involved, close junctions could also be preferential sites of intercellular ion flow. It is tempting to think that close and tight junctions are involved in the phenomena of contact inhibition which occurs in migrating embryonic tissues. . . . It seems likely from the work of Spratt and Haas (1961, 1962) that growth centers in localized areas of the epiblast (e.g., primitive streak) give rise to the migrating populations. As in the case of cultured cells, the direction of movement must be away from the densely populated center (primitive streak in this case), presumably because of the numerous contacts in this area. Cells touching the hypoblast are apparently immobilized by close and tight junctions in that region. The cells have to migrate laterally then. . . . It does seem reasonable to conclude that focal tight junctions between like or unlike cells in the early embryo are labile, primitive points of adhesion, suitable to a setting in which cells undergo constant remodeling. By their nonselective adhesiveness *in vitro* after disaggregation, younger tissues demonstrate the instability of their architecture. Whereas Townes and Holtfreter (1955) could not readily correlate with the stages of cellular differentiation what they took to be a change from indiscriminate to selective adhesiveness, the chronological sequence cannot be ignored. This so-called change could be the point at which the maculae adherens (desmosomes) develop, and this stage of tissue differentiation is probably also accompanied by increase of production of extracellular matrices in connective tissue and well developed basement laminae under epithelia and around the presumptive nerve and muscle parenchyma. In future studies of cell reaggregation, it will probably be helpful to distinguish at least two states of cell cohesion, the one involving primarily tight and close membrane junctions and the other, adhering junctions between membranes that are now coated with extracellular materials."[1] Within the epithelia, maculae adherentes (desmosomes) become particularly well developed at later stages and probably contribute to tissue-specific intercellular adhesion.

[1] R. L. Trelstad, E. D. Hay, and J. P. Revel, Cell Contact during Early Morphogenesis of the Chick Embryo, *Develop. Biol.*, **16**:78–106 (1966).

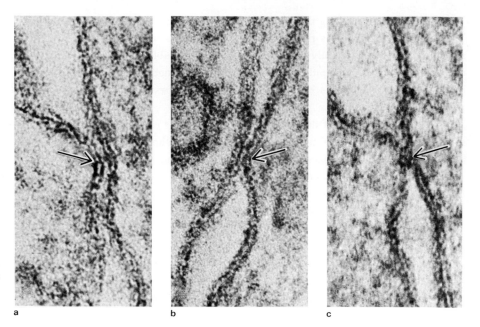

FIGURE 16.4 High-magnification electron micrographs showing junctions between dissimilar tissues in a stage-4 embryo near the primitive streak. (a) Close junction between plasmalemmas of mesoblast and presumptive endodermal cells (hypoblast). The arrow points to the outer leaflet of the trilaminar surface membrane of one of the cells. (b) Tight junction between mesoblast and epiblast. (c) Tight junction between mesoblast and hypoblast. Magnification: ×280,000. [*Courtesy of R. L. Trelstad, E. D. Hay, and J. P. Revel.*]

At least in the epithelial-mesenchymal interactions involved in early chick development, and probably generally, tight contacts between the cells of unlike tissues occur only during the events of primary induction, which is renowned for its lack of specificity. On the other hand, during the more specific interactions taking place later between the secondary mesenchyme and epithelia, the tissues are separated by the extracellular environment. Thus direct contact between unlike cells does not take place.

INTERCELLULAR COMMUNICATION

Adjoining cells communicate as well as adhere. Loewenstein's hypothesis concerning the flow of ions and molecules from cell to cell through opposing membrane regions that are sealed off from the surrounding environment by junctional insulation has already been described in Chapter 5. The concept that cells form junctions so structured as to allow direct flow of substances from one cell interior to another has implications in developing systems. Loewenstein develops the thesis that both cell communication (junctional communication) and interruption of such communication (uncoupling) play roles in tissue growth and differentiation; he is particularly concerned with the action of cell on cell or of tissue on tissue, i.e., with the machinery of differentiation. The concept here is that certain cells (inducers) produce diffusible substances which other cells do not produce, and that when such a substance reaches the latter type of cell it sets it into motion, directly or indirectly. This kind of communication is short-range and is effective only where interacting cells are closely joined. The signal substances for such induced differentiation may use the original cell junctions as diffusion paths, for they would supply a true multichanneled system that is both flexible and efficient; in other words, the cytoplasms plus the junctional membranes form the channel core, and the nonjunctional cell-surface membranes plus the perijunctional insulations form the channel walls. Such a system can bring forth patterns of differentiation with a precision as high as one cell diameter.

The hypothesis proposed is that *"junctional communication is instrumental in the flow of information concerning gene activity, that determines the initiation and maintenance of cellular differentiation; that interruption of this flow, junctional uncoupling, can also lead to differentiation under certain conditions."*[1] In support of this are the following considerations: There is a close surface contact (adhesion) between embryonic inducer and induced cell systems under normal conditions of development; there is extensive junctional communication between embryonic cells; in normally differentiated adult tissues, there is junctional communication between cells of the same kind (this appears to be the general rule in epithelial tissues; e.g., in normal mammalian liver, all parenchyma cells are communicating); the size range of particles permeating the junctional membranes in normal cells is such that a wide variety of metabolites (and other substances) can flow from cell interior to cell interior.

Loewenstein points out that so-called foreign inducers do not themselves enter into the chemical machinery of differentiation: "Agents so entirely different are likely to act through a common mechanism, and the evidence points to a common trigger mechanism. A piece of ectoderm left alone may develop into epidermis; if acted upon by one of the inducing agents, it will differentiate into neural tissue. An often-used figure in this relation is that the cell system is switched from one differentiation path to another. In the light of the present proposals, the cell junction provides just such a switch: in the on position, it establishes the information flow between inducer and target cells in the ectoderm that drives the system from ectodermal to epidermal differentiation; in the off position, it interrupts this flow, and the system runs toward neural differentiation."[2] It must be emphasized that this concept is primarily a good working hypothesis. As such it stimulates further experimental work, but as an explanation of certain phenomena it may or may not be valid.

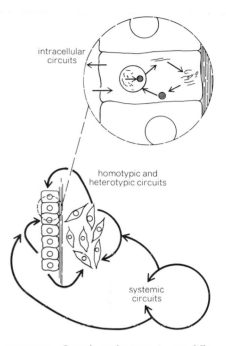

FIGURE 16.5 Control mechanisms in cytodifferentiation. [*After Grobstein, 1966.*]

intracellular circuits

homotypic and heterotypic circuits

systemic circuits

MAINTENANCE OF A TISSUE

The skin of vertebrates serves well as a model system of a tissue. It consists of two layers, the epidermis and the dermis, with a basement membrane between. The epidermis consists of a basal layer of columnar cells which are loosely united as

[1] W. R. Loewenstein, *Perspect. Biol. Med.,* **11**:267 (1967).
[2] *Ibid.,* p. 267.

an epithelium resting upon the basement membrane, together with a number of layers of progressively keratinized cells externally. All cell division takes place in the basal layer; cells above the basal neither divide nor synthesize the DNA which must precede mitosis. Following division, both daughter cells remain basal cells for an indefinite period; yet every time a basal cell divides, a nearby cell begins to move toward the skin surface. Once free of the basal layer, such cells change from columnar to cuboidal to flat and progressively produce the mixture of proteins that constitute keratin. At the same time they form stable connections, desmosomes, with their immediate neighbors. Consequently the epidermal cells external to the basal layer migrate toward the surface bonded together as a continuous sheet.

According to Bullough, who has long worked with mouse skin as an experimental system, "In mouse epidermis the plane of mitosis is usually such as to ensure that both daughter cells remain in the basal layer. It may be only the increasing pressure that ultimately squeezes cells, apparently at random, into the more distal layers where keratin synthesis begins. Thus in stratified epidermis the . . . choice between mitosis and keratin synthesis may be made primarily in terms of the position of the cell within the tissue, which may mean that it is taken in terms of the cell environment."[1]

There is evidence, however, that the cells of the basal layer of mouse skin are physiologically nonhomogeneous, even though similar in appearance. Studies of cell division initiated by plucking hair or wounding the skin show the presence of two cell populations, one with a G_2 period (the period between DNA synthesis and mitosis) of only a few hours, the other with a G_2 period of as long as 2 days. Whether these two populations represent the prospective persisting basal-cell population and the migrating keratinizing population, respectively, is undetermined. Environmental influence, however, is undeniable. In the presence of vitamin A, for example, the basal cells switch from a course leading to keratinization to the production of mucus or cilia, although the nature of the switch remains unknown.

On the assumption that the environment dictates gene expression in epidermal cells, Bullough suggests the hypothesis that a concentration gradient of some chemical messenger exists between the basal and the more superficial layers. Such a system might work as a negative feedback to suppress mitosis and to confine it to the basal layer. Working with this concept as a basis, Bullough and his colleagues have extracted an antimitotic chemical messenger from pig epidermis, which they have called the *epidermal chalone,* that appears to be of central importance in maintaining the stability of the system and seems to be a glycoprotein of relatively low molecular weight.

This substance is tissue-specific but not species-specific. Both in vivo and in vitro, mitosis in mouse epidermis can be suppressed by epidermal chalone extracted from a variety of mammals, including man, and from the skin of a codfish. Therefore it is not even class-specific; the cell mechanism which emits the message and that

[1] W. S. Bullough, "The Evolution of Differentiation," p. 112, Academic, 1967.

which receives it and acts upon it may accordingly be basically the same throughout the vertebrate kingdom. What is true of epidermis probably holds for every other type of tissue; that is, all are probably tissue-specific but without class distinction.

TISSUE ASSEMBLY

Single-cell suspensions prepared from kidney, liver, or skin of 8- to 14-day chick embryos, scrambled, recompacted, and transplanted to the chorioallantoic membrane of 8-day embryos, are able to give rise to remarkably complete and morphologically well-organized organs of the respective kinds, with the various tissue components having their normal mutual relations and functional activity. *The results reemphasize internal "self-organization" as one of the most basic problems in the study of development, in contradistinction to contemporary preoccupation with external inductions* (Weiss and Taylor, 1960).

For practical reasons chick embryo skin has been extensively used as experimental material. Chick embryos within the egg are more accessible than the mammalian fetus within the womb, and the chorioallantoic sac of the chick embryo serves as an excellent vascularized graft site for skin and other tissue explantations. In essential cell structure and cell turnover chick skin is similar to that of mammals, the main difference being the development of feather germs, or papillae, rather than hair follicles. The development of these two organized skin structures offers essentially the same opportunity for analysis.

AGE AND TISSUE DIFFERENTIATION

In tissues maintained in vitro in an appropriate culture medium, skin derived from a 5-day chick embryo failed to differentiate feather germs, whereas skin from a 6-day-old chick did so. Something occurs during the interval, although apparently not as a specific induction, for slightly younger skin grafted into the flank of a 3-day embryo differentiates ahead of its host skin, in spite of the 2-day difference in age, suggesting that all that is needed is sufficient exposure to the embryonic environment. Once this condition has been met, differentiation is seen to depend upon dermal-epidermal interactions.

This is shown by various transplantation experiments made by Rawles, who dissociated the two layers by gentle trypsin digestion, grafted them to the chorioallantoic membrane within the developing egg, and variously combined epidermis from one part of an embryo with dermis from another region. Back epidermis from a 8-day embryo placed against 13-day dermis of the normally scale-forming ankle region of the foot still developed feathers. Thus in the back epidermis-foot dermis combina-

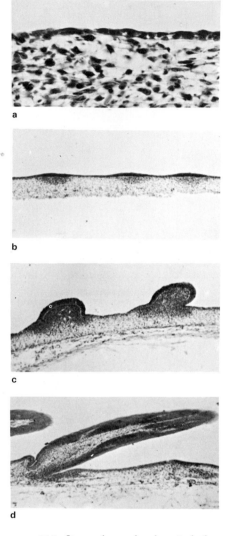

FIGURE 16.6 Stages of normal embryonic feather development (low-power magnification), showing (a) preplacodal stage, (b) placodal stage, (c) hump stage, and (d) elongation stage. [*Courtesy of B. Garber.*]

tion, the degree of determination already present in the younger, epidermal tissue is still able to dominate the system. On the other hand, beak dermis from 5- to 8-day chick embryos when placed against the same back epidermis induces a perfect beak. At 8 days the feather germs are already well developed, yet the commitment to further feather development is still reversible, and the epidermis can be remodeled to produce a different structure. These and other experiments show that the particular structure produced by the skin is primarily an epidermal construction but that the specific kind of structure is determined by the underlying dermis; and further that the commitment of the epidermis is alterable up to a certain developmental stage.

The normal development of the skin and feather has been well described, especially in a review by F. R. Lillie (1942) and in one by Wessells (1965). In brief, four stages are generally recognized:

Preplacodal stage In the 5-day embryo the ectoderm is a simple epithelial layer of cuboidal cells overlaid by flat peridermal cells, the underlying mesoderm being a loose mesenchyme network.

Placodal stage In the 7- and 8-day embryo the feather germs (primordia) first appear as localized nodes, or centers of dense cell populations, along the mid-dorsal line of the embryo, the ectodermal cells now becoming columnar and forming placodes while the underlying mesenchymal dermis condenses into dermal papillae. Other rows of feather germs later appear successively on either side of the primary row, i.e., the feather germs originate progressively both posteriorly and bilaterally.

Hump stage In the 8- to 10-day embryo the dermal component grows rapidly and raises the ectodermal component to form the feather bud.

Elongation stage The feather germ elongates, and condensed dermal papilla cells are evident at the feather base.

The capacity of dissociated and subsequently reaggregated embryonic cells to develop into organized structures changes with the age of the cells at the time of their dissociation, i.e., it changes as their differentiation progresses. Thus aggregates

SUMMARY OF THE CHANGES IN HISTOGENETIC POTENTIAL OF DISSOCIATED EMBRYONIC SKIN CELLS WITH ADVANCING DEVELOPMENTAL AGE*

Age in Days	Feathers	Stratification of Epithelium	Keratinization
6	+	−	−
8	+	−	−
10	+	+	+
12	−	+	+
14	−	+	+

*From Garber, 1967.

of cells dissociated from 5- to 8-day-old embryonic skin (preplacodal and placodal stages) form well-developed, typical feathers enclosed in large, thin-walled keratinized epithelial vesicles in the chorioallantoic membrane. Aggregates of cells dissociated from 10-day embryonic skin (hump stage) form imperfect feathers and skin, normal in cell types but poorly constructed. No feathers of any sort, only dense and extensive sheets of dermis and keratinized epidermis, form from aggregates of 12- and 14-day-old skin cells.

It is clear that with advancing development, dissociated skin cells exhibit a progressive loss of the capacity to form feathers, at the same time increasingly tending to produce fully keratinized skin. This is in keeping with the general concept that differentiating cells face exclusive choices, a path in one direction excluding another.

Yet age shows a dampening or inhibitory effect even at this level of cell differentiation. Mixed aggregates of chick skin cells of different embryonic age develop according to the nature of the mixture. Aggregated mixtures of skin cells from 6-day embryos and 14-day embryos form no feathers but only heavily keratinized skin sheets. Older skin cells function in this case as suppressors of feather morphogenesis.

TISSUE INTERACTION

Aggregates consisting of embryonic chick and mouse skin cells also show developmental capacities correlated with age. One of the first discoveries was that such aggregates give rise to bispecific chimeric sheets of heavily keratinized, stratified skin with mouse hair and sebaceous gland rudiments, but with no feather structure. It was further found that feather development is equally suppressed in cell aggregates consisting entirely of chick cells containing mixtures of embryonic skin cells with liver, heart, or lung cells. Suppression, therefore, must be caused by something other than the difference in genotype between chick and mouse. According to Garber et al., "The fact that not only heterotypic cells but also heterochronic differences between embryonic chick skin cells in aggregates could result in suppression of the morphogenetic program for feather formation made it important to re-examine the situation in interspecific combinations of skin cells; in particular to determine if the developmental state of the mouse cells was related to their capacity to suppress feather development in aggregates. . . . Cells from early embryonic mouse skin included in aggregates of chick skin cells do not suppress feather morphogenesis; on the contrary, they become incorporated into the feather structure and participate with the chick cells in the formation of chimeric feathers. Mouse skin cells from later embryos progressively acquire the capacity to suppress feather development in this system."[1]

[1] B. Garber, E. J. Kollar, and A. Moscona, Aggregation *in vivo* of Dissociated Cells, III: Effect of State of Differentiation of Cells on Feather Development in Hybrid Aggregates of Embryonic Mouse and Chick Skin Cells, *J. Exp. Zool.,* **168**:456 (1968).

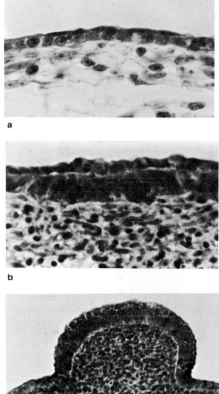

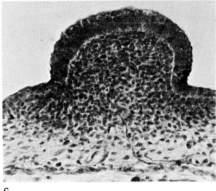

FIGURE 16.7 Stages of normal embryonic feather development, high-power details showing epidermal differentiation and dermal papilla (*DP*). (a) Preplacodal stage, 6-day skin. Two-layered epithelium, cuboidal cells covered by flat peridermal cells; mesenchyme forms a loose subjacent network; feather papillae are not yet defined. (b) Placodal stage, 8-day skin. Epithelium has palisaded into a thickened placode of columnar cells covered by a peridermal layer; dermal condensation beneath the placode forms a discrete dermal papilla. (c) Feather bud in hump stage, 10-day skin. Dermal papilla has greatly increased in volume and has elevated the placodal epithelium; blood vessels invade the papilla from the underlying connective tissue. [*Courtesy of B. Garber.*]

RESULTS OF COMBINATIONS OF CHICK AND MOUSE SKIN CELLS*

8-day chick epidermal and chick dermal cells	Reaggregated cells reconstructed typical feathers and nonkeratinized skin.
13- to 15-day mouse epidermal and mouse dermal cells	Reaggregated cells reconstructed hair and stratified keratinized epidermis.
14- to 15-day mouse skin cells commingled with chick skin cells (epidermis + dermis)	Feather development was suppressed, but hair and sheets of bispecific keratinized epidermis were produced.
13-day skin cells commingled with chick skin cells	Aggregates produced both hair and feathers, as well as chimeric, bispecific epidermis. Mouse cells frequently incorporated in feather structures.
Mouse dermal cells and mouse epidermal cells separately commingled with chick skin cells	Mouse epidermal cells contributed to epidermal elements of feathers. Mouse dermal cells formed groups at site of the feather dermal papillae.

* From Garber et al., 1968.

Three types of development occur in these bispecific aggregates, depending on the state of differentiation of the cells when they are first combined:

1   Cells of one species can suppress the developmental program typical of the other species.
2   Developmental programs characteristic of each species may be followed together and in a normal manner.
3   Cells of one species can participate in the developmental program typical of the other species.

With regard to the suppression of feather buds by older mouse skin cells, Garber et al. suggest that the mouse dermal cells occupy the site of the feather papilla and apparently prevent the entry of chick dermal cells into the feather rudiment, perhaps by blocking essential epidermal-mesenchymal metabolic exchanges, or by some effect of contact inhibition of movement. The results of this investigation further illustrate and stress the critical importance of communication among cells in regulating developmental expression at the level of structural organization.

COLLAGEN AND MORPHOGENESIS

Much of the pioneering work with tissue culture in relation to tissue pattern and morphogenesis has been that of Paul Weiss and his colleagues, beginning with his discovery that connective-tissue cells in culture tend to move along lines of least resistance in the culture media. Orientated tracts of protein subunits resulting, for example, from stress induced in the medium are followed by the mesenchymatous cells, i.e., through "contact guidance." This phenomenon is clearly of great importance with regard to formative movements of cells and tissues. It is a general feature of epithelial tissue, for example, that it requires a substrate over which to spread. Chick embryo epidermis when cultured alone fails to grow and loses its skinlike organization. Yet the normal organization is maintained if a suitable substratum, whether it be live dermis, frozen or killed dermis, or collagen gel, is present.

Collagen, and probably other proteins that are less abundant, may play an important role with regard to both cell movement and tissue induction. Moscona reports that in the morphogenesis of embryonic feather germs, the formation of dermal groupings is associated with the development of a highly regular pattern of birefringence in the dermis; this birefringence is due to a latticelike system of collagenous tracts along which the dermal cells become progressively aligned and grouped in regularly spaced sites.

Sheets of skin from the mid-dorsal area of 6-day embryos, maintained in organ culture, show that the first change in the dermis in connection with feather development is the appearance of cells elongated in an anteroposterior direction, while small clusters of cells appear marking the sites of the first row of dermal papillae. At the 8-day stage the entire dorsal feather area (field) is organized into a latticelike system of oriented dermal cells linking the sites of the dermal papillae. Epidermis alone shows a generalized meshwork, but dermis shows an additional fibrous pattern which first appears as a midline streak and then extends laterally, i.e., corresponding to the order of appearance of the feather germs. The pattern of the fibrous material and the pattern of feather buds are the same, both spatially and temporally. The fact that the enzyme collagenase which breaks down collagen inhibits the development of the fibrous material and of feather germs, as well as other evidence, indicates the collagenous nature of the fibrous lattice.

According to Wessells and Evans, however, there is no unique orientation of the collagen, and on the basis of current information it can only be hypothesized that the fibroblasts and some collagen bundles may orient between condensations in response to the same factors that determine feather-germ pattern on the embryo. Moreover, if microtubules are the agent by which cells elongate, as seems generally to be the case, then any action of extracellular environment upon this process must in some way influence the site and the orientation in which assembly of microtubules occurs.

Collagen has been implicated in reacting tissues besides the epidermis-dermis system, notably in the formation of kidney, pancreas, and salivary gland. Each of these involves an interaction between an epithelial tissue and mesenchyme. In these experiments of Grobstein and his associates Wessells and Rutter, the two components

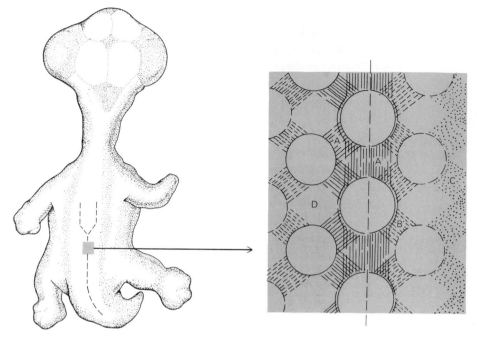

FIGURE 16.8 Skin in the vicinity of the square marked on the back of the chick embryo taken for examination. (*Left*) Dermal condensations at young stage; (*right*) dermal condensations at older stage. [*After Wessells and Evans, 1968.*]

in each case were separated by trypsin treatment, gently dissected, and cultured separately. Neither component alone can continue its characteristic development, but following recombination each pair is able to do so. Experiments consist of arranging one component of a system on one side of a membrane filter and the other component on the other side. Using filters of various porosity gives information concerning the size of molecules or particles that may diffuse through the filter.

The unit kidney rudiment of the mouse consists of a mesenchyme component derived from the nephrogenic mesenchyme, which gives rise to the glomerulus and secretory tubules, and a component from an outpocketing (ureteric bud) of the epithelial embryonic kidney duct (Wolffian duct), which gives rise to the collecting tubules and kidney pelvis. When the components are separated by the membrane filter, inductive action still occurs, i.e., the signals or stimuli pass through the filter. In fact, the nephrogenic mesenchyme will still produce secretory tubules if pieces of the dorsal side of the embryonic spinal cord are substituted for the normal derivative of the Wolffian duct. This is a morphogenetic response, therefore, to an inductor which is not tissue-specific, just as primary neural induction in the early embryo is not tissue-specific. After an initial response is made, continued exposure to the inductive agent is necessary to stabilize the event. This leads to the concept that a primary commitment, or determination, phase is followed by a secondary support

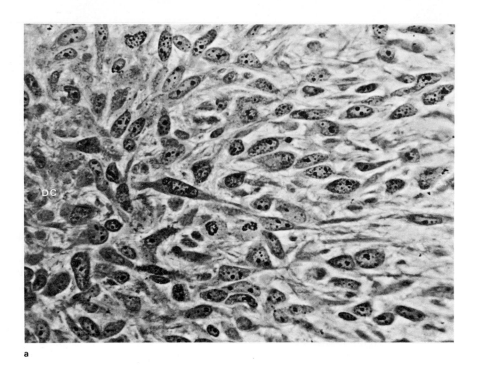

a

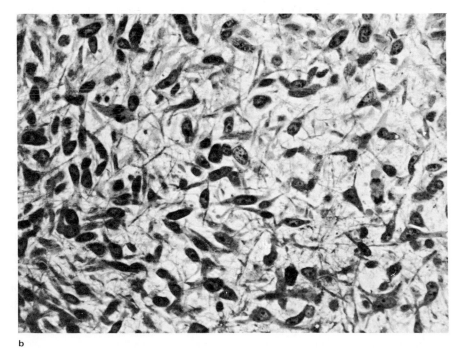

b

FIGURE 16.9 (a) A dermal cell condensation on the left, and oriented cells pointing toward a neighboring condensation. (b) An unoriented cell region equivalent to area *C* or *D* of Figure 16.8. A well-formed condensation is located below the lower left-hand corner of this field. [*Courtesy of N. K. Wessells.*]

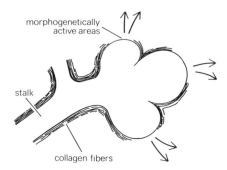

FIGURE 16.10 Location of collagen fibers around a developing salivary-gland unit. [*After Grobstein, 1966.*]

phase during which nonspecific mesodermal input is sufficient to permit tissue growth (Wessells, 1967).

Similarly, a pancreatic rudiment of the mouse embryo consists of an inner epithelial component and a surrounding mesenchymal component. When developing in culture, the epithelial component produces branchings with bulbous endings, within the mesenchyme, for about 2 days. Only after this morphogenetic process is virtually at an end is there any sign of specific pancreatic enzymes (e.g., amylase), the secretion of which then increases sharply and continues for some 5 days.

A concept in favor from time to time has been that cytodifferentiation is dependent on tissue mass, i.e., that it takes place because of increase in tissue size, but experiments combining tissue aggregates or pieces to form extra-large masses in culture have not resulted in greater extent or in acceleration of differentiation. On the other hand, fragmenting pieces of tissue below certain dimensions usually inhibits or postpones differentiation. Wessells (1967) has shown that such postponement of differentiation in subdivided pieces of pancreas epithelium is correlated with continuation of mitotic divisions, but that once cells reach a certain chronological age they differentiate, even in extraordinarily small masses, if undisturbed. In other words, for cytodifferentiation to occur, cells must be sufficiently mature and must not be stimulated to continue dividing.

In further studies collagen fibers were seen on the epithelial side of the filter membrane. Collagen has been considered to be primarily a secretion of mesenchyme cells, although in the formation of the basement membrane of amphibian skin the collagen fibrils first appear next to the epithelium and progress from there into the mesenchyme. However, in experiments in which tritium-labeled amino acid was given to the mesenchyme component but not to the epithelium (in this case, salivary gland mesenchyme and epithelium), labeled collagen was produced. Evidently collagen molecules are produced by the mesenchyme, pass through the filter membrane, and polymerize next to the epithelium. The resulting concept is that the formation of the basement membrane results from the interaction of two or more kinds of macromolecules produced on either side and combined in the intercellular space. This is important, because what may be true of the interaction between adjacent surfaces of tissues may also hold for interactions between adjacent or adjoining cells generally.

Several possibilities concerning control agencies for tissue differentiation exist, not necessarily excluding one another: intracellular circuits between nucleus and cytoplasm; intercellular junctional communication between adjoining cells; homotypic and heterotypic circuits between adjacent cells and tissues; and systemic circuits, particularly hormonal. In all but the first, the properties of the cell surface, both junctional and nonjunctional, the nature of intercellular material, and the sensitivity to diffusing substances play vital roles in the system of controls.

CONCEPTS

When tissues are disaggregated and then mingled together, unlike cells sort out and like cells adhere together, by means of random movement, until like meets like. Then contact inhibition causes movement to cease.

Unlike tissues take up specific positions relative to one another, i.e., internal or external, according to a specific hierarchy, which seems to be related to relative adhesiveness.

Homologous tissues, histologically alike, are compatible in chimeric assemblies when embryonic, even if they are from different classes.

Each specific tissue secretes a correspondingly tissue-specific inhibitor which is not class-specific.

Tight junctions form progressively between cells of specific tissues during development.

Junctional communication is instrumental in the flow of information concerning the gene activity that determines the initiation and maintenance of cellular differentiation.

Self-reorganization, seemingly based on a hierarchy of adhesiveness, is a dominant principle in the reconstruction of scrambled cell aggregations derived from organs and embryos.

Reorganization capacity of reaggregated embryonic cells changes with the age of the cells at the time of dissociation.

Cells of one species (genetic) can suppress the developmental program typical of another species, or they may participate in the developmental program typical of the other species, depending on the state of differentiation at the time of combination.

Extracellular matrices, such as collagen, may play a vital role with regard to tissue architecture and patterning.

Control agencies for tissue differentiation exist at several levels: intracellular circuits between nucleus and cytoplasm; circuits between adjoining cells; and systemic, particularly hormonal, circuits.

READINGS

BENNETT, M. V. L., and J. P. TRINKAUS, 1968. Electrical Coupling of Embryonic Cells by Way of Extracellular Space and Specialized Junctions, *Biol. Bull.*, **135**:415.

CURTIS, A. S. G., 1962. Cell Contact and Adhesion, *Biol. Rev.*, **37**:82–129.

ERICKSON, R. A., 1968. Inductive Interactions in the Development of the Mouse Metanephros, *J. Exp. Zool.*, **169**:33–42.

GARBER, B., 1967. Aggregation *in vivo* of Dissociated Cells, II: Role of Developmental Age in Tissue Reconstruction, *J. Exp. Zool.*, **164**:339–349.

———, E. J. KOLLAR, and A. MOSCONA, 1968. Aggregation *in vivo* of Dissociated Cells, III: Effect of State of Differentiation of Cells on Feather Development in Hybrid Aggregates of Embryonic Mouse and Chick Skin Cells, *J. Exp. Zool.*, **168**:455–464.

GROBSTEIN, C., 1966. What We Do Not Know about Differentiation, *Amer. Zool.*, **6**:89–95.

——, 1963. Microenvironmental Influences in Cytodifferentiation, in J. Allen (ed.), "The Nature of Biological Diversity," McGraw-Hill.

HOLTFRETER, J., 1939. Tissue Affinity, a Means of Embryonic Morphogenesis, in B. Willier and J. Oppenheimer (eds.), "Foundations of Experimental Embryology," Prentice-Hall.

MOSCONA, A., 1962. Analysis of Cell Recombinations in Experimental Synthesis of Tissues *in vitro*, *J. Cell. Comp. Physiol.*, **60**(suppl. 1):65–80.

——, 1957. The Development *in vitro* of Chimeric Aggregates of Dissociated Embryonic Chick and Mouse Cells, *Proc. Nat. Acad. Sci.*, **43**:184–194.

PATRICOLO, E., 1967. Differentiation of Aggregated Embryonic Cells of Amphibians (*Discoglossus pictus*), *Acta Embryol. Morphol. Exp.*, **10**:75–100.

—— and S. A. ROTH, 1964. Phases in Cell Aggregation and Tissue Reconstruction: An Approach to the Kinetics of Cell Aggregation, *J. Exp. Zool.*, **157**:327–338.

RASMUSSEN, H., 1970. Cell Communication, Calcium Ion, and Cyclic Adenosine Monophosphate, *Science.* **170**:404–412.

ROTH, S., 1968. Studies on Intercellular Adhesive Selectivity, *Develop. Biol.*, **18**:602–632.

SMITH, R. T., and R. A. GOOD (eds.), 1969. "Cellular Recognition," Appleton Century Crofts.

STEINBERG, M. S., 1970. Does Differential Adhesion Govern Self-assembly Processes in Histogenesis? Equilibrium Configurations and the Emergence of a Hierarchy among Populations of Embryonic Cells, *J. Exp. Zool.*, **173**:395–434.

TOWNES, P. L., and J. HOLTFRETER, 1955. Directed Movements and Selective Adhesion of Embryonic Amphibian Cells, *J. Exp. Zool.*, **128**:53–120.

TRELSTAD, R. L., E. D. HAY, and J. P. REVEL, 1966. Cell Contact during Early Morphogenesis of the Chick Embryo, *Develop. Biol.*, **16**:78–106.

TRINKAUS, J. P., 1969. "Cells into Organs," Prentice-Hall.

—— and J. P. LENTZ, 1964. Direct Observation of Type-specific Segregation in Mixed Cell Aggregates, *Develop. Biol.*, **9**:115–136.

UNSWORTH, B., and C. GROBSTEIN, 1970. Induction of Kidney Tubules in Mouse Metanephrogenic Mesenchyme by Various Embryonic Mesenchymal Tissues, *Develop. Biol.*, **21**:547–556.

WEISS, P., 1961. Ruling Principles in Cell Locomotion and Cell Aggregation, *Exp. Cell Res.*, **8**:260–281.

——, 1961. "The Biological Foundations of Wound Repair," The Harvey Lectures, ser. 55, Academic.

—— and A. C. TAYLOR, 1960. Reconstitution of Complete Organs from Single-cell Suspensions, *Proc. Nat. Acad. Sci.*, **46**:1177–1185.

WESSELLS, N. K., and J. H. COHN, 1967. Early Pancreas Organogenesis: Morphogenesis, Tissue Interactions, and Mass Effects, *Develop. Biol.*, **15**:237–270.

—— and J. EVANS, 1968. The Ultrastructure of Oriented Cells and Extracellular Materials between Developing Feathers, *Develop. Biol.*, **18**:42–61.

ZWILLING, E., 1960. Some Aspects of Differentiation: Disaggregation and Reaggregation of Early Chick Embryos, *Nat. Cancer Inst. Monogr.*, **2**:19–39.

CHAPTER SEVENTEEN

THE LIMB AS A DEVELOPING SYSTEM

The vertebrate limb has long been studied as a more or less independent developing system, both in embryonic development and in the process of regeneration. As these two related events show significant similarities and contrasts, they are discussed separately in this chapter.

DEVELOPMENT

The developing vertebrate limb has been intensively analyzed experimentally as a nearly independent interacting system of epithelial and mesenchymal components. The initiation stages have been studied both in salamander and chick embryos; tissue interaction has been more successfully analyzed in the developing wing of the chick; while reconstruction, or regeneration, of the differentiated limb has been mainly followed in larval and juvenile amphibians.

Embryonic Determination

The first sign of a developing amphibian limb is a small mound of tissue on the flank at the site of a prospective limb; it consists of two primary components, namely, an accumulation of mesenchyme cells overlain by a cap of ectoderm. This disc of tissue is large enough to be subdivided, rotated, or transplanted, before it develops into any sort of structure. The area involved, at least in amphibians, is circular and extends over 3½ somites in diameter, although the mesodermal somites as such do not contribute to limb development. In a normally developed limb, which is a complex and asymmetric structure, the anterior and posterior sides are clearly recognizable, and so are the dorsal and ventral, corresponding to the anteroposterior and dorsoventral axes of the body. In addition the limb has a proximodistal organization.

The important questions which arise concern the establishment of the several axial polarities, the nature of the limb disc itself in relation to its immediate surroundings, and the relative roles of ectoderm and mesoderm and their interaction. The pioneering experiments in this field were made by Ross Harrison during the first quarter of this century.

To begin with, as in virtually all developing systems and even though limb rudiments first appear relatively late in the development of the embryo, the area

CONTENTS
Development
  Embryonic determination
  Ectoderm-mesoderm system
  The ectodermal ridge
  Maintenance factor
  Axial growth sequence
  Cell death as a morphogenetic agent
  Mutations and morphogenesis
Reconstitution
  The blastema
  Dedifferentiation
  Redifferentiation
  Morphogenesis
  Neural trophic factor
Concepts
Readings

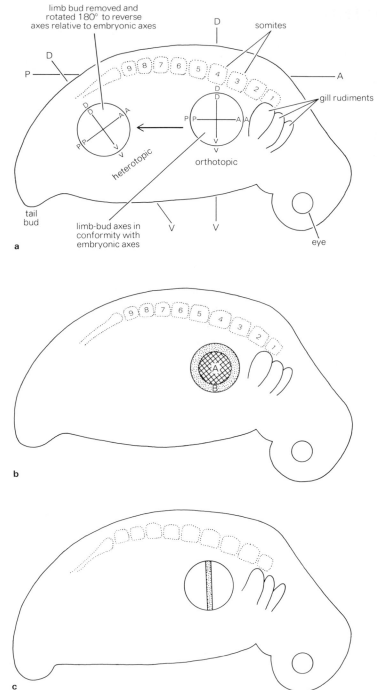

FIGURE 17.1 Experiments on limb-bud differentiation in salamander (*Ambystoma*) during tail-bud stage. **(a)** Transplantation of limb disc from normal (orthotopic) site to another (heterotopic) site. The limb differentiates normally. **(b)** Presumptive limb area (*A*) and limb field (*A* + *B*). If limb area is removed, peripheral-limb-field territory replaces it; if all of limb field is removed, no replacement occurs. **(c)** If limb disc is split and the two parts are prevented from refusing, two limbs develop. [*After Swett, 1937.*]

of tissue forming the limb disc constitutes an equipotential system, as defined by Driesch in 1905. This was shown by the following experiments on the earliest limb-bud stage.

1 If half a limb bud is destroyed, the remaining half gives rise to a completely normal limb.

2 If a limb bud is slit vertically into two or more segments, while remaining an integral part of the embryo, and the parts are prevented from fusing again by inserting a bit of membrane between them, each may develop into a complete limb.

3 If two limb buds are combined in harmonious orientation, a single limb develops which is large at first but soon is regulated to normal size.

The relation of limb polarities to those of the embryo as a whole is shown by inversion experiments. In these a limb-disc area was cut out, turned through 180°, and reimplanted, i.e., nothing was changed except that the anteroposterior and dorsoventral axes of the prospective limb were turned around. If these were fully determined at the time of operation, a limb should develop which is entirely normal except that it is turned backwards and upside down relative to the body. If there was no determination at this time, however, a limb should develop which is normal in every respect. In actuality a limb develops that is normal in that dorsal structure is still dorsal and ventral structure is still ventral, but anterior and posterior sides are interchanged. In other words the anteroposterior polarity of the disc tissue was already established, presumably as part of the primary axial polarity of the embryo as a whole, but the dorsoventral polarity was still reversible.

Ectoderm-Mesoderm System

The limb grows as a result of rapid multiplication of cells within the limb disc, rather than by migration of cells or tissue from outside the disc area. The limb-forming material is clearly localized in the mesoderm (mesenchyme), for no limb develops if the disc mesoderm is completely removed, leaving only the prospective limb ectoderm in place, nor does a limb develop where prospective limb ectoderm is transplanted to a new site, but a limb does develop where prospective limb mesoderm is transplanted beneath ectoderm in other places. Limb mesoderm transplanted to the flank with a covering of flank ectoderm gives rise to a limb with normal asymmetry if the mesoderm is oriented with its anterior side forward, but it gives rise to a limb with reversed asymmetry if the mesoderm is oriented with its anterior side facing backward. On the basis of these early experiments the concept of mesodermal determination of limb development seems to be a good working hypothesis. Nevertheless, an assumption that the prospective-limb ectoderm plays no important role is by no means justified. Isolated mesoderm from limb-disc stages cannot form limb structures;

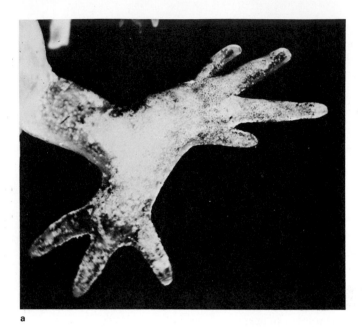
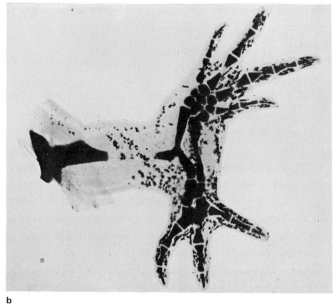

a   b

FIGURE 17.2   (a) Photo of live supernumerary three-
digit hand extending below normal hand, resulting
from ultraviolet irradiation of a developing right
forelimb of *Ambystoma* larva, and (b) stained section
of same, showing skeleton. [*Courtesy of E. G. Butler.*]

a covering of ectoderm is required, although the ectoderm from a variety of sources
is capable of cooperating with the mesoderm to this effect.

Although the earlier experimental analyses of the amphibian limb discs indicated
that the characteristic properties of limb development were mainly inherent in the
mesoderm, Balinsky concluded from his own experiments on limb induction that
limb-forming capacities are not limited to the normal limb areas. Various stimuli
applied to the flank mesoderm of salamander (newt) embryos between the prospective
anterior and posterior limb territories may cause the development of a heterotopic
(i.e., out-of-place) limb. He concluded that the activity of the prospective-limb mesoderm, when it is shifted to the new location, changes the overlying ectoderm into
specific limb epidermis, and that this change represents a biological reaction of the
epithelial component. The influences thus received by the ectoderm are then reflected
to the underlying mesoderm as return activities responsible for the morphogenetic
processes of the mesoderm in successive developmental stages.

The analysis of limb development and determination has shifted over the years
from amphibian material, in which limb rudiments although readily accessible are
inconveniently small, to the chick, which offers larger territories to work with but
has required more sophisticated techniques, mostly devised by Hamburger, whose
studies form the foundation for most of the modern work on limb development.
The shift in attention has also been from the initial limb-disc stages to the more
or less advanced limb bud. Contemporary work on wing buds is being done by
Saunders and Zwilling and their associates in the United States, and by Amprino

and Camosso in Italy, while Milaire in Belgium has extended it to the developing mammalian limb. The emphasis in all this work has been on the growth and individuation of territories, regulative processes, and the morphogenetic relationships between ectoderm and mesoderm of the limb bud.

The Ectodermal Ridge

In the chick the prospective mesoderm of the wing bud becomes finally localized at the two-somite stage. Grafting experiments show that the anteroposterior axis becomes determined at the five-somite stage, the dorsoventral axis at the 13-somite stage. The limb-forming areas, however, do not become morphologically distinguishable until the 14-somite stage. The wing rudiment is represented at this time by a slight condensation of mesenchyme, and a little later by a thickened ridge of ectodermal tissue.

According to the interpretation of Saunders and Zwilling, to quote the latter: "In this interacting system the mesoderm continues a sequential elaboration of distal limb structures under the influence of the ectoderm (ridge, cap) which, as a consequence, is responsible for distal outgrowth of the limb. If this system is deprived of the ectodermal ridge before all the presumptive limb elements are elaborated there is a halt in the formation of additional limb elements, but those which have been laid down may continue to develop their typical form and will grow (intercalary

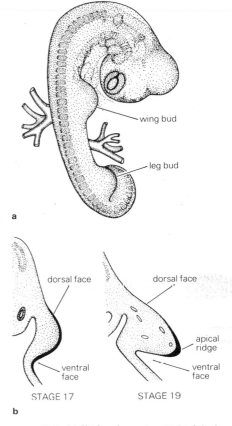

a

b

FIGURE 17.3 (a) Chick embryo, stage 21. Limb buds appear as prominent swellings on the body wall. (b) Cross sections of chick embryos, stages 17 and 19, at the wing-bud level. There are differences in thickness of the ectoderm between ventral and dorsal faces of the wing bud. [*After Amprino, 1965.*]

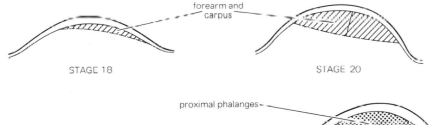

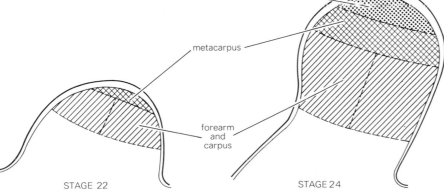

FIGURE 17.4 Maps of prospective wing segments in four embryonic stages. The apical ridge of the ectoderm is shown covering the mesodermal (shaded) tissues in each stage. [*After Amprino, 1965.*]

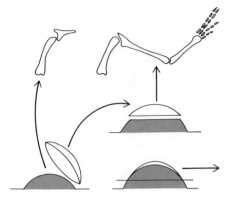

FIGURE 17.5 Induction by the apical cap. (*Left*) Differentiation of a limb bud after removal of the apical cap. Only the femur and a part of the tibia are differentiating. (*Right*) Graft of an apical cap on the basal part of the leg, after the distal half has been severed. The basal mesenchyme is induced by the apical cap to form the distal components. [*After Hampé*, J. Embryol. Exp. Morphol., **8**:247 (*1960*).]

growth) at nearly the usual rate. The result is a limb with deficient distal elements but with proximal parts that are quite normal. Exactly which distal parts will be missing depends on when the ridge is removed. The persistence of the ectodermal thickening as an active influence on continued elaboration of distal limb elements depends upon some factor (apical ectoderm maintenance factor) which is present in limb-bud mesoderm. . . . Little is known about the nature of this factor. . . . Limb type properties (i.e., fore-limb *vs.* hind-limb) are resident in the mesoderm."[1]

Several lines of evidence support the concept that the ectodermal ridge, as well as the mesoderm, has an important morphogenetic role. Zwilling separated ectoderm and mesoderm components, recombined them in different ways, and grafted the altered limb buds to chorioallantois. Mesoderm covered by nonlimb ectoderm fails to grow, while distal limb structures develop only when lateral limb-bud ectoderm, complete with an ectodermal ridge, is placed over the distal end of denuded limb-bud mesoderm. Saunders showed that when leg-bud mesoderm is added to the wing bud, by insertion between the stretched ectodermal ridge and the apical wing mesoderm, foot parts are formed at the tip, which are of normal size, whereas homologous wing parts are missing. The chimeric wing bud thus accommodates the added mesoderm and still forms a whole limb but of a different character.

Limb-bud combinations can be made between chicks and other species of birds. Leg mesoderm taken from an embryo of either Japanese quail or white Pekin duck and similarly implanted in a chick wing bud also produces foot parts at the tip of the wing host. Such parts, however, are smaller than the equivalent chick structures,

[1] E. Zwilling, Limb Morphogenesis, *Advance. Morphogenesis*, **1**:301–338 (1961).

FIGURE 17.6 (a) Large wedge-shaped sector (arrow), isolated from a donor wing bud (*bottom*), and implanted into a host wing bud. (b) Normal wing developed from the composite wing bud. [*After Amprino, 1965.*]

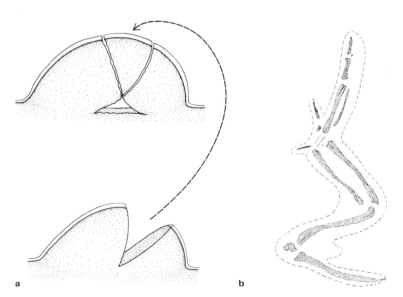

a   b

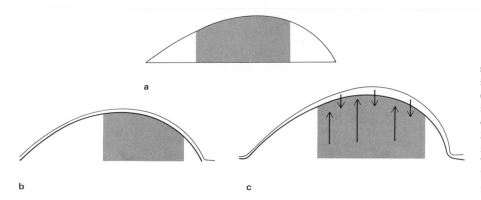

FIGURE 17.7   Distribution of the maintenance factor in the wing-bud mesoderm (shaded), according to (a) Zwilling and (b) Saunders et al. (c) Reciprocal interactions between limb-bud mesoderm and ectoderm, and vice versa, according to the Saunders-Zwilling theory. The activity of the mesodermal factor on the ridge is indicated by the lower arrows. The outgrowth activity exerted on the mesoderm by the activated and thickened apical ridge is represented by the upper arrows. [*After Amprino, 1965.*]

and some or all of the homologous wing parts are also produced. Apparently the sizes of labilely determined prospective regions of mesoderm and ectoderm must be in normal proportions relative to each other. Here the apical ectoderm presumably covers two smaller mesodermal prospective regions of different types and responds regionally and separately to each.

Comparable experiments have been made by combining limb-bud mesoderm with more than one ectodermal ridge. When all ectoderm, including the ridge, is removed from a limb bud and an ectodermal ridge from other limb buds is placed on each lateral surface, the outgrowths from the original distal end of the bud cease, but a new outgrowth and set of distal structures form in connection with each grafted ridge. On the other hand, if only part of a ridge is grafted, only a partial outgrowth forms, with but one or two digits. If, however, the ectodermal cap of a wing bud is rotated 180° about the proximodistal axis, two wings are induced in place of one.

Maintenance Factor

The Saunders-Zwilling theory postulates a mesodermal maintenance factor for the ectodermal ridge. When the ridge is separated from its underlying mesoderm by a very thin sheet of mica, or if limb mesoderm is replaced by nonlimb mesoderm, the ectodermal ridge degenerates. Conversely, thickened ectodermal ridge always appears wherever there is a limb-mesoderm out-pushing. Zwilling asks the crucial questions: "Does the ridge form independently and then induce mesodermal outgrowth or do the thickened regions of the ectoderm form in relation to some pattern present in the mesoderm? If the thickest regions from two or three ectodermal ridges are placed in tandem along the distal edge of one limb mesoblast do they all remain thick and active as outgrowth inducers or do they become modified under the influence of the mesoderm?" [1]

[1] *Ibid.,* p. 316.

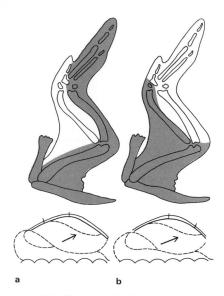

FIGURE 17.8 Effects on the development of the wing bud of the removal of (a) the cranial or (b) the caudal half of the apical ridge of ectoderm (lower schemes). Wing parts which form are shaded. [*After Saunders*, J. Exp. Zool., **108**:*363* *(1948)*.]

The result of experiments was that in every case the composite ectodermal ridge fused to form a single ridge, which gradually acquired a typical normal configuration that gave rise to a normal limb. Much the same results were obtained when limb-bud ectoderm was rotated 180° in relation to the mesoderm and when wing-bud ectoderm was placed on leg-bud mesoderm and vice versa. Ectodermal ridge pattern conformed to the expected normal mesodermal pattern. Moreover, essentially similar results were obtained when genetically normal ectodermal ridges were combined with mesoderm from limb buds of polydactyl mutants: a preaxial thickening of the ridge persists in regions where flattening normally occurs, and accessory distal structures develop there as in the limb buds of polydactyl controls.

Although such experimental findings seem to be conclusive, both the existence of a maintenance factor and the inductive influence of the ectodermal ridge have been questioned by the Italian workers. That the formation and persistence of the ridge depends on a mesodermal maintenance factor seems indisputable. The uncertainty lies rather in the origination of pattern, or individuation, which the American school regards as the result of interaction between the distal ectodermal ridge and the underlying mesoderm, possibly involving the basal membrane as a contributing agent; and which the Italian school regards as entirely the product of the mesoderm, although the presence of epidermis is still considered to be necessary for limb development. In general, however, there is a broad basis of agreement; much of the difference in opinion is subtle and related to the conceptual vagueness afflicting developmental biology as a whole. Thus, according to Milaire, "Although the existence of the apical ectodermal maintenance factor has been clearly demonstrated in recent experiments performed by Zwilling, the limited information on its spatial distribution, its mode of action, and its regulative abilities is not sufficient to provide an adequate explanation for all experimental or genetic modifications of limb morphogenesis."[1]

Axial Growth Sequence

The later course of limb development appears to be essentially the same in amphibians, birds, and mammals, and exhibits a precise proximodistal sequence of growth and differentiation. Older descriptions stated that the initial bud represented digital structures and that proximal structures were added behind. However, when mesodermal components of successively older limb buds were isolated and grown in the emptied eye orbits of older hosts, the mesoderm from younger buds formed proximal parts while more distal parts appeared only as older buds were used. Experiments consisting of marking the mesoderm of various parts of chick limb buds with insertions of small masses of fine carbon particles, and similarly marking both ectoderm and mesoderm with finely powdered colored chalk, and following the fate of the labeled regions, clearly demonstrate the general sequence: basal tissue differentiates first,

[1] J. Milaire, Aspects of Limb Morphogenesis in Mammals, in R. L. DeHaan and H. Ursprung (eds.), "Organogenesis," Holt, Rinehart and Winston, 1965.

and successively more distal regions of the limb are added as distal growth of the limb bud continues. This is also shown by experiments in which, for instance, a piece of tissue is cut out from the distal end of a leg bud and inserted in the distal end of a wing bud. In such a case, a normal wing develops except that a toe forms at the tip of one digit.

In the mammal the early development of the limb bud appears to be essentially like that of the chick, although it has been less studied experimentally because of the mammalian developmental circumstances. It has, however, been more intensively studied histogenetically, particularly with regard to the formation of skeleton. In the mouse embryo the first demonstrable activities connected with limb morphogenesis take place in the mesoderm, as a lateral crest of tissue, although ectoderm is still thin and shows no cytochemical pecularity. Yet even at this stage the limb mesoderm cannot grow and chondrify without ectodermal covering. By the eleventh day, the limb bud of a rat embryo is longer than it is wide and an apical ectodermal ridge is already present, covering the distal third of the bud. At this stage, femoral precartilage has already individuated, while more distally the presumptive mesoderm of the tibia and fibula has condensed as a common precartilaginous plate. In the distal third the mesoderm is still undifferentiated and forms a thick marginal layer, rich in RNA, beneath the ectodermal ridge, representing the presumptive material of the distal limb segment. A little later, as the result of simultaneous gradients of morphogenesis, different precartilaginous masses of mesoderm condense successively from the tarsal element to the distal phalanx in each digital ray.

On the twelfth day, the preskeletal mesoderm of the footplate appears as a series of five cartilaginous columns converging into a compact mass of cells proximally, and at each distal end each of these radiating precartilages is still continuous with the thick layer of marginal mesoderm from which it originates. According to Milaire, an early segregation occurs in undifferentiated growing mesoderm between the presumptive material of the soft tissues and the presumptive skeletal material within the developing limb. That is, presumptive muscle condenses in a thick cellular layer underlying the dorsal and ventral ectoderm, while the cartilaginous elements condense in a thick cellular layer underlying the dorsal and ventral ectoderm, while the cartilaginous elements condense in the central part of the bud. Later, the myogenic mesoderm actively moves close to the precartilaginous masses and finally divides into several morphological units. Myogenic mesoderm seems to require a morphogenetic influence from the ectoderm apart from the apical area. He concludes that the early events of limb morphogenesis in mammals appear as an interrupted series of interactions between ectoderm and mesoderm.

This conclusion of Milaire's is reached, however, with qualifications: "The precise chronological correlations between the developmental features occurring respectively in the mesoderm and the ectoderm show that, as a morphological entity, the mammalian ectodermal ridge is only concerned with the formation of the distal limb segment. Consequently, the presumptive mesoderm of the more proximal limb

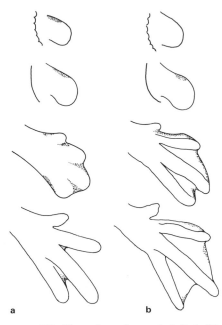

FIGURE 17.9 The pattern of necrosis (cell death) in leg primordia of (a) chick and (b) duck. [*After Saunders and Fallon, 1966.*]

territories must have originated under the influence of the apicoventral ectoderm before the latter has thickened into a ridgelike structure. This means that morphogenetic properties may be evinced by the ectoderm even if it does not show the morphological appearance of the ectodermal ridge. . . . Secondly, it is quite evident that, if we consider the morphogenetic role of the ectoderm as a nonspecific stimulation of mesodermal growth, the problem of the regional determination of the mesoderm still remains open."[1]

Cell Death as a Morphogenetic Agent

As limb development proceeds, whether in amphibian, bird, or mammal, a sculpturing process goes on, resulting in the particular form of the limb. This sculpturing is not entirely a matter of relative regional growth but is a truly erosive process involving cell death, i.e., necrosis. In the chick, an opaque patch of tissue appears in a 5-day limb bud at the site of the future knee joint. The patch consists of degenerating cells which will in some way be eliminated from the system.

Following this earlier discovery, Saunders has made an extensive study of cell death as a positive morphogenetic agent responsible for shaping a number of regions of the chick wing, such as the shoulder. Precartilage tissue, for example, may go on to produce cartilaginous matrix or may fail to do so and soon die, the two processes together helping to mold the form of the limb skeleton. Similarly, tissue regresses between the digits of the chick foot but remains as webbing in the duck foot. Yet tissue cells doomed to die if left in place may survive if transplanted to another site when moved before a critical stage, although they degenerate when moved a little later. Moreover, in the developing egg of a fly much the same process of necrotic phagocytosis sculpts out tissue between the foot structures. Cellular death as a regional phenomenon has been widely observed in embryonic development; it is clearly a useful process, at least inasmuch as it removes unwanted tissue whose substance usually can be reutilized elsewhere, as in the metamorphosis of many animals, though notably in amphibians and insects. Local removal of cellular material clearly facilitates many processes, such as separation of tissue layers, but how essential it may be to normal development is still somewhat in doubt. Death clocks and their genetic programming do seem to function in establishing shape in both insect and vertebrate morphogenesis.

Mutations and Morphogenesis

Limbs are particularly susceptible to developmental toxic agents, notoriously so in the case of thalidomide, which has caused so many mice and men to be born without limbs. This susceptibility appears to be due to the fact that a developing limb, from the moment of its inception as a limb disc and throughout the period of elongation

[1] *Ibid.,* p. 299.

of the limb bud, is a relatively rapidly growing mass of tissue. Anything that interferes, directly or indirectly, with the growth process will inhibit its development. Since growth of the limb bud occurs mainly at the distal end of the bud as long as the apical mesoderm remains undifferentiated, this part of the bud continues to be the most susceptible region. Analysis of such abnormal development, however, has been made mainly by studying the course of development of mutants where particular gene deficiencies produce the same abnormalities in abundant supply. There are abnormalities in the basic distal pattern, such as polydactyly, syndactyly, hypodactyly, etc., or no distal differentiation at all, as in brachypod limbs, and other anomalies where more or less of a limb is missing. Investigation of the development of these gene-dependent abnormalities, particularly experiments in which mesoderm or ectoderm from early limb buds of mutants is combined with normal ectoderm or mesoderm, throws some further light on the relative roles of ectoderm and mesoderm in normal limb development.

The extreme condition is seen in a "wingless" mutant in the chick. Experimental removal of the apical ectodermal ridge of a wing bud inhibits development. In the "wingless" mutant, the ridge is present early on the third day of incubation but then regresses, supporting the conclusion that limb development is truly dependent on the presence of the ectodermal ridge.

Polydactyl mutants have yielded more information. Limbs developing from buds consisting of polydactyl mesoderm and genetically normal ectoderm are polydactyl. In this case and also in control limb buds consisting of undisturbed ectoderm and mesoderm of the polydactyl mutant embryo, the first sign of departure from normal development is a relative enlargement of the bud apex. This excess outgrowth is associated with a more extensive apical ectodermal ridge. Marking experiments show that the apical region which gives rise to the excessive growth is the same in both normal and polydactyl forms until the third day. The enlargement of the apical ridge which then occurs is crucial: removal of one-third of the ridge from a 3-day limb bud results in a normal limb without polydactyly. Zwilling interprets these findings as indicating the following sequence of events: (1) factor in mutant mesoderm, (2) more extensive apical epidermal ridge, and (3) excessive distal outgrowth of limb.

In embryos with brachypod limb buds, a mutation is responsible for severe skeletal disturbances in the distal limb segment. All the primary morphogenetic events, however, are normal. The first genetic defect is seen as affecting late mesodermal activities leading to the formation of precartilage, i.e., to modification of the early processes of cellular organization as well as the ability of the chondroblasts to synthesize the mucoproteins of the basic cartilaginous substance. This has been interpreted as showing that genetic factors are still required to ensure the final organization of the mesoderm, a general conclusion that will be discussed later in a broader setting.

Another genetic approach has been to combine tissue from the limb buds of chick and duck legs to determine whether duck ectodermal ridge can induce outgrowth of duck mesoderm and whether the chick mesoderm can maintain such outgrowth.

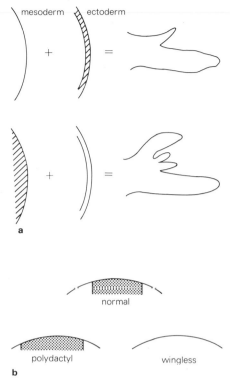

FIGURE 17.10 (a) Interchange of mesoderm and ectoderm between genetically normal and polydactyl (hatched areas) limb buds. The polydactyl condition developed from the combination of mutant mesoderm and normal ectoderm. (b) Probable distribution of apical ectoderm maintenance factor in normal, polydactyl, and wingless limb buds. [After Zwilling, 1956.]

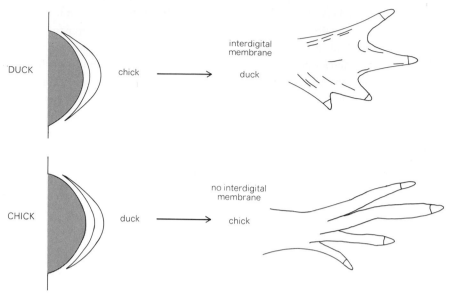

FIGURE 17.11 Specific induction of the mesodermal component, demonstrated by changing the apical caps between chick and duck embryos. The mesoderm of the duck induces an interdigital membrane, whereas none is induced by the mesoderm of the chick embryo. [*After Hampé*, J. Embryol. Exp. Morphol., **8**:248(1960).]

The answer is that ectoderm from either duck or chick can induce outgrowth of mesoderm in the other, and the mesoderm of either one can maintain the ectodermal ridge of the other, yielding perfect limbs in every case.

When the leg mesoderm of a duck is covered by chick ectoderm, a typical duck leg with webbed foot develops. In the converse experiment, when duck ectoderm covers chick mesoderm, webbing fails to develop. Both duck mesoderm and duck ectoderm normally seem to be involved in web formation.

To sum up, therefore, the morphogenetic factors so far demonstrated in the mesoderm and ectoderm of early limb buds contribute to our understanding of limb morphogenesis but remain insufficient to account for the complete, normal development of such a structure.

RECONSTITUTION

Virtually all the phenomena and problems associated with development generally, especially those associated with the vertebrate embryo, are inherent in the production of a limb. The development of limb buds, as we have seen, more or less parallels the early development of the embryo as a whole, beginning with an initial and relatively undetermined stage, i.e., the spherical egg or the circular limb disc, with little more than primary polarity, and developing mainly as the result of interactions between epithelial (epidermal) and mesenchymal (mesodermal) tissues, leading to tissue differentiation and pattern elaboration. A comparable event is seen in the reconstitution or regeneration of a part from the whole at later stages of growth and development,

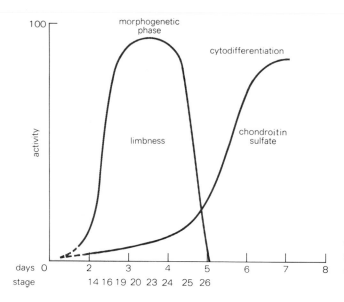

FIGURE 17.12 Relative activities of both the morphogenetic phase and the synthesis of chondroitin sulfate against a time axis. [*After Zwilling, 1968.*]

even in fully grown and mature organisms. The process of reconstitution of the vertebrate limb accordingly offers further opportunity to analyze the morphogenetic and histogenetic events responsible for the creation of such a structure. The situation differs from that of embryonic limb development inasmuch as the stump from which a new limb, following amputation, is produced already consists of differentiated cells and tissues. The following account, therefore, extends the analysis of limb development already given in this chapter and also serves as an introduction to the general phenomenon of regeneration.

Limb regeneration has been studied mostly in amphibians, particularly in salamanders of various ages. In these forms the limbs are readily regenerated throughout life, although more rapidly when the amphibian is young and small. In fact no other vertebrates exhibit such a wide capacity for regeneration of missing parts, including tail and snout and even the eye to some extent.

The Blastema

When a limb is amputated, a process of restoration begins immediately, in which three phases are recognized: a period of wound healing, a period of blastema formation, and a period of differentiation. Wound healing consists essentially of epidermal cells migrating from the basal layer of the adjacent epidermis toward the center of the wound. Active migration of this epithelial layer of cells continues until the wound is closed. Cells accumulate beneath this newly formed epithelial covering, and the combined cap of epithelial and subjacent cell mass is known as the *blastema*. As such,

FIGURE 17.13 Two concepts of regional organization in the limb regenerate. Four successive stages of regeneration are shown, from left to right. **(a)** The youngest blastema is depicted as being morphogenetically neutral. Limb pattern arises as a result of determinative events proceeding from stump distalwards. **(b)** The youngest blastema has distal differentiation tendencies. Proximal differentiation tendencies appear later than distal ones. [*After Faber, 1965.*]

"limb field"

apical field

basal field

a

b

stump ▦ proximal differentiation tendencies ▥ distal differentiation tendencies ∧∧

it has the general properties associated with a limb bud or an egg insofar as it has the potentiality of growing and differentiating into a highly organized unit structure.

The production of the blastema continues for some time before developmental events become discernible. Epidermal cells continue to accumulate at the apical region, possibly through further migration but certainly in part through mitotic division of cells initially forming the cap. Mesenchymatous cells progressively accumulate beneath the cap, so that an epithelium-mesenchyme reacting system is established, resembling that of a limb bud or a feather bud but on a much larger scale.

The general questions which arise are already familiar. What is the role of the newly formed ectoderm, and what is the role of the mesodermal elements in the subsequent processes of differentiation and organization? What is the developmental sequence? And so on. The new questions relate to the fact that the blastema cells derive from mature epidermis and from internal tissue that contained fully differentiated muscle, nerve, and cartilage, together with connective-tissue cells (fibroblasts or mesenchyme).

Dedifferentiation

The source of the mass of mesenchymatous cells comprising the bulk of the blastema has been a problem plaguing analysis of limb regeneration for many years. Following amputation, the residual differentiated tissue of the stump, mostly cartilage and muscle, undergoes extreme disintegration and apparent dedifferentiation. At the same time the mesenchyme cells beneath the epidermal cap increase in number enormously.

The problem has been to ascertain the connection between these two events. Opposing interpretations, not entirely exclusive of one another, have been variously adopted. The points at issue have been questions not of plausibility but of fact. Possible interpretations are:

1   Is the apparent dedifferentiation of mature muscle and cartilage cells a true return to an undifferentiated state? If so, they can obviously give rise to a mass of undifferentiated blastema cells. It has been extraordinarily difficult, however, to establish whether or not true dedifferentiation and subsequent redifferentiation may actually take place.

2   Are the disassociating and otherwise changing cartilage and muscle cells degenerating and in fact dying, allowing their products to stimulate and support rapid multiplication of any unspecialized mesenchyme cells present? A dependent population replacement of this sort would be effective and is known to occur in regenerative processes in some other forms (e.g., ascidians). In this event no transformation of cells is called for, only differential multiplication.

3   Do the cartilage and muscle cells merely appear to dedifferentiate, i.e., lose their distinctive histological character, but in actuality retain their fundamental character and as such contribute to the blastema? This phenomenon of pseudo-dedifferentiation is commonly seen in tissue culture and has been termed *modulation* by Weiss. If this is the case, however, in a blastema, then we are faced with another problem, because the implication is that modulated cartilage, muscle, and other cells, having been mixed together, subsequently sort out according to their kind, i.e., like to like, and also become reassembled in the new configurations of the developing limb structures.

The long-standing difficulty has been the failure in following the history of individual cells during the crucial period, that is, in tracing individual cartilage or muscle cells from the original stump tissues to whatever their destiny may be, or in determining the source of the mesenchyme cells with any precision. Electron-microscopic studies support the contention that the differentiated cells of the stump do become morphologically undifferentiated cells in the blastema. What is less clear is whether they redifferentiate into new cell types.

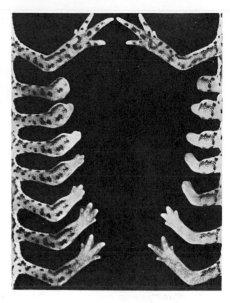

FIGURE 17.14   Regeneration of forelimb of salamander. (*Left*) Amputation below elbow. (*Right*) Amputation through upper arm. [*Courtesy of R. J. Goss.*]

Redifferentiation

When the stumps of amputated limbs are treated with tritiated thymidine and the tissues are fixed at later intervals and radioautographed, the pattern of incorporation of the thymidine indicates that DNA synthesis begins 4 to 5 days after amputation in all differentiating tissues within 1 mm of the wound. In animals injected with thymidine before amputation, only the epidermis of the limb shows incorporation of the tracer. When this is followed by amputation, the labeled epidermis migrates over the wound surface and forms a labeled apical cap, which remains labeled throughout blastema formation. None of the labeled epidermal cells contribute to the internal blastema.

Such evidence as there is of this kind shows that epidermal cells of the cap do not contribute to the mesenchymal blastema; that blood cells, for instance, that appear at an early stage in the limb do not derive from the blood-forming organs of the body; and that the dedifferentiating tissues of the stump, more or less by default, give rise to the unspecialized blastema cells. Evidence that cartilage cells can dedifferentiate and subsequently redifferentiate into muscle cells, or vice versa, is supplied by labeling, or marking, experiments of a different kind. The question, of course, is a general one of vital significance: can the daughter cells of a highly differentiated vertebrate cell, following mitosis, exhibit a variety of differentiations, or must they perpetuate the parental type? Can they truly dedifferentiate and redifferentiate, or only modulate?

The crucial experiment consisted of grafting triploid salamander cells of a known differentiated type into the limbs of diploid hosts, amputating the limb, allowing limb regeneration to occur, and then examining the regenerate to see whether the triploid cells had given rise to cell types other than that of the original graft cells. The grafts consisted of pure cartilage, pure muscle, cartilage plus perichondrium, and epidermis, i.e., the various types typically present in the stump of an amputated limb, but introduced separately into the hosts. Host limbs were previously exposed to x-irradiation to discourage host cells from participating in the regeneration process when the host limb was amputated. Triploid cells are recognizable as having three nucleoli in the interphase nucleus instead of two.

When the graft tissue consists of pure cartilage or of cartilage plus perichondrium, the marker appears subsequently in the regenerated limb in the following cell types: cartilage, perichondrium, the connective tissue of joints, and in fibroblasts, but not in muscle; when pure muscle is grafted, the marker appears in all the above cell types and in muscle cells as well; no regeneration occurs when pure epidermis is grafted. Similar experiments, made independently but employing both tritiated thymidine and triploidy (used either independently to mark graft and host cells differentially, or simultaneously to label the same cell) have provided cross-checks on any uncertainties in either method alone. Again, clean cartilage grafts give rise to morphologically dedifferentiated blastema cells, which redifferentiate almost exclusively into chondrocytes that retain their specific character through at least five divisions; and, again, muscle tissue contributes to both muscle and cartilage, *though it is highly significant* that muscle tissue consists of connective-tissue cells as well as contractile elements.

The question whether real dedifferentiation takes place, with subsequent redifferentiation along another line, still receives a somewhat dusty answer. Yet the mass of morphologically undifferentiated cells in a young blastema can be completely disrupted with a fine needle and still give rise to a normally regenerated limb. What is truly evident is that the cone-shaped mass of blastema cells develops into the amputated structure, whatever the removed structure may be, no more and no less; and that redifferentiation, according to position, and relocation, by means of selective reassortment, may both be at work.

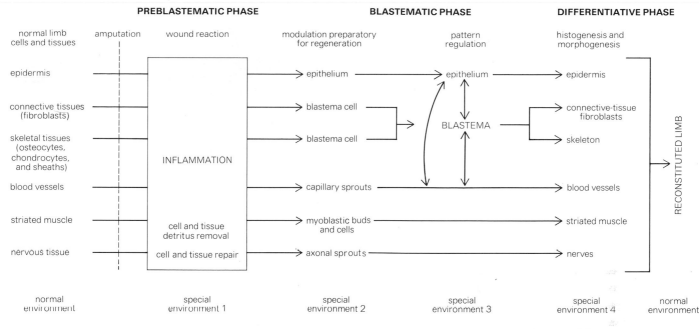

FIGURE 17.15   A concept of the regeneration process in amphibian amputated appendages. With the exception of the final reconstitution of the amputated appendage, this diagram could also serve as a summary of repair processes in all vertebrates. [*After Schmidt, 1968.*]

Morphogenesis

The limb blastema passes through several definable stages during the course of regeneration. Morphologically, in the adult newt at 20°C, 15 days after amputation, the blastema is filled with undifferentiated cells; by 20 days the blastema has become a *cone*; the *palette* stage (a flattened cone) is reached at 25 days; the *notch* stage, representing the first sign of digits, is reached at 30 to 35 days; and from then on digital pattern becomes progressively in evidence, the precartilaginous skeleton condenses, and a complete limb is present by 75 days.

The cone stage is of particular interest morphogenetically because not only do the various tissues and structures differentiate within the regeneration outgrowth, but they form in continuity with the stump tissues to produce an anatomically and functionally complete limb. Only those structures distal to the plane of amputation are formed, and a long-standing explanation has been that the differentiated tissues of the limb stump induce the undifferentiated blastema to redifferentiate the lost parts in conformity with their own organization. The failure of young blastemas to continue their differentiation when transplanted to foreign sites and the ability of older blastemas to do so have supported this concept.

The validity of this concept has been tested by removing cone-, palette-, and notch-stage blastemas and grafting them to the body (the dorsal fin) of the larval salamander, with or without a stump. Whether the stump is included or not, whole blastemas are able to self-organize into all the skeletal and muscular components

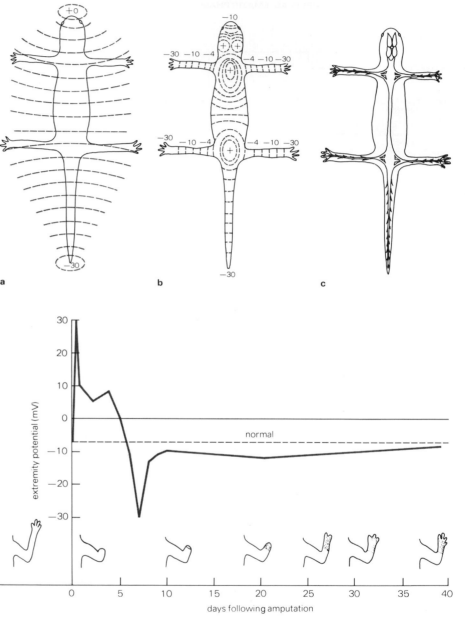

FIGURE 17.16 Bioelectric field of adult newt. (a) Total body potential, with tip-of-nose reference electrode and tip-of-tail recording electrode; animals are immersed in fluid. (b) Field plot found using two-electrode equipotential line-plotting technique on an animal kept moist but not immersed in fluid. (c) Gross arrangements of the central nervous system in the adult newt. [*After Becker, 1961.*]

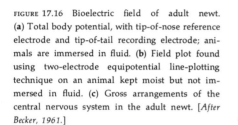

FIGURE 17.17 Voltage changes in adult-newt forelimb during regeneration. Zero point on horizontal scale immediately precedes amputation. Very high peak in positive voltage curve is concurrent with limb amputation. Secondary positive peak is observed on the fifth day. Negative maximum is reached on the eighth day. Negative bias is maintained until limb regeneration is complete. Normal limb voltage is about −10 mV (interrupted line) at point of amputation. Lower figures show schematically the stages of regeneration. [*After Becker, 1961.*]

of the lost limb distal to the level of amputation. Distal parts of blastemas grafted in the same way develop hand structure; proximal parts develop the more proximal parts. In other words, the limb blastema as early as the cone stage is a self-organizing system already imbued with a pattern representing the prospective regional structure.

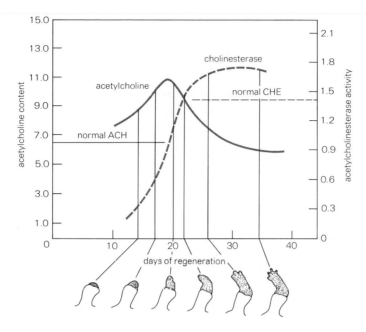

FIGURE 17.18   Comparison of acetylcholine content and acetylcholinesterase activity during the course of regeneration of forelimbs of adult newts. [*After Singer, David, and Arkowitz,* J. Embryol. Exp. Morphol., **8**:98 (1960).]

Neural Trophic Factor

Regeneration of the limb blastema of vertebrates, particularly amphibians, and of regeneration blastemas of various but by no means all invertebrates is normally dependent upon a critical supply of nerves at a very early stage. Many nerve endings in fact enter the epidermal cap that covers the wound, and make close, synapse-like contacts with epidermal cells. However, if a limb is first denervated and then amputated, or if nerves are by any means blocked from penetrating the epidermis, no regenerative outgrowth occurs and the stump tissues merely undergo degenerative changes. Yet once the process of dedifferentiation and blastema formation has taken place, interference with the nerve supply no longer has any effect. The influence of nerves is something other than nerve function as such and is not due to acetylcholine. The agent is known as the *neural trophic factor.* What it is and how it works are still unsettled, although the establishment of neural-epidermal junctions seems to be an essential condition in such nerve-dependent regenerations.

Regeneration nerve dependency, however, is not universal. Embryonic limb buds, for instance, comparable to though smaller than regeneration blastemas, develop without initial neural stimulation. Similarly, salamander limbs that have been denervated and maintained nerveless for a month or more by repeated denervation at 5-day intervals, acquire the capacity to regenerate amputated limbs. In other words the dependency is not absolute. Yet neural dependency is not limited to the establishment of limb blastemas of vertebrates.

CONCEPTS

Embryonic limb discs, like eggs, are at first equipotential systems and may be regarded as individuation fields.

Polarity of the system is derivative, the anteroposterior axis being derived from or imposed by the axial polarity of the embryo, the dorsoventral axis being acquired later in keeping with the later development of the embryonic dorsoventral polarity.

Ectoderm and mesoderm comprise an interacting system, the mesoderm determining regional character relative to the host embryo, i.e., limbness and also forelimbness as distinct from hind-limbness, the ectoderm exhibiting the equipotential property.

Limb development exhibits a proximodistal time sequence of differentiation.

Organized or controlled cell death serves as a morphogenetic tissue-sculpturing procedure.

Mutant forms show that the ectoderm is mainly inductive and that the mesoderm is a maintenance agent.

The regeneration blastema, like the limb disc, is initially an equipotential system.

During blastema-cell accumulation, mesodermal-cell dedifferentiation and cell modulation both occur and contribute, whereas epidermal cells remain true to type.

Specific stump character determines the particular nature of adjoining regenerating tissue, and only those structures distal to the plane of section regenerate.

Blastema development represents a self-organizing system imbued with a pattern representing specific regional structure.

Blastema development is normally but not absolutely nerve-dependent, involving a neural-growth factor rather than nervous stimulation. The distal epidermis of the blastema is the target tissue.

READINGS

AMPRINO, R., 1965. Aspects of Limb Morphogenesis in the Chicken, in R. L. DeHaan and H. Ursprung (eds.), "Organogenesis," Holt, Rinehart and Winston.

BALINSKY, B. I., 1956. A New Theory of Limb Induction, *Proc. Nat. Acad. Sci.*, **42**:781–785.

BECKER, B. O., 1961. The Bioelectric Factors in Amphibian Limb Regenerates, *J. Bone Joint Surg.*, **43-A**:643–656.

BUTLER, E. G., and H. F. BLUM, 1963. Supernumerary Limbs of Urodele Larva Resulting from Localized Ultraviolet Light, *Develop. Biol.*, **7**:218–233.

FABER, J., 1965. Autonomous Morphogenetic Activities of the Amphibian Regeneration Blastema, in V. Kiortis and H. A. L. Trampusch (eds.), "Regeneration in Animals and Related Problems," North-Holland Publishing Company, Amsterdam.

FLICKINGER, R. A., 1967. Biochemical Aspects of Regeneration, in R. Weber (ed.), "The Biochemistry of Animal Development," vol. II, Academic.

GLÜCKSMANN, A., 1951. Cell Deaths in Normal Vertebrate Ontogeny, *Biol. Rev.,* **25**:59–86.

GOETWICK, P. F., and V. K. ABBOT, 1964. Studies in Limb Morphogenesis, *J. Exp. Zool.,* **155**:161–170.

GOSS, R. J., 1968. "Principles of Regeneration," Academic.

——, 1965. Regeneration of Vertebrate Appendages, *Advance. Morphogenesis,* **1**:103–152.

——, 1965. Metabolic Patterns in Limb Development and Regeneration, in R. L. DeHaan and H. Ursprung (eds.), "Organogenesis," Holt, Rinehart and Winston.

MILAIRE, J., 1965. Aspects of Limb Morphogenesis in Mammals, in R. L. DeHaan and H. Ursprung (eds.), "Organogenesis," Holt, Rinehart and Winston.

NICHOLAS, J. S., 1955. Regeneration in Vertebrates, in B. Willier, P. Weiss, and V. Hamburger (eds.), "Analysis of Development," Saunders.

SAUNDERS, J. W., and J. F. FALLON, 1966. Cell Death in Morphogenesis, in M. Locke (ed.), "Major Problems in Developmental Biology," 25th Symposium, Society of Developmental Biology, Academic.

SCHMIDT, A. J., 1968. "Cellular Biology of Vertebrate Regeneration and Repair," The University of Chicago Press.

SINGER, J., 1965. A Theory of the Trophic Nervous Control of Amphibian Limb Regeneration, in V. Kiortis and H. A. L. Trampusch (eds.), "Regeneration in Animals and Related Problems," North-Holland Publishing Company, Amsterdam.

SMITH, S. D., 1970. Effects of Electrical Fields upon Regeneration in the Metazoa, *Amer. Zool.,* **10**:133–140.

STEEN, T. P., 1970. Origin and Differentiative Capacities of Cells in the Blastema of the Regenerating Salamander Limb, *Amer. Zool.,* **10**:119–132.

STOCUM, D. L., 1968. The Urodele Limb Regeneration Blastema: A Self-organizing System, *Develop. Biol.* **18**:441–456, 457–480.

SWETT, F. H., 1937. Determination of Limb Axes, *Quart. Rev. Biol.,* 12:322–339.

THORNTON, C. S., 1968. Amphibian Limb Regeneration, *Advance. Morphogenesis,* **7**:205–250.

TRAMPUSCH, H. A. L., and A. E. HARREBOMÉE, 1965. Dedifferentiation a Prerequisite of Regeneration in V. Kiortis and H. A. L. Trampusch (eds.), "Regeneration in Animals and Related Problems," North-Holland Publishing Company, Amsterdam.

WHITTEN, J., 1969. Cell Death during Early Morphogenesis: Parallels between Insect Limb and Vertebrate Limb Development, *Science,* **163**:1456–1457.

ZWILLING, E., 1968. Morphogenetic Phases in Development, in M. Locke, "The Emergence of Order in Biological Systems," 27th Symposium, Society of Developmental Biology, Academic.

——, 1961. Limb Morphogenesis, *Advance. Morphogenesis,* **1**:301–338.

—— and L. A. HANSBOROUGH, 1956. Interactions between Limb Bud Ectoderm and Mesoderm in the Chick Embryo, III: Experiments with Polydactylous Limbs, *J. Exp. Zool.,* **132**:219–339.

CHAPTER EIGHTEEN

MORPHOGENESIS OF THE VERTEBRATE EYE

Of all the organs of the vertebrate body, apart from the brain, the eye presents the greatest complexity, particularly in terms of cell diversity, tissues, and parts unified to form an optical instrument of amazing proficiency and efficiency. The course of development has been fairly well worked out, although it is replete with problems at every stage. The evolution of the vertebrate eye, however, and therefore the evolution of the development of the eye have as yet been hardly considered. The challenge is great.

The overall developmental event is well described by Coulombre as follows: "During the development of the vertebrate eye a large number of tissues assemble in such a manner that their size, shape, orientation, and relative positions meet the precise geometrical tolerances required by the optical function of this organ. The cells which make up these tissues are contributed by the ectoderm and mesoderm. They become highly specialized to fulfill the diverse functional requirements of the eye. For example, both the cellular and extracellular portions of the dioptric media (cornea, lens, vitreous body) become highly transparent. The intrinsic musculature of the eye (protractor lentis, retractor lentis or ciliary musculature) is appropriately oriented and attached to alter the position or shape of the lens and to focus the visual image in the plane of the retina. The outer tips of the visual cells are specialized to transform the light energy of these images into coded trains of nerve impulses. The neural retina, which is a peripherally developed portion of the central nervous system, develops a cytoarchitectural organization, enabling it to reorganize the output of the visual cells into a form suitable for transmission to the brain. It is important that the eye maintain constant shape during visual function. The scleral and corneal cells construct an outer eye wall which maintains its shape in the face of intraocular pressure and the pull of the extrinsic muscles."[1]

The morphogenesis of the whole can be subdivided into several phases: a first phase in which inductive effects separate the specific material of a particular rudiment, such as the retinal rudiment within the neural plate and the lens and cornea within the surface ectoderm; a second phase during which cell differentiation occurs within each rudiment, e.g., the differentiation of optic-vesicle cells into pigment and retinal epithelia, and the differentiation of the lens; and a third phase, of a supracellular character, concerned with the formation of the ciliary body and corneal curvature, the differentiation and growth of sclera and cornea, and the general growth of the eye and proportional growth of the retina and pigment epithelia.

CONTENTS

Determination of the eye rudiment

Lens induction

Lens regeneration

Whole-eye regeneration

Development of retinal architecture

Specificity of neuronal connections

Concepts

Readings

[1] A. J. Coulombre, The Eye, in R. L. DeHaan and H. Ursprung (eds.), "Organogenesis," p. 219, Holt, Rinehart and Winston, 1965.

FIGURE 18.1 Development of brain and sense organs of fish embryo.

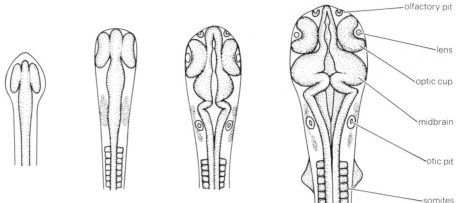

DETERMINATION OF THE EYE RUDIMENT

The presumptive retina begins to evaginate from the lateral walls of the forebrain to form the optic vesicles at about the time that the neural folds meet in the midline. The first question concerns the events that precede this evaginative process: What determines the establishment of the pair of prospective retinal outgrowths, and when does this occur? Vital-staining experiments have shown that the material destined to become the optic vesicles and their optic stalks and chiasma region lies well forward in the neural plate. The same procedure has also shown that the area from which the future lens develops lies outside the presumptive neural plate, lateral to and in front of the presumptive-eye rudiment. Primarily, the eye develops as the result of these two areas being brought together and interacting. Identification of these areas, however, is merely a mapping process which says nothing concerning the state of the tissues i.e., whether or not they are still indifferent, whether in some way they have yet to be determined.

One method that shows when retinal determination becomes irreversibly established is to constrict a developing egg in the sagittal plane—the plane of the embryonic axis. When this is done, an embryo develops as a double-headed monster, with two brains, two pairs of eyes, etc. This holds true in the newt up to the commencement of gastrulation, so that until this stage, determination of eye and lens is labile or absent, and the determinative process may accordingly be assigned to the gastrulation or early neurulation periods.

A number of experiments have been made in this connection, with regard to the eye rudiment proper. One is to explant the presumptive-eye rudiment of the open-neural-plate stage into the flank of another embryo to test for its capacity for self-differentiation. When this is done, an optic cup develops in the new position, although only if the explant consists of both neural-plate and archenteric-roof tissue.

Other experiments made with the early neurula stage consisted of excising a rectangular piece of neural plate and subjacent mesoderm and replacing it after rotating

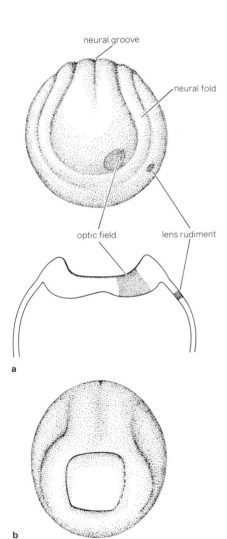

FIGURE 18.2 (a) The position of the presumptive optic cup and presumptive lens in the neurula stage. (b) Neurula with a rectangular piece of the neural plate excised and inverted. [*After Spemann, 1938.*]

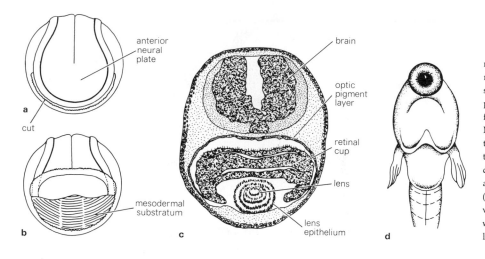

FIGURE 18.3 Cyclopia in the newt, produced by mechanical defect. Excision of the mesodermal substratum of the anterior part of the medullary plate in the early neurula stage has entailed median fusion of the eyes into a single ventral eye. (a) Neurula prepared for operation by cutting around the anterior border of the medullary plate. (b) Anterior part of medullary plate is lifted. The mesodermal substratum thus exposed is then excised, and the medullary flap is put back in position. (c) Cross section through the head, showing the ventral eye with a single median lens. (d) Fish larva with cyclopean eye induced by magnesium (or lithium) chloride treatment of early embryo.

through 180°. These classical experiments of Spemann involved cutting transversely through the prospective-eye rudiment, thereby leaving a portion in place and displacing the other portion to a more posterior location on the opposite side. Optic cups develop from each portion, the size depending on how the prospective area is divided—so that diagonal pairs resulting from a divided rudiment are complementary in size. Accordingly the optic-cup rudiment has already been determined in the early neurula, to the extent of its general destiny to form optic cup; yet it is still labile with regard to any regional determination within it, since any part when isolated attempts to form a whole. This labile state continues for some time.

In certain circumstances one median eye (cyclopia) forms, instead of a pair of laterally placed eyes. The presumptive-eye area, as already stated, is located in a median position in the extreme anterior part of the neural plate. In normal development the area divides into right and left components, the definitive optic-vesicle areas. Cyclopia appears, however, if developing eggs are chemically treated with lithium chloride, alcohol, chloretone, etc., or if the two lateral parts of the prospective retinal area fail to become separated by an intruding wedge of median neural plate and mesoderm. Thus excision of the mesodermal substratum of the anterior region of the neural plate in the early neurula stage allows the prospective-eye rudiments to remain as one, and a single large median eye develops.

Intimate association with mesenchyme is a general condition, however, for continuing development of the optic vesicle. When vesicles evaginating from the neural tube are isolated in vitro without access to the matrix of mesenchyme which normally surrounds them, they attain only a very rudimentary level of organization.

Under normal circumstances the optic vesicle continues to grow, and its distal region, after making contact with the lateral head epidermis, invaginates. By this means a single-layered vesicle transforms into the two-layered optic cup, the inner layer becoming the retina and the outer layer becoming the pigmented layer of the eye.

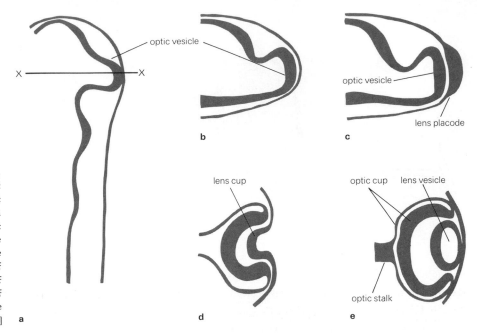

FIGURE 18.4 Development of the optic cup, lens, and cornea. (a) Dorsal aspect of the embryo at the stage at which the optic vesicle makes contact with the overlying ectoderm. (b) Transverse section through the level X–X at the same stage as that represented in (a). (c) Transverse section at the same level following induction of the lens placode by the tip of the optic vesicle. (d) Invagination of the optic cup and optic vesicle. (e) Separation of the lens vesicle from the surface, and reunion of the surface ectoderm to form the presumptive anterior corneal epithelium. [*After Coulombre, 1965.*]

For a time during early development the prospective retinal layer can form pigmented epithelium and prospective pigmented epithelium can form retina, as shown by reversal experiments and by isolation of the two components in vitro. The retinal, or sensory, layer soon loses its capacity to form pigmented epithelium, but the pigmented layer retains the capacity to produce retina almost indefinitely, at least in the case of amphibians.

Not only does the laterally expanding optic vesicle require adjacent mesenchyme for normal development, but the tip of the optic vesicle, i.e., the prospective sensory retina, must make contact with the surface ectoderm. Otherwise it fails to invaginate to form the optic cup and tends to form pigmented epithelium instead of neural retina. Only that part of the vesicle wall actually adjoining the ectoderm undergoes the invaginative process. Conversely only that part of the surface ectoderm in contact with the distal part of the optic vesicle invaginates to form the lens, and commonly, but not in all species, a lens fails to form if no approximation is made. An interacting system consisting of the neuroectoderm and the surface ectoderm evidently exists. Ectodermal cells overlying the tip of the vesicle elongate to form a thickened disc, the lens placode. Somewhat later the placode invaginates to form the lens cup, a process which is independent of the simultaneous invagination of the underlying optic vesicle. As the lens cup deepens, it progressively pinches off until it forms the lens vesicle. By then the marginal ectoderm has replaced the original area of the lens ectoderm and forms the corneal epithelium.

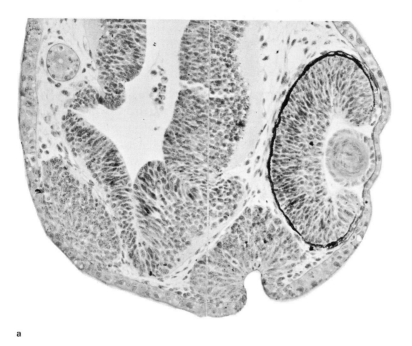

a

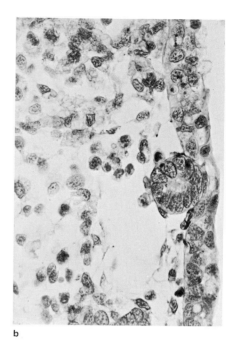

b

FIGURE 18.5 (a) Transverse section of a 12-mm salamander larva in which a lens has developed in the absence of the retina. The embryo was reared at 13°C until the early neurula stage, when the right retinal anlage was excised. (b) Section of an explant of presumptive-lens epidermis combined with the subjacent entodermal wall of the archenteron and the anterior portion of the lateral-plate mesoderm. A lens has been induced from the epidermis. [Courtesy of A. G. Jacobson]

LENS INDUCTION

The investigation of lens induction has had a confusing history. Spemann and later investigators reported different results from experiments depending on which species frog was used. In *Rana sylvaticus* and *Rana palustris*, transplanted optic vesicles induce a lens from the overlying ectoderm of any part of the body. In *Rana esculentes*, only the normal prospective-lens ectoderm responds to the presence of the optic vesicle, although the *R. esculentes* vesicle evokes lens development anywhere in *R. sylvaticus* and *R. palustris*, indicating that the vesicles are generally inductive but that ectodermal competence is variable. In *R. catesbiani* the lens is completely dependent even though the ectoderm is generally responsive, and in *R. esculentes* and *R. fusca* the lens develops in the normal position even in the absence of the optic vesicle. Results such as these show that:

1 All optic vesicles are lens-inductive, even to ectoderm of other species, genus, or order.

2 A lens can develop independently of induction by optic vesicle.

3 Induction may be necessary for lens development.

4 Ectodermal competence may be local or general.

To sum up, most regions of the early embryonic ectoderm are competent to form lens when properly stimulated, but this competence rapidly becomes restricted to that region overlying the optic vesicle or optic cup. The competent region is induced

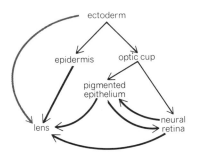

FIGURE 18.6 A schematic presentation of the "competencies" of optic tissues during early development.

to become lens by a succession of interactions, first with the underlying endoderm of the foregut, then with portions of the presumptive-heart mesoderm, and finally with the tip of the optic vesicle.

In some species the endodermal and mesodermal influences are alone sufficient in strength and duration to initiate ectodermal lens differentiation. In others they at least prepare the ectoderm by lowering the threshold to optic-vesicle induction. In all, the optic vesicle plays an important role, whether or not it is the principal inductor, in determining the final phase and in aligning the lens precisely with the rest of the eye. The nature of the inductive influences remains unknown; embryonic optic vesicle and guinea pig thymus, for instance, can also induce and sustain lens development in competent ectoderm. The terminal induction requires proximity between vesicle and ectoderm, although not full cell contact. In fact an acellular layer develops between the two epithelia which is several microns thick and results from fusion of the basement membranes of the two layers; nevertheless the intervening distance is very small, and the barrier has been shown experimentally to pose no difficulty to the diffusion of molecular substances.

Even when lens ectoderm has been committed to lens differentiation, the commitment is by no means an all-or-none affair, and removal of the lens at various stages of development into a neutral environment such as the body cavity shows that it only gradually becomes independent of the optic cup. This influence persists even in the adult in some salamanders. In any case, as the lens vesicle pinches off from the ectoderm, the basement membrane of the placode envelops the vesicle and gives rise to the lens capsule, while the cells constituting the back of the lens vesicle continue to elongate and, under the influence of the neural retina, produce the lens fibers. As the fibers grow, they obliterate the lens cavity, while cells of the front side of the vesicle form the lens epithelium.

The dependency and regulative capacity of the lens have been well described by Coulombre, who states that this continuing influence is shown "during normal development of the chick embryo by surgically reversing the lens at 5 days of incubation so that its epithelium faces the vitreous body. The epithelial cells which have been turned toward the retina immediately begin to elongate and form lens fibers. The original group of lens fibers, now situated beneath the cornea, stops growing. The polarity of the equator becomes reversed so that fibers are now added to the surface of the newly forming fiber mass on the vitreal side, while epithelial cells are added on the corneal side to reconstitute a new, appropriately positioned, but incomplete lens epithelium. . . . Thus the size and shape of the lens must ultimately be controlled by factors which control the number, size, and shape of the lens cells. These regulatory factors are of two types: retinal factors necessary not only for growth of the lens, but also for the orientation of lens fibers within the lens; and lens-inhibitory factors possibly arising from the lens."[1]

_____
[1] Ibid., pp. 230, 231.

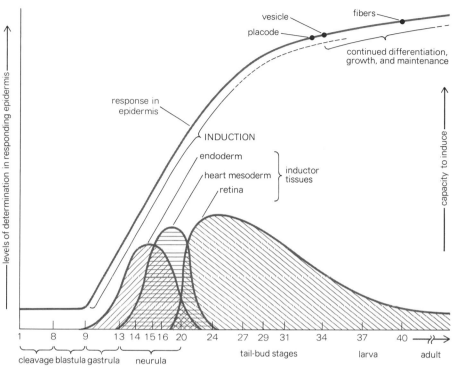

FIGURE 18.7 Lens induction in a salamander. The abscissa, representing time, is marked off in arbitrary stage numbers and names. The relative amount of time between stages is illustrated; actual time varies greatly with the temperature. At 17°C, the time between stages 9 and 40 is 3 weeks in the West Coast newt. The ordinate for the response curve is logarithmic, the level of response being a function of the sum of all past inductions. The ordinate representing the capacity of inductor tissue to induce is linear. [*After A. G. Jacobson, 1966.*]

The lens exerts its own influence on the development of the eye. It is principally responsible for the induction and maintenance of the cornea, although this capacity is gradually lost as development proceeds, while anterior corneal epithelium becomes progressively more independent of its inductor. The cornea normally arises only over the embryonic eye and is of a size and shape appropriate to the eye. It consists of three types of cells, together with nerve fibers and a very small number of extracellular components produced by fibroblasts, mainly collagen fibrils and mucopolysaccharides which become deposited in the interfibrillar space. These few components, again according to Coulombre, are "assembled in time and space in such a manner that the cornea develops a large number of strikingly different functional characteristics (avascularity, tensile strength, deturgescence, a tissue-specific population of ions, regenerative capacity, a characteristic interference pattern in polarized light, an appropriate refractive index, a precisely controlled curvature that contributes to its refractory power, and transparency). Simultaneous development of such diverse properties sets strict limits on the manner and sequence in which the corneal components can be compatibly assembled during development."[1]

The lens also controls the accumulation of the material forming the expanding vitreous body, and this in turn generates mechanical forces important in the

[1] *Ibid.*, p. 231.

construction of the skeletal wall of the eye, the sclera. Scleral cartilage develops in the first place from mesenchyme that condenses around the expanding optic cup. Although the inductor of scleral cartilage has not been identified, circumstantial evidence suggests that the neural retina or the pigmented epithelium, or both, may be the inductors; for the neural tube is known to induce vertebral cartilage, while scleral cartilage corresponds precisely in area with the underlying pigmented epithelium of the retina. It is of general interest that the cells of precartilaginous scleral mesenchyme, once their fate has been determined, form flat plates of cartilage in tissue culture, after being disaggregated and reaggregated, whereas similarly treated precartilaginous mesenchyme from chick embryo limbs form rods of cartilage. Such properties assume great significance in connection with processes of regeneration.

LENS REGENERATION

A lens is readily regenerated in some vertebrates, notably in salamanders, though rarely in frogs and toads. This is in keeping with the exceptional capacity of urodele amphibians, especially species of the newt, *Trituris*, to regenerate whatever part may have been lost, even the front part of the head, including the jaws. The great interest given to the regeneration of the lens comes from the discovery made by Colucci and Wolff in the late nineteenth century that it regenerates from the dorsal rim of the iris, a source remarkably different from its epidermal origin in the embryo.

The capacity to regenerate a lens in this way is generally absent during embryonic development and first appears in the young larval stages when tissues have already attained a high degree of functional differentiation. In the newt, removal of the lens from the larval or adult stages is followed by regeneration of a lens from the iris epithelium. The principal histological events following such removal are:

1 Increase in height of the cells of the inner layer of the dorsal iris
2 Separation of the inner and outer layers
3 Depigmentation of the inner layer and of the pupillary margin of the iris
4 Cell proliferation at the margin and a consequent downgrowth of an unpigmented epithelial vesicle
5 Formation of a lens nucleus through elongation of cells of the back wall of the vesicle and associated lens-fiber formation, together with secondary fiber formation from the equatorial zone
6 Detachment of the new lens from the dorsal edge of the iris

This is admittedly a special case of regeneration, but it calls attention in a dramatic way to the fact that the same structure may be produced in different ways by embryonic and differentiated tissues, even though, once initiated, the rudiment may follow much the same developmental course. Also the question arises concerning what normally holds the regenerative impulse in check, in this case while the original lens is still in place.

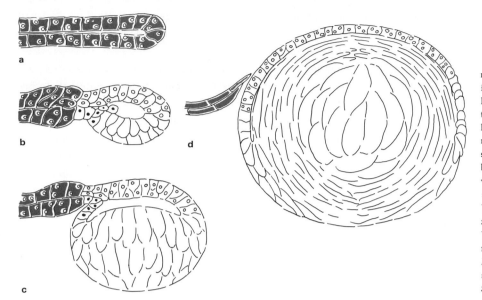

FIGURE 18.8 Histological location of the proliferating zone at the representative stages of Wolffian lens-regeneration. Each figure shows a section through the mid-dorsal pupillary margin of the lens regenerate, oriented perpendicular to the main body axis. The cornea (external "anterior") side is above, the retina (internal "posterior") side below, the dorsal side toward the left, and the ventral side toward the right. (a) Regeneration stage II; (b) stage V; (c) stage VIII; (d) stage XI. White circles denote cells in the proliferating zone; black circles, cells which are in proliferation. White cells are depigmented; black cells are pigmented. Lines indicate cell boundaries. [*After Papaconstantinou*, Science, **156**:338 (1967). *Copyright American Association for the Advancement of Science.*]

As a rule a regenerative process follows injury to the tissue. Here, lens regeneration occurs without injury to the iris, since it still occurs if the lens is removed through the roof of the mouth instead of through the cornea. Moreover, if a lens is removed and replaced, no regeneration occurs. Neither does it occur if the lens of an adult is replaced by a much smaller lens from a younger individual. Instead, the small lens grows rapidly to the size appropriate to the host. Evidently a lens has an influence, presumably chemical, which inhibits adjacent tissue from regenerating a lens as long as it is present. However, more than the absence of a lens inhibitory factor is required. Iris tissue transplanted to head connective tissue or any other region regenerates a lens only when retina is also present. When iris and retina are transplanted together and they round up to form a vesicle, a lens then develops. Lens, retina, iris, and vitreous mass thus appear to maintain an active mutual influence, the harmony of which is expressed in the normal structure of the eye.

WHOLE-EYE REGENERATION

The capacity of the urodele eye to regenerate completely following removal of all except a very small remnant has long been known. As a rule, cornea, sclera, and retina are renewed from corresponding components of the eye fragment as a process of coordinated self-assembly. The whole event is a spectacular performance, but the particular interest is in the regeneration of the sensory retina following its surgical removal. This layer readily regenerates from the adjoining pigmented epithelium and can be forced or flooded out of the vitreous chamber through a slit in the eyeball. A new retinal layer is then replaced by the pigmented epithelium, a phenomenon thoroughly investigated by Stone and his students.

Cells of the pigment layer round up and enlarge, undergo depigmentation, and multiply through mitosis. Subsequently some of the daughter cells differentiate into a multilayered sensory retina which establishes functional connections with the brain, while other cells redifferentiate and re-form a typical pigment epithelium. The same procedure is seen when the sensory retina degenerates following transplanting or replanting of the eyeball. In fact, any separation of the pigment layer from the sensory layer stimulates the pigment layer to form secondary retinal tissue. All that seems to be necessary is removal of inhibitory influence, for Stone found that pieces of pigment epithelium always give rise to sensory retina when grafted into the pupillary space following removal of the lens. Here is an unequivocal case of a highly differentiated cell redifferentiating in another way.

DEVELOPMENT OF RETINAL ARCHITECTURE

The differentiation and establishment of the impressive cytoarchitecture of the retina have been studied more in the development of the embryonic eye, where exactly known temporal stages are readily available, than in reconstitution from the functional pigmented layer. During early phases of development, cell division is confined mainly to the margin of the optic cup. In embryos, radioautographic analysis of retinas labeled with tritiated thymidine shows that immature retinal cells migrate to the outer surface of the layer as they divide, and the daughter cells then migrate back to their appropriate locations within the differentiating retinal layers. Increase in thickness and complexity is accompanied by waves of cell death, which are probably related in some way to the orderly establishment of the precise neurosensory network. The innermost layers of the retina (i.e., those nearest the center of the eye) differentiate first and the sensory layer of rods and cones last. Once differentiation has begun, the intricate and interrelated processes leading to the final cytoarchitecture apparently comprise a self-directing and self-sustaining complex.

The sequence of retinal-cell differentiations is as follows:

1 Glial cells, which become Müller's fibers and will envelop the retinal neurons, develop cytoplasmic processes connecting the inner and outer surfaces of the sensory retina; the terminals of these processes combine to form the internal and external so-called limiting membranes.

2 Ganglion cells, which form the innermost layer of neurons, send out axons, which together grow to form the optic nerve by way of the optic stalk. The dendritic processes of the ganglion cells form synapses with axons from the next layer to differentiate, i.e., the inner nuclear layer, which contains at least three kinds of neurons.

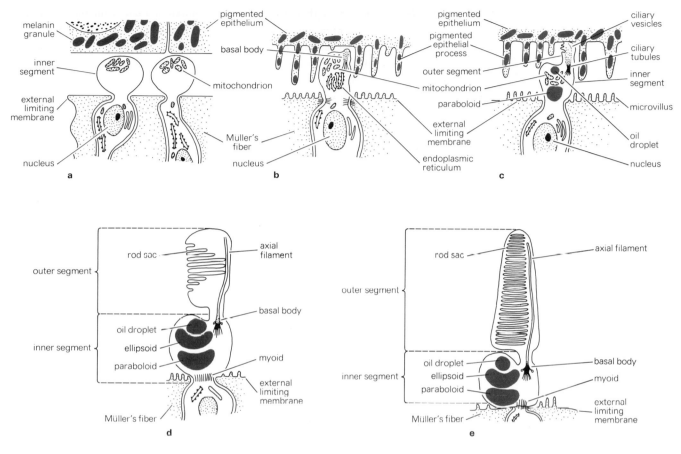

FIGURE 18.9 Rod and cone cytogenesis, illustrating the successive stages in the maturation of the cone. [*After Coulombre, 1961.*]

3   The outer nuclear layer, which is the outermost layer of the neural retina, is the last to differentiate, after local cell division ceases. Their axons synapse with the dendrites of the bipolar cells of the inner nuclear layer.

4   The light-sensitive rods and cones develop from the outer ends of the cells of the outer nuclear layer, each cell producing a cytoplasmic bud at its outer end, which protrudes through a pore in the external limiting membrane. Here again, further differentiation is sequential and centrifugal. The inner segments of the rods and cones are first formed, and from them the outer segments later arise. Various organelles are formed. One, a basal body or centriole, elaborates a noncontractile cilium which has nine peripheral filaments but no central filaments, and is otherwise typical. It takes part in the development of the outer segments of the rods and cones. Rod or cone sacs fold in from the plasma membrane close to the tip of the cilium, which may induce them, and become arranged like a stack of coins.

5 Finally, pigmented epithelial cells, whether differentiating from the outer layer of the embryonic optic cup or from the reconstituted layer during retinal regeneration, extend fine cytoplasmic processes which interdigitate with the emerging outer segments of the rods and cones, thus locking the two layers together.

SPECIFICITY OF NEURONAL CONNECTIONS

Some years ago Sperry (1944) showed that following section of the optic nerve, the axons from the retinal ganglion cells grow back through the sheath of the degenerated optic nerve and reestablish neuronal connections with the optic tectum of the midbrain. There is not only a reestablishment of the original point-to-point connections between the retinal map and the tectal map, but also a reestablishment in depth with the

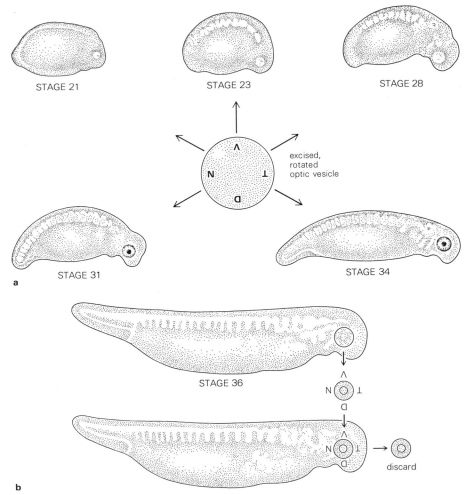

FIGURE 18.10 (a) Early optic vesicle (central figure) of Harrison's stages 21, 23, 28, 31, and 34 of *Ambystoma punctatum* embryos (indicated by arrows); in an embryo of each stage the vesicle is excised, rotated 180°, and reimplanted. D (dorsal), V (ventral), N (nasal), and T (temporal) poles are reversed. All hosts later showed normal vision through the grafted eye regardless of orientation. (b) Harrison stage 36 of *A. punctatum* in which the early optic cup is excised, rotated 180°, and transplanted to a new host. Later, the visuomotor reactions of the host are reversed. [*After Stone, 1960.*]

several layers of the exquisitely organized tectal neuronal structure, both in frogs and fish. Sperry also found that when the eye was reversed, with reversal of its dorsoventral and anteroposterior axes, the visual fields were also reversed, following optic-nerve regeneration. In other words, regenerating nerve fibers found their way back to the same specific points in the optic tectum, even though they consequently recorded inverted vision. Once this phenomenon was confirmed, it became logically necessary, it seems, to generalize the concept that certain specific local properties of retinal neurons establish functionally effective connections only with tectal neurons of corresponding local specificities.

In a recent summary, M. Jacobson states: "At an early stage of their development the ganglion cells of the retina are pluripotent with respect to the connections that they can form in the brain, but at a later stage the central connections of each ganglion cell become uniquely specified. In order to discover the time at which this change occurs, I inverted the eye of the toad *Xenopus* at various stages of development before connections had formed between the retina and midbrain tectum, and then mapped the connections after they were formed. These experiments show that, if the left eye is inverted before the early tailbud larval stage 29, its connections with the tectum are the same as the point to point connections of the normal right eye. Before larval stage 29 the retinal ganglion cells are unspecified and will form connections in the tectum which are appropriate to the new position of the eye."[1] Specification of retinal ganglion cells occurs first in the anteroposterior axis of the retina of stage 30. After this the ganglion cells have the information to form their connections in the correct order in the axis of the tectum. During the next 5 to 10 hours of development the ganglion cells become specified in the dorsoventral axis of the retina and are able to form correct connections in the mediolateral axis of the tectum. Other experiments of Jacobson's have shown that the retinotectal connections of *Xenopus* are completely determined after the cessation of DNA synthesis and mitosis in the ganglion cells, and about a day before their axons start growing from the eye to the tectum.

Homologous maps of neuronal specificities in the retina and tectum appear to be laid down independently, in parallel, so that connections form as the result of specific affinities between neurons located at homologous positions in the retinal and tectal maps. When the optic nerve of the goldfish is cut, for example, the optic fibers grow back to their correct places in the tectum; if regeneration of fibers from a small region of the retina is prevented, the corresponding region of the tectum remains unconnected. Further, in the adult frog or fish, Sperry has shown that the regeneration of optic-nerve fibers to the correct places in the tectum is not affected by either function or experience; while if one bundle of the optic tract is displaced from its normal pathway during regeneration, the fibers, as they continue to grow, regain their proper channel and continue to their correct terminations. He suggests that the

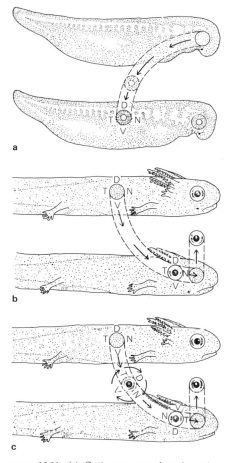

FIGURE 18.11 (a) Optic cup transplanted to the side of body where it will develop and differentiate independently of any connections. (b) The eye, developed on the larval body wall, excised and transplanted to denuded orbit of another host of same age. Eye is normally oriented. When the regenerated optic nerve connects with the brain for first time, normal visuomotor reactions are established through the graft. (c) Eye, developed on larval body wall, excised, rotated 180°, and transplanted to denuded orbit of another host of the same age. When regenerated optic nerve connects with the brain for the first time, reversed visuomotor reactions are established through the graft. [*After Stone, 1960.*]

[1] M. Jacobson, Development of Specific Neuronal Connections, *Science,* **163:**545–546 (1969). Copyright 1969 by the American Association for the Advancement of Science.

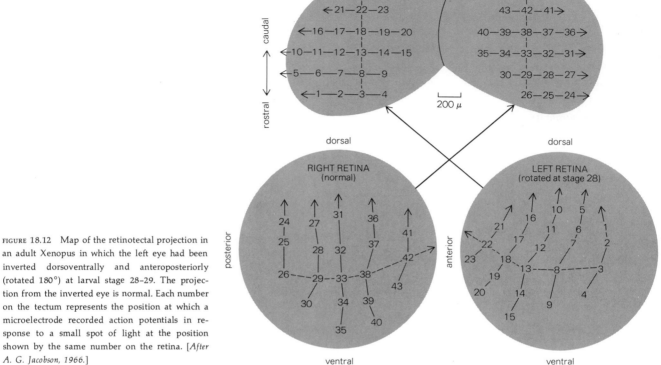

FIGURE 18.12   Map of the retinotectal projection in an adult Xenopus in which the left eye had been inverted dorsoventrally and anteroposteriorly (rotated 180°) at larval stage 28–29. The projection from the inverted eye is normal. Each number on the tectum represents the position at which a microelectrode recorded action potentials in response to a small spot of light at the position shown by the same number on the retina. [*After A. G. Jacobson, 1966.*]

proper matching of optic fibers with their respective tectal terminations depends on a system of local specification of retinal and tectal areas by cell-bound markers, determined by cytochemical gradients operating over the extent of the retina and tectum. Complementary or matching markers, arising independently in the retina and tectum, then provide a mechanism for linking specific optic fibers with the corresponding tectal cells.

Comparable experiments with chick embryos, by DeLong and Coulombre, support such an interpretation, although the embryologic and cellular mechanisms underlying specific fiber orientations in the various experiments remain entirely hypothetical. "The retinal and tectal markers must have some means of mutual recognition. The nature of the postulated markers is not known. . . . One possible mechanism is that pioneering fibers, searching randomly, may find matching fields by trial-and-error, with secondary enlargement of the successful tract by 'selective fasciculation' (Weiss, 1950) and atrophy of the unsuccessful pioneering fibers. The available data give no evidence of such a process. . . . Further, the fascicles clearly do not follow any random or meandering path; they course in a straight and direct

line. . . . This strongly suggests that the fibers are accurately guided throughout their course."[1]

However this may be, there is clearly a specific active matching of retinal fibers and tectal areas in the brain, which may indicate a chemical individuality associated with each small area, possibly with each cell, in both retina and tectum. This conclusion has been questioned on the ground that the differences in the complexity of brains of various orders of mammals are not reflected in corresponding differences in the DNA content of the cell nuclei. The diploid cell nucleus of both man and rat contains close to 10^{-12} g of DNA in the form of a double strand containing about 10^{10} nucleotides per strand. This is capable of coding for a few million different kinds of proteins, a large number but insufficient for labeling every neuron with a different protein.

Whatever the causal relationship with the genome may be, it is unlikely to be direct, and the retinotectal specificity question becomes part of the general problem of neuronal connectivity. This is an immense area of inquiry, seemingly incomprehensible in its complexity and requiring monumental accumulation of data; it is beyond the range of the present discussion. Nevertheless, it is significant that in normal eye development (of *Xenopus*), replication of DNA ceases just before the ganglion cells are specified in the anteroposterior axis of the retina. M. Jacobson suggests, since as a general rule in differentiating cells DNA replication stops before specific macromolecular synthesis appears to start, that neuronal specification involves synthesis of specific macromolecules, and that there is a stepwise specification of progressively finer details of the pattern of neuronal connections, beginning before the outgrowth of neuronal processes.

[1] G. R. DeLong and J. Coulombre, The Specificity of Retino-tectal Connections Studied by Retinal Grafts onto the Optic Tectum in Chick Embryos, *Develop. Biol.*, **16**:529 (1967).

CONCEPTS

All the phases and problems associated with the development of the whole organism are also characteristic of the development of an organ.

The principal stages are a phase of inductive separation of rudiment material, a labile phase, a still-labile phase of early determination, and a phase of supracellular control.

Morphogenesis precedes histogenesis and diversified cytodifferentiation.

Inductor tissues act and react upon one another, as in the optic-vesicle–lens system.

Inductors are generally not species-specific and not even class-specific, whereas competence is typically highly specific.

Reconstitution of an organ, in whole or in part, takes different courses according to whether the initiating system is embryonic or is already differentiated.

Point-to-point cell specificity is exhibited in reconstitution of neuronal and possibly other types of patterned cell associations.

READINGS

ADELMANN, H. B., 1936. Problems of Cyclopia, *Quart. Rev. Biol.*, **11**:161–182; 284–304.

COULOMBRE, A. J., 1965. The Eye, in R. L. DeHaan and H. Ursprung (eds.), "Organogenesis," Holt, Rinehart and Winston.

———, 1965. Problems in Corneal Morphogenesis, *Advance. Morphogenesis*, **4**:81–109.

———, 1961. Cytology of the Developing Eye, *Int. Rev. Cytol.*, **11**:161–194.

COWAN, W. M., A. H. MARTIN, and E. WENGER, 1968. Mitotic Patterns in the Optic Tectum of the Chick during Normal Development and after Early Removal of the Optic Vesicle, *J. Exp. Zool.*, **169**:71–92.

DELONG, G. R., and A. J. COULOMBRE, 1967. The Specificity of Retino-tectal Connections Studied by Retinal Grafts onto the Optic Tectum in Chick Embryos, *Develop. Biol.*, **16**:513–531.

HASEGAWA, M., 1968. Restitution of the Eye after Removal of the Retina and the Lens in the Newt, *Triturus pyrrhogaster, Embryologia*, **4**:1–32.

JACOBSON, A. G., 1966. Inductive Processes in Embryonic Development, *Science*, **152**:25–34.

JACOBSON, M., 1968. Cessation of DNA Synthesis in Retinal Ganglion Cells Correlated with the Time of Specificity of Their Central Connections, *Develop. Biol.*, **17**:219–232.

REYER, R. W., 1962. Regeneration in the Amphibian Eye, in D. Rudnick (ed.), "Regeneration," 21st Growth Symposium, Ronald.

SPEMANN, H., 1938. "Embryonic Development and Induction," Yale University Press.

SPERRY, R. W., 1963. Chemoaffinity in the Orderly Growth of Nerve Fiber Patterns and Connections, *Proc. Nat. Acad. Sci.*, **50**:703–712.

STONE, L. S., 1960. Polarization of the Retina and Development of Vision, *J. Exp. Zool.*, **145**:85–93.

———, 1959. Regeneration of the Retina, Iris and Lens, in C. S. Thornton (ed.), "Regeneration in Vertebrates," The University of Chicago Press.

STROEVA, O. G., 1960. Experimental Analysis of Eye Morphogenesis in Mammals, *J. Embryol. Exp. Morphol.*, **8**:349–368.

SZÉKELY, G., 1966. Embryonic Determinations of Neural Connections, *Advance. Morphogenesis*, **5**:181–221.

TWITTY, V., 1955. Eye, in B. Willier, P. Weiss, and V. Hamburger (eds.), "Analysis of Development," Saunders.

CHAPTER NINETEEN
METAMORPHOSIS, HORMONES, AND GENES

The developmental system, seen as a continuously maintained although progressively elaborating and changing organization, the very molecules of which are in perpetual flux as long as life persists, is a reality whose essence is elusive. Yet any change in the genetic information appears as some sort of change in the developmental outcome, and any substance to which cells and tissues are sensitive can alter the course in some degree. Hormones, which are themselves the products of development, have such a role and act on cells and tissues in various ways to modify or modulate the timing and direction of events. Yet no more than the so-called inductors, organizers, and organ-forming substances of the embryo can they be considered to be truly developmentally instructive.

CONTENTS
Metamorphic transformation
  Metamorphosis and evolution
  Perspective on metamorphosis
Amphibian metamorphosis
  Hormonal control
  Threshold and rate of response
Insect metamorphosis
  Molting and the molting hormone
  Juvenile hormone and transformation
  Extent and timing of metamorphosis
  Pupal reorganization
Hormones and gene activation
  Puffing and genic information
  Hormonal control of puffing
  Correlation within the genome
  Integration within cells and tissues
Concepts
Readings

METAMORPHIC TRANSFORMATION

The most spectacular shifts in the precise control of growth and form in a rapidly changing developmental system, in both animals and plants, are the metamorphoses seen in the developmental cycle of many marine invertebrates, in the holometabolous insects and anuran amphibians, and in the initiation of flowering in the higher plants. These events are both challenging and potentially enlightening. Attention focuses mainly on the nature of the target tissue on the one hand and on the nature of the triggering, usually hormonal, agents on the other. It should be kept in mind, however, that hormonal activity in a developing system is possible only in organisms already sufficiently developed to produce hormones. Hormones are unlikely to be present in embryonic stages, nor are they likely to be operative at the time of metamorphosis of the small larval organisms of marine invertebrates such as sea urchins and ascidians.

Metamorphosis is a widespread developmental phenomenon which is usually associated with a dramatic change in habitat and consequent way of life, such as the change from a planktonic to a benthic existence in the sea urchin, from an aquatic to a terrestrial existence in frogs and toads, and from nonflying to a flying existence in insects. Such changes in environment and activities demand equally rapid transformations of the structure and function of the living machinery.

Transformation during development is typical of most animals. In many, where no radical change in life style occurs, the course of development may be a gradual, progressive change from one condition to another. Or the change may be sudden but not so superficially remarkable that it goes by the name of metamorphosis. Such is the shift from fetal structures and adaptations to postnatal life in human beings and other mammals. Metamorphic change during the developmental cycle is an acceleration or condensation of essentially the same basic processes characteristic of most forms of development. Primarily it consists of the differential destruction of certain tissues, accompanied by an increase in growth and differentiation of other tissues. The phenomenon of regional growth and differentiation associated with local cell death in developing limbs comes into this category.

As a rule, metamorphic change is associated with and highlights the phenomenon of duality in the development of most kinds of eggs.

Metamorphosis and Evolution

The general significance of the more specialized types of marine larvae has long been debated. An early, generally discarded interpretation is that they represent ontogenetic relics of ancient ancestral types. Eggs, being conservative, do retain old ways of doing things, but the generally accepted current belief is that echinoderm larvae, ascidian tadpoles, and nemertean larvae are evolutionary inventions interpolated into an originally more direct developmental cycle. The various special types of insect larvae are similarly regarded. On the other hand amphibian larvae, both anuran and urodele, are almost certainly retentions of an ancestral type of larva, with some modern improvements, retained because of persisting advantages, for amphibians, of continuing to develop and grow in their primitive freshwater environment.

Each group presents a developmental-evolutionary problem. In the first, which includes the marine invertebrates and the insects, the nature of the egg and the genome have been changed from that associated with a relatively more direct development to a more complicated course. The development of the egg is forced into a new path to yield a viable organism strikingly different from the adult; this is accomplished without harming the capacity to develop the adult organization. In the second group, the amphibians, the genome (and the egg as a whole) has evolved so as to produce a terrestrial tetrapod in place of an ancestral lobe-finned fish, but has done so without changing the particular character of the early freshwater vertebrate type of egg, which follows the old developmental path for a considerable developmental distance.

In both groups a dual system of development is very evident. In both groups the egg exhibits a specific early organization that relates to the development of the first of the two organismal stages, but in one group this special organization has been added to the primary developmental process leading to the mature adult, and in the other group a comparable organization has been retained as an inheritance of an ancient type. In terms of development, the two situations have much in common, although they are different enough to call for separate discussion.

Several fundamental developmental, and evolutionary, problems accordingly confront us. The development of an organism adapted to two very different environments at different times during its life span implies the operation of two sets of environmental selection pressures. Whatever such selection for functional phenotypes may have been, an internal developmental selection of genotypes may well have been the primary factor in determining evolutionary direction. In the examples of metamorphic life cycles just mentioned, at least two effective genotypes, i.e., two virtually independent sets, exist together in the same genome.

As Curt Stern concludes, in his essay already alluded to, "The versatility of any given genic constitution is amazing. The genotype can give rise to all the types of cells which make up the differentiated organism, and can arrange and synthesize the cells into many diverse organs and organ systems. The same genetic constitution can produce first a caterpillar and then a butterfly, first a tadpole and then a frog, first a polyp and then a jellyfish. There is no difference in the genes of an egg which develops into the large female of the marine worm *Bonellia* or into the microscopic, semiparasitic male which lives inside the body of the female. The genes in the egg of a termite may lead to the appearance of queens, or soldiers, or workers. Within one individual the action of the genes may result in structural adaptations which change weak muscles into strong, cause one kidney to do the work of two, or rebuild in adjusted fashion the structures of broken bones."[1]

Perspective on Metamorphosis

The phenomenon of metamorphosis presents the most basic of all questions, the genetic control of development. It also presents other questions that are more readily approached. Metamorphosis is essentially the elimination of one phenotype and its replacement with another, all as part of the same continuum that we recognize as the course of development of an egg. Since the metamorphic event must necessarily occur at an appropriate time in the cycle of an organism that has to cope with its own growth and with environmental change, timing is crucial. Therefore controllable triggering agents have evolved which have a differential effect on the target tissues. The questions concerning the identity of these agents have been the most readily investigated. Those concerning their means of operation have been less satisfactorily answered.

The details of the transformative event itself have in most cases been reported fairly well, although in no case well enough to be fully understandable. Moreover, the *means* of transformation should not be confused with the *course* of transformation; each may vary according to circumstances and still yield essentially the same result. Yet the real basis for the differential response to a triggering agent, although seemingly clear in a few cases, remains elusive. Most of the attention concerning the nature of metamorphosis and its control has been give to the metamorphosis of amphibians and insects. Nevertheless, a brief account of metamorphic development as seen in certain marine invertebrates may serve to give some perspective:

1 In the life cycle of the jellyfish *Aurelia,* an example already described in Chapter 7, the polyp stage undergoes a succession of transverse epidermal constrictions

[1] C. Stern, Two or Three Bristles, *Amer. Sci.,* **42**:246 (1954).

under the triggering influence of environmental iodine. Each such segment constricts off and simultaneously develops the primary organization of the medusa. A profound change is induced; a series of parts gives rise to a series of new wholes of a very different character, but there is no significant destruction of old tissue. Yet this is metamorphosis. What property of the polyp is affected by the agent, and what actually is taking place? We have no answers.

2   In hydroids such as *Tubularia* and in ascidians such as *Clavelina,* exposure to high temperatures or other adverse conditions results in dissolution or resorption of the comparatively highly differentiated tissues comprising the hydranth and thorax, respectively, but leaves the relatively unspecialized tissues of the stalk unaffected. Subsequent redevelopment or reconstitution of the functional organism usually occurs. This is replacement rather than metamorphosis, but we see differential susceptibility to the tissue environment: in both cases the more differentiated and specialized cells are destroyed and the least specialized cells survive.

3   Metamorphosis is usually profound, or cataclysmic, in the more highly specialized marine larvae, involving extensive destruction of the functional and precociously differentiated larval tissues, and the survival and continued development of the less differentiated tissues destined to give rise to the adult-type organism. The metamorphic event in the sea urchin and ascidian larvae, for example, seems to be triggered by internal factors, although at a stage too early for hormonal intervention. The onset of metamorphosis, however, can be precociously induced by exposure to potentially toxic agents, particularly to traces of metals. Very dilute solutions of mercuric chloride in seawater bring about precocious metamorphosis of advanced pluteus larvae of sea urchins, the larval tissues being more affected by the poison than is the urchin rudiment developing within them. In various ascidian tadpole larvae, precocious metamorphosis is induced by similar exposure to apparently nontoxic concentrations of iodine, iron, copper, and aluminum salts. In all these cases the highly differentiated larval tissue is resorbed and the less differentiated prospective adult tissue survives.

4   Metamorphosis is a complex of destructive and constructive events responding to an overall, pervasive change of a little understood nature. Such a complex is indicated in ascidian metamorphosis, for example, by a major activation of a new set of enzymes, a general migration of mesenchyme cells through the epidermis everywhere into the tunic, and, in certain cases, a change from red to yellowish color throughout the body, all coinciding with the onset of tail resorption.

Considering these cases together, several general conclusions can be made:

1   There is an invisible but presumably substantial basis for organization, the character of which can be transformed by external chemical agents.

2   Highly differentiated tissues are generally much more susceptible to adverse conditions than are relatively unspecialized tissues.

3   The presence of an established organization tends to retard or inhibit the development of another potential organization.

4   The destruction or resorption of an established organization releases the potential of any latent organization and at the same time may supply the latter with growth-stimulating nourishment.

5   Metamorphosis may be regarded as a general response to the attainment or establishment of a set of organismal circumstances, whether regarded as genomic, metabolic, or cellular, that is new or different from the set previously in control.

AMPHIBIAN METAMORPHOSIS

Frogs, toads, and, in a less spectacular way, urodeles transform from an aquatic gill-breathing larva to a lung-breathing terrestrial animal. The general course of development through this life cycle is well known. Two transformations take place:

1   Certain adaptive structures formed during embryonic development, namely, the ventral suckers and the external gills of the anuran tadpole and the balancers of the urodele larva, are resorbed during early functional life. These are merely resorptions of precociously formed structures, not a part of the true metamorphic event, which occurs much later. They disappear when they have served their purpose.

2   Almost every organ system of the frog undergoes alterations during metamorphosis. Before the onset of metamorphosis, the tadpole, a vegetarian, is a fully aquatic creature with well-developed gills, a long flattened tail, lidless eyes, horny rasping teeth, and a long coiled intestine. After the completion of metamorphosis the froglet, no heavier and usually somewhat lighter than the tadpole that gave rise to it, is a lung-breathing creature with no tail, with well-developed limbs, with eyelids, and with teeth, gut, and other structures associated with its carnivorous habits. Even structures which persist into the adult undergo changes. For instance, the skin thickens, becomes more glandular, attains an outer keratinized layer, and acquires a characteristic pattern of pigmentation, while the brain becomes more highly differentiated. At the cellular level, cell modifications are evident in eyelids, limbs, lungs, tongue, eardrum, operculum, skin, liver, pancreas, and intestine. Horny teeth are lost, and the tail and the larval cloaca atrophy. Probably no cell or tissue or organ remains entirely unaffected.

At the same time many physiological and biochemical changes take place. According to Frieden, the biochemical alterations may be considered to have direct adaptive value or to serve as a basis for morphological, chemical, or other changes which have adaptive value relating to the transition from freshwater to land: "Among the most important adaptive changes are the shift from ammonotelism to ureotelism [i.e., from the excretion of ammonia to the excretion of urea], the increase in serum albumin and other serum proteins, the alteration in the properties and biosynthesis of hemoglobins. The development of certain digestive enzymes and the augmentation of respiration also contribute to the success of the differentiation process. During

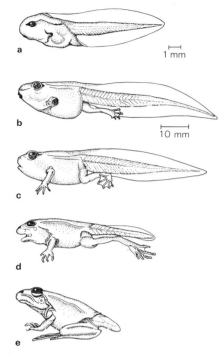

FIGURE 19.1   Typical anuran metamorphosis (Rana pipiens). (a) Premetamorphic tadpole. (b) Prometamorphic tadpole (growth of hindlimbs). (c) Onset of the metamorphic climax (eruption of forelimbs, retraction of tail fin). (d,e) Climax stages, showing gradual appearance of the froglike organization.

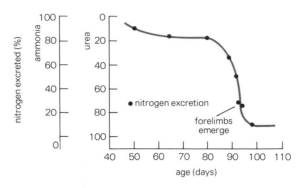

FIGURE 19.2   The transition from ammonia to urea excretion in *Rana temporaria* tadpoles. [*After Frieden, 1961.*]

metamorphosis, there are many additional important chemical developments which may be secondary to the primary morphological or cytological transformations which aid in the adjustment to land. These include alterations in carbohydrate, lipid, nucleic acid, and nitrogen metabolism. Major modifications in water balance, visual pigments (vitamin A), pigmentation, and tail metabolism are also observed. Finally, there is a partial mobilization of the enzyme machinery to promote the metamorphic process and the colonization of the land."[1]

Hormonal Control

Frogs metamorphose after various periods of growth, according to the species, and it was early recognized that the attainment of a critical species-specific size, rather than the duration of growth, was crucial. Thus, in nature, bullfrog tadpoles metamorphose at the end of the third summer-growth season in the north but at the end of the second in the south, after having attained a certain size, the time required depending on the mean environmental conditions. And it has also long been known that addition of iodine to the water or feeding with thyroid-gland tissue causes metamorphosis to occur earlier, at a smaller size, while elimination of iodine from the diet postpones it. From the first, therefore, the iodine-containing thyroxine of the thyroid gland has been studied intensively with regard to the timing of metamorphosis and the metabolic effect of the hormone. Any vertebrate hormone, however, is a component in a complex interacting hormonal system, in which pituitary hormones are particularly involved.

So far, hormone activity in development has been approached in several ways, and with regard as much to the development and differentiation of the vertebrate reproductive and other systems as to the questions of metamorphosis.

1   The initial concern is to show that particular hormones influence and participate in the control of differentiation of various target structures, by administration of excess hormone or removal of its source. In the case of thyroxine in relation to

[1] E. Frieden, Biochemical Adaptation and Anuran Metamorphosis, *Amer. Zool.*, **1**:115–150 (1961).

amphibian metamorphic events the task of proving this relationship has been mostly accomplished.

2 A more analytical concern is to identify the specific developmental processes that are subject to hormonal interference during critical periods of development.

3 It is necessary to interpret the hormone action in terms of cellular mechanisms.

4 Finally, and prospectively, efforts are being made to fit the effects of hormones on target tissues with the central doctrine that differentiation reflects the emergence of selective protein synthesis preceded by differential repression or derepression of the coding genes.

In studying frog metamorphosis, therefore, we are confronted with a complex hormonal situation and a seemingly infinite variety of responding or nonresponding target tissues. The main event, however, can be stated, oversimply, as the general response of the tadpole tissues and organization to a critical threshold of thyroid hormones in the circulation. The hormone threshold normally varies with the species. On the one hand, the secretory activity of the thyroid gland is itself under pituitary hormonal control, and its relative rate of growth is to some degree variable from species to species. On the other hand, tissue sensitivity presumably also varies to some extent according to the species. Present analytical investigations are mainly concerned with the effects of varying threshold levels of thyroid hormones and their analogs, and with differences in susceptibility among the multitude of responding tissues.

The hormonal situation has been investigated during many years by Etkin, who long ago emphasized that *spacing* of metamorphic events depends on the concentration of thyroid hormone, while the *sequence* of the events is inherent in the tissues. Most tissues respond when a critical threshold is reached, but some respond earlier and some later. Thyroid hormone concentration in the blood and tissues of the tadpole gradually increases during the last two-thirds of larval life up to the phase of metamorphic climax, and then undergoes a fairly sudden drop. At a critical point in the development of the tadpole, some factor, presumably controlled by a genetic mechanism, renders the hypothalamus sensitive to the low level of thyroid hormone already circulating in the blood.

The neurosecretory apparatus of the hypothalamus responds by secreting a thyrotropin-releasing factor (TRF) which stimulates the anterior pituitary to secrete a thyroid-stimulating hormone (TSH), which turns on the orderly increase of thyroid secretion. This increase in thyroid hormone then trips the orderly sequence of tissue changes that transform the tadpole into a frog. If the part of the hypothalamus attached to the pituitary is removed, metamorphosis proceeds through the early stages but stops abruptly at the climax. The presence of the hypothalamus seems to be necessary for the completion of metamorphosis. Isolation from the pituitary inhibits the hypothalamus from attaining a high level of TSH activity. Nevertheless, animals thus inhibited continue to grow and may achieve gigantic size.

Etkin and coworkers have shown that another pituitary hormone, prolactin, is also involved (as an inhibitor) in the overall control of metamorphosis. In his own

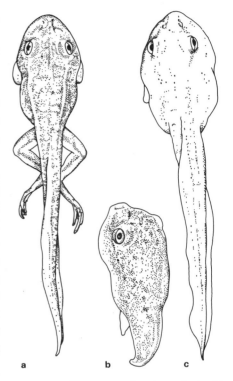

FIGURE 19.3 Hormones and metamorphosis. (a) Normal metamorphic stage. (b) Premature metamorphosis following exposure of young tadpole to thyroxine. (c) Inhibition of metamorphosis following removal of either the thyroid or the pituitary gland.

FIGURE 19.4 The interaction of endocrine factors in determining the time and pattern of anuran metamorphosis. In the early premetamorphic period the thyroxine level is very low; it remains so until just before prometamorphosis begins. At this time, the mechanism of the hypothalamic thyrotropin-releasing factor (TRF) becomes sensitive to positive thyroxine feedback, thereby initiating prometamorphosis. The increase in TRF provoked by the action of the initial thyroxine level on the hypothalamus stimulates increased release of TSH (thyroid-stimulating hormone), which acts in turn to raise the thyroxine level. This leads to a spiraling action which raises the thyroxine level and thereby induces prometamorphosis, with its characteristic sequence of changes. The positive feedback cycle leads to maximal activation of the pituitary-thyroid axis, thereby bringing on metamorphic climax. During early premetamorphosis, prolactin is produced at a high rate (dark shading). With the activation of the hypothalamus, the production of prolactin drops, under the inhibitory influence of hypothalamic activity. As the level of TSH rises during prometamorphosis, that of prolactin decreases. The growth rate of the animal therefore falls, and the metamorphosis-restraining activity of prolactin diminishes. Thus the premetamorphic period, in which growth is active and metamorphosis is inhibited, is characterized by the predominance of prolactin over TSH. The reverse situation exists during metamorphosis. The time of shift in hormone balance is determined by the initiation of positive thyroid feedback to the hypothalamus. This varies greatly among species. The pattern of change during metamorphosis is regulated by the pattern of the feedback buildup and is much the same in most anurans. [*After Etkin, 1968.*]

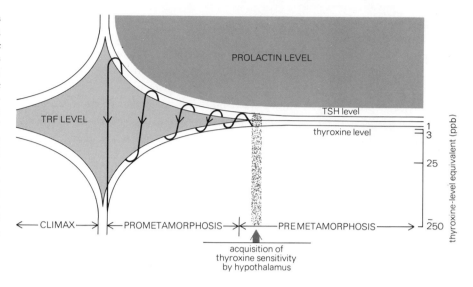

words, "At the level of endocrine action we again see that developmental control is effected by a balance between inhibition and disinhibition rather than by simple stimulation. The pituitary-thyroid axis is kept at a low level of activity in the growth phase of the tadpole's development by negative feedback, and at metamorphosis its activation is brought about by hypothalamic action. . . . The complexity of the push-pull type of interaction in governing metamorphosis is further emphasized by the discovery of the role of prolactin as a thyroid antagonist in amphibian development. Whether this substance acts at the peripheral level or as a goitrogen or in both ways its role again emphasizes that development is controlled by a dynamic balance of plus and minus factors." [1] Etkin calls attention to a fascinating aspect of this particular interaction, which is the manner in which it appears to have been exploited as the mechanism for the evolution of a "second metamorphosis" in the common newt.

This urodele, after a typical first metamorphosis that produces the land form, and after a period of growth on land, then undergoes a second metamorphosis for the return to water in order to breed. This second metamorphosis has been shown to be under the influence of prolactin, presumably resulting from a shift in the balance between the pituitary factors, prolactin, and TSH, rather than from an activation of one of them. In other words, the first metamorphosis is induced by a shift in favor of TSH, the second metamorphosis by a return to the predominance of prolactin.

Threshold and Rate of Response

According to Kollros, metamorphosis involves alteration of larval tissues which normally acquire sensitivity to thyroid hormones long before significant quantities

[1] W. Etkin, "Metamorphosis," p. 341, Appleton-Century-Crofts, 1968.

of such hormones are released into the circulation. In general, structures that meta-morphose early are more sensitive than those that undergo change later. Whether this variation is related to differences in threshold values or to differences in total thyroxine requirements has been in debate. However, by using greatly lowered hormone concentrations for long periods of time, up to a year, Kollros has shown that different tissues have different thresholds. Successive metamorphic events have different and somewhat higher thresholds from preceding ones. The threshold for the onset of corneal reflex, for example, is lower than that for perforation of the operculum, and the threshold for perforation of the operculum is lower than the threshold for gill absorption.

Not only do such experiments establish the existence of individual thresholds for different target organs, but they also support the concept of progressively increasing thresholds at different stages of development of a particular organ. In the case of the hind limb, for instance, long exposure to very low concentration of thyroxine promotes differentiation to the stage of three digits. A higher level of hormone concentration is required for the development of five toes, and a still higher one for the formation of interdigital webs and toe pads.

Thus limb development, inhibited during most of the growth period of the tadpole, apparently needs a progressive booster at a certain stage. This may be associated with the enormous growth that anuran hind limbs finally undergo during metamorphosis, compared with urodele limbs, which remain relatively small and require only developmental release. In the frog, however, as hormone concentrations increase, tissue responses become progressively more rapid, until maximum rates of change are attained. In effect, at high concentrations, all metamorphic events become crowded together, and the time sequence is disturbed; the tail begins to resorb before limbs become well developed, so that without providing means of locomotion and other essential factors, the transformation leads to death.

Difference in tissue sensitivity is seen in a striking form in an early experiment in which an eye cup was transplanted to a tadpole tail, where it differentiated. During metamorphosis the eye moved forward as the tail resorbed and finally came to rest in the sacral region when metamorphosis was complete. Transplanted limbs and transplanted kidney tumor are likewise unaffected by the degenerative processes in the surrounding tail tissue.

Perhaps the most striking single feature of the hormonal control of amphibian metamorphosis is that a single hormone, the low-molecular-weight compound thyroxine, evokes multiple responses from diversified tissues. Responses are specific although the inducer varies only quantitatively. Moreover, they are both constructive and destructive, depending on the target tissue. Thus in response to tri-iodothyronine, a companion hormone of thyroxine, biosynthesis of nucleic acid is decreased in the tail but increased in the liver. Similarly, thyroxine induces rapid aging and destruction of the red blood cells of the metamorphosing tadpole which carry tadpole-type hemoglobin, while simultaneously (or later) it stimulates the development of cells

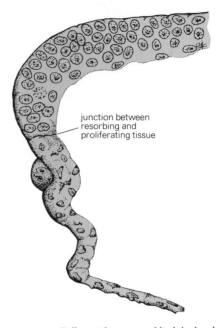

junction between resorbing and proliferating tissue

FIGURE 19.5 Differential response of limb-bud and gill tissue to thyroxine; the line separating the two districts, one growing, the other degenerating, is very sharp.

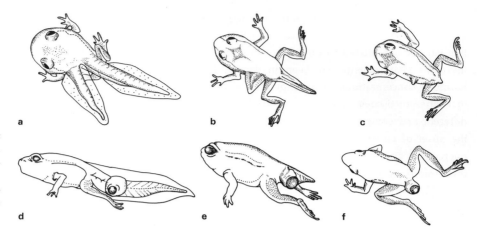

FIGURE 19.6 Organ specificity of metamorphic responses in tadpoles. (**a–c**) Tail tip transplanted to the trunk region undergoes atrophy simultaneously with the host's tail. [*After Geigy, Rev. Suisse Zool.* **48**:*483 (1941)*.] (**d–f**) Eye cup transplanted to the tail remains unaffected by the regressing tail tissue. [*After Schwind, 1933.*]

that synthesize the adult frog type of hemoglobin exclusively. The agent is the same, but the response is cell death and cell proliferation, respectively, in the two erythrocytic cell populations.

The problem, therefore, is: What general property of thyroid hormone is responsible for these effects? The association with level of basal metabolism in higher

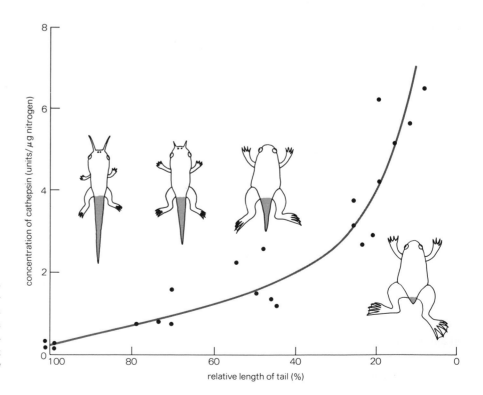

FIGURE 19.7 Regression of tadpole tail, in metamorphosis of *Xenopus laevis*, is accomplished by lysosomal digestion of cells. As metamorphosis proceeds, the enzyme concentration increases (the absolute amount of enzyme remaining constant). Eventually the stub contains almost nothing but lysosomal enzymes, and it falls off. [*Partly after Etkin, and Weber.*]

vertebrates suggests that an increase in oxidative metabolism may be the discriminating factor. However, there is no increase in oxygen uptake during spontaneous metamorphosis. On the other hand, a flood of metabolites, resulting from tissue destruction, is immediately available for rapid protein and nucleic acid synthesis.

The process of tail resorption has been investigated in isolated tails or pieces thereof, of *Xenopus* in particular. Isolated tails respond to thyroxine. Involution always begins at the very tip, and the typical pattern of metamorphosis follows; namely, thickening of the epidermis, migration of pigment cells, and involution of notochord, neural tube, and muscle cells. A surprising feature is that an initial *synthesis* of certain tail enzymes appears to *precede* the dissolution of the tail. Tail resorption apparently requires RNA and protein synthesis before the demolitive processes begin. There is evidence, however, that the tadpole anterior pituitary releases a factor that interferes with the action of thyroxine on tail-disc resorption in vitro, suggesting that an antagonism exists between a pituitary growth factor, possibly a prolactin or a growth-hormone-like molecule, and thyroid hormone, and that this interaction takes place peripherally through direct interactions at the site of the target cells.

INSECT METAMORPHOSIS

Insect and amphibian development have much in common and much that is different. In both cases an egg develops along certain general lines characteristic of eggs as a whole, and in both cases during the life cycle of certain groups a profound metamorphosis takes place. A dual developmental system is evident in both. Otherwise the eggs, the course of development, and the functional organisms are about as different in the two classes as it is possible to be. The amphibian egg and larva are, in their general organization, clearly a retention of an ancient ancestral type, serving only while development proceeds in the original freshwater environment. Amphibian metamorphosis is in essence a release or activation of the genomic set underlying the adult organization, which requires for its expression a mass of tissue which is minimal but greater than that of the egg. Concomitantly the genomic support for the larval organization is withdrawn and specifically larval tissues resorb.

Insects, in both egg and adult form, have become fully terrestrial. Special larval forms, especially those of holometabolous insects, are relatively new evolutionary inventions. This holometabolous transformation is surely as radical an alteration of form as graces the animal kingdom. According to Snodgrass (1954), the extraordinary fact is that a caterpillar hatches from the egg of a butterfly, the subsequent change of the caterpillar into the butterfly being merely the return of the metamorphosed young to the form of its parents. The transformation of the caterpillar is a visible event reenacted with each generation. The change of the young butterfly into a caterpillar has been accomplished gradually through the past evolutionary history. As Snodgrass emphasizes, the larva of a holometabolous insect cannot be regarded

FIGURE 19.8 Life history of the *Cecropia* silkworm.
(a) First-, third-, and fifth-instar larvae. (b) First-
instar larva hatching from egg. (c) Larval-pupal
molt within cocoon. (d) Pupa within cocoon. (e)
Adult male. [*Courtesy of L. I. Gilbert.*]

as an embryonic form prematurely exposed, or as a recapitulation of the evolutionary past. The true metamorphic characters are the structures found in the young insect—without counterpart in adult evolution—which are discarded as the adult develops. The legs of a caterpillar, for example, are not primitive structures representing an early step in the development of the adult legs, because the legs of the adult exist within the larva as distinct structures, segregated off as imaginal discs. The larval legs must be recognized as relatively modern and autonomous organs, which were evolved for the special conditions of larval life and which are sacrificed in the return toward the ancestral form.

The eggs of the large *Cecropia* silkworm, for example, hatch as larvae after 10 days and subsequently molt four times as they grow some 5,000-fold to mature as fifth-instar larvae, all the while transforming leaves into silkworm. They then enter pupation, the pupa being enclosed in a cocoon. Williams summarizes the event as follows: "The completion of the cocoon signals the beginning of a new and even more remarkable sequence of events. On the third day after a cocoon is finished, a great wave of death and destruction sweeps over the internal organs of the caterpillar. The specialized larval tissues break down. But meanwhile, certain more or less discrete clusters of cells, tucked away here and there in the body, begin to grow rapidly, nourishing themselves on the breakdown products of the dead and dying larval tissues. These are the imaginal discs which throughout larval life have been slowly enlarging within the caterpillar. Their spurt of growth now shapes the organism according to a new plan." [1] New organs arise from the discs. Also, some less specialized larval tissues, such as the epidermal layer of the abdomen, are transformed directly into pupal tissues. Pupation is followed by a developmental standstill, a diapause, lasting 8 months, a device that allows the pupa to survive the winter and emerge at an appropriate time the following spring.

Then comes a second period of intense morphogenetic activity. "The result is a predictable pattern of death and birth at the cellular level as the specialized tissues of the pupa make way for the equally specialized tissues of the adult moth. Spectacular changes occur throughout all parts of the insect: in the head, the formation of compound eyes and feather-like antennae; in the thorax the molding of legs, wings, and flight muscles; in the abdomen, the shaping of genitalia and, internally, the exorbitant growth of ovaries and testes. And in the newly formed skin we can witness the strangest behavior of all—the extrusion and transformation of tens of thousands of individual cells into the colorful but lifeless scales so typical of moths and butterflies. After three weeks of adult development, the process is complete. The full-fledged moth escapes from the cocoon and unfurls its wings." [2]

Insects have been studied mainly with regard to their external features and structures, which consist of cuticle laid down by a single layer of epidermal cells.

[1] Carroll Williams, "Nature of Biological Diversity," p. 251, McGraw-Hill, 1961.
[2] *Ibid.*

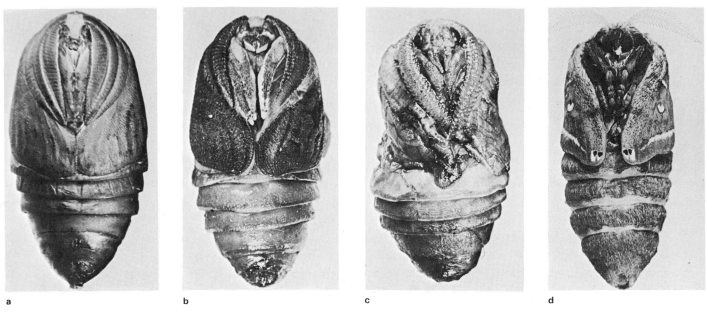

a   b   c   d

FIGURE 19.9 Advanced stages of pupation of *Cecropia* silkworm. (**a,b**) Female and male pupae. (**c,d**) Later stage of female and male pupae, respectively. [*Courtesy of L. I. Gilbert.*]

Internal organ systems are as important to the insect as to any other animal but are even less amenable to experimental developmental or genetic studies. It seems, therefore, that in the insect the epidermis is the chief agent of morphogenesis, although our ignorance may be bliss and seeing is believing. However this may be, the general characters of the cuticle change at successive stages of growth.

Molting and the Molting Hormone

Since the cuticle is a mosaic made up of the contributions of each individual epidermal cell, the growth activity of each cell, at each stage of the whole developmental history, is firmly registered in the characters of the little patch of cuticle that it lays down. And since the cells comprising the single epidermal sheet not only lay down cuticle, a process incompatible with cell division, but are also responsible for growth and change of form, which requires cell multiplication, these two processes do not occur at the same time. Accordingly the growth of insects takes place in cycles, in which mitosis and cellular growth alternate with the deposition of a new cuticle and the shedding of the old, i.e., in molting cycles. Therefore, as growth proceeds, insects periodically discard the old cuticular structures and produce a new set, irrespective of whether, apart from growth, the new and the old need to be alike. All insects are subject to this pattern of growth and its restraints.

The sequence of events is under precise hormonal control. In brief, certain stimuli associated with the state of nourishment cause the brain to discharge the "brain hormone" from the neurosecretory cells of the pars intercerebralis. This hormone in turn activates the endocrine organ known as the *prothoracic gland,* which then secretes the molting hormone, *ecdysone.*

Ecdysone is a cyclic compound of small molecular size. Just as the same thyroid hormones are produced in all classes of vertebrates, the hormone ecdysone is produced by all insects and other arthropods that have been investigated. Molting cycles are characteristic of all classes of arthropods, and ecdysone extracted from insects causes molting in shrimps, while ecdysone from shrimps causes molting in insects. It is an activator of a nonspecific type.

Differential response must be a property of the target tissues. For example, many of the cells of the epidermis are highly differentiated and are no longer capable of division or of renewal of the cell cycle, while other cells lying between the groups of specialized cells remain dormant or undifferentiated.

Under the impact of ecdysone, at each molt, the old cuticle is loosened and thrown off, the specialized epidermal cells are resorbed or discarded, and the ordinary epidermal cells which have been stimulated to grow and divide give rise to new cuticle and to new groups of specialized cells. In fact, ecdysone acts directly upon those cells that are in a state of dormancy. Within a few hours the nucleolus is enlarging, RNA begins to accumulate in the cytoplasm, and mitochondria enlarge and multiply by subdivision. By the time the old cuticle is thrown off, the renewed epidermal growth has been virtually completed, with locally expanded regions folded compactly while awaiting release from cuticular confinement. Remarkably, the muscles of the larva, e.g., of *Rhodnius,* dedifferentiate between molts and redifferentiate muscle fibers shortly before the next molt occurs. To a degree the organism is renewed on each occasion.

As a general but not absolute rule, molting occurs each time the volume of the growing animal doubles. Such is the cyclic basis of growth which underlies the more spectacular changes commonly seen in the life history of many insects (and crustaceans).

Juvenile Hormone and Transformation

The growth and molting cycles of an insect are more specifically controlled, however, by a third hormone, the juvenile hormone, secreted by the corpora allata, which are endocrine glands located near the brain. In a negative sense this hormone controls the onset of metamorphosis. A general pattern of hormonal control in insects accordingly emerges, although there are differences among insect groups.

In holometabolous insects, whether fruit flies or giant saturniid silk moths, the life cycle consists of usually four or five larval molts, a larval-pupal molt, and a pupal-adult molt. The larval-pupal-adult transformations constitute an extensive metamorphosis. In the silkworms metamorphosis is usually interrupted soon after pupation by a prolonged pupal diapause, during which development usually ceases.

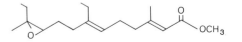

FIGURE 19.10 Molecular backbone of juvenile hormone.

Sexual maturation, which is a part of metamorphosis, occurs during the construction of the adult during pupation. Accordingly, six stages follow the hatching of the embryo, namely, four (in some, five) larval stages (or instars), the pupation phase, and the mature adult. The intermittent molting, with its discard of the old and replacement with the new, together with the changes that can be rung with the triple-hormone system, allows control both of phase duration and of remodeling during the span of individual life. In the absence of ecdysone, for instance, the insect lapses into a state of developmental standstill. When ecdysone is again secreted, diapause is terminated, and growth is resumed.

The dual nature of the development of the egg is evident and important. On the principle of first come, first served, the embryonic development leads directly to the formation and function of the insect larva—whether it be the so-called nymph of a grasshopper (resembling its parent except dimensionally and in the absence of wings), the dragonfly nymph (remarkably different from the adult in being adapted to aquatic life), or the grub or caterpillar (which must be entirely remodeled as a pupa).

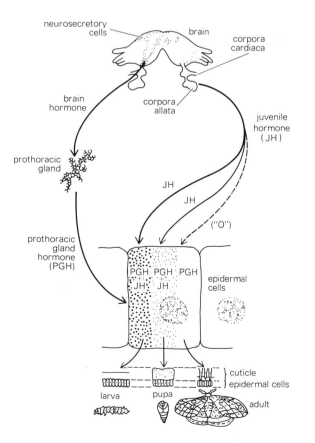

FIGURE 19.11   The endocrine control of growth and molting in the *Cecropia* silkworm. [*After Schneiderman and Gilbert, 1959.*]

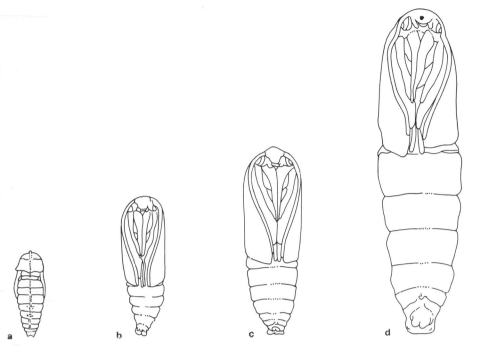

FIGURE 19.12 Experiments on effect of molting hormone on pupa size. (a,b) Dwarf pupae of moth, resulting from the removal of the corpus allatum (source of the molting hormone) from third- and fourth-instar larvae. (c) A normal pupa. (d) A giant pupa produced by implanting an extra corpus allatum from a young larva into one which had already reached the stage at which it would normally pupate.

In every case the developing system, having been diverted to some degree during embryonic development, moves toward expression of the adult organization. The adult organization is latent but ready-to-go, so to speak. The hormonal controls operate to determine when the major event occurs, to what extent development can be temporarily arrested, and how long juvenile states may be prolonged. For example, in the presence of the prothoracic gland hormone (PGH), juvenile hormone (JH) promotes larval development or maintains the *status quo* and so prevents metamorphosis. The presence of JH in an immature insect, whether larva or pupa, ensures that when the immature insect molts it retains its larval or pupal characters and does not differentiate into an adult. When the insect molts, in the absence of JH, it differentiates into an adult. In other words, withdrawal of JH initiates metamorphosis. Conversely, implantation of the JH-secreting gland from a second-stage larva to a fourth-stage larva maintains the larval state, in effect inhibiting metamorphosis, so that larval growth continues and a giant larva is produced.

Quantitative differences are as significant as the extremes of presence and absence, as is seen in the response of epidermal cells to JH. When the larval epidermis molts in response to ecdysone, the response varies as follows:

1   In the presence of a high concentration of JH, the cells secrete larval cuticle.

2   In the presence of a low concentration of JH, the epidermal cells secrete a pupal cuticle.

buds or discs, and be permitted at last to fulfill their destiny. Within the confines of the pupal epidermis the broken-down tissues yield a veritable nutritive soup. This promotes massive proliferation and reconstructive processes by the less specialized cells that survive the cellular massacre or mass suicide, whichever it may be.

Two main types of larval tissue have been recognized with regard to growth and metamorphosis. Many tissues grow by increase in cell size and undergo disintegration at metamorphosis, the corresponding adult tissue being formed from imaginal discs that were not functional parts of the larval tissue. In the second class, tissues grow by cell division and are carried over, with or without modification, into the adult. Some other tissues behave in a manner intermediate between these two methods.

The muscle system is a case in point. In some of the more primitive orders of the endopterygote group, the majority of larval muscles are carried over into the adult stage. On the other hand in the thorax of the honeybee, for example, no larval muscles remain unchanged, although most of them are associated with the development of adult muscles. In general, metamorphosis involves the destruction of some larval muscles, the rebuilding of others, and the formation of muscles which were never represented in the larva. In muscles that disappear, the cross-striations are lost, fiber bundles lose their connections and separate, nuclear membranes disappear, and nuclei degenerate, and only when the tissue is disintegrating do phagocytes pick up the pieces. The rebuilding of a muscle from larva to adult involves the replacement of large nuclei (which have multiplied by amitosis) by smaller nuclei, which multiply by mitosis and have been sheltered in the larval muscle. At metamorphosis the large nuclei degenerate, and each of the small nuclei becomes enclosed in a small bag of myoplasm, thus forming the myoblasts, which construct new adult muscles.

A comparable situation is seen in the reconstruction of the intestine, in the mosquito, for example. Larval growth is accomplished solely by increase in cell size. At metamorphosis the adult tissue is formed by the simultaneous division, reduction in size, and increase in number of these same larval cells. The great increase in size of the intestinal cells, associated with larval growth as a whole, is accompanied by replication of chromosomes in the individual cell, with the largest cells having as many as 32 sets of chromosomes. During metamorphosis these cells divide until daughter cells are produced having the normal diploid sets. The new cell population gives rise to the new intestine of adult character.

Other types of tissues have cells which grow large but do not divide during larval growth, although the tissues may be augmented by growth of associated small reserve cells. These tissues also undergo excessive chromatin replication but of a different sort. Such cells do not subdivide at metamorphosis; characteristically they suffer cell death, and they may or may not be replaced, depending on organ function. The best known and most intensively studied are the cells of the salivary gland of dipteran flies, although it must be emphasized that this type of cell is typical of

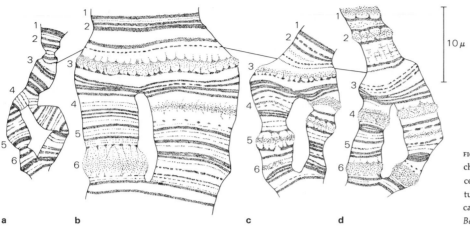

FIGURE 19.14 A short section of one of the chromosomes of the midge, *Chironomus*, from a cell of (a) intestine, (b) salivary gland, (c) excretory tubule, (d) rectum. Six prominent bands (numbered) can be recognized in each chromosome. [*After Beermann, 1958.*]

differentiated dipteran larval tissues in general. Chromosomes are not replicated as such, i.e., the diploid condition persists, but daughter chromatids align themselves side by side to form compact, ribbon-like chromosomes up to 16,000 times larger than usual. Within each chromatid is a much-coiled structure with a periodic pattern; daughter chromatids pair "gene for gene" and are consequently large and banded. Salivary glands, midgut, rectum, seminal vesicles, and Malpighian tubules all have cells with giant chromosomes of this type. In *Drosophila* there are 5,072 crossbands. Investigation of the effect of hormones or other agents on the nucleus during the developmental cycle is therefore facilitated. Before entering a discussion of this, however, some further comment on the significance of giant cells is appropriate.

Growth of larval insects by means of cell enlargement rather than by cell proliferation is characteristic of the more extreme forms of holometabolous development. This type of growth is essentially an overall enlargement of the larval organization first established during the development of the embryo, without possible disruption of its cellular basis by cell multiplication. The meaning of this form of growth in animals has had little discussion, but it is significant that a comparable condition is characteristic of other cases where perpetuation of a distinctive larval organization is seen:

1 Among tunicates, the predominating class of ascidians produce a tadpole larva (with a structure described earlier, on page 295) that has a brief existence, metamorphoses, and then grows into the adult. The tunicate class of larvaceans, however, consists of small animals of the tadpole-larva type that never metamorphose but become sexually mature, free-swimming adults. This condition is apparently attained by "freezing" the larval structure, i.e., the tail contains no more than 20 notochord cells and two bands of 10 muscle cells each, while the heart in some cases consists of only two contractile cells. Yet considerable growth occurs, so

that the mature organism consists of a very small number of exceptionally large cells.

2 Among amphibians, certain kinds, the perennibranchiate salamanders, grow to sexual maturity as permanently aquatic creatures that retain larval features, such as external gills, and do not undergo metamorphosis. As in larvacean tunicates and dipteran insects, the persisting, growing larval condition is correlated with constituent giant tissue cells.

Maintenance of pattern or organization is, at least to some extent, associated with suppression of cell division in differentiated tissues; conversely, change of cellular organization is associated with cell multiplication.

HORMONES AND GENE ACTIVATION

The giant chromosomes of the salivary glands have opened the door to a visualization of the process of gene activation, particularly in relation to metamorphic events. Just before pupation the salivary glands show a sharp change in their synthetic activity, secreting a brownish fluid in place of their previously clear secretion. At the same time striking changes occur in the pattern of puffing of the giant chromosomes. Most of the puffs that had been prominent during larval life now collapse, the DNA strands seeming to fold back into the chromosome to re-form compact bands. Meanwhile new puffs form at a series of other specific loci.

Pupation itself is the response to an increasing concentration of ecdysone and a decreasing concentration of juvenile hormone. When ecdysone is injected into an immature larva, therefore, conditions are established that favor premature pupation, an event that proceeds to take place. In this experiment, 2 hours after the injection of ecdysone, the pattern of puffing in the salivary glands shows a changeover from the larval to the pupal type. Accordingly, the primary effect of ecdysone is to alter the activity of specific genes, and this action is documented in the giant chromosomes. Cells are made to undertake synthetic acts accompanied by derepression and utilization of fresh genetic information.

Puffing and Genic Information

The correlation between ecdysone level and the puffing of the giant chromosomes indicates some sort of connection but says little concerning its nature. Does ecdysone, or any other hormone, exert its effect partly or wholly by intervening with the pattern of information retrieval from the genome? By what mechanism do the relatively small and simple molecules exert such complicated and fundamental effects? How is the transition from the use of one set of information to another instigated in all cells of an organism?

The premise is that most if not all features of a cell are determined by its population of protein molecules and that any fundamental changes in a cell, therefore, might be expected to involve alterations in the composition or activity of its protein pool. In view of the process of DNA $\rightarrow$ RNA $\rightarrow$ protein information transfer, the protein complement of a given cell should reflect that set of cistrons in the genome whose code message is being transliterated. Therefore if only part of the information stored in the genome is used at a given time, the most efficient way of doing this would be to allow some cistrons to form messenger RNA while others are kept inactive. Such a procedure should manifest itself in autoradiographic experiments as a discontinuity of RNA-precursor incorporation along the axis of the chromosome.

Experiments on the giant chromosomes of cells of salivary gland excretory tubules and on the midgut of dipterans, using tritium-labeled RNA-precursor molecules and autoradiography, show that some regions incorporate more labeled precursors than others, and that the synthesized material is RNA. They also show that for the same *organ* taken from animals at the same developmental stage, *this pattern is constant.* Accordingly RNA synthesis is discontinuous along the axis of the giant chromosomes. The active regions are seen as puffs, made up of DNA, RNA, and protein, the DNA being much less tightly coiled than elsewhere.

The RNA produced in the puffs appears to migrate through the nuclear sap and into the cytoplasm in the form of ribonucleoprotein granules, partly ribosomal. By comparing puffing patterns from different stages of development and from different tissues, one can determine whether genomic-activity patterns change during development and whether they are tissue-specific.

Puffs appear and disappear, ranging from barely discernible to 20 bands. They begin to form at a single band but progressively incorporate neighboring bands, which suggests that during puff development more and more cistrons are stimulated to deliver their information in a sequential way. About 10 percent of the giant chromosome bands are puffed at any one time. Drastic changes in the pattern of puffing occur shortly before the first signs of molt preparation, while during the last larval instar, preceding pupation, 15 percent of the chromosomal bands are puffed at the same time.

Kroeger (1968) concludes that those puffs which are present at all times, change their activity little if at all during metamorphosis, and are shared by many or all tissues, are those involved in cellular functions common to all cells, such as respiration, carbohydrate metabolism, and so forth. They form by far the larger group of puffs. In contrast, the smaller group of puffs that appear and disappear in preparation for a molt may mediate the less common chemical reactions characteristic of a specific step in metamorphosis. It seems, therefore, that the genome exerts permanent control over the chemical and morphological events that occur in the cytoplasm during metamorphosis.

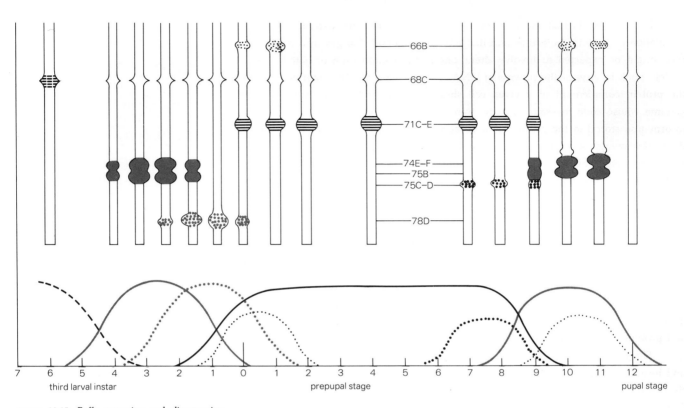

FIGURE 19.15 Puffs appearing and disappearing during the third larval instar and the prepupal stage at the base of chromosome arm 111 L of *Drosophila melanogaster* salivary glands. Numbers indicate hours before or after puparium formation. [*After Becker,* Chromosoma, *10:654 (1959), by permission of Springer-Verlag.*]

Hormonal Control of Puffing

The question is raised whether ecdysone and juvenile hormone exert their effects partly or entirely through intervention with the patterns of information retrieval. What is their mechanism of action? Studies have been made on isolated cells and tissue in culture employing ecdysone at a critical level as the test substance. Little has been done with juvenile hormone. It is postulated that ecdysone causes a cell to accumulate potassium ions, since the primary set of puffs can also be induced without ecdysone by increasing intracellular or intranuclear concentrations of K^+. Accordingly the stimulation of K^+ uptake by cell and by nucleus may be the primary effect of ecdysone. Results are summarized as follows:

1 Within 1 minute after application, the electrical-potential difference across the nuclear and cell membranes begins to rise and continues to do so for about 12 minutes, until it reaches a value of about 15 mV (millivolts).

2 Within 15 minutes, isolated nuclei from epidermal cells show a general increase in RNA synthesis.

3 Within 15 to 30 minutes, the "primary set" of puffs appears.

4   Within 60 minutes, the resistance of the nuclear membrane has risen to about twice its original level.

5   Within 65 to 72 hours, the "secondary set" of puffs appears.

6   This is followed by a series of events such as enlargement of nucleoli, swelling of mitochondria, and stimulation of mitosis that precede the actual molt.

Although the appearance of the primary set of puffs is the first easily demonstrated effect of ecdysone, during normal development this event is preceded by the disappearance of a number of puffs in a specific sequence. These disappearing puffs can be made to reappear by an increase in Na⁺ of a salivary-gland cell, suggesting that a lowering of intracellular Na⁺ causes these puffs to disappear and that this is another effect of ecdysone in low concentrations. Extreme sensitivity to electrolytic disturbance of the intracellular and intranuclear environment is evident. Clever (1965) suggests that mRNA synthesized in the puffs of the primary set directs the synthesis of proteins that alter transport or permeability functions of the cell in such a way that the intracellular solute undergoes further alterations, and that these changes then cause the loci of the secondary set to become puffed. He concludes that molting depends not on the activation of a *pattern* of genes (such as $A + B + C \cdots$) but on activation of a *sequence* of genes ($A \rightarrow B \rightarrow C \rightarrow \cdots$), of which only the first is hormonally activated. However, these are only guidelines.

Provisionally, therefore, we may say that the hormonal message acts at two sites. At the *cell membrane* it causes an alteration of the intracellular and intranuclear

FIGURE 19.16 Probable ecdysone and possible juvenile hormone action. Arrows pointing to transport systems indicate that effectiveness of these systems is modified; this might also be achieved by changes in membrane permeability, leaving the transport systems themselves unaffected. [*After Kroeger, 1968.*]

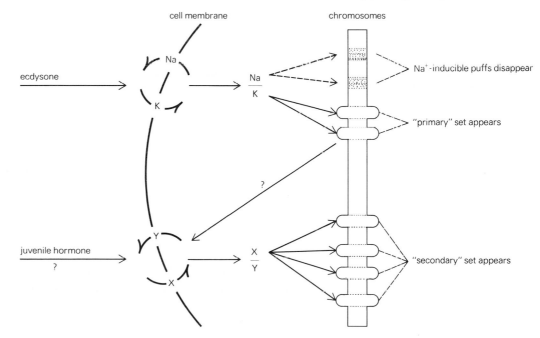

electrolyte media. At the *chromosomal* level this shift in milieu is translated into a differential pattern of mRNA synthesis, involving an alteration of the molecular configuration of histones. Histones are bound to DNA by salt linkages and are therefore susceptible to electrolyte change.

Correlation within the Genome

The following questions are crucial:

1   Which part of the cell is dispensable, and which part is necessary for the formation of those coherent patterns which occur during development and which may be termed "strategic patterns"?
2   At what site or by what interaction of cell parts is this "strategy" elaborated? Is coordination a purely nuclear process, arising, for example, by an interaction of chromosomal sites, or do cytoplasm and gene cooperate in activating any single site on the chromosome?

Since transplantation of cytoplasm of nerve tissue from the excretory (Malpighian) tube into the cytoplasm of salivary-gland cells does *not* alter the puffing pattern of the host cell but rolls it back to a more juvenile pattern, tissue-specific patterns are *not* elicited by the cytoplasm.

The fact that isolated salivary-gland nuclei can also undergo a similar rollback shows that a cell nucleus devoid of all observable cytoplasm can form a more juvenile pattern and implies that "strategy" is purely nuclear. Moreover, when as many as three-quarters of the chromosomes or chromosome parts are removed from isolated salivary-gland cells, rollback changes still occur, showing that "strategy" does not involve mutual interaction of chromosomes.

Integration within Cells and Tissues

Communication between cells has already been presented in terms of Loewenstein's concept of junctional membranes and associated ionic coupling. These offer the possibility of ion-distribution patterns and consequently of gene-activation patterns. If free communication between cells is impaired experimentally (or by natural developmental events), new patterns, subcenters, duplications, and multiplication of pattern elements should follow. In other words, ion-distribution patterns can be expected to display features characteristic of phenomena usually categorized under the term *morphogenetic field.* This term may well be replaced by an expression such as "field of gene-activation regulation." If ions are the vehicles of this activity control, as also indicated by L. G. and L. C. Barth's work on neural induction in amphibian embryos and Burnett's on *Hydra,* the persisting problems of regulative development, pattern formation, and related phenomena may be explained in physical and chemical terms.

CONCEPTS

Metamorphic Transformation

An invisible but real substantial basis underlies visible organization, and its character can be transformed by external chemical or physical agents.

Metamorphosis is a process associated with the existence of a dual system of development.

The presence of an established organization tends to retard or inhibit the development of another potential organization.

The destruction or resorption of an established organization releases the potential of a latent organization and simultaneously supplies growth-stimulating nourishment.

Highly differentiated, nondividing cells are relatively highly susceptible to all agents that interfere with enzymatic processes.

Metamorphosis is a general response to the establishment of a set of organismal controls that is new or different from the set previously in control.

Amphibian Metamorphosis

Metamorphosis is a general response to a critical threshold of circulating thyroid hormones.

Spacing of metamorphic events depends on the concentration of hormone; sequence of metamorphic events is inherent in the tissues.

Hormonal control is exerted through hormonal interplay among hypothalmic, anterior pituitary, and thyroid hormones.

Sensitivity to hormonal influence is variable among tissues, although not according to histological type.

Later-responding tissues have higher sensitivity thresholds than early-responding tissues.

A single molecular species evokes varying and multiple tissue responses according to its concentration and duration of action.

The thyroid hormone probably intervenes in protein synthesis at the level of transcription.

Insect Metamorphosis

The complex life cycle of many insects is an expression of the great versatility of genic constitution.

Reconstitutive changes occur during every molting period that are fundamentally similar to but less profound than those associated with metamorphosis.

Larval stages represent mainly a device whereby the initial mass of the egg is increased to one appropriate to the adaptive requirement of the mature phenotype.

The timing of metamorphic events is controlled by a push-pull relationship between inhibiting and disinhibiting hormones.

A circadian biological clock is involved in the temporal control of embryonic, larval, and metamorphic development as a whole.

Cell division, though not chromatid replication, is generally suppressed in the cells of tissues destined to become destroyed during metamorphosis, and is maintained as long as the larval state persists, larval growth accordingly being the result of increase in cell size rather than cell number.

Hormones and Gene Activation

The genome continuously releases a flow of messenger (mRNA) molecules that determine many and possibly all functional capacities of a cell.

Metamorphosis proceeds by an alteration in the composition of this flow.

Alteration in composition of flow is in part instigated by ecdysone. Ecdysone and possibly juvenile hormone alter the intranuclear electrolyte balance.

The genome is highly sensitive to alterations in the electrolyte composition of the nuclear sap, probably because single-ion species locally and specifically interfere with the repressive activity of histones.

Development appears as a progressive liberation of the intranuclear milieu from its surroundings. Consequently, groups of genes become sequentially activated and inactivated.

Intercellular communication at the genic level is effected by transmission of electrolytes and other substances, directly by way of junctional membranes and indirectly by hormonal circulation.

READINGS

BARRINGTON, E. J. W., 1968. Metamorphosis in the Lower Chordates, in W. Etkin and L. Gilbert (eds), "Metamorphosis," Appleton-Century-Crofts.

——, 1961. Metamorphic Processes in Fish and Lampreys, *Amer. Zool.*, 1:97–106.

BERRILL, N. J., 1955. "The Origin of Vertebrates," Oxford University Press.

CLEVER, U., 1965. Chromosomal Changes Associated with Differentiation, in "Genetic Control of Differentiation," Brookhaven National Laboratory, Symposium in Biology, no. 18.

CLONEY, R. A., 1961. Observations on the Mechanism of Tail Resorption in Ascidians, *Amer. Zool.*, 1:67–88.

COHEN, P. P., 1970. Biochemical Differentiation during Amphibian Metamorphosis, *Science,* **168**:533–543.

ETKIN, W., 1968. Hormonal Control of Amphibian Metamorphosis, in W. Etkin and L. Gilbert (eds.), "Metamorphosis," Appleton-Century-Crofts.

——, 1966. How a Tadpole Becomes a Frog, *Sci. Amer.,* May.

FRIEDEN. E., 1961. Biochemical Adaptation and Anuran Metamorphosis, *Amer. Zool.,* **1**:115–150.

GILBERT, L. I., and H. A. SCHNEIDERMAN, 1961. Some Biochemical Aspects of Insect Metamorphosis, *Amer. Zool.,* **1**:11–52.

HARKER, J. E., 1965. The Effect of Photoperiod on the Developmental Rate of *Drosophila* Pupae, *J. Exp. Biol.,* **43**:411–423.

HENSEN, H., 1946. The Theoretical Aspect of Insect Metamorphosis, *Biol. Rev.,* **21**:1–14.

KALTENBACH, J. C., 1968. Nature of Hormone Action in Amphibian Metamorphosis, in W. Etkin and L. Gilbert (eds.), "Metamorphosis," Appleton-Century-Crofts.

KEMP, N. E., 1963. Metamorphic Changes of Dermis in Skin of Frog Larvae Exposed to Thyroxin, *Develop. Biol.,* **7**:244–254.

KOLLROS, J. J., 1961. Mechanisms of Amphibian Metamorphosis, *Amer. Zool.,* **1**:107–114.

KROEGER, H., 1968. Gene Activities during Insect Metamorphosis and Their Control by Hormones, in W. Etkin and L. Gilbert (eds.), "Metamorphosis," Appleton-Century-Crofts.

LYNN, W. G., 1961. Types of Amphibian Metamorphosis, *Amer. Zool.,* **1**:151–162.

SCHNEIDERMAN, H. A., 1967. Insect Surgery, in F. W. Witt and N. K. Wessells (eds.), "Methods in Developmental Biology," Crowell.

SCHWIND, J. L., 1933. Tissue Specificity at the Time of Metamorphosis in Frog Larvae, *J. Exp. Zool.,* **66**:1–14.

WEBER, R., 1967. Biochemistry of Amphibian Metamorphosis, in R. Weber (ed.), "Biochemistry of Animal Development," vol. II, Academic.

WIIITTEN, J., 1968. Metamorphic Changes in Insects, in W. Etkin and L. Gilbert (eds.), "Metamorphosis," Appleton-Century-Crofts.

WHYTE, L. L., 1960. Developmental Selections and Mutations, *Science,* **132**:954, 1694.

WIGGLESWORTH, V. B., 1966. Hormonal Regulation of Differentiation in Insects, in W. Beermann (ed.), "Cell Differentiation and Morphogenesis," North-Holland Publishing Company, Amsterdam.

——, 1959. Metamorphosis and Differentiation, *Sci. Amer.,* February.

WILLIAMS, C. M., 1963. Differentiation and Morphogenesis in Insects, in J. Allen (ed.), "Nature of Biological Diversity," McGraw-Hill.

CHAPTER TWENTY
GENES, PREPATTERNS, AND DETERMINATION

The phenomena of determination and differentiation confront us with two fundamental questions: (1) What causes apparently genetically identical cells to become different in structure and function? (2) How do different kinds of cells become arranged in orderly spatial patterns? Both questions, which we now take up, emphasize the approach from a lower to a higher level in the hierarchical system. Insect development is particularly well suited to analysis of the second question, especially the development of *Drosophila,* since its genetics and experimental procedures have been so well worked out. Insects vary greatly, however, in life histories, and the often complex changes that occur clearly have a developmental basis. Even so, the developmental and life-history differences appear to be variations on a theme rather than basically different procedures, for the insect egg is a special type with certain general features common to all. Very briefly, the course of early development is as follows.

CONTENTS
Organization centers
Extrinsic control of regional development
Determination in imaginal discs
Prepatterns
Reaggregation and assembly
Initiation of discs
Transdetermination
Cellular diversification
Genetic control of developmental pathways
Concepts
Readings

ORGANIZATION CENTERS

Insect eggs, even the smallest, are relatively rich in yolk. The nucleus lies in a central position, embedded in a small island of cytoplasm from which fine strands extend to the periphery of the egg, where they form the cortical layer of cytoplasm. This layer is called the *periplasm.* As the nucleus undergoes successive divisions, the daughter nuclei move apart, in which stage they have been termed *energids.* The process continues through eight divisions, by which time the nuclei enter the periplasm, and cell boundaries form between them for the first time. The single layer of cells now surrounding the yolk is the blastoderm. Subsequently a germ band appears in the blastoderm as the first visible differentiation of the embryo. It first appears in the presumptive prothorax region of the embryo-to-be and continues to differentiate anteriorly and posteriorly from this region. Two interacting centers appear to be responsible for the differentiation of the germ band: (1) an activation center located at the posterior pole, which interacts with the cleavage nuclei; and (2) a differentiation center in the presumptive prothorax region, which may be regarded as a center of morphodynamic movement that provides the stimulus for the aggregation of cells that form the germ band. The series of reactions evoked by the activation center is accordingly essential for the function of the differentiation center, which is the focal point for all subsequent processes of differentiation.

The eggs of insects vary greatly among the different insect groups, ranging from eggs with great regulatory powers (e.g., those of the dragonfly, which can be

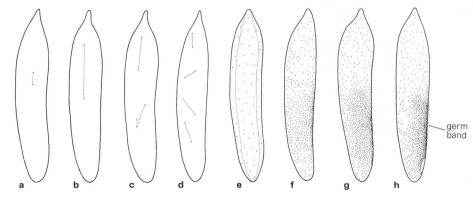

FIGURE 20.1 Early development of the dragonfly *Platycnemis*, showing the first three divisions of the nucleus, the spreading of the daughter nuclei through the egg cytoplasm (the two sister nuclei from a division are joined by a dotted line), the multiplication of the nuclei to form a blastoderm, and their aggregation to produce the germ band. [*After Seidel*, Arch. Entwickl.-mech., 119:322 (1929).]

FIGURE 20.2 Regulation in the insect egg. (a) Normal embryo of the dragonfly *Platycnemis pennipes*, seen from the left side, and dwarf embryo, obtained by partial constriction of the egg at the four-nucleus stage. The dwarf is normally proportioned and developed, and its organs have arisen from regions the presumptive fate of which was quite different; their fates were therefore not irreversibly determined at the stage operated upon, and regulation has been possible. (b) The operation of the formation center. If a very small part of the posterior of the egg is constricted off at an early stage, an embryo can develop. (c) If the constriction lies a little further forward, no embryo forms. (d) After the formation of the blastoderm, an embryo is formed even if the constriction lies well forward, since the formation center has by that time completed its action. [*After Seidel*, Arch. Entwickl.-mech., 119:322 (1929).]

made experimentally to produce twin embryos, by dividing the germ band longitudinally during early cleavage) to those of dipterans (e.g., *Drosophila*, in which determination of the presumptive embryonic parts is already set at the time of fertilization).

In mosaic eggs of this sort, typical of higher Diptera and of Lepidoptera (i.e., holometabolous insects that undergo drastic, complete metamorphosis), the main regions of the body appear to be already mapped out in the cortical plasma at the time of laying, before the egg nucleus has begun to divide. Since elimination of some of the migrating nuclei by ultraviolet irradiation at a sufficiently early stage of development does not affect morphogenesis, their place being taken by other nuclei arriving in the cortical region, the subsequent fate of the cleavage cells is evidently determined by the regional nature of the cortical plasma. The nuclei are clearly equivalent. The general conclusions are that in the absence of nuclei in the cortical region, the cortical plasma or periplasm has become chemically differentiated and that the nuclei are later subject to this chemodifferentiation. In the egg of a hemimetabolous insect such as the dragonfly, however, which undergoes an incomplete

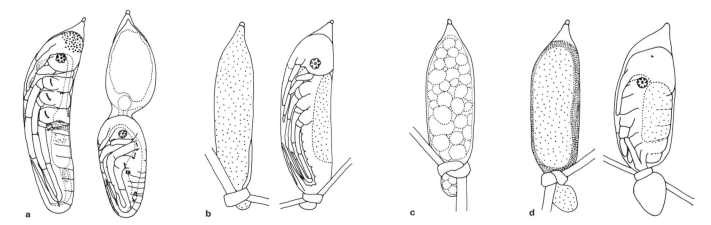

metamorphosis, the determinative chemodifferentiation of the cortex is not attained until some time after the cleavage nuclei have reached the cortical layer. If a substantial portion of the anterior region of the egg at this stage is removed by ligation, a well-proportioned but reduced embryo differentiates in the remaining posterior part of the egg.

EXTRINSIC CONTROL OF REGIONAL DEVELOPMENT

During the embryonic development of holometabolous insects such as *Drosophila* a remarkable segregation takes place. One population of embryonic cells becomes determined for forming the larval body, i.e., the cells differentiate so that at the time of hatching, the organs and systems of the first-instar larva (the stage before the first molt) are ready to function. A second system is set apart within this primary embryonic system as small isolated territories, the imaginal discs, which individually undergo some growth during larval existence but do not otherwise proceed with further development until metamorphosis. Differentiation is postponed in this second system, but cell division is not, for the discs which are so small in the embryo that they are hardly recognizable, each consist of thousands of cells when they reach their final size at the beginning of metamorphosis.

At metamorphosis most of the larval organs break down within the pupal case and their cells disintegrate. Simultaneously the cells of the imaginal discs begin to differentiate into specific cell types, and the cells of each disc collectively develop a structure characteristic of the location of the disc on the body wall. Thus three pairs of anterior discs together form the head (with its special mouth parts and sense organs), three pairs of thoracic discs give rise to the six legs, two pairs of more dorsal thoracic discs form the wings and specialized wing rudiments (halteres), and a posterior bilobed disc gives rise to the genital structure. Each disc accordingly develops into a structural complex that eventually connects with neighboring complexes, and together they construct the complete adult. Each disc is therefore specifically determined with regard to its destiny relative to the whole organism and this determination must occur some time before the differentiation of the disc becomes evident.

DETERMINATION IN IMAGINAL DISCS

Experiments, variously employing mutants, high temperature, and ether, show that the prospective fate of a particular imaginal disc becomes set during the first few hours of embryonic development. Thus bithorax-type flies with two pairs of wings may be produced, resulting from the formation of a second mesothorax body segment instead of a metathorax (the metathorax produces *halteres*, stubby wing rudiments with gyroscopic function). By the end of the fifth embryonic hour, determination of

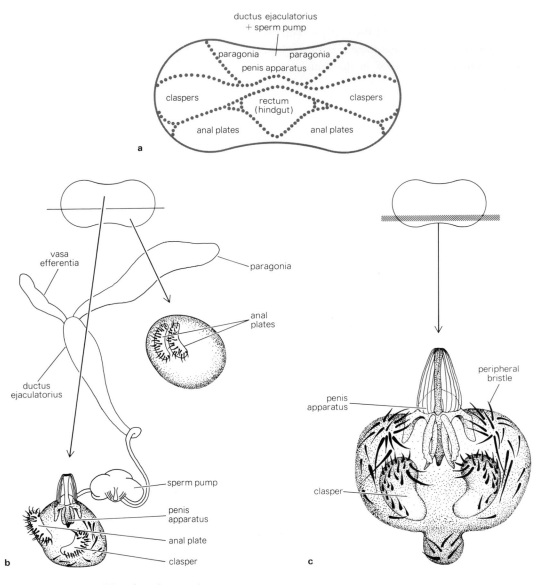

FIGURE 20.3 **(a)** A map of the anlagen (presumptive areas) in a male genital disc of *Drosophila melanogaster*. **(b)** Result of a fragmentation experiment; below, the metamorphosed structures obtained by the anterior half *(left)* and the posterior half *(right)*. **(c)** Result of a localized irradiation indicated by hatching; below, the metamorphosed implant in which no anal plates are differentiated. [*After Hadorn, 1965.*].

the nature of the third thoracic segment of the body, which will become either mesothorax or metathorax, has already been made. Whatever the nature of the determination switch may be, it can be operated by both genic and environmental factors.

The final differentiation of the discs beginning with the onset of metamorphosis is controlled by hormones. Throughout larval life the discs are exposed to a high level of juvenile hormone secreted by the corpora allata, which has the combined effect of sustaining the larval organization and inhibiting disc differentiation. When metamorphosis begins, the level of this hormone drops and disc differentiation is allowed to proceed, each according to its initial determination. The specificity of this determination is at least sufficient to allow localization in the leg disc, for example, of such organs as transverse rows of bristles, claws, groups of sensilla, and even a single bristle in the trochanter. The primordia for the segments of the adult leg are arranged in concentric annuli in the disc, and these annuli are converted to tubular structures when the disc everts.

Imaginal discs can be cut out of an insect larva with a fine tungsten needle and then implanted by means of a micropipette into the body cavity of other larvae, i.e., tissue culture in vivo. The hosts pupate, and the implanted discs also undergo differentiation, each according to its original fate, i.e., an eye disc gives rise to an eye, a leg disc to a leg, etc. Such discs therefore lend themselves to a variety of experimental procedures.

To begin with, while an implanted disc usually differentiates only according to its original prospective fate, under certain conditions it can give rise to more and different structures, although generally as duplication of structures typical of the particular disc. Each remains district-specific, but within the district significant regulation can take place. Even so, regional determination is apparent within a particular disc well before disc activation begins. A leg disc, for instance, contains cell subdistricts that develop into claws, tarsal parts, tibia, femur, trochanter, coxa, and adjoining parts of the thorax, each such district having bristles and hairs in specific numbers and patterns. Similarly the genital disc (male) is a mosaic of subdistricts more or less determined as prospective sperm pump, penis, anal plates, and hind gut, and each of these in turn is a mosaic of still smaller subdistricts. If the disc is subdivided before any differentiation becomes evident, however, each part reorganizes so as to form a harmonious whole of proportionate size.

PREPATTERNS

Pattern determination is evident at several levels of organization in a developing insect egg such as that of *Drosophila*. At the highest level, the imaginal discs occupying the various sites of the epidermis of the embryo collectively exhibit a topographical pattern relative to one another and to the whole. They are precisely located. Each such disc, presumably initially indifferent, acquires a "prepattern," a pattern of invisible determination within the district; and so on down to the cytodifferentiation of each constituent cell.

The term *prepattern*, established some years ago by Curt Stern, has been defined as a convenient expression of all the real chemical and physical factors which are distributed in a nonrandom fashion in a tissue, and which account for the nonrandom local initiation of differentiation. According to Ursprung, "After the term 'prepattern' had been created, it was often attempted to compare it to other embryological terms,

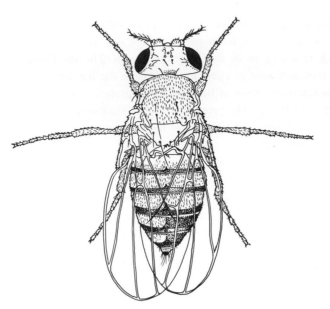

FIGURE 20.4 A gynandromorph of *Drosophila*. The female parts of the body are recognizable by their straight bristles, the male parts by their singed bristles. Note the male foreleg *(right)* with its "sex comb," and the dark coloration of the male tip of the abdomen *(right)*. [*After Stern, 1954.*]

such as morphogenetic fields. That latter term is based on experimental evidence; we ascribe field character to a tissue if a fragment of it will restore, under experimental conditions, what the intact piece would have done under normal conditions. The undifferentiated genital disc, for example, is composed of a number of morphogenetic fields, as demonstrated by the fragmentation experiments. Yet, without doing any experiment, it can be stated that the disc is prepatterned for a typical bristle arrangement. The term prepattern thus is much broader than the field-term, and it is a conceptual one that does not require, *per se*, experimental support." [1]

Prepattern studies have been carried out in connection with a number of differentiating structures in *Drosophila*, though particularly with the genital discs of gynandromorphs, i.e., individuals consisting of a mosaic of male and female genetically determined tissues. The genetic sex of the male or the female cells represents a natural marker. When the dividing line between tissues of the two types passes through the genital disc, female tissue forms female parts, male tissue forms male parts, in both cases regardless of the size of the tissue of the opposite sex. Accordingly, a genital disc seems to contain a prepattern for both male and female organs. The final differentiation is dependent on the male or female genotype of the respective cells, the prepattern itself being invariant.

In a series of experiments, by Kroeger, a different technique was used in which the fore- and hind-wing imaginal discs of the moth *Ephestia* were cut out, threaded, and tied together with a silk thread produced by the moth caterpillar, implanted into

[1] H. Ursprung, Development and Genetics of Pattern, *Amer. Zool.,* **3**:84–85 (1963).

a host, and later analyzed after differentiation. Particular attention was given to the nature of the complex hinges of the wings. In every case a mosaic hinge developed, containing both forewing and hind-wing structures, the two discs always forming an integrated hinge. If much forewing developed, little hind-wing material formed, and vice versa. Apparently the two discs grow together to form a compound rudiment, and thereafter together they lay down a new prepattern for wing hinges. According to Kroeger, imaginal discs from different body segments carry identical prepatterns, the determination of their cells being responsible for the segment-specific differentiation.

In other words, "The prepattern that gives rise to the forewing part of the hinge is identical to the one which forms the hindwing part of the hinge. Therefore, only when the final pattern is differentiated do the differences in determination begin to interact with the prepattern. The resulting pattern is the consequence of a differential response to an identical prepattern. The term 'determination' in this case denotes the competence of mosaic tissue to construct a wing hinge, while 'prepattern' signifies the site of differentiation which appears to be identical in each disc." [1] The overall situation appears to be essentially the same as in xenoplastic grafting in amphibians, in which competent tissue of the frog embryo reacts specifically to particular sites in salamander embryos, and vice versa, the grafted tissue in each case responding in its own way to location in a general system shared by the two orders.

REAGGREGATION AND ASSEMBLY

In another line of investigation, principally by Hadorn and by Ursprung, wing and leg discs of *Drosophila* were removed, disaggregated by means of trypsin, reaggregated in vitro for a short time, and then injected into a common host larva. Genetic markers were used so that the origin of the cells could be recognized; e.g., wild-type cells were mixed with yellow cells from a mutant which also carried a gene for multiple-wing hairs, that is, three or more hairs were in every cell of the wing lamella instead of one as in the wild-type strain. Similarly, in the case of leg-disc experiments, cells of the yellow mutant were mixed with those of a dark ebony mutant. Subsequent development or differentiation of the mixed aggregates, at the time of host metamorphosis, showed that in each case a perfect organ is produced by the assembly of randomly mixed cells, i.e., giving rise to a wing or leg, respectively. Yellow-leg cells and ebony-leg cells, for instance, produce leg parts which are an ebony and yellow mosaic.

According to Hadorn, those cells in the implanted mixture that carry a particular differentiation program recognize one another and move about until they come together to form precisely the indicated structure; whereas cells from different sub-

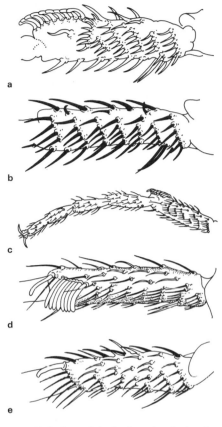

FIGURE 20.5 Parts of the forelegs of male, female, and gynandric drosophilas. Male bristles (which are yellow and singed, or forked) are drawn in outline. Female bristles (which are not yellow, and are straight) are drawn in black. (a) First tarsal segment of male. (b) First tarsal segment of female. (c) Mosaic tarsus: male anterior areas *(right)* with sex comb, female posterior areas. (d) Mosaic first tarsal segment. A female bristle within a gap of the sex comb. (e) Mosaic first tarsal segment. A single tooth is shown formed by male tissue. [*After Stern, 1954.*]

[1] K. C. Sondhi, The Biological Foundations of Animal Patterns, *Quart. Rev. Biol.,* **38**:311 (1963).

FIGURE 20.6 Reaggregating cells originating from dissociated wild-type (black) and y;mwh (white) wing imaginal discs give rise to a normal bristle pattern consisting of wild-type (black) and y;mwh bristles and hairs (white). Note the horizontal row of slender bristles above, the row of heavy teeth, the row of slender bristles arranged at larger intervals, and the hair zone below. [*After Ursprung and Hadorn,* Develop. Biol., 4:40 (1962).]

districts of a disc or from different discs never combine but separate from one another. A cell from a leg disc never joins one from a wing disc, nor do wing-base cells collaborate with wing-tip cells to form a base or a tip. Thus every cell in an imaginal disc appears to be individually determined for a specific differentiation long before any morphological differences can be detected.

In combination experiments, employing the same two genetic markers, cells originating from two or more wing discs of different genotypes were mixed and reaggregated in vitro for a short while, and then injected into a common host larva. The cells readily grew together and differentiated a mosaic wing lamella consisting of both wild and yellow types: multiple-wing-hair cells. More importantly, the heavy bristles which are produced are not arranged at random but are in a typical pattern identical to the one of a normal wing, consisting of a row of heavy teeth, another row of slender bristles, and a third row of slender bristles arranged at greater intervals. In the experimentally produced patterns, the rows are formed of yellow and black bristles, indicating that they are composed of bristle cells from different imaginal discs. Alternative explanations have been suggested:

1 A prepattern is originally present in each disc, and, following dissociation and reaggregation, the appropriate cells of each disc sort out and find one another by cell affinity, a process requiring cell movements which have not been observed.

2 A new prepattern is organized in the reaggregated tissue, causing cells originating anywhere in a disc to form bristles if they lie within the new margin-forming area.

The latter explanation is favored.

According to Ursprung, "If we think of differential gene regulation as a means of the local initiation of differentiation, the prepattern would necessarily comprise gene regulators. These in turn would be direct or indirect gene products themselves, and as a part of the prepattern, not be distributed at random, and therefore represent a pattern. Their formation would again require the non-random distribution of gene regulators, in other words, earlier prepattern. Development could then be characterized as a time hierarchy of prepatterns and patterns, every pattern being the prepattern for the next pattern, and the first prepattern consisting of the organization of the egg."[1]

[1] H. Ursprung. *loc. cit.*

INITIATION OF DISCS

The discovery that imaginal discs are essentially determined with regard to their prospective fate in somewhere between 2.5 and 4 hours of development raises the question of their origination. Do they arise in the earliest stages of embryonic development from single epidermal cells or from groups of cells? Is the single cell the source of a specific cell population, or does the initial definitive area usually embrace a group of cells, large or small, and therefore relate to a supracellular agency of some kind? The technical problem here is a familiar one, namely, to trace particular cell populations to their source or to the structures they ultimately form. Concerning the so-called determinate-type eggs of many invertebrates, the early studies of cell lineage yielded descriptive mappings of regional cellular differentiations, from the uncleaved egg onward.

The same principle has been applied by different methods to the development of *Drosophila* in relation to the pattern of differentiation. The technique employed is the x-ray induction of somatic chromosomal crossing-over, whereby single cells in the imaginal discs are caused to become homozygous for recessive marker genes that remain heterozygous elsewhere in the body. These marked cells give rise to generations of daughter cells which are also homozygous for the marker, and the clones thus produced are recognizable as patches of mutant tissue on the surface of the adult, notably as clones bearing both the bristle markers, yellow and singed.

The center of attention in these experiments was the differentiation of the leg, and the most striking feature of the cell lineage of leg discs is that clonally related cells come to occupy long narrow stripes in the longitudinal direction of the leg. The numerous somatic crossing-over experiments indicate that the number of cells initially destined to give rise to the leg is about 20, and that these cells are present as a group of prospective leg cells in developing eggs less than 3 hours old. This group of cells apparently remains static during the remainder of the embryonic period; further cell division does not commence until the second half of the first larval instar. Results show that determination of the leg disc into regions does not occur until after the third larval instar. These findings suggest that a certain area of the egg surface is destined to give rise to the leg disc (similarly with all other types of disc) and that this localization of prospective leg or other organ tissue is independent of the particular cellular state at any given time.

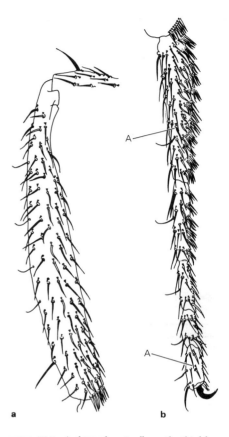

FIGURE 20.7   A clone of *y sn*[3] cells on the third leg (shown from two sides) in which somatic crossing-over was induced by irradiation at 12 hours after oviposition. The clone extends into the distal part of the femur, and into the tibia, basitarsus, and second to fifth tarsal segments. At *A*, the bristle and its bract are of different genotypes. [*After Bryant and Schneiderman, 1969.*]

TRANSDETERMINATION

In the foregoing experiments, predetermined but undifferentiated disc tissue implanted into a larval host received full exposure to the insect metamorphic hormone during the pupation of the host. In further experiments, pieces of imaginal discs

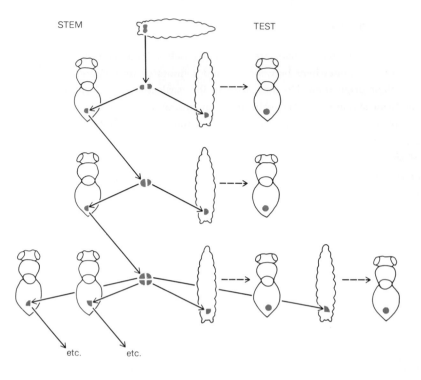

STEM   TEST

etc.   etc.

FIGURE 20.8 Method used for permanent cultures in vivo. Stem-line fragments grow but remain undifferentiated; test fragments pass metamorphosis and develop into adult structures. [*After Hadorn, 1965.*]

were inserted directly into the abdomen of the adult fly. Ordinarily the disc cells would stop dividing and begin to differentiate as soon as the host pupation begins. In this circumstance they continue to grow and divide indefinitely, as long as a sample of proliferating cells is transferred to a new adult abdomen every 2 weeks, i.e., as long as they are regularly subcultured as in typical tissue-culture procedure. By this means the cultured disc cells live for years, from fly to fly, without undergoing any observable differentiation. Since cell populations cultured in this way have been maintained for more than 6 years, passing through close to 200 adult fly generations, there have been ample time and opportunity to check on cell potentialities.

The starting point of these experiments was a male genital disc. Half of such a disc, the stem piece, is implanted in the abdomen of an adult fly, and the other half, the test piece, is implanted in a larva where it will differentiate during pupation. After 2 weeks the stem piece is recovered and divided into two; one piece is again inserted in an adult abdomen as the stem piece, and the other into a larva as the test piece, to see if it still differentiates as before. At first the test pieces continue to differentiate into typical genital structures. Then, as time and transfers continue, test pieces develop into leg parts or head organs, switching from the original determined path to another, a change which has been termed *transdetermination*.

Transdetermination can proceed from state to state. In a particular transfer series, for instance, genital-disc cells gave rise to head and leg structures in the eighth transfer generation, or after about 4 months. After the thirteenth transfer they gave rise to wings. After nineteen transfers, a thoracic type of structure appeared. In other words, genital-disc cells may give rise to head or leg cells, which may give rise to wing cells, which may give rise to thoracic cells, but apparently the genital cells cannot transform directly to wing cells. Moreover the transdetermination may become stabilized at any stage and need not proceed to a specific end point.

The underlying mechanism is not known, except for the fact that the frequency of transdetermination is directly correlated with the rate of cell proliferation. Sequences of diverse differentiations arise even in clones, i.e., cell populations derived from a single cell. Obviously a switch from a determinative process responsible for genital-complex differentiation to one responsible for the differentiation of a leg complex or for any comparable change must operate in some way different from a switch, say, that controls whether a mesenchyme cell differentiates as a muscle cell or as a cartilage cell.

CELLULAR DIVERSIFICATION

The problem of control of differentiation pervades all levels of organization. It is seen in the diverse differentiation of the body segments of an insect and any other segmented type of animal; it is seen in the specific course of differentiation of an individual cell; and it is seen in a simple but nonetheless tantalizing form in the cellular diversification within cell clusters of the integument of insects and other creatures.

The patterning of differentiation centers presents a problem distinct from that of differentiation within a center. Both are evident in the development of the integument of insects and have been studied both in *Drosophila* and particularly, by Wigglesworth, in the bug *Rhodnius*. With regard to the first problem, we are dealing with what may be regarded as a morphogenetic field, where the field is first divided by one patterning process into a small number of large regions and is then subdivided by a second process into a larger number of smaller regions. The development of the genital disc of *Drosophila*, for instance, has a stepwise character of this kind. In general, in patterns where structures are arranged in linear series, the two patterning processes need to be separated in time. However, two patterning processes can occur simultaneously but along different axes, one determining the number of rows and the second the number of structures in each row. Thus in the integument of *Rhodnius*, on which the only structure visible in the adult is a series of transverse ripples, Locke (1967) has shown the presence of both an anteroposterior gradient and of a side-to-side gradient.

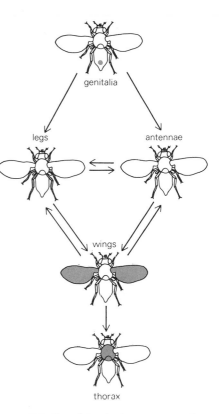

FIGURE 20.9 Transdetermination sequence undergone by seven kinds of imaginal-disc cells is shown by arrows. Genital cells, for example, may change into leg or antenna cells, whereas leg and antenna cells may become labial or wing cells. In most instances the final transdetermination is from wing cell to thorax cell; the change to thorax appears to be irreversible.

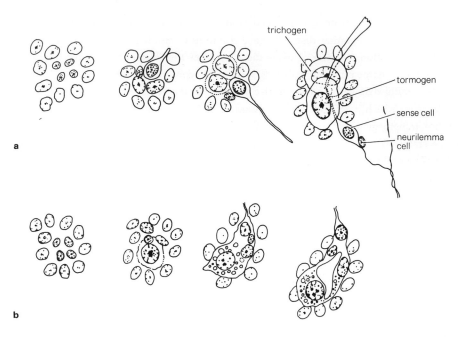

FIGURE 20.10   (a) Four stages in the development of a sensory hair, showing four small cells becoming differentiated into the trichogen, tormogen, sense cell with inwardly growing axon, and neurilemma cell. (b) Four stages in the development of a dermal gland by the differentiation of four small cells. [*After Wigglesworth, 1966.*]

The surface of *Rhodnius* and of other insects is covered with numerous sensory bristles or hairs and skin glands arranged in a patchwork of areas determined for different activities, each area having its own distinctive pattern of specialized cells. In a particular region of the abdomen, for instance, the sensory bristles are fairly evenly distributed, each being separated from its neighbors by an extent of about 15 ordinary epidermal cells. After each larval molt, when the old chitinous integument has been cast off and the epidermis expands (in keeping with the body as a whole), new sensory hairs appear wherever the existing bristles are most widely separated, and never close to an existing bristle, suggesting that existing hairs inhibit the appearance of new hairs within a certain distance.

Each new bristle complex is made up of four cells formed by division from a single epidermal cell. Each of the four daughter cells differentiates into a particular type of differentiated cell: a bristle-forming cell, a socket-forming cell, a sense cell which connects the bristle base to the central nervous system, and an insulating cell investing the sense cell and axon. Similarly dermal glands are produced by the division of single epidermal cells into groups of four diversely differentiated cells, one cell enlarging as the main glandular cell and the other three contributing to the sheath and duct. The two four-cell complexes—sensory hair and gland—may be homologous, with only a small change in local environmental conditions needed to deflect the development one way or the other.

However this may be, the phenomenon is both striking and challenging. What determines whether a particular epidermal cell remains undisturbed or undergoes two mitotic divisions culminating in extreme cytodifferentiations? In any such group of four cells, what leads to the specific diversity of specializations, what determines whether this set or that set of single-cell differentiations is started on its course? A possible clue is seen in the neuroblasts of the grasshopper embryo, which give rise by repeated cell division to a succession of ganglion cells. When these cells are already in metaphase it is possible, as shown by Carlson (1952), to rotate the spindle by means of a microdissection needle through 180° so that the chromosomes which would have remained in the neuroblast now go to the ganglion cell. Yet it makes no difference with regard to which cell becomes which. It is the cytoplasm that determines which will be the ganglion cell and which will remain as the continuing neuroblast.

According to Wigglesworth a transforming or inducing factor is already present in the epidermal cells. A developing sensory hair or dermal gland is seen as taking up this hypothetical factor and so draining it away from the surrounding cells that they are thereby deprived of the factor and are inhibited. This could explain the original spacing but does not account for the interpolation of new centers as the epidermis grows.

A comparable view has been expressed by Curt Stern (1956) following an account of tissue interactions in genetic mosaics of bristle pattern in *Drosophila*: "The differentiations in mosaics show that the prepattern and the realized pattern of bristles do not stand in a fixed relation to each other. The prepattern which is the prerequisite for the formation of an individual bristle defines a peculiarity which extends over a certain region and has its 'peak' in a more restricted area. In this it resembles a morphogenetic field and its gradients. The realized pattern is a derived one. It depends on (1) the prepattern, (2) the genetic competence of the tissues and (3) the developmental interactions which are set in motion as soon as either differentiation of a bristle or lack of differentiation has been determined. If a bristle organ once begins to form it exerts a suppressing action on its surroundings while, if it fails to begin, differentiation may occur at unusual places.

"The nature of the prepattern remains unknown. There is evidence that the patterned differentiation of imaginal discs can occur independently of their eversion to form adult surface structures, although cases have been reported of apparent regulation after eversion. Likewise the evocation of the differentiation of a bristle apparatus may conceivably consist in as unspecific a process as localized stimulation of hypodermal cells to increase mass, an increase which may automatically entail the sequence of events which truly result in differentiation."[1]

---

[1] C. Stern, The Genetic Control of Developmental Competence and Morphogenetic Tissue Interactions in Genetic Mosaics, *Arch. Entwickl.-mech.,* **149**:22,24 (1956).

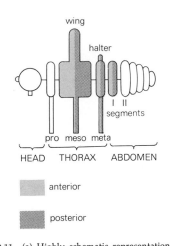

a

anterior

posterior

FIGURE 20.11　**(a)** Highly schematic representation of the pattern of body segments in the wild-type fly. **(b)** Chromosome map of the bithorax pseudo-allelic series, showing the location and names of five of the principal types of mutant genes and the body segmentation pattern associated with each. [*After Lewis, 1963.*]

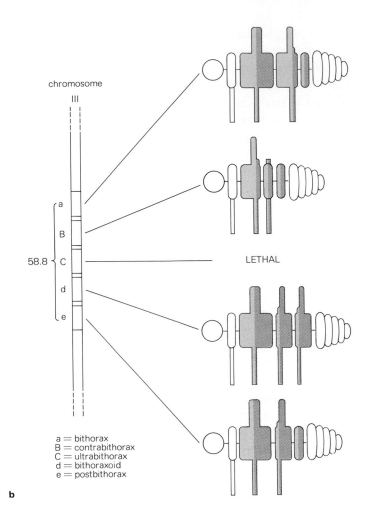

a = bithorax
B = contrabithorax
C = ultrabithorax
d = bithoraxoid
e = postbithorax

b

GENETIC CONTROL OF DEVELOPMENTAL PATHWAYS

The thorax of all insects consists of three segments, the prothorax, mesothorax, and metathorax. The prothorax is always wingless. In fully winged types, both the mesothorax and metathorax develop wings. In dipterans, only the mesothorax develops functional wings, the metathorax normally developing a pair of halteres with gyroscopic function but clearly homologous with wings. In certain mutants of *Drosophila,* the bithorax series, wings form on the metathorax in place of halteres, which might be regarded as a reversion to an ancestral type since the halteres of dipteran insects are without doubt specialized forms of once-functional wings. In certain mutants in the series, the first abdominal segment, normally wingless and legless, shows trans-

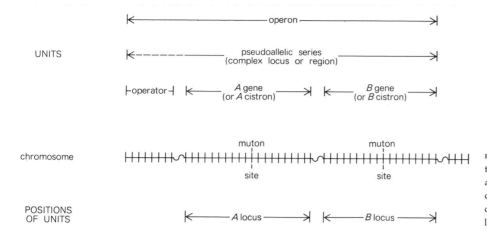

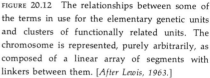

FIGURE 20.12 The relationships between some of the terms in use for the elementary genetic units and clusters of functionally related units. The chromosome is represented, purely arbitrarily, as composed of a linear array of segments with linkers between them. [*After Lewis, 1963.*]

formation toward the thoracic type and develops rudimentary to fully formed legs together with small, partially winglike halteres on the dorsal side. These variations, with their known genetic basis, offer exceptional material for cytogenetic analysis of control of differentiation at a high level of organization, an analysis notably undertaken by Lewis, employing the concept that cis-trans position has an effect within functional genes, particularly with regard to efficiency in synthesis.

The positions of genes relative to one another activate a specific gene at earlier or later times during development. This could be a very important form of control. Influence of functions of neighboring genes on one another also requires that they be not too far from one another in the genome. Lewis has worked with a series of pseudoallelic genes in *Drosophila*, involving intragenic recombination and crossing-over. In his own words, there is a stable-position effect which seems to depend on the presence here and there in the chromosomes of clusters of functionally interrelated genes or pseudoalleles. One way of detecting the stable type of position effect is to compare cis and trans heterozygotes for two such closely linked genes, say *a* and *b*.

In the cis form, *ab/* + +, the mutant genes are together in the same chromosome and their wild-type alleles are in the opposite chromosome.

In the trans heterozygote, *a* + */* + *b*, the mutant genes are in opposite chromosomes, as are also the wild-type alleles.

A cis-trans position effect is said to occur if the two heterozygotes differ phenotypically. In general, the cis form is the wild type, or nearly so, whereas the trans form is mutant in phenotype.

Two or more mutant genes which exhibit a cis-trans position effect are sometimes said to belong to the same functional unit, or cistron. Thus the two recessive mutants, *a* and *b*, are said to belong to the same cistron if the trans form of the

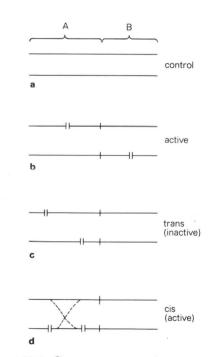

FIGURE 20.13 Cis-trans arrangement of mutations. (**a**) Heterozygote with two intact genomes. (**b**) Heterozygote with two mutations located in different cistrons *A* and *B*. (**c**) Heterozygote with two mutations located in the same cistron *A*. (**d**) Result of the cross-over that took place in (**c**). [*After Staehelin*, Biochem. An. Develop., I:463 (1965).]

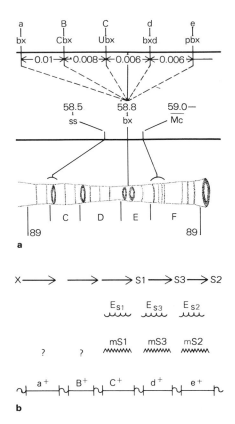

a

b

FIGURE 20.14 (a) Correspondences between the linkage map and salivary gland chromosome map of the bithorax region of the third chromosome. (b) Features of the sequential reaction and of the operon models as applied to the bithorax pseudoallelic series. The function of the wild-type alleles of the a and B genes is left undefined, as discussed in the text. B^+ may correspond to the operator region (see page 17); or the a^+ and B^+ genes may be elaborating enzymes involved in steps which precede the production of the hypothetical bithorax substances (S1). It is assumed that the C^+, d^+, and e^+ genes, at least, produce specific messenger RNA's (mS1, etc.), which in turn specify the sequence of amino acids in enzymes (E_{S1}, etc.). On the sequential reaction (but not the operon) model it is also necessary to assume that enzyme synthesis occurs in the vicinity of the chromosome. [After Lewis, 1963.]

heterozygote, or $a + / + b$, has a mutant phenotype. Since in practice the double mutant ab must be derived by recombination, the cis-trans test is intimately tied to a recombination test.

In *Drosophila* the bithorax pseudoallelic series comprises five groups of phenotypically distinguishable mutants, each group having a separate locus on the chromosome. These have been designated a_1, B_1, C_1, d, and e, corresponding to their order in the chromosome, the capital letters signifying dominant mutants. On the other hand, the various bithorax mutant phenotypes may be specified in terms of transformations of anterior and posterior portions of certain body segments into structures resembling the corresponding parts of homologous segments. The affected segments are the mesothorax, the metathorax, and the first abdominal segment. The line of separation between anterior and posterior portions is well defined for wing and thorax mutant effects, less well defined for halteres and legs, and not defined in the abdominal segment.

Results of recombination experiments are summarized by Lewis as follows:

1 The a mutant causes the *anterior portion* of the metathorax and its appendages to be transformed into structures closely resembling the mesothorax (i.e., type 1 transformation). The degree of transformation depends on the particular a allele present (four being known), and an extreme form of this gene (bx^3) effects almost complete transformation.

2 The e mutant is complementary to the a mutant in the sense that it causes the *posterior portion* of the metathorax to resemble the corresponding portion of the mesothorax (i.e., type 2 transformation).

3 The d mutant has the type 2 effect *plus* a thoraciclike modification of AB_1 in varying degree, depending on which of three mutant alleles is involved (i.e., type 3 transformation).

4 The B mutant (of x-ray origin) is dominant and viable when homozygous, and causes a metathoraciclike modification of the posterior mesothorax (i.e., type 4 transformation).

5 The C mutant, dominant phenotype in heterozygote but lethal in homozygote, causes enlargement of the distal segment of the halteres (i.e., type 5 transformation).

These results indicate that although the functioning of all the genes of a cluster is closely coordinated, at least three (type 1, type 2, type 3) mutant interactions vary more or less independently of one another, while gene-dosage studies show fairly conclusively that each of these three mutant-induced transformations is the result of a loss of gene function. A developmental abnormality associated with a mutant gene therefore presumably arises because of a loss of some substance that is elaborated by the wild-type allele during the course of normal development.

According to Lewis, these position effects can be interpreted in a reasonably consistent way on the basis of either a sequential reaction model involving enzymatic reactions occurring in proximity to the chromosome, or a model which uses the operon concept of Jacob and Monod, or some combination of the two. He suggests that possibly the bithorax genes are regulated by an inducible type of operonic system, and that the inducer in the bithorax case is a substance which diffuses with difficulty between cells, being at first uniformly distributed in the embryo. As development proceeds, a gradient in the inducer arises, perhaps as a result of an anteroposterior gradient in mitotic division rates. "The bithorax genes evidently exploit and amplify this gradient by producing a whole set of new substances that repress certain systems of cellular differentiation and thereby allow other systems to come into play. Possibly in this way the bithorax genes control the pathway of development of certain major body segments of the fly."[1] Specific alleles, however, may either change the extent or type of a field (or a gradient) or they may change the reactivity of cells to an existing unchanged field.

Gradient systems undoubtedly exist and probably in multiple form such as single gradients, double or opposing gradients, and stimulator-inhibitor gradients. They may play a very vital role, perhaps as the most important general agency, in the development of multicellular organisms, or we may be asking too much of too little. Nevertheless, they represent a useful and suggestive working hypothesis, and the gradient concept and that of genic regulation may in combination be most fertile.

[1] E. B. Lewis, Genetic Control and Regulation of Developmental Pathways, in M. Locke (ed.), "The Role of Chromosomes in Development," 23d Symposium, Society of Developmental Biology, Academic, 1964.

CONCEPTS

The early development of the insect egg involves an interaction between an activation center and a differentiation center.

The primary organization first appears as a regionally patterned chemodetermination of the cortical plasma. It takes place before zygote division in determinate (mosaic) eggs but after cortical nucleation in regulative insect eggs.

The developing insect egg is a dual system consisting (1) of a patterned distribution of cell groups or imaginal discs representing the adult organization, which remain inhibited and undifferentiated until the onset of metamorphosis; and (2) of a precociously differentiating residuum which gives rise to the larva.

The developmentally inhibited discs are initially determined according to their location in the embryo system as a whole, each being the presumptive rudiment for a part specific to its location.

Discs, apart from location specificity, are at first equipotential systems with regard to their respective determinations. In this they resemble amphibian limb buds and limb blastemas.

Regional determination within a disc occurs before disc activation, while, subsequently, districts give rise to subdistricts in a progressive or epigenetic manner.

Prepatterns represent all the real chemical and physical factors that are distributed in a nonrandom fashion in a tissue and account for the nonrandom local initiation of differentiation.

Imaginal discs from different body segments carry identical prepatterns, the determination or specific competence of their cells being responsible for segment-specific differentiation.

Aggregations of randomly mixed wild-type and mutant disc cells produce harmoniously developed structures characteristic of the particular disc type.

Disc cells from different disc types do not collaborate but instead form separate aggregations that develop their respective types of structure.

In determinate-type insect eggs, discs are essentially determined with regard to their prospective fate at about 3 hours after the onset of development, when each consists of a small cell population.

Predetermined disc tissue maintained as continually growing but undifferentiated "in vivo cultures," during a long period of time, may lose the initial type of determination and transform into virtually any other type, a process known as *transdetermination.*

Specific differentiation of cells, leading to diversification among adjoining cells, is determined by supracellular or extracellular factors.

Mutant genes, such as bithorax, exploit and amplify an inducer gradient by producing new substances that repress certain systems of cellular differentiation and thereby allow other systems to come into play.

Specific alleles may either change the extent or type of a field or gradient or change the reactivity of cells to respond to an existing unchanged field.

READINGS

BAKER, W. K., 1967. A Clonal System of Differential Gene Activity in *Drosophila, Develop. Biol.,* **16**:1–17.

——, 1963. Genetic Control of Pigment Differentiation in Somatic Cells, *Amer. Zool.,* **3**:57–69.

BECKER, H. J., 1966. Genetics and Variegation Mosaics in the Eye of *Drosophila,* in A. Moscona and A. Monroy (eds.), "Current Topics in Developmental Biology," Academic.

BEERMANN, W., 1966. "Cell Differentiation and Morphogenesis," North-Holland Publishing Company, Amsterdam.

BODENSTEIN, D., 1955. Progressive Differentiation: Insects, in B. Willier, P. Weiss, and V. Hamburger (eds.), "Analysis of Development," Saunders.

——, 1953. Embryonic Development, Postembryonic Development, in K. D. Roeder (ed.), "Insect Physiology," Wiley.

BRYANT, P. J., and H. A. SCHNEIDERMAN, 1969. Cell Lineage, Growth, and Determination in the Imaginal Leg Discs of *Drosophila melanogaster, Develop. Biol.,* **20**:263–290.

CARLSON, J. G., 1952. Microdissection Studies of the Dividing Neuroblast of the Grasshopper, *Chortophaga viridifasciata, Chromosoma,* **5**:199–220.

CASPARI, E., 1941. The Morphology and Development of the Wing Pattern of Lepidoptera, *Quart. Rev. Biol.,* **16**:249–273.

GEHRING, W., 1967. Clonal Analysis of Determination Dynamics in Cultures of Imaginal Discs in *Drosophila melanogaster, Develop. Biol.,* **16**:438–456.

HADORN, E., 1968. Transdetermination in Cells, *Sci. Amer.,* **219**:110–123.

——, 1966. Dynamics of Determination, in M. Locke (ed.), "Major Problems in Developmental Biology," 25th Symposium, Society of Developmental Biology, Academic.

——, 1965. Problems of Determination and Transdetermination, in "Genetic Control of Differentiation," Brookhaven National Laboratory, Symposium in Biology, no. 18, pp. 148–161.

KROEGER, H., 1968. Gene Activities during Insect Metamorphosis and Their Control by Hormones, in W. Etkin and G. I. Gilbert (eds.) "Metamorphosis," Appleton-Century-Crofts.

LEWIS, E. B., 1964. Genetic Control and Regulation of Developmental Pathways, in M. Locke (ed.), "The Role of Chromosomes in Development," 23d Symposium, Society of Developmental Biology, Academic.

——, 1963. Genes and Developmental Pathways, *Amer. Zool.,* **3**:53–56.

LOCKE, M., 1967. The Developmental Pattern in the Integument of Insects, *Advance. Morphogenesis,* **6**:33–88.

SCHULTZ, J., 1965. Genes, Differentiation, and Animal Development, in "Genetic Control of Differentiation," Brookhaven National Symposium in Biology, no. 18, pp. 116–147.

SONDHI, K. C., 1963. The Biological Foundations of Animal Patterns, *Quart. Rev. Biol.,* **38**:289–327.

STERN, C., 1956. The Genetic Control of Developmental Competence and Morphogenetic Tissue Interactions in Genetic Mosaics, *Arch. Entwickl.-mech.,* **149**:1–26.

——, 1954. Two or Three Bristles, *Amer. Sci.,* **42**:213–247.

URSPRUNG, H., 1967. *In vivo* Culture of *Drosophila* Imaginal Discs, in F. W. Wilt and N. K. Wessells (eds.), "Methods of Developmental Biology," Crowell.

——, 1966. The Formation of Patterns in Development, in M. Locke (ed.), "Major Problems in Developmental Biology," 25th Symposium, Society of Developmental Biology, Academic.

——, 1963. Development and Genetics of Pattern, *Amer. Zool.,* **3**:71–86.

WADDINGTON, C. H., 1962. "New Patterns in Genetics and Development," Columbia University Press.

WIGGLESWORTH, V. B., 1966. Hormonal Regulation of Differentiation in Insects, in W. Beermann (ed.), "Cell Differentiation and Morphogenesis," North-Holland Publishing Company, Amsterdam.

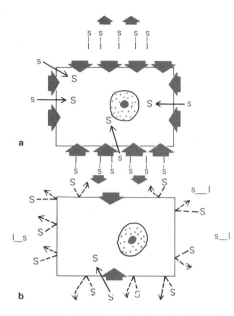

FIGURE 21.1 Model suggesting how stimulator and inhibitor may act at the cellular level. The stimulator (large arrows) becomes associated with the cell membrane and alters permeability. The inhibitor (I) attaches to a substrate (S) in intercellular spaces. When the inhibitor leaks to the culture medium, it carries the attached substrate with it. In (a) the stimulator and inhibitor are concentrated around a cell. Although substrate is withdrawn in large amounts by the inhibitor, the cell is still an efficient "gatherer" of substrate because of the large amount of stimulator present. Thus, the substrate enters the cell. In (b), although substrate is available, little of it can enter the cell because of scarcity of stimulator. Thus the stimulator inhibitor ratio determines the amount of substrate which will enter the cell. [*After Burnett*, Amer. Natur., **100**:165–190 (1966).]

the bulk of the cistrons, would consist of general-cell-type set determiners, with subsidiary sets for further switches, while another level would be concerned with supracellular or intercellular determiners, over and above the first two.

The differentiation process as seen in the pattern of accumulation of specific proteins in the developing mouse and rat pancreas has been intensively investigated by Rutter, Wessells, and others, with regard to both exocrine (digestive enzyme-secreting) and endocrine (insulin-secreting) cells. They recognize four levels of differentiation, namely, the undifferentiated state, the protodifferentiated state, the differentiated state, and modulation.

1 The primary regulatory event consists of the conversion of the undifferentiated cell to a cell with essentially pancreatic character. At this stage specific pancreatic proteins are present at detectable levels, even before the onset of histogenesis, suggesting that this event may "uncover" all the genes required for complete pancreatic differentiation. Such an event could involve major changes at the genetic level, with profound developmental consequences.

2 The secondary regulatory event involves conversion from the protodifferentiated state to the differentiated state. Studies of cell cultures show that embryonic mesenchyme or tissue extract is needed for about 2 days before this event, if it is to take place. During this period cells undergo extensive proliferation, and then proliferation ceases in those cells which are undergoing differentiation. Specific messenger RNA must already be available, since specific proteins were present in the preceding stage. The synthesis of ribosomal RNA, the assembly of functioning ribosomes, or the development of the endoplasmic reticulum could initiate the secondary regulatory event. Rutter suggests that, in fact, the strong linkage between cessation of cell proliferation, development of intracellular structure, and increased synthesis of protein could have a common cause.

3 The tertiary regulatory event is believed to be a mechanism distinct from the others, although it is not clear whether the observed alterations in specific enzyme activity among the differentiated cells (i.e., changes in relative protein concentration) reflect changes in the levels of the production of specific messenger RNA in the cell (transcription) or changes in the efficiency of ribosomal function in protein synthesis (translation). These are changes that occur late in development and may be regarded as *modulations* of the differentiated state in response to extracellular factors such as metabolites or hormones.

Experiments employing actinomycin D show that the synthesis of various specific proteins is inhibited to different degrees, indicating that the messenger RNA's that encode the various proteins are synthesized at slightly different times and not as a single set. Other experiments suggest that some event that occurs only in dividing cells or is facilitated by DNA synthesis must precede the changes in RNA synthesis that bring about the new pattern of protein synthesis. The most striking feature of the differentiating pancreatic cells is that many complex regulatory events are apparently coordinated at a few distinct times. Hundreds of genes must be in-

CHAPTER TWENTY-ONE

CYTODIFFERENTIATION

The phenomenon of cell differentiation confronts developmental biologists with two major but closely related problems: (1) how a cell of a higher organism acquires its structural and functional features as it develops, and (2) how the various cell types become different from one another during development. The molecular viewpoint is that if polypeptide sequences are made at the right time and in the right amounts, the developing organization will take care of itself. Processes of self-assembly create progressively ordered structure and associated function out of the building blocks available at a given moment, alterations being possible at any time. However this may be, we have a general problem and a specific problem. All cells that are not caught up in the treadmill cycle of one cell division after another are likely to undergo differentiation in some degree, which may be reversible. Most cells in any multicellular organism, however, proceed along one path of specialized differentiation or another. A muscle cell has contractile proteins, red blood cells form hemoglobin, plasma cells make antibodies, lens cells make crystalline lens proteins, etc. In each case these proteins are cell-specific and may be many thousand times higher in concentration than in other cells of the adult organism.

CONTENTS

Levels of differentiation
Differentiation of the lens
Tissue maintenance and replacement
Stem cells
Microenvironments and mass effects
Cell fusion and hybrid enzymes
Genotype-phenotype relationship
Concepts
Readings

LEVELS OF DIFFERENTIATION

Enzymes and specific proteins in differentiating systems fall into three categories:
1 Fundamental structures and metabolic pathways that exist in all cells
2 Secondary metabolic pathways found in the cells of a number of tissues
3 Characteristic cellular function which occurs in single, specific cell types
Since the relative concentration of proteins in all categories changes during differentiation, the overall process must therefore involve the activation or modulation of the activity of many genes.

An outstanding feature of cell differentiation in a multicellular organism, whether a hydra or a mammal, is that there is no continuous spectrum of types ranging from one type through fine gradations to an alternate type, except in the sense of an undifferentiated or partially differentiated cell becoming fully differentiated. Each cell type falls into a clearly recognizable discrete category, sharply set off from other cell types. This seems to imply, on a genetic level, that as one set of genes is turned on, another set in some linked way is turned off. In these terms, sets of genes of the first order determine the general cell category, such as the mesenchyme family; and sets of genes of the second order determine the cell type within the general category. The essential part of the genome (DNA) would relate to the cell as a cell, i.e., primary and universal. A large part of the genome, perhaps

volved in the two major differentiation phases. These genes do not change as a coordinated set during the transition from one stage of differentiation to the next, but rather as a number of small sets regulated together and more or less synchronized during a certain period of development.

For differentiation of cells to occur in culture, at least in the case of pancreas cells, growth and the presence of embryo extract or mesoderm are both necessary, although it is not known whether the extract and mesoderm participate actively in cytodifferentiation or act by their ability to support mitosis. Mitosis is clearly important to growth; yet some of the division cycles may be a prerequisite for the differentiation itself. Two examples in adult systems are cited by Wessells:

1 Cell division seems to be a necessary step before mammary cells secrete casein in response to hormone stimulation.

2 Immunologically competent cells, after being challenged by an antigen, divide before antibody production begins.

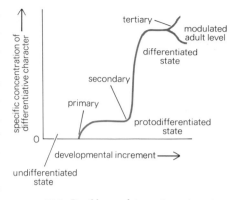

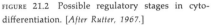

FIGURE 21.2 Possible regulatory stages in cyto-differentiation. [*After Rutter, 1967.*]

DIFFERENTIATION OF THE LENS

The early appearance of proteins in embryonic development can be detected by electrophoresis and by immunological techniques. The latter, however, can detect only antigenic determinants on a protein molecule, and it is possible that positive immunological reactions are obtained from antigenic determinants or protein subunits at a stage in synthesis of a protein before final conjugation and tertiary folding have been completed.

In chick embryonic lens differentiation, the first adult lens antigen appears at the lens-placode stage, and its appearance is followed by that of three other antigens during the formation of the lens vesicle and the development of the first lens fibers, while a sixth appears 10 days after incubation and a seventh at hatching. The first lens antigen, representing the first chemically detectable onset of lens differentiation, therefore appears shortly after contact between the optic vesicle and the presumptive-lens ectoderm. It appears before the ectodermal cells show any loss of vacuolization, nuclear orientation toward the vesicle, or palisade appearance. A unique structural protein, called *delta crystallin*, is the first protein to appear during lens development. Later, it can no longer be detected in the layer of epithelial cells, but synthesis of alpha crystallin begins in the lens-fiber cells and gradually spreads; synthesis of beta crystallins begins even later. Therefore there seems to be a continuous regulation of synthesis of the special lens proteins during the growth and differentiation of cells in the lens, involving regulation of DNA, all classes of RNA, and protein synthesis.

Morphologically the mature lens is an avascular tissue composed of an outer, single layer of epithelial cells; a zone of cellular elongation, or equatorial region, composed of cells which are in the process of developing into fiber cells; and the inner fiber cells. After the embryonic lens has been formed, fiber cells are con-

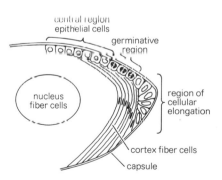

FIGURE 21.3 The lens of the adult vertebrate. The lens is surrounded by an external noncellular capsule. Beneath the capsule are the lens epithelial cells. In the peripheral area is the transitional region of cellular elongation, where the epithelial cells begin to elongate into fiber cells: The fiber cells that are newly laid down constitute the cortex region; those laid down during the early growth period of the lens compose the nucleus region of the adult lens. [*After Papaconstantinou, 1967.*]

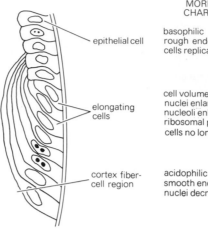

	MORPHOLOGICAL CHARACTERISTICS	BIOCHEMICAL CHARACTERISTICS

epithelial cell	basophilic
	rough endoplasmic reticulum
	cells replicate
		α, β-crystallin synthesis
		    inhibited by actinomycin
		oxidative metabolism
		    is efficient
		calf: LDH-5 > LDH-1
		adult: LDH-1 > LDH-5

elongating cells	cell volume increases
	nuclei enlarge
	nucleoli enlarge
	ribosomal population increases
	cells no longer replicate
		initiation of γ-crystallin
		    synthesis
		α, β, γ-crystallin synthesis
		    inhibited by actinomycin
		transition from LDH-5
		    to LDH-1 enhanced

cortex fiber-cell region	acidophilic smooth endoplasmic reticulum nuclei decrease in size	ribosomes break down mRNA for crystallins is stabilized DNA is metabolically inactive actinomycin stimulates crystallin synthesis LDH-1 > LDH-5 active aerobic glycolysis

FIGURE 21.4 The region of cellular elongation in the vertebrate lens. The major morphological and biochemical characteristics associated with lens-cell differentiation are listed and are discussed in detail in the text. [After *Papaconstantinou, 1967.*]

tinuously laid down throughout life in the zone of cellular elongation, where cuboidal epithelial cells continually transform to the elongated fiber cell characterized by gamma-crystallin protein. Finally, since the fiber cell loses its replicative activity, it essentially enters a permanent stationary phase and can only proceed to death. In contrast to the fiber cells the epithelial cells of the central region retain their ability to replicate and are in a reversible stationary phase.

It seems, therefore, that the potential for synthesizing gamma crystallins is inherent in the genome of the cell; yet this part of the genome is nonfunctional in the epithelial cell. Can these genes be activated without bringing about a simultaneous cellular elongation, nuclear inactivation and loss of cellular replication, stabilization of messenger RNA, and breakdown of the ribosomes, all of which occur during the normal differentiation process?

During regeneration of the lens from the iris epithelium following lens removal, cells become activated for protein synthesis even before depigmentation of the dorsal iris is completed. During the depigmentation phase, the ribosomes in the cytoplasm are mostly single. Later, when the cells are multiplying, the ribosomes cluster as polysomes of various sizes. Fiber differentiation begins when the clusters are middle-sized. In older fiber cells the clusters are fewer again. Gamma crystallin appears only during the terminal phase of the cell cycle, after the cells have stopped dividing, and there is some evidence that gamma crystallins may be synthesized in the nucleus as well as in the cytoplasm.

The situation in the lens is typical of that in many other tissues in the body. Each kind of tissue represents a cell population of a certain sort. Each kind must in some way be maintained throughout the life of the organism, and, in fact, aging and death of the organism as a whole may be essentailly the progressive, inevitable, col-

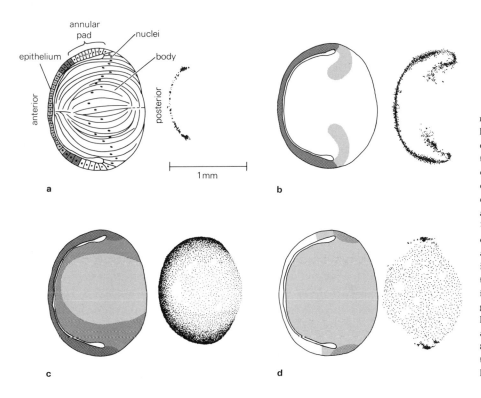

FIGURE 21.5 Autoradiographs of 12-day chick lenses. The diagrammatic sketches on the left of each autoradiograph represent histological sections showing the parts of the lens. The body consists of a central core of cells surrounded by a cortex. The intensities of stippling over the parts correspond to the distribution of grains in the autoradiograms. (a) Pattern of incorporation of ^{14}C-thymidine. Radioactivity is restricted to the epithelium and is concentrated in a ring of cells anterior to the annular pad. (b) ^{14}C-Uridine is incorporated by cells in the epithelium and by those in the cortex of the body, but not by cells in the core. (c) ^{14}C-Leucine is incorporated in greatest degree by cells of the epithelium, to a lesser extent by cells of the cortex of the body, and least by cells of the core. (d) ^{14}C-Leucine after 8 hours in actinomycin D is incorporated uniformly throughout the body but not at all in the epithelium. [*After Reeder and Bell,* Science, **150**:71 (1965).]

lective failing of the tissue-maintenance mechanisms. In the lens new fiber cells continually form from the peripheral equatorial zone of elongating epithelial cells, while these in turn derive from the adjoining germinative region of the lens epithelium. Thus fiber cells are systematically laid down, layer upon layer, throughout life. Eventually those fiber cells forming the central or nucleus fiber cells of the lens lose their cell nuclei and remain permanently entombed in the center of the lens. Inevitably the lens ages and dies at the center, although it is renewed at the edge. Its life course and destiny therefore depend on the relative rates of progression toward cell death and of peripheral cell birth, and this situation is complicated by the fact that the aging or dying cells in the center, or their equally aging protein products, cannot be sloughed off to make way for new cells or material.

The time for the cell cycle of the lens epithelial cells in newborn mice has been calculated to be 56 hours, the G_1 phase taking up three-quarters of this time. In the 3-day-old mouse most of the cells undergo DNA synthesis, but in the 12-day-old mouse only the cells of the germinative epithelium do so. The differentiation phase, however, is very long. The time taken for cells moving from the germinative region to differentiate into an elongated fiber cell is 6 months. In the chick the length of time required for cells of the lens ring, or annular pad, to develop into a fully differentiated fiber cell is 2 years.

TISSUE MAINTENANCE AND REPLACEMENT

Accordingly we may conclude, tentatively, that the *rate* of differentiation in a particular type of cell is itself under strict regulative control, so that the differentiated cell persists for a period in keeping with the special circumstances of its existence and the life span of the parental organism as a whole. The lens situation is a rather special case because of the spatial restrictions on the lens as a dioptic body. The cells of many other tissues, however, follow the path of differentiation to inevitable death. In almost all cases, therefore, the survival time of the differentiated or differentiating cell and the mechanism for replacement are vitally important to the maintenance of the tissue and the welfare of the organism.

In general, tissues fall into one of several categories:

1 Cells remain essentially undifferentiated and persist in the stationary, or G_0, phase until activated by growth or repair requirements, or by hormones. As such they may serve simple functions such as forming membrane, ciliated or not.

2 Cells may become large and differentiated in a general sense, with multiple function, for example, liver cells of vertebrates, the large epithelial cells of hydras, and most chlorophyll-containing plant cells. Such generally differentiated cells typically retain the capacity for mitotic cell division and therefore constitute tissues that are innately self-maintaining.

3 Cells enter the differentiating and aging pathway. Up to a certain stage along this course they may be turned back, by hormones for example, to the mitotic cycle. As a rule they proceed beyond the point of no return, have a life span much shorter than the organism to which they belong, and must be replaced. They therefore form a constituent part of tissues which typically consist of fully undifferentiated replacement cells, differentiating cells, fully differentiated cells, and dying cells. An outstanding example, and the subject of much investigation, is the red blood cell system, in which the death and replacement rates are extremely high. This is true also of the epithelium of the intestinal villus.

4 Cells may become highly differentiated but have neither the capacity to undergo mitosis nor a replacement source, such as the nerve cells of the higher vertebrates. If they fail to persist in fully functional condition as long as the rest of the animal, senility ensues. In any case the organism possesses a waning population.

Each organ and every organ system has its own characteristic mixture of cells and tissues consisting of some or all of the above categories. The overall situation in the mammal (rat) has been investigated and reviewed by Leblond (1964). He classifies cell populations within the organism as static, expanding, or renewing.

Static populations are homogeneous groups of cells in which no mitotic activity can be detected and in which the DNA content remains constant. Neurons, for instance, come in this category, although other neural cells (spongioblasts) do proliferate but the mitotic frequency declines with time.

Expanding populations exhibit scattered mitoses sufficient to account for the observed increase in total DNA content, as in pancreatic exocrine tissue, kidney tubule, and thyroid follicle.

Renewing populations exhibit intensive cell proliferation, producing cells in numbers far exceeding those required to account for any increase in the DNA content. In these populations the high rate of cell production is balanced by cell loss. Cell loss may occur either by cell attrition (as in the digestive tract, pulmonary epithelium, and epidermis) or by cell emigration [as in the blood-forming tissue (hematopoietic), reticuloendothelial system, and seminiferous tubules]. In the renewing populations, mitotic activity is restricted to a stem-cell line frequently set apart from the specialized cells of the tissue.

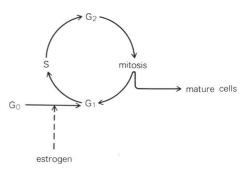

FIGURE 21.6 Hormonal stimulation of differentiation: mitosis precedes terminal differentiation process. During diestrous, most of the cells are in what is called a G_0 phase—i.e., they can synthesize DNA and subsequently undergo mitosis, but will not do so until the appropriate output of ovarian estrogen is attained. When this level of estrogen is reached, there is a transition to a presynthetic or postmitotic interphase (G_1), that is, a transition to cells that will synthesize DNA (S phase) in the near future. These cells then enter the S phase and subsequently divide.

SOME EFFECTS OF ESTROGEN ON UTERINE CELLS*

Increase in total cell protein

Increase in transport of amino acids into cell

Increase in protein synthesis activity per unit amount of polyribosomes

Increased synthesis of new ribosomes

Alteration of amounts of nuclear protein to nucleus

Increased amount of polyribosomes per cell

Increase in nucleolar mass and number

Increase in activity of two RNA polymerases

Increase in synthesis of contractile proteins

Imbibition of water

Increased synthesis of many phospholipids

Increased *de novo* synthesis of purines (dependent on new enzyme synthesis)

Alteration in membrane excitability

Alteration in glucose metabolism

Increase in synthesis of various mucopolysaccharides

* After Britten and Davidson, 1969.

STEM CELLS

Stem cells are typically "undifferentiated" cells that serve as a source for specialized, nonproliferative tissue throughout the body. There are many kinds of stem cells, and although a stem cell may give rise to multiple derivative types, a particular stem cell, even though undifferentiated, generally belongs to a certain basic category. In other words, some degree of commitment has already been made. Blood-forming tissue is a good example. This has been studied mainly in the chick, where the red blood cells, or erythrocytes, retain their nucleus throughout cell life, and in the mammal, where the erythrocyte nucleus degenerates and disappears during cell

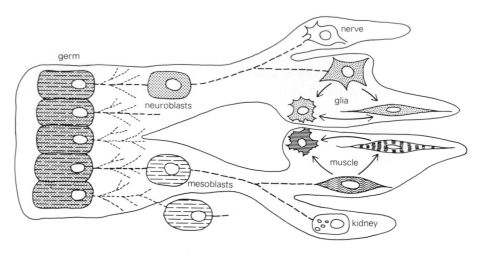

FIGURE 21.7 Differentiation of cells in steps, from unspecialized embryonic cells through stem cells such as neuroblasts and mesoblasts to differentiated nerve, glia, muscle, and kidney cells. [*After Weiss, 1953.*]

maturation. Bone marrow, at least in the developed organism, is the main blood-forming, or erythropoietic, tissue.

In the bone marrow an undifferentiated stem cell receives a stimulus which directs it toward the erythrocytic series. This may be called *determination,* for it is an all-or-none phenomenon. The stem cell, which is but one constituent in a highly mixed cell population, is thought to be a small, round cell commonly found in several organs. *Erythropoietin* is the name given to the determiner substance, known to be a protein. Once determined, the stem cell becomes a proerythroblast, following which a series of steps occurs known as *maturation.* A similar sequence was seen earlier in gametogenesis. In spermatogenesis and oogenesis the spermatogonia and oogonia, respectively, represent the stem cells, and a comparable series of steps is seen in the subsequent process of differentiation.

In red blood cells, when a cell leaves the compartment of the stem cells, which it does when it becomes a proerythroblast, it undergoes four cell divisions and six definite stages of morphological maturation. This period corresponds to the appearance of new proteins, of lipoproteins and nucleoproteins, of carbohydrates, and of enzymes relating to multiplication and growth on the one hand and to the synthesis of hemoglobin on the other.

Erythropoietin is a hormone, obtainable from plasma, which serves as a useful experimental tool. It can induce erythropoiesis in tissues which, although normally erythropoietic, have been totally suppressed with regard to red blood cell formation. According to Bessis (1967), it acts on stem cells by initiating transcription of previously repressed genetic loci, i.e., it derepresses. In vitro it causes rapid messenger RNA synthesis, hemoglobin synthesis, glucosamine incorporation, and cellular iron accumulation in a nonheme form. With regard to the accumulation of iron, there appear to be two possibilities:

1 Stimulation of iron accumulation by marrow cells, stimulated by erythropoietin, requires the prior synthesis of a protein, which may be needed to ensure an adequate intracellular supply of iron for heme synthesis.

2 Globin may play a key role in the regulation of hemoglobin synthesis. Heme is a feedback inhibitor of its own synthesis. Globin acts as a scavenger of heme and channels the whole system toward hemoglobin synthesis.

After 3 days, following the four cell divisions, the cytoplasm of the erythroblast has become filled with hemoglobin. It then expels its nucleus (in mammals) and enters the circulation as a reticulocyte. The reticulocyte ripens during the next 2 to 3 days and becomes an adult erythrocyte. In man the erythrocyte lives for about 120 days and is then phagocytosed. In both mammal and chick there is a stepwise reduction of transcriptional activity during differentiation, and even in the chick, where the erythrocyte nucleus is retained, there is the total arrest of RNA synthesis in the mature cell.

MICROENVIRONMENTS AND MASS EFFECTS

The basal layer of cells in the epidermis consists of actively proliferating cells which are cytologically and biochemically unspecialized. Individual cells of the basal layer lose their attachment to the basement membrane and appear to be forced up by the crowding into the more superficial layers of the epidermis. Only then do they begin to synthesize keratin.

The basal-layer cells are stem cells. Their differentiation into keratinizing cells is clearly a response to the change in their environmental circumstances, i.e., they become separated by the remaining basal layer from their original contact with the basement membrane and the subepidermal chemical medium. At the same time, their fate is not immediately determined, since exposure of the cells to vitamin A results in differentiation of the keratinizing epidermis into a mucoid-secreting epithelium.

The cell microenvironment, therefore, has a great deal to do with the particular direction of differentiation taken by an unspecialized cell. Experiments with isolated, undifferentiated vertebrate pigment cells by Wilde (1961) are significant. In no case did a single isolated cell undergo differentiation in a minute hanging drop of nutrient culture medium. But when two cells are present together in a microdrop, one undergoes differentiation. Apparently there is an exchange of cues into the microenvironment, extending to the neighbor cell, so that one of the two cells is able to initiate differentiative processes.

Similar contrasting cell systems, differing only by one cell, responded in a variety of ways, but in each case the single isolated cell merely survived, while one cell of the double isolate differentiated. As the number of cells isolated together increases up to 15, the rate, degree, and numbers of cells differentiating also increase. It was concluded that an exchange of a metabolic nature takes place between cells and medium, and the altered medium is thus made more suitable for support of

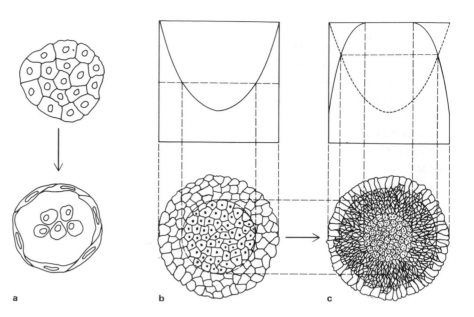

FIGURE 21.8   Gradient systems within a cell cluster. (a) An equipotential cell cluster segregates into an outer envelope and an inner residual cluster, the two components now having different microenvironments. (b) A resultant gradient of composite conditions relating to inward and outward diffusion of metabolites. (c) Changes in cell type and cell shape resulting from release of reactant products of cells responding to the initial diffusion gradients. [*After Weiss, 1953.*]

a   b   c

cellular differentiation in the isolated small groups of cells. Such cues may be electrolyte gradients, particular amino acids in varying amounts, or other substances, leaking through the cell membrane. The differentiation of the vertebrate pigment cell, for instance, might then be a response to nuclear clues at one period, to microenvironmental ones at another period, and at still another time to conversion through "infective differentiation."

A model system of this sort has been described by Weiss (1953), under the heading of "field effects," although in this case the cells of a cluster were in actual contact with one another. In this system the starting condition is a simple equipotential cell cluster, bordering either on cells of some other type or on some other medium. As a result of interaction along the free surface, a certain course of reactions will be activated in the cells that share in the surface; this entails the release from those cells of substances or activities that spread inward with a gradient from their source. Further, metabolites which are produced by the whole cell mass and which can diffuse into the outer medium will tend to be more concentrated in the center. We may then assume that the cells of the group can react to a particular constellation or combination of such factors by a differentiating step. If the threshold for this reaction lies at a certain level along the gradient, cells at a lower level remain undifferentiated, whereas in the central area all cells are switched to an altered course.

The differentiation of cartilage seems to be a good example of such a system. The first histological sign of chondrification in the precartilaginous mesenchyme of a developing embryo is the close apposition of rounding cells to form precartilaginous mesenchyme. The cell-mass effect and cell cohesion appear to be the most important

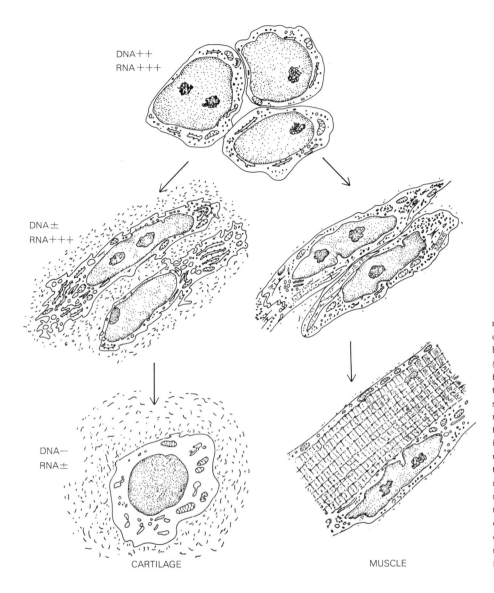

DNA++
RNA+++

DNA±
RNA+++

DNA—
RNA±

CARTILAGE

MUSCLE

FIGURE 21.9 Changes in fine structure of mesenchymal cells as they aggregate within the limb blastema *(top)* and differentiate into cartilage *(left)* and muscle *(right)*. In cartilage, the newly formed differentiated products are destined for the extracellular compartment. They are synthesized and secreted by the granular endoplasmic reticulum and Golgi complex. In muscle, the differentiated proteins (myofibrils) are intracellular in location. They seem to be synthesized by free ribosomes in the cytoplasm without the intervention of the granular endoplasmic reticulum. The muscle cell contains only a few profiles of granular reticulum and has a smaller Golgi apparatus than the cartilage cell. It later acquires a special kind of smooth-surfaced reticulum. Also depicted in the diagram are the relative amounts of DNA and RNA synthesized by the cells in the developmental stages illustrated. [*After Hay, 1965.*]

factors among the various recognized conditions for cartilage formation. Holtfreter (1968) reports that amphibian neural-crest cells in culture emigrate and disperse as a predominantly single layer of flattened cells. The majority differentiate into polymorphic mesenchymal cells, while others become pigment cells or neurons. But within this layer of spreading cells, matrix-encapsulated cartilage cells appear, this cytodifferentiation being associated with an aggregation and piling up of the cells into nodules, with cell numbers varying from about ten to several hundred.

Chondrogenesis has attracted much attention because the massive secretion and chemical complexity of cartilaginous matrix constitute a challenging and available target for biochemical investigation. The problem of elucidation, however, has been unexpectedly difficult. Under suitable nutritive conditions and given sufficient time, somite tissue will form cartilage without additional stimulation. Lash has shown, however, that somite tissue produces cartilage sooner and in greater amount if exposed to notochord or spinal cord, or to extracts thereof. The essential feature is the synthesis of chondroitin sulfate. In the 3-day embryos, various tissues besides cartilage are able to synthesize this substance, but in older embryos (10 days) this synthesis occurs mainly in cartilage, indicating that such synthesis becomes restricted during development. Lash suggests that when somites are exposed to various inducers in cultures, e.g., to notochord and spinal cord extracts, no new metabolic pattern is elicited but a preexisting pattern seems to be enhanced.

A similar opinion has been expressed by Holtzer, who suggests that differentiation of somite cells into chondrocytes involves a sequence of permissive events, whereby only covertly differentiated cells are capable of interpreting the message carried by an inducer. These data indicate that it is unlikely that a single chemical entity can trigger the complex process of cartilage formation, and it is probable that many stimuli act in a synchronized and concerted fashion. Once full commitment has been attained, however, it seems never to be truly lost. Dedifferentiated chondrocytes in cell culture may return to a functional state even after a long period of dedifferentiation.

CELL FUSION AND HYBRID ENZYMES

The development or differentiation of striated vertebrate muscle has also attracted much interest. In the fully differentiated state, striated muscle consists of large multinucleate myotubes containing long myofibrils lattice-spaced in the sarcoplasm. The first question concerning myogenesis is whether a myotube develops from a single myoblast whose nucleus undergoes successive divisions without involving cytoplasmic cleavage, or whether large numbers of myoblasts fuse together, each contributing its nucleus to the giant cell thus formed.

One approach to the problem, as in the investigation of other tissues, has been the study of embryonic muscle cells, or myoblasts, in culture. Such cultures show two distinct phases: during the first period the cells attach and spread and then multiply rapidly, forming a confluent sheet of cells within a few days. As confluency is reached, long ribbonlike cells appear which resemble myotubes. Konigsberg succeeded in deriving such a cell colony from a single isolated myoblast. Each such colony eventually contained large numbers of multinuclear, cross-striated muscle fibers similar to those observed in mass cultures and in embryonic muscle, and apparently formed by myoblast fusion. None of the myoblasts obtained from muscle-cell colonies will form clones, i.e., give rise to new single-cell strain colonies, unless a conditioned medium is provided. This conditioning appears to be primarily a function of fibro-

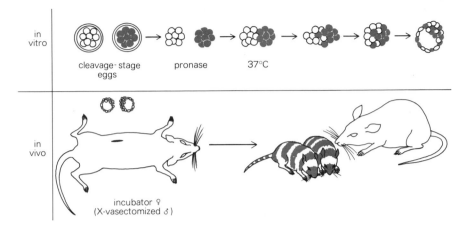

blasts secreting collagen. The latent uncertainty in this type of experiment is caused by doubt as to whether cell behavior is the same in the body as it is outside in the culture medium.

The conclusion that the multinucleate myotube in the body is produced by nuclear divisions and accompanying growth as a single cell, already accepted, has received confirmation by experimental procedures of a very different sort, employed by Mintz and involving the creation of a new kind of laboratory animal, the *mosaic mouse,* or multimouse. In this procedure, fertilized mouse eggs or cleavage-stage embryos are first deprived of their containing noncellular zona pellucida by digestion with pronase, and the blastomeres from two mouse embryos are then placed in contact in a siliconed dish. Without further ado the partners fuse completely in about 1 hour and rapidly round up to form a single morula. This morula or the blastocyst which it subsequently becomes is then implanted into a foster mother. When fusions are made at the eight-cell stage, almost 100 percent normal blastocysts develop, of quadruple parentage. To follow contributions made by cells from each egg, blastomeres from one member are prelabeled for autoradiographic analysis by incubation with tritiated thymidine before fusion. Eggs, or blastomeres, of unrelated mouse strains fuse easily and have yielded hundreds of mosaic embryos which have gone on to become healthy, breeding adults.

With regard to the muscle-tube situation, such mice offer a biochemical means of detecting cell fusions at the differentiated tissue level. If each myotube contains nuclei descended from an original mononucleate myoblast, all myotubes will be genetically alike. If the multinucleate condition results from the fusion of myoblasts, the myotubes of mosaic mice will contain nuclei from both parental strains. If the latter is the case, then hybrid enzymes should be found in addition to the two pure-strain isozymes.

Analysis shows that hybrid enzymes exist in the skeletal muscle, and not in any other tissues. Therefore the myotubes are formed in vivo by myoblast fusion.

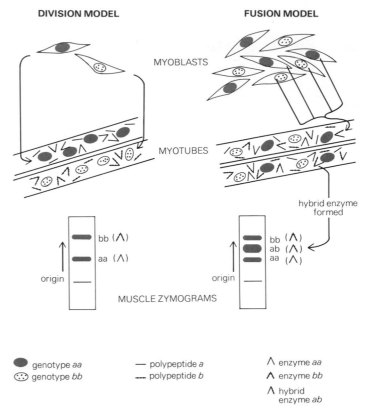

FIGURE 21.11 Expected isozyme results in allophenic mice, on the "division" versus "fusion" models of skeletal-muscle development. When homozygous cells of different NADP-isocitrate dehydrogenase genotypes coexist, heterocaryons would result in the event of myoblast fusion, and hybrid enzyme could be formed. Enzyme molecules are represented as dimers formed in the cytoplasm from polypeptide subunits. [*After Mintz and Baker, 1967.*]

Accordingly, the hybrid enzyme found in skeletal muscle must be produced within the multinucleated cell itself. All in vivo evidence points to retention of nuclear individuality in muscle. Without nuclear fusion, an intracellular hybrid enzyme would require genetic information from both kinds of nuclei. Mintz and Baker conclude that the hybrid enzyme in mosaic mice is made in a two-step sequence: first monomeric subunits or polypeptides of each of the pure types are separately synthesized on templates received from each nucleus, and then the two kinds of subunits are assembled in the cytoplasm into the final macromolecule. In other words, the genetic material codes independently for each polypeptide, not for the completed enzyme as a single unit.

GENOTYPE-PHENOTYPE RELATIONSHIP

Employing the same kind of mosaic, or allophenic, mice, Mintz has undertaken a series of projects relating to pigmentary patterns in the coats of various mouse strains. The distribution of melanocytes has long been recognized as an indicator of supracellular organization, as shown by the many studies on pigment-cell migrations and patterns in amphibian neural-crest experiments reported by Twitty (1949). An intact organism is necessary for experimental studies of the genotype-phenotype relationship, and the multimouse serves well in this connection. The differentiation problem is that a single genome produces quite distinctive subpopulations of pigment cells which are arranged in a geometry peculiar to the genotype. In brief, the results, or conclusions, are reported as follows:

1 Analyses of the mosaics have disclosed that all melanocytes in the coats are clonally derived from a fixed, small number of primordial melanoblasts. Genes which determine that these cells will be melanoblasts first become active in them at a definable time in early embryonic life. Gene activity remains stabilized through many successive cell generations, during which proliferation occurs in a similar manner in all clones, unless selection is made.

2 Many (possibly most) genotypes can cause the *primordial* melanoblast population in an individual to consist of two or more different cellular phenotypes, as a result of several possible kinds of genetic control mechanisms. When such diverse cells coexist, their primary distribution is not random with respect to one another but follows a given arrangement so that, after the cloning is complete, a reproducible color pattern is found. Alternative products of only one pigmentary locus appear sufficient to mediate cell recognitions and to dictate preferential association among early melanoblasts.

3 Selection plays a major role in normal development of all tissues and systematically leads to gene-specific modifications of clonal histories and, therefore, of the final, total, or organismic phenotype (e.g., color pattern).

Regulator genes are increasingly being invoked as the controlling agents of cell-differentiation processes and, by projection, as controlling agents of supracellular developmental events. With regard to production of specific organization of cell protein, for example, the protein molecule characteristic of the myeloma tumor, a single gene (the master) is assumed to control the constant part of the myeloma protein molecule, and five alternating genes (the slaves) are thought to control the variable parts of the molecule. It is also suggested, by Ephrussi, that there are two aspects of somatic-cell genetic analysis, namely, "caryotype" and "epigenotype," the latter encompassing problems of embryonic differentiation. He proposes a division into "household" and "luxury" functions, subject to different regulatory mechanisms.

However this may be, the basic problem remains, as Waddington has emphasized in commenting on the theory of gene regulation for higher cells recently proposed by Britten and Davidson: "We need a mechanism that accounts not only for gene activation or derepression in such instances as the puffing of particular salivary bands after treatment with ecdysone or a changed ionic medium; the synthesis of hemoglobin following erythropoietin; the development of a drosophila imaginal disc into adult structures after the action of pupation hormones; and so on. We also have to show what has happened previously to determine which particular bands will puff;

why erythropoietin stimulates hemoglobin synthesis in determined blood cells but not in other cells; and why the cells of eye imaginal bud develop into adult eye cells and those of other discs into other structures, even many generations after this determination occurred. This implies that we need a double action control mechanism, with one action concerned with determination and the second with activation."[1]

[1] C. H. Waddington, *Science*, **166**:639 (1969). Copyright 1969 by the American Association for the Advancement of Science.

CONCEPTS

Three levels of differentiation are evident in cell specialization, namely, fundamental (common to all cells), secondary (common to a number of cell types), and specific (characteristic of one type).

Overt differentiation follows the terminal division of cells.

Cell populations may be static (with neither cell loss nor cell replacement), expanding (with new cells accounting for growth), or renewing (with new cells continually replacing cells lost).

Stem cells, which are characteristic of renewing cell populations, are undifferentiated cells that become stimulated to take a particular pathway of differentiation.

Chemical agents act as determinants, either as hormones reaching a stem-cell population, or as components of a system established by cell clustering.

Diversity of differentiation is generally in direct ratio to cellular mass.

Simple external signals may serve to change the state of differentiation of higher cell types.

A given state of differentiation tends to require the integrated activation of a very large number of noncontiguous genes.

Repetitive sequences are transcribed in differentiated cells according to cell type-specific patterns.

READINGS

BESSIS, M., 1967. Morphology of the Different Stages of the Cells of the Erythrocyte Series in Mammals, in "Morphological and Biochemical Aspects of Cytodifferentiation," *Exp. Biol. Med.*, 1:220–243, Karger.

BRITTEN, R. J., and E. H. DAVIDSON, 1969. Gene Regulation for Higher Cells: A Theory, *Science*, **165**:349–357.

EPHRUSSI, B., and M. C. WEISS, 1967. Regulation of the Cell Cycle in Mammalian Cells: Inferences and Speculations Based on Observations of Interspecific Somatic Hybrids, *Develop. Biol.*, suppl. 1, 136–169.

HAY, E. D., 1965. Metabolic Patterns in Limb Development and Regeneration, in R. L. DeHaan and H. Ursprung (eds.), "Organogenesis," Holt, Rinehart and Winston.

HOLTFRETER, H., 1968. Mesenchyme and Epithelia in Inductive and Morphogenetic Processes, in R. Fleischmajer (ed.), "Epithelial-Mesenchymal Interactions," Williams & Wilkins.

HOLTZER, H., 1960. Aspects of Chondrogenesis and Myogenesis, in D. Rudnick (ed.), "Synthesis of Molecular and Cellular Structure," 19th Growth Symposium, Ronald.

—— and J. ABBOTT, 1968. Oscillations of the Chondrogenic Phenotype in Vitro, in H. Ursprung (ed.), "Stability of the Differentiated State," Springer.

KONIGSBERG, I. R., 1965. Aspects of Cytodifferentiation of Skeletal Muscle, in R. L. DeHaan and H. Ursprung (eds.), "Organogenesis," Holt, Rinehart and Winston.

LASH, J., 1968. Phenotypic Expression and Differentiation: In Vitro Chondrogenesis, in H. Ursprung (ed.), "Results and Problems in Cell Differentiation," Springer-Verlag.

—— 1963. Tissue Interactions and Specific Metabolic Response: Chondrogenic Induction and Differentiation, in M. Locke (ed.), "Cytodifferentiation and Macromolecular Synthesis," 22d Growth Symposium, Academic.

LAVIETES, B. B., 1970. Cellular Interaction and Chondrogenesis in Vitro, *Develop. Biol.*, **21**: 584–610.

LEBLOND, C. P., 1964. Classification of Cell Populations on the Basis of Their Proliferative Behavior, *Nat. Cancer Inst. Monogr.*, **14**:119–150.

—— and B. E. WALKER, 1956. Renewal of Cell Populations, *Physiol. Rev.*, **36**:255–276.

MARKERT, C. L., 1965. Mechanisms of Cell Differentiation, in J. A. Moore (ed.), "Ideas in Modern Biology," Natural History Press.

MINTZ, B., 1969. Differentiation *in Vivo* and *in Vitro,* in "In Vitro," Williams & Wilkins.

——, 1967. Gene Control of Mammalian Pigmentary Differentiation, I: Clonal Origin of Melanocytes, *Proc. Nat. Acad. Sci.*, **58**:344–355.

—— and W. D. BAKER, 1967. Normal Mammalian Muscle Differentiation and Gene Control of Isocitrate Dehydrogenase Synthesis; *Proc. Nat. Acad. Sci.*, **58**:592–598.

PAPACONSTANTINOU, J., 1967. Molecular Aspects of Lens Differentiation, *Science*, **156**:338–346.

RUTTER, W. J., W. D. BALL, W. S. BRADSHAW, W. R. CLARK, and T. G. SANDERS, 1967. Levels of Regulation in Cytodifferentiation, In "Morphological and Biochemical Aspects of Cytodifferentiation," *Exp. Biol. Med.*, **1**:110–124, Karger.

TWITTY, V. C., 1949. "The Developmental Analysis of Amphibian Pigmentation," 9th Symposium of Soc. Develop. and Growth, *Growth,* suppl. to vol. 8.

WEISS, P., 1953. Some Introductory Remarks on the Cellular Basis of Differentiation, *J. Embryol. Exp. Morphol.*, **1**:181–211.

WESSELLS, N. K., and J. W. COHEN, 1967. Early Pancreatic Organogenesis: Morphogenesis, Tissue Interactions, and Mass, *Develop. Biol.*, **6**:279–310.

—— and W. J. RUTTER, 1969. Phases in Cell Differentiation, *Sci. Amer.*, March.

WHITTAKER, J. R., 1968. Translational Competition as a Possible Basis of Modulation in Retinal Pigment Cell Cultures, *J. Exp. Zool.*, **169**:143–159.

WILDE, C. E., 1961. The Differentiation of Vertebrate Pigment Cells, *Advance. Morphogenesis*, **1**:267–300.

WILT, F. W., 1967. The Control of Embryonic Hemoglobin Synthesis, *Advance. Morphogenesis*, **6**:89–125.

ZALAK, S. E., and T. YAMADA, 1967. The Cell Cycle during Lens Regeneration, *J. Exp. Zool.*, **165**:385–393.

CHAPTER TWENTY-TWO

MALIGNANCY, DIFFERENTIATION, AND DEVELOPMENT

MALIGNANCY AND DIFFERENTIATION

The development of a multicellular organism consists of cell multiplication, growth, cytodifferentiation, and morphogenesis proceeding together to an organized end. To some extent each of these component processes is independent of the others and has been studied as such so far as possible. In this connection malignant tumors represent an exceptional opportunity to study some component processes in the absence of others. A malignant tumor is the result of unrestricted cell growth and cell multiplication, comparable to tissue culture in vitro but taking place within the body, without response to the normal growth-controlling agencies. Typically such a tumor consists of a single prevailing type of cell which may or may not retain obvious characteristics of the tissue from which it arose. In teratomas, a special class of malignant tumors, various cell types are evident and, compared with the process of development as a whole and apart from the quality of malignancy, only the phenomenon of morphogenesis appears to be lacking. Consequently, malignant growth may be seen as the inverse of organized development, and the study of one may throw light upon the other. In this final discussion, the problem of malignant growth, whether caused by a virus or by some other agent, is examined in some detail; this is followed by a brief account of certain special aspects of early vertebrate development.

Carcinogens and Cell Division

Any tissue in the body that is capable of cell division may become cancerous, with varying degrees of malignancy. The problems involved seem endless, although inquiry falls into a few main categories: the nature of the instigating agents, the changes produced within the cells, and the changed behavior of the cells with regard to normal tissues. The clinical picture does not concern us here. The general inquiry into the nature of the cancerous transformation, however, has much in common with the search for the "organizer" of the vertebrate embryo. In both cases the initial efforts concerned the nature of the causative agents and their common property. In both cases the effective agents were found to be numerous and to have no recognizably significant physical or chemical properties in common. In both cases, therefore, interest turns from the nature of the agents to the nature of the response and to whatever aspect of the cell machinery may be responsible for the response. The action of the inducing agents, however, remains vitally important. In spite of their

CONTENTS

Malignancy and differentiation
   Carcinogens and cell division
   The malignant transformation
   Chromosomes and centrioles
   Viral induction of tumors
   The cell membrane
   Growth-promoting factors
   Nonviral carcinogenesis
   Tumor progression
   Genetic basis of susceptibility
   Teratomatous development
Cells and development
   Cell transformations
   The labile state
Concepts
Readings

diversity, each in some way interferes with the precise controls of cell growth and multiplication. Yet it does not follow that the control system is affected in the same way by all, except insofar as all the agents may be disruptive.

Accordingly, at one end of a chain of events are the various carcinogenic agents. They include ionizing radiations that penetrate a cell, namely, ultraviolet radiation, x-rays, and the various radiations emitted from radioactive elements. They include a host of chemical compounds, particularly the polycyclic hydrocarbons, which belong to the same general class as the sex hormones and cholesterol, with dibenzanthracene, benzopyrene, and methylcholanthrene leading the way. And they include the viruses that invade and replicate within animal and plant cells and in many specific cases induce tumorous growth.

The remainder of the chain is the cell itself. Is the carcinogenic effect, whatever the agent, exerted at the DNA level, at the cell membrane, or on enzymatic processes between? The answer may vary according to the agent and the type and developmental age of the cell.

The Malignant Transformation

To begin with, what is the operative malignant transformation? It is essentially the behavior of cells with regard to growth, division, and movement, particularly in relation to normal cells.

FIGURE 22.1   Normal cells in tissue culture. (a) Cells tend to adhere to one another or, (b) when crowded, to form a mosaiclike arrangement referred to as *pavement*. [*Courtesy of R. Dulbecco.*]

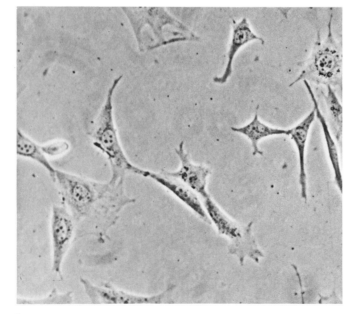

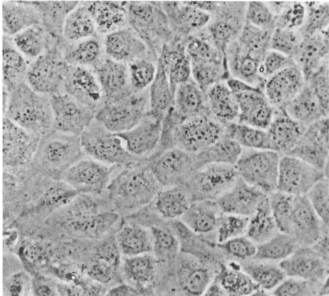

a

b

In normal tissue repair, as in a simple skin wound, fibroblasts secrete a new collagenous base to re-form the dermis. The epidermal cells then begin to close the wound. Within 12 hours after wounding, the epidermal cells nearest the wound lose their organized arrangement, become shapeless and develop ruffled borders much like those seen in actively moving cells in tissue culture, and travel quickly into the wound. When the leading edges of the sheets of rapidly proliferating epidermal cells meet, the edges fuse, all migratory activity ceases, and each cell regains its normal identity. As the thickness of this layer of cells is restored, the rate of division among the epidermal cells also returns to what it was before the injury. Contact inhibition of cell movement and of cell division is evident. Similarly in tissue cultures of normal epithelial cells, cells continue to divide until they establish contact with one another, at which point they stop dividing. In contrast to this process, epidermal cells, for instance, in the early stages of cancerous transformation begin to grow and divide relatively rapidly under the influence of agents such as ultraviolet light or dibenzanthracene, and eventually they penetrate the basement membrane to invade subjacent tissues as a truly malignant carcinoma, with no evidence of any contact inhibition of either cell division or movement.

Two main categories of tumors are recognized: (1) carcinomas which derive from epithelial tissues and generally retain at least some vestige of their epithelial character, although known as adenocarcinomas when the epithelial character is fully retained, and (2) sarcomas of connective-tissue origin, such as those derived from

FIGURE 22.2 Cells transformed by viruses generally overlap one another and form irregular patterns. [*Courtesy of R. Dulbecco.*]

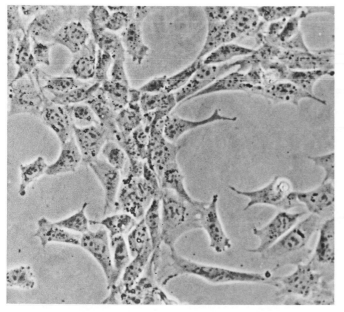

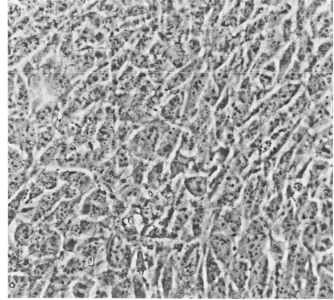

a   b

fibroblasts, chrondroblasts, etc., in fact, most mesodermal tissues that are not epithelial. Two other types are lymphomas, as in leukemia, in which the cells are dispersed, and teratomas, which exhibit manifold cell differentiations otherwise associated with embryonic development.

We have, therefore, two fundamental questions relating to the effect of the carcinogenic agents on the cell: What exactly is the nature of the change, in terms of the cell membrane, which results in the failure of the cell to respond in a normal manner to contact with other cells? How is this change brought about in terms of the cell as a whole?

The *capacity* for growth, division, and movement is much the same for normal and for malignant cells, at least in many cases. In tissue culture, normal cells when fully released from inhibitory factors, grow and divide at much the same rate as do malignant cells, as long as certain conditions are met. The difference is that in the body, cells grow at a maximal rate only in unusual circumstances and stop when normal tissue structure is restored, whereas malignant cells do so indefinitely, as though the body were only a culture medium serving as a nutritive support. In fact, most kinds of tumors in the body give off small clusters of cells from their periphery, as metastases, which circulate until they lodge in some other region, where secondary growths become established. The behavior of normal cells and that of malignant cells are two sides of the same coin, and any progress in understanding the one must throw light on the other.

The basic question is that of cause and effect. Do cancer cells multiply because they continue to grow? Do they continue to grow because their division mechanism runs out of control? Do they continue to grow and divide because their social behavior, mediated by the cell membrane, is radically changed so that external inhibitory agencies are no longer effective? Is abnormal membrane response the result of genic change, whether innate or virally produced? If so, is it at the translation or transcription level? Does change in the membrane affect the genome? Can change in membrane structure be self-perpetuating? And so on. The field is wide open, and were it not for the devastating effect of cancer as a fatal disease in human beings, we might enjoy the challenge at leisure for a long time to come. We may take the time in any case, for while we know much, we understand little. The theories are many. We have devised cancer theories involving chromosome aberration, somatic mutation, enzyme deletion, mitochondria, centrioles, anaerobic glycolysis, lysogeny, and others, on different occasions and depending on what was fashionable in biology at the time. The several aspects of the malignant cell and tissues, therefore, must be examined separately.

Chromosomes and Centrioles

It is a fact, nevertheless, that chromosomal abnormality is a common feature of many tumors, usually associated with abnormal mitoses. Tripolar and tetrapolar mitoses are frequent in many kinds of tumors, with inevitable irregular distribution of the two sets of chromosomes among the three or four progeny. As a rule divisions of this sort are associated with the deeper parts of a tumor and may be regarded as a product of an unhealthy internal environment. All such cells are probably capable of few further divisions at the most and are on the path to death. Vigorously growing, peripheral cancer cells typically divide by regular bipolar mitosis.

Another possibility is that centriolar misbehavior may be a characteristic, if not a causal agent, of some tumors. Among nondividing somatic animal cells, the two centrioles of a single cell are close to each other and are regularly disposed so that their long axes form a right angle. Schafer (1969) asks whether the constant geometry of centriolar ultrastructure and the constant orthogonal relationship of the two centrioles in a single cell is part of a larger intercellular pattern of spatial order among these tiny organelles. The spatial distribution of centrioles has been examined in a variety of normal tissues in fish, mice, and man, in certain normal mouse cells exposed *in situ* to an artificial electromagnetic field, and in a variety of spontaneous human tumor cells.

In single nondividing cancer cells, particularly in a lethal human esophageal tumor, the two centrioles were separated by long random distances, were randomly oriented toward each other instead of being mutually perpendicular, and exhibited no connecting fine structure. To whatever extent ordered growth and reproduction of normal cells relate to a regular 90° angulation between the long axes of the two centrioles of each cell and to their close constant proximity, to that same extent these relations are absent in at least the tumor that has been most closely examined. The suggestion is made that centrioles interact electromagnetically, and that such a centriolar function may have much to do with the inexorable continuum that has characterized the diverse evolution of cellular life. Cancer may represent a discontinuity in that continuum, evident as a defect in integrity of the centriolar angle.

Viral Induction of Tumors

The experimental attack on cancer as a problem of cell proliferation and behavior began early in this century, and the literature long ago reached monumental proportions. Cancer-inducing chemicals, especially the polycyclic aromatic hydrocarbons already mentioned, at first seemed to be the most promising tools, but they have complex chemical effects on a large number of cell constituents, and, even on the assumption that they cause cancer by inducing somatic mutations, the total complexity of the cytoplasmic and genic situation becomes too great. Attention has turned mainly to the study of viruses in relation to malignancy, not so much as a possible general cause but as a genetic tool of comparative simplicity. Viruses were in fact first reported in tumors as early as 1908, when extracts were found to transmit leukemia through successive passages from fowl to fowl; and in the pioneering work of Rous, begun in 1910, on the propagation of virus-induced sarcomas. The virus causing chicken sarcoma (Rous sarcoma virus, or RSV) has been maintained now for about 60 years and is still busily investigated in many countries. Similarly a number of rat and mouse tumors, not necessarily virus-induced, have been maintained for study almost as long.

When cell-free extracts of tumors are injected into hosts of the same species, not only is malignancy transmitted but so is the tumor tissue type. Such is the case in birds, mice, and frogs. For example, in the chick an extract of fowl chondrosarcoma gives rise to chondrosarcoma when injected into another fowl; fowl endothelioma extract acts upon endothelial cells. Rous sarcoma extracts, injected into the chick, induce sarcomas and lymphosarcomas, particularly of the liver and spleen. The same extract injected into ducklings evokes no response. Only if massive injections are made into the ducklings, within a day or two of hatching, are tumors induced. In this case widespread tumors of the skin and digestive tract, etc., appear in the duckling, i.e., epithelial carcinomas are produced rather than mesodermal sarcomas; the agent has undergone change in the transfer and in fact can no longer readily induce tumors in the chick.

In recent years mammalian virus-induced tumor-cell strains have been intensively studied, particularly those associated with small, DNA-containing viruses called the *polyoma virus* and *simian virus 40* (SV 40). This last virus, originally found in monkey kidney cells, is seemingly harmless in the native monkey species but induces malignant, neoplastic growths in young hamsters. Both these viruses induce cancer when they are inoculated into newborn rodents, particularly hamsters, rats, and in the case of the polyoma virus, mice. They are generally referred to as the *small papava viruses.*

The DNA of polyoma virus, for instance, contains about 5,000 nucleotide pairs and can code for proteins containing about 1,600 amino acids, more than one-third of which are needed for building the viral coat. The number of genetic functions performed by the virus in infected cells is accordingly very small, probably between two and four, and there is good chance therefore of identifying the functions that affect the growth-regulatory mechanisms of the cell.

Two types of host cell are studied in tissue culture in connection with each virus: (1) the "productive" host cell, in which the virus multiplies unchecked within the cell until the cell is killed; and (2) the "transformable" host cell, in which the virus causes little or no productive infection but induces changes similar to those of cancer cells. For transformation studies, cell clones derived from a single cell and therefore uniform in composition are employed. Thus the effect of the virus can be studied without interference from other forms of cellular variation, and transformed cells can readily be compared with normal cells of the same clone.

When suitable clone cells are exposed to viruses of this sort, many viral particles are taken up intact by the cells and accumulate around the nucleus. Most of the particles remain inert, but some lose their protein coat, and their naked DNA core enters the nucleus. Various experimental results show that the transforming process is caused by the incorporation of viral DNA into the cell nucleus and that it is not caused by the viral protein coat. Mutation in the viral genetic material can abolish the ability of the virus to transform the host cell. For example, in a temperature-sensitive mutant line of polyoma virus, the virus can cause either transformation or production, depending on the kind of cell it infects, at a temperature of 31°C; but at 39°C, the effect of the mutation shows up, the virus becomes inactive, and the host cells remain unchanged.

A striking effect in cells infected with polyoma virus, which normally have a very low rate of DNA synthesis, is an induction of cellular DNA synthesis. A very active incorporation of labeled precursors into DNA begins a few hours after infection and leads to synthesis of DNA which is about two-thirds cellular and one-third viral. When the DNA synthesis begins, three enzymes involved in DNA synthesis also become active. In normal, uninfected, crowded cultures the activity of these enzymes is slight, but the activity is much greater in less crowded, uninfected cultures where cell growth and division are relatively rapid. In the infected cultures, even when crowded, activity of these enzymes is high. In other words, virus infection induces a set of activities which are high in growing cells such as regenerating liver, and center around DNA synthesis. According to Dulbecco, the important point is that the DNA growth complex is induced because the virus removes the growth inhibition caused by crowdedness, and therefore neutralizes the regulatory mechanism that inhibits the growth of the cells in crowded cultures.

How does this DNA induction occur? The effect of specific inhibitors of protein or RNA synthesis throws some light on the subject. In the first place, the induction is blocked when protein synthesis is inhibited by puromycin, indicating that the enzyme activity requires the synthesis of new protein. Secondly puromycin inhibits DNA synthesis. The induction of the DNA complex therefore appears to consist of a simultaneous activation of several genes, some responsible for DNA synthesis and others for enzyme synthesis. The effect of various inhibitors of protein synthesis indicates that a regulator protein is involved that is required for the functioning of the DNA growth complex. The virus may induce the DNA complex either by producing its own regulatory protein or by influencing the cellular regulatory gene.

Is induction of the DNA growth complex actually a part of the process of transformation of cells toward the malignant state? The answer seems to be in the affirmative; if this is so, the viral gene causing the induction should persist in the transformed cells and play an essential role. Cells in which such induction is permanent would continue to synthesize DNA irrespective of regulatory influences of the environment, such as crowding or contact situations, and presumably would multiply without restraint.

What is the relationship between induction of DNA synthesis and the function of the viral gene causing the surface changes in the transformed cells? Different viral genes may be responsible, but Dulbecco considers that the DNA induction may well be an indirect consequence of alterations produced in the surface membrane by the same gene. "The reason for suggesting this hypothesis is that the surface of the normal cell has a key role in regulating cell movement and probably also multiplication by cell to cell contact; thus, the surface is like a sensor and may control the regulator gene. The insertion of a virus-coded component, by making the cell surface incapable of sensing the normal regulatory signals of the environment, would cause the cellular regulator gene, and consequently the whole growth machinery of the cell, to be turned on permanently."[1] That is, the system that regulates cell growth may be rather simple, involving the cell surface and a regulator gene.

The Cell Membrane

The most obvious visible property of cancerous cells and the one that gives them their malignant quality is their invasiveness of other tissues, which appears to be primarily a property of the cell membrane. Any theory of cancer based on genic change of any kind, whether somatic chromosomal aberration or somatic gene mutation (which is essentially a modified version of Boveri's original idea, progressively refined as the years go by), must account for the change in membrane behavior. Genic change leading to loss or addition of enzyme might well result in vital change in membrane property. It is a question of evidence. Opinions vary between two extremes. Rubin (1966), for instance, states that when we say that a virus is or is not present in a cell, we are saying precious little about what it does to cause the cell to alter its behavior. He considers that the contribution of the somatic mutation hypothesis is even flimsier. Yet the cancerous behavior of a cell is perpetuated in its progeny, and it seems almost self-evident that there is a change in the genetic material of the cell. He raises the question of whether there are ways of making hereditary changes in cells without altering the DNA of the cell, and he looks to the cell membrane for the answer.

Both the direction and extent of cell movement appear to be controlled at the cell surface, and both are responsive to the nature of the surface contacted by the cell. Carter (1965) has presented strong evidence that contact inhibition is the result of the greater strength of adhesion of a cell to its substratum than to another cell, and that the direction of movement of cells on a gradient of differential adhesion is toward the region of strongest adhesion.

Cells escape from contact inhibition when they become malignant, as a result of decrease in differential adhesiveness, with the following consequences:

1 The movement of a malignant cell is not obstructed or diverted when the cell is confronted by a normal cell.
2 The movement of normal cells, in tissue culture, is obstructed when they are confronted by malignant cells.

The fine structure of the cell membrane, however, is near the limits of resolution by the electron microscope, while any analysis that involves membrane disrup-

---

[1] R. Dulbecco, Mechanisms of Cell Transformation by Polyoma Virus, *Perspect. Biol. Med.*, **10**:304 (1966).

tion is self-defeating, since the question concerns the precise nature of the macromolecular structure. Whatever the precise molecular organization may be, there is little doubt that the membrane is made up of macromolecular units in a state of flux and in equilibrium with the immediate environment. In certain circumstances, when growth rate is high or the medium is in short supply with regard to building blocks, the membrane appears to be under stress and able to maintain its integrity only with some difficulty. It tends to become leaky even in normal cells. In malignant cells there is evidence that the membrane is not only very leaky but is relatively thin and active, and fragile.

Growth-promoting Factors

In relatively dense populations of normal cells cultured in a protein-free medium, the medium itself gains material which has the capacity to enhance the growth of cell populations that are growing slowly either because the cells are too sparse or because they are too crowded. In such a protein-free medium cells are apparently under stress, membrane structure may be difficult to maintain, and leakage results. Normally leakage is not known to occur, even under the condition known as density-dependent inhibition.

A powerful growth-promoting factor, however, leaks from cancerous cells, and the available evidence suggests that it is a protein closely related to the growth-promoting factors in normal cells. According to Rubin, "When chick embryo cells in culture are exposed to high concentrations of the Bryan strain of Rous sarcoma virus (RSV), only a fraction of the cells are infected with RSV. The remaining cells are infected with the helper virus (RAU) which occurs in the stock and which interferes with RSV infection. Within a day, scattered cells undergo a malignant transformation which is characterized by rounding up of the cells, disorganization of the typical whorled pattern of fibroblastic growth, and persistent rapid multiplication of cells under crowded conditions that would otherwise slow their growth. Most cells in the culture, necessarily including many not infected with RSV, take on the appearance and behavior of transformed cells by day 4 or 5 after infection. This suggests that the cells infected by RSV release material which stimulates continued multiplication (overgrowth) among the cells not infected by RSV as well as those infected."[1]

In other words, the cancer cells appear to make excessive amounts of the overgrowth-stimulating factor and to let it escape by leakage through the cell membrane. After it has leaked out, it acts on the surfaces of neighboring cells. This is shown by adding fluid from the original infected cell cultures to crowded cell cultures that have not been infected with the cancer virus. The result is a dramatic spurt in growth and cell multiplication in the normal cells, which also become abnormal in

[1] H. Rubin, Growth-stimulating Factor Isolated from Rous Sarcoma Cells, *Science,* **167**:1271 (1970). Copyright 1970 by the American Association for the Advancement of Science.

shape. The cells in a culture exposed in this way double their number in 65 hours but return to normal behavior and shape after about 100 hours. The factor responsible for the change is not the virus itself, for the fluid remains effective after centrifugation more than sufficient to separate out the virus particles. Moreover the growth factor, responsible for the enhanced growth of the normal cells, does not in itself bring about a permanent malignant change.

Further analysis shows that the increase in intracellular activity of treated cells is contemporaneous with the malignant transformation of the cells, whereas the release of the overgrowth factor into the medium seems to be a later consequence related to the increased leakiness of the Rous sarcoma cells. The overgrowth-stimulating factor could therefore be the agent which effects the release of RSV-infected cells from the density-dependent inhibition. The stimulating effect can be mimicked with traces of trypsin or pronase, and since peptidases undergo a marked increase in many tumors and since trypsin unmasks tumor-type agglutinins in normal cells, it is possible that the factor is a peptidase or protease. It is significant that the mesenchymal growth factor isolated from mouse submaxillary glands has both peptidase and esterase activities and that the nerve-growth factor isolated from the same source has similar enzymatic activities.

In fact, Levi-Montalcini and Cohen reported some time ago that the potent nerve-growth factor extracted from snake venoms duplicates the effects of a "protein fraction" of mouse sarcomas in vivo and the effect of living sarcomas in vitro, although the purest fraction of snake venom obtained was 1,000 times as effective as their purest fraction of mouse tumor.

The most clear-cut case of tumor induction by a virus is that of the lens. Undisturbed lens epithelium in man and experimental animals has never been observed to undergo neoplastic change. Lens epithelium of the hamster, however, undergoes malignant transformation when infected with simian virus 40. The special interest here is that lens epithelium constitutes a population of a single-cell type, and can be isolated without contamination by any other tissue elements, since the lens is surrounded by a noncellular laminated capsule. It therefore serves as a useful model for studying viral neoplastic transformation. Cells in the tumors become spindle and epithelioid types, showing numerous mitoses, together with multi-nucleate giant cells in moderate numbers, but have no structures resembling lens fibers. Portions of such tumors are transplantable and can be carried through many generations of host hamsters without alteration in morphologic characteristics or spontaneous regression of the tumors.

Nonviral Carcinogenesis

Apart from the action of viruses as neoplastic agents, the known carcinogenetic agents represent a confusing multiplicity of means. The various ionizing radiations intervene in cytoplasmic and nuclear activities in a variety of ways, none well enough known. The chemical carcinogens may be as diverse in effect as in composition. Only a brief presentation of some of the possibilities can be given here.

1 The physical and chemical properties, particularly the electronic structure, of polycyclic hydrocarbons and of purines and pyrimidines have been theoretically related to possible cellular receptors, proteins, or nucleic acids. The initial reaction would be physical attraction followed by enzymatic bond formation.

2 A weakening of hydrogen bonds would result in a mispairing in the DNA transcription process. Benzopyrene, a potent carcinogen, when activated photochemically or by radiant energy, reacts chiefly with nucleoprotein rather than with DNA alone or with histone. A lower-energy barrier may exist after strand opening and rotation in DNA, i.e., a carcinogen may open and activate a restricted part of of the genome in a deep groove of double-stranded DNA covered by a protein chain.

3 Most chemical carcinogens require an activating mechanism. The polycyclic hydrocarbons were often considered as directly acting, but there may well be an oxidized intermediate ionic structure which can react in a number of ways. The still-tentative general conclusions, however, may be simply stated, namely, that chemical carcinogens are ultimately converted to electrophilic reactants, and that DNA is the primary molecular target with a variety of carcinogens.

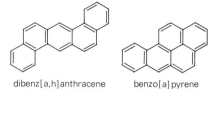

dibenz[a,h]anthracene   benzo[a]pyrene

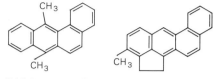

7,12-dimethylbenz [a]-   3-methylcholanthrene
anthracene

FIGURE 22.3 Molecular backbone of polycyclic hydrocarbon carcinogens.

A hypothesis concerning the operation of polycyclic hydrocarbons as carcinogens has been stated by Cromwell (1965) as follows: "The carcinogen initiates an infectious DNA through a dual interaction with the cell nucleus. The nucleoprotein complexes with the polycyclic compound and the DNA is laid bare for further attack. Intercalation of the large flat polycycles between base pairs in the more disordered stretched sections of the DNA chain follows. . . . Such specific complex formation might mask part of the nucleic acid chain (some of the bases) in the case of compounds with critical steric requirements leading to interference with the replication process for the normal DNA and alteration in the assembling of amino acids in the nucleoprotein by RNA. Thus, a new cell nucleus emerges. When it happens that there is no controlling immune response available from the host, uncontrolled growth begins and neoplasia is started which no longer requires the presence of the original carcinogen for its propagation. It is also possible that the carcinogen plays an important role in the suppression of the normal immune response expected from the host to deal with the appearance of strange cell nuclei. Complex formation between the carcinogen and specific antigens might result in their loss by the animal at this critical time, allowing for the uncontrolled growth of the abnormal cell."[1] This hypothesis accounts for tumor induction but hardly explains the well-established fact that mouse tumors originally induced by a carcinogen such as dibenzanthracene can be carried through more than 500 generations in mouse hosts without any change in tumor character and long after any trace of

[1] H. N. Cromwell, Chemical Carcinogens, Carcinogenesis and Carcinostasis, *Amer. Sci.*, **53**:231 (1965).

the carcinogenic chemical could conceivably be present. If it is valid, then some further change is required to fix the hereditary character of the new cell nucleus.

Tumor Progression

Tumor development in the body is a progressive process involving several phases and considerable intercellular competition, with the fastest growing cells eventually dominating.

Initiation, therefore, depends primarily on a specific pattern of damage to those genes involved in cellular homeostasis, perhaps a single regulator gene, concerned with the activation of the cell cycle. In addition to the direct or indirect action of viruses and nonviral carcinogens, part of the necessary damage may be inherited or may result from spontaneous gene mutations, such as those known to occur in many kinds of nondividing mammalian cells as the organism ages, possibly as the result of ^{14}C decay within DNA molecules. Initiation, however, evokes no distinctive change and, in particular, incites no cell growth and proliferation. Some further event is generally needed to release them from the normal tissue control.

Promotion, unlike initiation, is usually slowly acting and may be reversible. The change from dormancy to progressive growth may depend merely on a simple nonspecific stimulus to cell division, as seen in wounds. Continuous tissue repair resulting from chronic irritation and the presence of initiation agents such as chemical carcinogens usually present in tobacco tar or oily smog, comprise a model system for the production of active malignancy. Hormone-dependent tissues, such as mammary gland, are especially vulnerable.

Progression consists of the final steps leading to faster and more independent growth. What happens at each step is more than enhancement of the previous state; it is a wholly new event, resulting in a new type of cancer cell distinct from the one which gave rise to it. Stable, irreversible hereditable changes occur, not necessarily of chromosomal nature, during which the original histological character of the tissue becomes progressively lost. This is most evident when growth of a primary tumor changes to metastatic proliferation.

For example, a naturally occurring tumor of the cottontail rabbit that is widespread through the Midwestern United States is a warty growth, or papilloma, capable of progressing to carcinoma. Fleshy, keratinized, pigmented tumors of the skin grow all over the body. Virus extracted from these tumors when applied to the scratched skin of either cottontail or domestic rabbits induces tumors similar to those of the tumor from which it is obtained, the first evidence of such tumors becoming visible 2 or 3 weeks after inoculation. Discovered by Shope in 1931, it is known as the Shope rabbit papilloma virus, a mammalian counterpart to the Rous chicken sarcoma virus.

In neither species do they regress. Yet although the tumors look alike in the two species, the papillomas induced in cottontails usually contain virus that can in

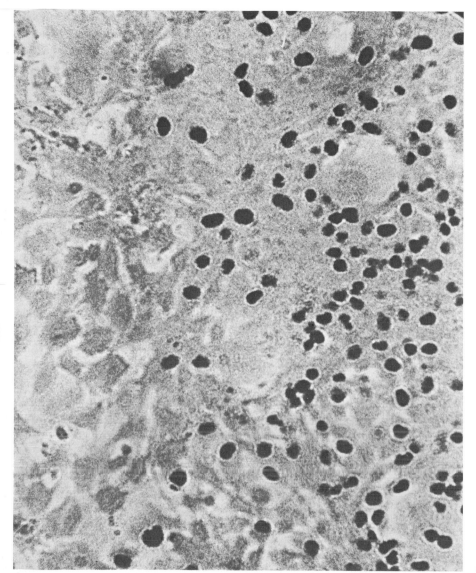

FIGURE 22.4 Autoradiograph of the edge of a ³H-thymidine-labeled Rous sarcoma focus. The region with all the black (labeled) nuclei contains infected and transformed cells. The unlabeled region contains normal cells. This is a graphic illustration of the capacity of virus-transformed cells to continue growing while their sister uninfected cells are inhibited from growing. [*Courtesy of H. Rubin.*]

turn infect and cause tumor formation, while those induced in domestic rabbits contain no demonstrable virus or infectivity, and cannot be propagated beyond the first experiment. This is a regularly reproducible phenomenon. Yet, despite the seeming absence of infective papilloma virus, a viral antigen is nevertheless present in the noninfective domestic-rabbit tumors, which suggested that a 'masked' virus might be present. More recent work by Shope and his coworkers indicates that "in cottontail papillomas, virus seemingly exists in two forms: one, the complete mature infective virus composed of nucleic acid and protein; and the other, immature virus,

Teratomas are tumors, mostly of germ-cell origin, composed initially of undifferentiated embryonal-type cells. Some of these cells may differentiate, giving rise to malignant teratocarcinomas consisting of both embryonal and mature tissues. Teratomas may be simple in composition, or they may contain almost all the different kinds of tissues of the body. In some cases so-called embryoid bodies are produced which resemble embryonic stages of the parental species. Teratomas occur in all classes of vertebrates, with the possible exception of cyclostomes, although the incidence varies greatly from species to species even within a single class. Occasionally they arise from the pineal or the adrenal gland and elsewhere, but all such are rare compared with those associated with the testes and ovaries. Most studies have been made on teratomas of the testes, either experimentally induced in fowl and mice or spontaneously arising in strain 129 mice at about 12 days of gestation. The degree of malignancy is extremely variable and depends mainly on which cell type becomes fully malignant. They can be transmitted experimentally.

The common germinal origin of teratomas is demonstrated by the experimental means of induction. They are readily induced in mice, occasionally in birds, by the injection of copper or zinc salts, with or without gonadotropic hormones, at times when the diploid spermatogonia are actively dividing. If the genital ridges of the male fetus of susceptible mouse strains are transplanted to the testis of adults, about 75 percent develop into testes with multiple teratomal centers. However teratomas can also be induced in mature testes.

There is strong evidence that all teratomas such as these arise from pluripotent undifferentiated cells. In fetal mice the tumors consist at first of undifferentiated cells of a generally embryonic character; they become histologically more complex as the animals age. Direct evidence that adult-type tissues in mouse teratomas are derived by differentiation of pluripotent embryonal stem cells comes from an elegant in vivo cloning technique designed by Kleinsmith and Pierce (1964). They dissociated cells of small embryoid bodies, taken from a transplanted teratoma of strain 129, and transplanted them directly into mice. Forty-three tumors grew from single stem cells, composed of as many as fourteen well-differentiated tissues in addition to the original embryonal carcinoma type of cell. In fact single embryonal carcinoma cells are clearly multipotent and capable of producing most of the somatic tissues of the body.

We may, therefore, make the following conclusions:
1　Spermatogonia are multipotent undifferentiated cells.
2　They are normally destined to proliferate and differentiate as spermatocytes and finally as spermatozoa.
3　They are able, when proliferating as teratoma cells, to differentiate to form most of the specialized somatic-cell types characteristic of the species.

As a rule the embryonal carcinomas first formed are composed of cells that are not only multipotent but also highly malignant, although most of the differentiated tissues they give rise to are generally benign. What we see, in fact, is an initial malignant change in the spermatogonial cell which transforms it from a well-behaved constituent cell of the germinal epithelium into an unrestrainedly multiplying and invasive cancer. But this very process in effect sets the sperm mother cell free from its original directives under circumstances that enable it, as a multipotent cell of unlimited proliferative capacity, to express its full histogenetic potential. It is a natural experiment and a remarkable performance.

What can such a cell do apart from perpetuating the original undifferentiated but malignant type of cell? This depends on circumstances we know little about, for the composition of teratomas is extremely variable. Some teratomas may be mostly embryonal, either entirely undifferentiated or with neural tissue. Most of them contain notochord, together with respiratory, alimentary, and glandular epithelia, and cartilage, bone, and marrow. Many teratomas have hair and sebaceous glands. Various degrees of tissue organization may be seen in them. Some transplantable, i.e., malignant, teratomas may consist entirely of myoblastic cells, although at first consisting of a variety of tissues; others may become entirely neural. Microenvironmental influences seem to be responsible for their development. These influences may be no more than local variations in electrolyte concentrations within each population of cells. The undifferentiated, multipotent cells need little directing in order for them to give rise to the multiplicity of cell types characteristic of the body, but to form an embryo as distinct from an assortment of tissues, supracellular organization of another order is required.

Two theories concerning the origin of teratomas have been favored. According to one, teratomas originate from germ cells. The other is that they originate from embryonic totipotent cells that have escaped the influence of embryonic organizers. The former interpretation has been adopted in the foregoing account. Supporting evidence comes from the discovery that mice that are sterile because of a genetically determined absence of primordial germ cells do not develop teratomas, whereas their normal litter mates do. This means that the tumors *are* derived from primordial germ cells.

On the other hand Stevens, who has made intensive studies of spontaneous teratomas, has also found supporting evidence for the alternative theory (Stevens, 1970). In preliminary experiments two-cell (1-day) eggs of mouse strain 129 were implanted in adult testes. Development was normal for about a week, after which the embryos became disorganized mixtures of embryonic, immature, and adult tissues, for as long as 60 days. The delay in onset of differentiation of some cells was attributed to disruption of normal intercellular relationships. When 3- and 6-day mouse embryos were similarly implanted, a comparable mixture of undifferentiated, immature, and adult tissues arose. Several of the embryos, however, gave rise to serially transplantable tumors, indistinguishable from transplantable teratomas derived from spontaneous teratomas of the same strain of mouse. They were apparently derived from undifferentiated stem cells. As in spontaneous mouse teratomas, embryoid bodies are produced when tumor cells are grafted intraperitoneally.

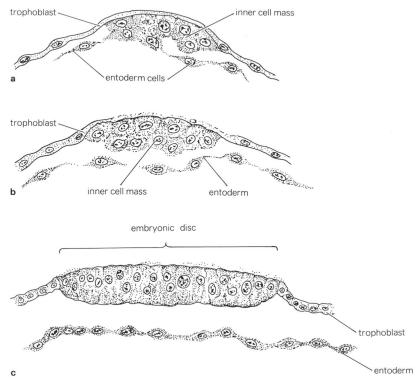

FIGURE 22.6 Sections of pig blastocysts showing the first appearance and subsequent rapid extension of the entoderm. [*After Patten, "Foundations of Embryology," 2d ed., p. 124. Copyright 1964, McGraw-Hill Book Company. Used by permission.*]

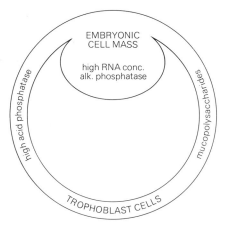

FIGURE 22.7 Biochemical patterns in the mammalian blastocyst. [*After Deuchar, 1965.*]

Typical teratomas accordingly can arise either from primordial germ cells or from a disorganized population of undifferentiated embryonic cells. The distinction between these two kinds of stem cells therefore appears to be purely semantic.

Embryoid bodies have been investigated, and the literature on them has been reviewed, by Stevens (1967), and by many others. The debated question is whether they are truly comparable to mammalian embryos of any stage. They arise not directly from the spermatogonia but from undifferentiated embryonal carcinoma stem cells of the tumor. They lack typical organization, but then so do normally fertilized rat and mouse eggs when grafted to many extrauterine sites. On the other hand, grafts from transplanted embryoid bodies derived from a transplanted testicular teratoma of a strain 129 mouse develop formations remarkably like early mouse embryos that have undergone morphogenesis. According to Stevens they form neural folds, amnion, coelomic epithelium, somite material, and yolk sac, all in proper spatial relationships to one another. When some sublines of transplantable teratomas are grafted into the peritoneal cavity, thousands of free-floating embryoid bodies similar to mouse embryos 5 and 6 days of age appear in the peritoneal fluid. He states that mouse embryoid bodies not only look like normal mouse embryos but have similar embryonic potency. He concludes that they are homologous to embryos.

CELLS AND DEVELOPMENT

More and more, it seems that much of the cellular and physiological paraphernalia associated with the egg, together with special features of the egg itself, are primarily devices and boosters that make it possible for the basic, undifferentiated cell to carry out its developmental potentiality.

Cell Transformations

The cell-transformation hypothesis has been evoked as an explanation of cell-differentiation processes observed in the reconstitution of amphibian limbs and as an explanation of transdetermination in long-term cultures of imaginal discs. In connection with cell clusters formed by reaggregation of dissociated tissue cells, the concept of *critical mass* has been put forward on several occasions. Trinkaus and Groves (1955) state: "Several lines of evidence indicate that with standard culture conditions a mass of cells must exceed a certain minimal size before differentiation may occur *in vitro*. . . . When this cell mass reaches the necessary size, it will begin to differentiate, if it is located in the proper region in the explant. It is assumed that with the onset of differentiation this mass becomes an active center, specifically influencing surrounding cells to follow suit. As these surrounding cells begin to differentiate they, in turn, will influence cells peripheral to them. This process continues until a large area composed of many cells is involved. It ceases when cells are reached which have already passed under the influence of another active center. This hypothetical phenomenon resembles an embryonic field."[1] The concept of critical mass, which we may extend to area or extent, since the shape of a coherent cell population is also important, takes on further significance in relation to size and shape of the embryonic blastoderm and the inner cell mass.

The question has been more recently examined by Hiroka Holtfreter (1965) with regard to self-differentiation and regulation of variously sized fragments of the isolated dorsal area of the amphibian gastrula. Using the early gastrula of the newt and the axolotl as donors, she cut out the whole dorsal area, comprising prospective chordamesoderm and some endoderm, and subdivided it into smaller fragments according to precise schemes of fractionation. The sets of fractions thus produced ranged in steps from one-half to one–twenty-fourth of the original standard area. The isolates were then cultured to determine their potentiality.

The overall diversity and frequency of the differentiations decreased drastically with the progressive fractionation of the standard area until, in the one–twenty-fourth fractions, only four different cell types were produced. This decline was associated with a gradual simplification and loss of the histo- and organotypical patterning of the explants, so that the smallest fragments developed merely into amorphous groups

---

[1] J. P. Trinkaus and P. Groves, Differentiation in Culture of Mixed Aggregates of Dissociated Cells, *Proc. Nat. Acad. Sci.*, **41**:794 (1955).

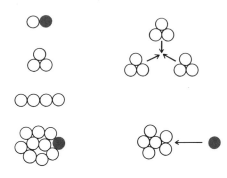

FIGURE 22.8 Experimental combinations of mid-cleavage eggs. Each circle represents an entire, denuded egg. Black circles are eggs preincubated in ³H-thymidine before aggregation. Arrows show a second step in aggregation. [*After Mintz*, Proc. Nat. Acad. Sci., **58**:344 *(1967)*.]

of cells, whereas undivided standard pieces even exhibited the bilateral and antero-posterior configuration of the corresponding tissues characteristic of a normal organism. In fact the explants of larger size formed a much greater variety of tissues than normally arise from the same area when it is left within the context of the embryo.

The developing situation in early mammalian development, particularly with regard to the questions of lability and determination, has been intensively and elegantly investigated by Mintz, employing the "multimouse" procedure already described: "The question to be explored was not whether topographical inhomogeneities exist at fertilization. Unequal distributions of several inhomogeneities exist at fertilization. Unequal distributions of several components have been reported (Dalcq, 1957; Mulnard, 1961). Rather, the dilemma may be summarized as follows: is there a 'patterning' in the oocyte which is *causally* responsible for the onset of morphogenesis, and therefore for establishment of inner cell mass, as against trophoblast, in the blastocyst? The answer will not, in itself, reveal the basic mechanisms of the beginnings of differentaition, but it will indicate whether it is more fruitful to look for critical information at the onset of development, in cytoplasmic gradients, or at a much later time, possibly as the dividing cells become packaged into smaller units which must then have disparate microenvironments.

"To determine whether a causal relationship obtains between egg topography and morphogenesis, it is necessary to conduct an *operational* test which would unambiguously reveal whether, in the face of a demonstrable perturbation of any 'pre-pattern' in the egg (either of detected or unidentified nature), normal embryogenesis still occurs. The methods of egg aggregation lent themselves favourably to such an investigation, and in the culture environment and development of 'rearranged' embryos could be directly observed.

"The results clearly established in a positive manner the extreme lability of the embryo before the blastocyst stage. . . . A summary follows.

1 "Blastomere movements were observable in living eggs by combining two denuded cleaving eggs with characteristically different types and different degrees of cytoplasmic granularity. Blastomere migrations appeared to be at random, rather than selectively orientated, and neither complete intermingling nor ordered employment of the two types of cells occurred. This is consistent with the fact that morulae with highly unusual arrangements of cells, which were occasionally recovered from pregnant females, developed into normal blastocysts during later implantation.

2 "Synchronous pairs of eggs developed into unitary blastocysts following aggregation at any time during cleavage, from the two-cell stage onward. This was true even of some pairs of morulae with clear signs of cavities, i.e., early blastulas.

3 "Increasing numbers of midcleavage eggs were assembled in clusters. As many as 16 have been united and have formed one enormous blastocyst. These large balstocysts at first have a correspondingly large inner cell mass.

4 "Aggregation experiments in which the initial stages of aggregation were retarded, consequently decreasing the time for any possible sorting out among the cells, resulted in normal blastocyst formation. The timetable for development to the blastocyst stage remains constant, whatever the stage at which fusion occurs.

5 "Individual eggs were preincubated with tritiated thymidine during the S-period close to the eight-cell stage, until nuclei of all cells were labeled in the DNA. The labeled eggs were then made to adhere to unlabeled eggs, and allowed to develop further. Radioautographs showed that cells from an entire egg often remained in relative proximity, as a clump or streak, but that many kinds of distribution occurred, and no specific pattern was evident."[1]

These experiments and others, considered together, strongly support the view that the mouse egg (and probably every mammalian egg) is morphogenetically labile throughout the cleavage phase. This is perhaps indicated also by the fact that giant blastocysts subsequently raised in the uterus of a foster mother give rise to embryos of normal size at birth, although the time and mechanism of such adjustment are not known.

Mammalian oocytes, which normally mature in the ovary, can be induced to mature in a culture medium. Even human oocytes, obtained from ovaries, have been brought to maturation, fertilized, and cultured in vitro to the blastocyst stage, by R. G. Edwards and colleagues (1970). Such mammalian eggs may then be introduced into the uterus, where they implant in the normal manner. In the mouse, when a single labeled cell from a donor embryo (inner cell mass of the blastocyst) is injected by means of a micropipette into another mouse blastocyst, large parts of the developing host embryo are seen to be colonized by descendants of the labeled cell. This implies that very few cells in the host blastocyst are truly embryonic, for otherwise a single additional cell could not contribute so much to the future embryo, and accordingly embryonic determination is not initiated until the blastocyst is well established.

The observations on mouse development, together with those on twinning in the armadillo, in birds, and in bony fish, suggest that the labile or undetermined (with regard to embryo formation) state of the egg in all meroblastic vertebrate eggs, including the mammal, persists throughout the cleavage period, up to the time that a visible embryonic axis or primitive streak begins to form on the blastoderm or the inner cell mass. In other words, in this general type of egg, cleavage primarily yields a mass of undifferentiated, uncommitted cells which serves simply as a territory within which one or more embryonic axes may become established.

This conclusion is borne out by phenomena seen in the development of so-called annual fishes. These are fish which have a very short life span and which lay eggs that undergo cleavage but proceed with embryo formation and further development only after a period of developmental arrest lasting a minimum of several days.

[1] B. Mintz, "Ciba Foundation Symposium on Preimplantation Stages of Pregnancy," p. 195, 1965.

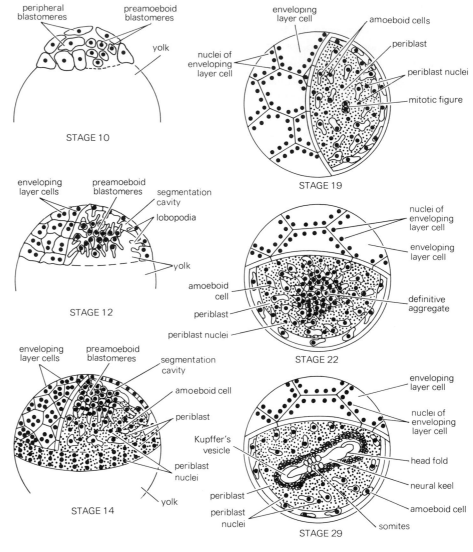

FIGURE 22.9 Dispersion and reaggregation in *Austrofundulus myersi*. Chorion removed; surface view of the egg with enveloping layer intact on one side; on the other side, the enveloping layer is removed to illustrate underlying amoeboid cells and periblast; periblast stippled; cell numbers reduced for clarity; oil droplets not illustrated. [*After Wourms, 1967.*]

Normal development of annual fish eggs is characterized by a phase of complete dispersion and subsequent reaggregation of amoeboid blastomeres which is interposed between cleavage and embryogenesis.

According to Wourms, working with the eggs of *Austrofundulus myersi*, two populations of blastomeres form during cleavage and segregate at the flat, hollow blastula stage, in the typical manner of teleostean development. By the early flat blastula stage there are flattened periblast cells adjacent to the yolk continuous with a surface-enveloping layer, and a compact mass of amoeboid, more or less spherical blastomeres. As epiboly commences, the amoeboid blastomeres consolidate into a mass, then migrate away from that mass as individual cells, and move into the space formed when the enveloping layer and periblast advance over the yolk. When epiboly is completed, the amoeboid blastomeres of the 3-day egg are randomly distributed. At this stage no germ ring or embryonic shield is present; this would be an obvious impossibility. Both cell contact and, presumably, consequent intercellular communication are essential for a cell population to express collective potentiality.

In the late dispersed stage, cell contacts increase in number and duration, until in the 4-day egg an aggregation of cells is found in a single region, an ellipsoid mass at one pole. By 10 days a solid neural ridge has formed in the aggregate, while development of a typical teleostean embryo follows. The development of five other genera of annual fishes appears to be essentially similar. The reaggregation mass may therefore be considered identical to the embryonic shield of nonannual teleosts.

The disaggregation experiments on sea urchin embryos of Guidice (1962–1967) may be relevant here. The cells of embryos dissociated at the early pluteus stage reassociate as small masses and reconstitute pluteuslike larvae. The process consists, however, of a reconstitution of a blastula, or large vesicle, which does not invaginate but acquires a new gut by the enlargement and confluence of intracellular vesicles within the inner cells, which arise a few hours after a solid aggregate has formed.

The Labile State

From all the foregoing it is evident that the prospective embryo-forming territory of the blastoderm of fish and bird and the inner cell mass of the mammal are in a completely labile state. No determinative processes leading to embryo formation are either detectable or apparently possible during blastulation. Thus the condition at the time of embryo initiation is merely a coherent cellular sheet or mass consisting of a very large number of cells which are individually already of approximately the same size as the less differentiated types of somatic cell typical of the species. The situation is that of a disc of undifferentiated multipotent cells, one or two layers deep, that are numerous enough to initiate an embryonic axial thickening and are nutritively supported for further growth and cell multiplication.

The egg is still a specialized cell, but the specialization is now primarily the storage of yolk to serve for later nutrition of the system and as a suitable base for the developing tissue, or it is a device for implanting the cortical cell mass in an appropriate parental environment. However this may be, the starting point for embryonic development is a cell territory of certain minimal size and cell number, in which the cell population can express its collective developmental potentiality.

This prospective development territory, whether it be the blastoderm of a fish or chick at a certain stage or the inner cell mass of the mammalian blastocyst, may be compared with the imaginal disc of an insect. Both consist of undifferentiated cells, both are capable of developmental expression of structured pattern under permissive circumstances, but the imaginal disc already carries a regional prepattern derived from its location within the total embryonic system at the time of its formation, whereas the comparable disc of the vertebrate has not been subject to any such organization and can therefore start from scratch. In other words we see in this type of vertebrate development the capacity of a group of undifferentiated cells *as cells* to undergo development, as distinct from the development of the egg cell *as an egg* characteristic of the precociously determined eggs described earlier. This recalls the capacity of comparable somatic cell territories of some invertebrate animals to undergo complete and direct development to the mature adult, either as buds or as fragments, naturally or operationally isolated from the parent, described in earlier chapters.

We are thus led in almost full circle from our first discussion of the capacity of populations of coherent unspecialized cells to reconstitute the whole of which they originally were a part. The special phenomena associated with the formation of an egg as a reproductive device, and with the particular events associated with the precocious development of many features of embryonic development, tend to distract attention from the latent capacity of the undifferentiated cell simply as a cell to express the full phenotype. What is usually lacking, except in eggs, are the conditions necessary to the initiation and sustaining of such development. The two basic questions concerning the development of the multicellular organism therefore remain. What are the guidelines that organize, or enable cell populations to self-organize, into the shape and structure, dynamically maintained, of the organism as a whole? And what are the microenvironmental cues that underlie the diverse differentiation of cells within a cell population?

CONCEPTS

The cell membrane of malignant cells is basically different from that of normal cells, in its leakiness, its fragility, and its immunity to contact inhibition.

The growth potential of normal and malignant cells in culture is the same, but in the body growth-inhibitory controls are effective in the one but not in the other.

Cell-free extracts of malignant tumors transmit malignancy to normal tissues by virtue of specific viruses.

Growth-promoting factors, specific for tissue type, are induced by viral infection, which can enhance growth of normal cells but are not themselves inductors of malignancy.

Chemical carcinogens induce cell transformation leading to malignancy by interfering with the nucleoprotein complex of the cell nucleus.

Tumor development is progressive, involving successive changes, at first reversible but finally irreversible, in cell character.

Specific tissue susceptibility to tumor-inducing viruses lies in the genome of the target cells.

Prospective male germ cells, before the onset of gametogenesis, can develop into histologically diversified, embryolike bodies, given adequate activation (malignancy) and supportive circumstances (body tissues or cavities).

Reaggregated blastomeres generally retain the capacity for normal development.

Eggs with direct development are essentially undifferentiated cells supplied with nutritive and/or protective devices that establish necessary conditions for such development.

READINGS

ABERCROMBIE, M., J. HEAYSMAN, and H. KARTHAUSER, 1957. Social Behavior of Cells in Tissue Culture, III. Mutual Influence of Sarcoma Cells and Fibroblasts, *Exp. Cell Res.,* **13**:279–291.

AMBROSE, E. J., and F. J. C. ROE (eds.), 1966. "The Biology of Cancer," Van Nostrand.

BERRILL, N. J., 1943. Malignancy in Relation to Organization and Differentiation, *Physiol. Rev.,* **23**:101–123.

BIESELE, J. J., 1956. Tissue Culture and Cancer, *Sci. Amer.,* October.

BULLOUGH, W. S., 1967. "The Evolution of Differentiation," Academic.

BURNETT, A., 1968. The Multipotent Cell and the Tumor Problem, in H. Ursprung (ed.), "Results and Problems in Cell Differentiation," Springer-Verlag.

CARTER, S. B., 1965. Principles of Motility: The Direction of Cell Movement and Cancer Invasion, *Nature,* **208**:1183–1187.

CROMWELL, H. N., 1965. Chemical Carcinogens, Carcinogenesis and Carcinostasis, *Amer. Sci.,* **53**:213–236.

CURTIS, H. J., 1966. "The Biological Mechanisms of Aging," Charles C Thomas.

DEUCHAR, E. M., 1965. Biochemical Patterns in Early Developmental Stages of Vertebrates, in R. Weber (ed.), "The Biochemistry of Animal Development," vol. I, Academic.

DULBECCO, R., 1969. Cell Transformation by Viruses, *Science,* **166**:962–968.

——, 1967. The Induction of Cancer by Viruses, *Sci. Amer.,* April.

EBERT, J. D., and F. W. WILT, 1960. Animal Viruses in Embryos, *Quart. Rev. Biol.,* **35**:261–312.

EDWARDS, R. G., and R. E. FOWLER, 1970. Human Embryos in the Laboratory, *Amer. Sci.,* December.

FOULDS, L., 1963. Some Problems of Differentiation and Integration in Neoplasia, in Harris (ed.), "Biological Organization at the Cellular and Supercellular Level," Academic.

GROSS, R. J., 1964. "Adaptive Growth," Academic.

GREEN, D. E., and J. F. PERDUE, 1961. Membranes as Expressions of Repeating Units, *Proc. Nat. Acad. Sci.,* **55**:1295–1302.

GUIDICE, G., 1962. Restitution of Whole Larvae from Disaggregated, Cells of Sea Urchin Embryos, *Develop. Biol.,* **5**:402–411.

HIEGER, I., 1961. "Carcinogenesis," Academic.

HOLTFRETER, H. B., 1965. Differentiation Capacity of Spemann's Organizer Investigated in Explants of Diminishing Size, Thesis no. 65-12,008, University Microfilms, Ann Arbor.

HUXLEY, J. S., 1956–1957. Cancer Biology, I: Comparative and Genetic; II: Viral and Epigenetic, *Biol. Rev.,* **31**:474–514; **32**:1–37.

KLEINSMITH, L. J., and G. B. PIERCE, 1964. Multipotentiality of Single Embryonal Carcinoma Cells, in A. Monroy and A. A. Moscona (eds.) "Current Topics in Developmental Biology," pp. 223–246, Academic.

LEVI-MONTALCINI, R., and S. COHEN, 1956. In Vitro and in Vivo Effects of a Nerve Growth-stimulating Factor Isolated from Snake Venom, *Proc. Nat. Acad. Sci.,* **42**:695–699.

MILLONIG, G., and G. GUIDICE. 1967. Electron Microscopic Study of the Reaggregation of Cells Dissociated from Sea Urchin Embryos, *Develop., Biol.,* **15**:91–101.

MINTZ, B., and G. SLEMMER, 1969. Gene Control of Neoplasia, I: Genetic Mosaicism in Normal and Preneoplastic Glands of Allophenic Mice, *J. Nat. Cancer Inst.,* **45**:87–95.

MIZELL, M., 1969. "Biology of Amphibian Tumors," Springer-Verlag.

RASMUSSEN, H., 1970. Cell Communication, Calcium Ion, and Cyclic Adenosine Monophosphate, *Science,* **170**:404–412.

ROSS, R., 1969. Wound Healing, *Sci. Amer.,* June.

ROUS, PEYTON, 1967. The Challenge to Man of the Neoplastic Cell, *Science,* **157**:24–28.

RUBIN, H., 1970. Growth-stimulating Factor Isolated from Rous Sarcoma Cells, *Science,* **167**:1271–1272.

——, 1966. Facts and Theory about the Cell Surface in Carcinogenesis, in M. Locke (ed.), "Major Problems in Developmental Biology," 25th Symposium, Society of Developmental Biology, Academic.

SCHAFER, P., 1969. Centrioles of a Human Cancer: Intercellular Order and Disorder, *Science,* **164**:1300–1303.

SHOPE, R. E., 1966. Evolutionary Episodes in the Concept of Viral Oncogenesis, *Perspect. Biol. Med.,* **9**:258–274.

SMITH, R. T., and R. A. GOOD (eds.), 1969. "Cellular Recognition," Appleton Century Crofts.

STEVENS, L. C., 1970. The Development of Transplantable Teratocarcinomas from Intratesticular Grafts of Pre- and Postimplantation Mouse Embryos, *Develop. Biol.,* **21**:364–382.

——, 1967. The Biology of Teratomas, *Advance. Morphogenesis,* **7**:1–32.

TRINKAUS, J. P., and P. GROVES, 1955. Differentiation in Culture of Mixed Aggregates of Dissociated Cells, *Proc. Nat. Acad. Sci.,* **41**:787–795.

WEISS, L., 1967. "The Cell Periphery, Metastasis and other Contact Phenomena," North-Holland Publishing Company, Amsterdam.

WEITLAND, A. M., and G. S. GREENWALD, 1968. Influence of Estrogen and Progesterone on the Incorporation of 35S Methionine by Blastocysts in Ovariectomized Mice, *J. Exp. Zool.,* **169**:463–469.

WILLMER, E. N., 1961. Steroids and Cell Surface, *Biol. Rev.,* **36**:368–398.

WOURMS, J. P., 1967. Annual Fishes, in "Methods in Developmental Biology," Crowell.

INDEX*

Abalone, 221

Abnormal mitoses in tumors, 494

Accessory cells, 206, 207

Acetabularia, 58, 90–94, 103

Acetylcholine, 107, 142

Acetylcholinesterase, 142

Acrasin, 107

Acrosomal filament, 222

  membrane, 224

  tubule, 222, 230

Acrosome, 196, 220, 222, 224

Acrosome reaction, 220

Actin, 46

Actinomycin, 70, 98, 204, 263, 297, 474

Actinosphaerium, 49–51

Activation of egg, 217–237

Activation center, 453, 469

Activation-transformation hypothesis, 359

Actomyosin, 37, 46, 49, 51

Adaptation, 190

Adaptive changes, 189

Adenocarcinomas, 272, 493

Adenosine monophosphate, 142

Adenosine triphosphate (ATP), 46, 57, 188

Adhesion of blastomeres, 306

Adhesive suckers, 190

African frog (see *Xenopus*)

African violet, 151

Age and tissue differentiation, 375

Agglutination of sperm, 219

Aggregation, 2, 106

  of amoebae, 108

  (*See also* Reaggregation)

Aging:

  of collagen, 26

  and competence, 354

  of unfertilized eggs, 320

Albumen, 215

Algae:

  Acetabularia, 89–98

  blue-green, 13

  brown, 153, 154

  development of egg, 196, 242–244

  life cycle, 154

Allantois, 325

Allophenic mice, 505

Alpha helix of polypeptide, 47

Alternation of generations, 156, 157

Alternative differentiation, 159

Amaroecium, 300

Ambystoma, 285

  (*See also* Salamander)

Amino acid sequence, 8, 14, 23, 34, 226, 258

Amino acids, 297

Amnion, 325

Amniotic folds, 328

Amoeba, 54, 58, 279

Amoeboid blastomeres in annual fishes, 512

Amoeboid movement, 59, 114, 117

Amphiaster, 231

Amphibia, 190

  developmental stages, 310

  egg, 283

    cortex, 285

    inversion, 286

    rotation, 284

    size, 201, 247

  embryonic axis, 350, 351

  exogastrula, 320, 343

  eye reconstitution, 418–421

  gastrulation, 321

  gene activation at gastrulation, 344

  induction, 344–353

Amphibia:

  ionic environment and tissue differentiation, 352

  limb development, 386–388

  limb regeneration, 396–403

  metamorphosis, 427–449

  neural-crest cell migration, 332

  neurulation, 331–337

  nucleus transfer, 269–271

  oogenesis, 210–214

  polyploidy and cell size, 269

  polyspermy, 233

  renal tumor, 270–272

  sperm track, 284–285

  tissue competence, 253–255

  tissue spreading, 319

  vital staining maps, 312, 313

Amphioxus, 189, 207, 247, 283, 294, 307–309, 329, 332, 342

  blastomere recombination, 342

  blastomere separation, 251

  egg size, 247

  gastrulation, 309

Amprino, R., 388

Anfinsen, C. B., 21

Animal-vegetal axis, 239, 245, 254–256

Animalization, 256, 257

Annelid, 143

  (*See also* Polychaete)

Annual fishes, development of, 511

Antheridia, 156

Antibody production, 475

Anticlinal division, 173

Antifertilizin, 220

Antigen-antibody reactions, 220, 336

Antigenic determinants, 475

Antimitotic agent, colchicine, 49

*Does not include references to readings or footnotes.

Antimitotic chemical messenger, 374

Antimycin, 297

Apical cell of meristem, 168–170
dome, 177
dominance, 172
ectodermal ridge, 393
growth center, 179
meristem, 165–171, 179

Arbacia, 298

Archegonium, 156, 218

Archenteron, 254, 256, 308, 309, 316, 320, 335, 343, 345
roof, 358

Arginine, 214

Aristotle, 4

Artemia, 265, 266

Ascaris, 245, 246, 249, 253, 267, 274, 287

Ascidia, 290, 293, 297

Ascidian, 126
asexual development, 299
blastomere segregation, 341–342
cleavage planes, 292
cytodifferentiation, 293
determination and cleavage, 290
egg size, 247, 294
egg symmetrization, 281
follicle cells, 290
gastrulation, 295
neurulation, 294
ooplasmic movements, 282
regeneration, 126

Ascidian tadpole larva, 293, 299, 337

Ascidiella, 247, 283, 287

Ascidiozooid, 299

Asexual development, 159, 166, 171, 299
ascidian, 299, 300
Aurelia, 161–162
carrot, 150, 151
fern, 156

Asexual development:
Hydra, 131–132
hydroid, 159, 165–180
plant meristem, 165
polychaete, 126, 145
spruce, 172
teratoma, 506
tobacco, 150, 151
tunicate, 299

Aspergillus, 20

Assembly:
of protein, 24–29, 34
self- (*see* Self-assembly)

Asterias, 306

Asters, 66, 67

Astral organization, 204

ATP, 46, 57, 188

Aurelia, 161, 247, 425

Austin, C. R., 217

Austrofundulus, 512

Autodifferentiation, 359

Auxin, 152, 175, 176

Available-space theory, 168

Axial filament, 196, 198

Axial gradient, 125, 142, 146

Axial growth sequence, 392

Axillary bud, 174, 179

Axis of symmetry, 326

Axolotl (*Necturus*), 334, 356

Axonal flow, 59

Axoneme growth, 49–52

Axopod, 49

Bacterial DNA, 16, 17, 30
disassembly and reassembly, 48
flagella, 48
viruses, 30

Baer, E. Von, 4

Baker, W. K., 486

Balancers, salamander, 356

Balanoglossids, 217, 221

Balinský, B. I., 388

Ball, E., 171

Ballard, W. W., 322

Barbulanympha, 64

Barth, L. C., 351, 448

Barth, L. G., 351, 448

Basal bodies, 52
basal disc (hydroid), 127, 133, 140
metabolism, 432

Basal-layer cells, 481

Basement (basal) membrane, 28, 90, 104, 215, 374, 382, 393, 412

Basement membrane and carcinoma, 493

Begonia, 151

Benzanthracene, 492

Benzopyrene, 492, 501

Beroë (ctenophore), 280

Berrill, N. J., 139

Bessis, M., 480

Bilateral symmetry, 254, 281, 283, 315

Bioelectric fields, 147, 402

Bioelectrical potentials, 143

Biogenetic law, 5

Biological clock, 98, 441, 450

Biological crystallizations, 28

Bipolar differentiation, 240, 241

Bispecific aggregates, 378

Bithorax pseudoallelic mutants, 455, 468

Blastema, 397
development, 404
formation, 397
morphogenesis, 401

Blastocoel, 254, 304, 308, 312, 329

Blastocyst, 328, 329, 485, 511, 513
fusion, 485

Blastoderm, 313, 321, 325, 453, 511, 513
fusion, 322, 325
gastrulation, 321

Blastoderm, subdivision, 322, 325

Blastodisc, 66, 313

Blastokinin, 329

Blastomeres, 238–240, 246, 253, 303, 313
adhesion of, 306
dispersal and reassociation, 513
induction by, 341
isolation, 250, 253, 287, 293, 341, 346
rotation, 342
size, 303

Blastoporal lip, 345, 358

Blastopore, 315, 316, 324, 346

Blastula, 238, 239, 254, 262, 263, 269, 272, 304, 305, 310
expansion, 312
nuclei, 272

Blastulation, 263, 513
in meroblastic vertebrates, 313

Blepharisma, 78

Blocks to polyspermy, 232, 233

Boke, N. J., 178

Bonellia, 425

Bonner, J. T., 4, 107

Botryllus, 299

Boveri, T., 245, 256, 267, 498

Brachypod limbs, 395

Brain graft, 143

Brain hormone of insects, 436

Brien, P., 127, 129

Briggs, R. W., 269

Brine shrimp, 265–266

Britten, R. J., 487

Brønstedt, H. V., 126

Brummett, A. R., 322

Bryozoans, 155

Bud, hydra, 159

Bud sequence in hydra, 162

Budding zone, 140

Bufo, 247

Bullough, W. S., 144, 374

Burnett, A., 130, 133, 139, 145, 448

Butyric acid, 229

Camosso, M., 389

Campbell, R. D., 130

Cancer, 349, 495

Cancer-inducing chemicals, 495

Cap formation in *Acetabularia*, 93–95

Capacitation (fertilization), 219

Carbon particles, as tracers, 171, 393

Carcinogens, 491, 500, 501

Carcinoma, 493, 496, 504
embryonal stem cells, 508

Carlson, J. G., 465

Carpels, 179

Carrot, cell culture and development, 150, 151

Carter, S. B., 498

Cartilage, 72, 400, 482

Casein, 475

Cataphoresis, 244

Cecropia, 435, 440

Cell, 3, 35
adhesion, 114, 115, 136, 304, 318, 334, 366, 368
affinities, 303
aggregation, 114
clones, 496
communication, 118
contact, 308
inhibition, 115, 119, 371, 378, 493, 498
cycle, 57–75, 266, 475, 477
death as a morphogenetic agent, 394, 404, 435, 442
differentiation, 71, 132, 136
division, 74, 151, 174, 176

Cell:
division: and differentiation, 71
planes of, 174
elongation, 174
enlargement, 175–176
evolution, 9
growth, 59, 74, 93, 136, 151
regulator, 498
guidance, 379
inclusion, 244
ligands, 120
lineage, 287
membrane, 40, 60
leakage, 498
and malignancy, 498
migration, 136, 324, 326, 332
movements, 365
number, 127, 133, 159, 182, 191, 296, 303, 307, 308, 313, 325, 444
population-density gradient, 326
populations, 488
categories of, 479
reassembly, 54
shape, 305, 337
during mitosis, 303
size, 58, 74, 93, 103, 151, 182, 190, 268, 444

Cell-transformation hypothesis, 509

Cellular diversification, 463

Cellular slime molds, 68

Cellulose, 91
microfibrils, 90, 104

Central bodies, 64
dogma, 15
tubules, 51

Centrifugation, 246, 279, 290
and polarity, 244

Centriole, 53, 64, 67, 196, 198, 204, 231, 494
abnormal relationship in tumor cells, 495

Centriole:
  and activation, 231
Centrolecithal, egg, 206, 265
Centrosome, 67, 230
Cephalic structures, 345
Ceratodus, 285
Chabry, L., 287
Chaetopterus, 185, 288, 290
Chalkley, H. W., 139
Chalones, 144, 374
Chambers, Robert, 66, 279
Chaos (amoeba), 89
Chemical carcinogens, 501
  diffusion theory, 142
  gradient systems, 119, 121, 125
Chemodetermination, 469
Chemodifferentiation, 455
Chemotaxis, 218
Chick blastoderm, 314
  eye development, 410–417
  gastrulation, 321
  limb development, 389–398
  sperm, 196
  twinning, 322–323
Child, C. M., 125, 126, 170, 255, 257
Chimeras, 130
Chimeric assemblies, 383
  wing bud, 390
Chlorophyll precursor, 44
  synthesis, 45
Chloroplast, 37, 43, 96, 97
  DNA, 44, 45, 96
  lamellae, 40
  membrane, 45
Cholesterol, 492
Chondrification, 482
Chondroblasts, 369, 395
Chondrocytes, 400, 484
Chondrogenesis, 483
Chondroitin sulfate, 484

Chondrosarcoma, 496
Chordamesoderm, 312, 316, 324, 354
Chordamesodermal plate, 316, 335
Chordaneuroplasm, 282
Chordates, 189
Chorioallantoic membrane, 375
Chorion, 215
Chromomeres, 275
Chromosomal puffing, 274
  hormonal control, 444–448
Chromosome diminution, 246, 267
Cilia, 51
Ciliary membrane, 51
Ciona, 281, 283, 290, 293
Circadian cycle, 68
Cis-trans position, 467
Cistron, 77, 445, 474
Clavelina, 126, 426
Cleavage, 54, 61, 65, 183, 185, 229, 267, 280, 303, 337
  asters, 234
  and blastula formation, 303
  course, 261
  and development, 237
  and differentiation, 251
  furrow, 66, 204, 280
  pattern, 246, 280–288, 295, 301
  plane, 239, 245, 251, 281, 304, 311
    and embryonic axis, 346
  rate, 237, 238, 296
Clement, A. C., 204, 290
Clever, U., 447
Cliona, 117
Clones and cloning, 72, 127, 461
Cloney, R. A., 337
Clymenella, 143
Cnidaria (*see* Coelenterata)

Cnidoblasts, 131, 133, 136
Cnidocytes, 133
Coacervates, 10
Coconut milk, 150
Coelenterata (Cnidaria), 129, 155, 161
  Aurelia strobilation, 161–162
  egg size, 247
  Hydra, 126, 141
  Rathkea, medusa budding, 160
  Tubularia regeneration, 126–127, 142–143
  oocyte, 206
Coelom, 254, 332
Coelomic sacs, 308
Cohen, S., 500
Collagen, 24–27, 90, 379, 382, 485
  aging, 26, 27
  fibril periodicity, 24
  fibrils, 413
  and morphogenesis, 379
  polymerization, 26
  reassembly, 26
Collagenase, 379
Collagenous endoskeleton, 132
Colon bacillus (*Escherichia*), 106
Colonial organism, 300
Colucci, V. L., 414
Colwin, A. L., 222, 223
Colwin, L. H., 222, 223
Competence, 353, 357, 359
  and aging, 354
Conditional development, 159
Conjugation, 9, 163, 217
Conklin, E. G., 183, 280, 281
Contact guidance, 379
  inhibition, 115, 119, 371, 383, 493, 498
    loss in malignant cell cultures, 498
Contractile filaments, 46

Contracting ring theory of cleavage, 66

Copulation, 219

Cornea, 412

Corneal epithelium, 410, 413

Corpora allata, 437

Corpus tissue, 174

Cortex of cell, 78, 80, 93
  egg, 279, 290
  and nucleus as an interacting
    system, 84

Cortical activation, 234
  differentiation, 78
  granules, 223, 226
  inheritance, 84
  localization, 280
  organization of cell, 99
    of eggs, 220, 301
  pattern, 80, 82, 99
  reassembly, 81
  segregation, 281
  ultrastructure, 279

Corticotypes, 87

Coulombre, A. J., 407, 412, 413, 420

Cowden, R. R., 296, 297

Crepidula, 220

Critical mass concept, 509

Cromwell, H. N., 501

Cross-fertilization, 215

Ctenophore, cortical plasm and
  cleavage pattern of, 280

Culmination process in cellular
  slime mold, 106–111

Cultured imaginal disc cells, 462

Curtis, A. S. G., 347

Cycad sperm cell, 198

Cyclic adenosine monophosphate
  (AMP), 142

Cyclopia, 409

Cytochrome system, 42

Cytodifferentiation, 4, 57, 71, 74,
  237, 238, 292, 382, 457, 465–
  474

Cytokinesis, 67

Cytokinins, 175–176

Cytoplasmic DNA, 214

Cytoplasmic streaming, 97

Cytoskeleton, 102

Dalcq, A., 282, 510

Dalyell, J. G., 126

Dan, K., 64, 304

Dandelion, 151

Danielli, J. F., 54

Daphnia, 268

Darwin, C., 4

Davidson, E. H., 344, 487

Dedifferentiated blastema cells, 400

Dedifferentiation, 71, 399

Degenerating cells, 394

DeLong, G. R., 420

Delta crystallin, 475

Dentalium, 288

Deoxyribonucleic acid (see DNA)

Dermis, 373, 376
  reformation, 493

Desmosome, 102, 116, 131, 136,
  304, 369, 374, 377

Determinate (mosaic) eggs, 247,
  286

Determination and cleavage, 290
  of eye rudiment, 408
  in imaginal discs, 455
  of limb axes, 385

Development of feather germs, 375

Developmental selection of genotypes, 424

Developmental timetable of mouse,
  328

Diapause, 438

Dibenzanthracene, 493, 501

Dictyostelium (cellular slime mold),
  105

Differential gene alteration, 274
  multiplication of cells, 399

Differentiation:
  of aggregation mass, 109
  of imaginal discs, 456
  in mixed cell aggregates, 377
  of optic vesicle, 408
  of spermatozoon, 199

Diffusion gradient, 107, 218, 359

Direct and indirect development,
  188

Disaggregation (disassociation),
  117, 365, 414

Disc activation, 470

Disulfide linkages, 21

Division and cell differentiation,
  74

DNA (deoxyribonucleic acid):
  assembly, 35
  bacterial (see Bacterial DNA)
  basal body, 52
  and carcinogenic target, 492, 501
  cell growth, 59
  chloroplast, 44, 45, 96
  code, 15–17, 99
  content relative to cell size, 89
  cytoplasmic, 214
  histone binding, 274, 448
  length of thread, 63
  in mammalian ovum, 328
  mass per cell, 61
  mitochondrial, 43, 44
  nerve cell growth, 59
  in oogenesis, 213
  plastid, 45
  polymerase, 15, 37
  polyoma virus, 496
  regulation, 475

DNA (deoxyribonucleic acid):
  replication during cell differen-
    tiation, 421
  in sperm nucleus, 196
  strand-folding following puffing,
    444, 445
  synthesis, 15, 474–475
    during amphibian develop-
      ment, 276
    in cell cycle, 37, 64, 70–72, 261,
      477, 511
    following fertilization, 264
    during limb regeneration, 399
    in malignant cell cycle, 497
    and retinotectal connections,
      419
  and tumor induction, 501
  viral, 30, 497
Dormin, 176
Dorsal lip of blastopore, 324, 343
Double gradient system, 255, 259
Doublet cells (*Paramecium*), 85
Driesch, H., 5, 250, 253, 255, 267,
  387
Drosophila, 5, 209, 274, 441–468
Dual function of fertilization, 217
Duality of egg development, 183,
  191, 298, 428, 438
Dulbecco, R., 497

Ecdysone, 437, 444, 446, 487
Echinoderms, 189, 221, 227
  (*See also* Sea urchin; Starfish)
Echiuroids, 217
Ecker, R. E., 213
Ecteinascidia, 247
Ectoderm, 316
Ectoderm-mesoderm system in
  limb development, 387
Ectodermal competence, 411
"Ectodermal-fold" technique, 354

Ectodermal ridge, 389, 391, 395
Ectoplacental cone, 328
Edds, M. V., 191
Egg, 193, 216
  aggregation, 510
  centriole, 231, 234
  cortex, 224, 227, 279, 290, 300,
    317, 346
  cytoplasmic organization, 281
  envelopes, 214
  establishment of polar axis, 242
  fixation of axis, 285
  inversion, 286
  localization patterns, 282
  numbers, 200
  polarity, 240
  pronucleus, 272
  size, 188, 189, 191, 200, 201, 206,
    210, 283, 294, 310
  as specialized cell, 201, 514
Electrical cataphoresis, 250
  communication, 119
  fields, 119, 126, 144, 241, 402
  polarity, 126, 170
Electrochemical communication,
  250
Electrolyte balance and induction,
  352
Ellis, C. H., 265
Embryo sac, 157–158
Embryoid teratomas, 508
Embryonal carcinomas, 506
Embryonic axis, 281, 286, 322, 324,
  325, 511
  fields, 355, 359
  limb discs, 404
  shield, 313, 322
Emergent evolution, 7
Endocytosis, 60
Endoderm, 312, 316
Endomesoderm, 305, 344

Endoplasmic reticulum, 21, 37, 38,
  63, 103, 208, 210, 474
  reassembly, 63
Endoskeleton, 132
Endosperm nucleus, 158
Energids, 453
Energy, 34
Enteromorpha, 153
Enteron, 333
Entosphenus, 247
Enucleate embryos, 344
Enzymes, 13, 24, 29, 43, 221, 486
  assembly, 22–24, 29
  cholinesterase, 107
  hybrid, 486
  hydrolytic, 37, 107
  lactate dehydrogenase, 24
  lysosome, 36
  membrane assemblies, 42
  mitochondrial, 43
  population, 29, 59
  repressible, 18
  specific, 113
  systems, 42
  virus assembly, 45
Ephestia, 458
Ephrussi, B., 487
Epiblast, 326, 369
Epibolic growth, 316
Epiboly, 313, 321, 512
Epidermal wound healing, 318
Epidermis, 127, 130, 140, 144, 313,
  372, 373, 376
Epidermis-dermis system, 379
Epigenetic development, 9, 11, 45
Epithelial-mesenchymal interac-
  tion, 372, 398
Epitheliomuscular cells, 131, 132
Epistylis, 101
Equipotential system, 387, 404, 469
Equivalence of nuclei, 267

Erythroblast, 481

Erythrocyte, 71, 479

Erythrocytic cell populations, 432

Erythropoietin, 480

Escherichia, 16, 106

Estrogen, 330

    effect on uterine cells, 478

Etkin, W., 429, 430

Eucaryote cell, 10, 13, 35, 55, 103

Euglena, 68

Euplotes, 84

Evaginations, 303

Evans, J., 379

Evocation, 341

Evocator, 348

Evolution of the cell, 9

Exogastrulation, 256, 320, 343

Expansion theory of cleavage, 66

Experimental activation, 229

Extension:

    of neural plates, 335

    of notochord, 332

Extent of metamorphosis, 440

Extracellular collagenous material, 146

Extracellular enzyme, phospho-diesterase, 109

Extracellular material, 102, 120, 132, 383

Extraembryonic blastoderm, 329

Eye, morphogenetic phases, 408

Fallopian tube, 219

Fankhauser, G., 268

Fate maps, 313

Feather germs, 376

    morphogenesis, 377

    primordia, 376

Female gametes, 155, 199

    pronucleus, 204, 217, 230, 234

Fern, 156, 168, 171, 218

    apical meristem analysis, 171, 172

    cycle, 154

    embryo, 157

    gametes, 196, 198

    spore development, 156

Ferritin, 60

Fertilizability of eggs, 219

Fertilization, 217, 223, 262, 264, 284, 285

    cone, 222, 225, 230

    membrane, 217, 223, 225, 233

    specificity of, 220

Fertilizin, 218

"Fertilizin" theory, 220

Fibroblasts, 485

Field concept, 169

First proteins, 262

Fischberg, M., 270

Fish, 190

    blastulation, 313

    development rate and tempera-ture, 187

    egg types, 209, 281

    fertilization, 285

    gastrulation, 321

    sperm, 196

    twinning, 322, 323

Fission:

    flatworms, 149

    protists, 84, 89

Flagellin, 48

Flagellum, 51, 222

Flatworm (planaria):

    centriolar orientation and cleavage, 290

    cleavage, 189, 240

    egg size, 247

    regeneration, 126, 144, 166

Flavoproteins, 42

Flax, 151

Floral apex, 178, 179

Floral hormone, 177

Floral meristem, 178

Floral pattern, 178

Floral primordia, 177–178

Florigen, 176, 177

Flowering apex, 165

Flowering plants, 157

Follicle cells, 206, 208, 211, 214, 215

Foreign inductors, 348

Form maintenance, 127

Frieden, E., 428

Frog (*see* Amphibia)

Frog kidney tumor, 270

Fruiting body (cellular slime mold), 112

Fucus, 153, 242, 251, 284

Fundulus, 322

Fusion:

    of mammalian eggs, 485

    of oocytes, 206

    of pronuclei, 217

Galston, A. W., 177

Galtsoff, P. S., 117

Gametes, 138, 149, 153, 195, 218

Gametogenesis, 153

Gametophyte, 153, 158

Gamma crystallin, 476

Garber, B., 377, 378

Gastric column, 133

Gastrodermis, 127, 131, 133, 140

Gastrula, 240, 254, 269, 305

Gastrulation, 253, 263, 268–269, 274–276, 293, 295, 305, 329, 344, 347, 352, 354

    convergence, 322

Gene activation during gastrulation, 344

Genes, 17
  operator, 17
  regulator, 17
  structural, 17
Genetic basis of cancer suscepti-
  bility, 505
Genetic control of developmental
  pathways, 31–33, 112, 466–
  469, 484–487
Genetic markers, 459
Genetic spiral (plant meristem),
  171
Germ band, 453
Germ-cell determination, 137
  differentiation, 147
Germ cells, 147
  and teratomas, 507
Germ line, 139
Germ layers, 312, 316, 322, 328,
  337
Germ ring, 321
Germinal vesicle, 204, 213
  rupture of, 281, 283
Giant chromosomes, 214, 443, 444
Giant insect larvae, 440
Gibberellins, 175–176
Gill absorption, 431
Globin, 481
Glucosamine incorporation, 480
Glucose polymerization, 113
Glutamine, 258
Glutinants (nematocysts), 136
Glycoprotein, 26
Golgi apparatus, 38, 136, 198, 199,
  210, 211
Gonadotropic hormones, 506
Gonyaulax, 68–70
Gradient concept in sea urchin
  development, 256
Gradient systems, 147
Gradient theory, 140, 253
Grana, 43, 45

Gray, J., 217
Gray crescent, 282, 312, 315, 320,
  346–348
Greenwald, G. S., 330
Griseofulvin (antimitotic agent),
  64
Grobstein, C., 67, 121, 379
Growth, 185, 202, 205
  Hydra, 131, 136, 139
  buds, 131
  and metamorphosis, 442
  oocytes, 205
  regulator, 172
Growth-rate gradient of meristem,
  170
Growth-stimulating agent in hy-
  dra, 140
Growth zone of hydra, 130
Guidice, G., 513
Gurdon, J. B., 270, 272
Gustafson, T., 304, 308
Gynandromorphs, 458

Hadorn, E., 459
Haeckel, E., 5
Haliotis, 221
Halteres, 455, 466
Hamburger, V., 388
Hämmerling, J., 93
Harker, J. E., 441
Harmonious equipotential sys-
  tem, 259
Harrison, R. G., 58, 385
Harrison stages, salamander, 310
Harvey, W., 4
Hay, E. D., 369, 371
Haynes, L., 134
Head-tail axis, 326
Heart cells, 119
Heliozoans, 49

Hemimetabolous insects, 454
Hemoglobin in metamorphosing
  tadpole, 431
Hemoglobin synthesis, 480, 481
Hensen's node (primitive knot),
  324, 348
Herbst, C., 250, 256
Hierarchy of cell and tissue adhe-
  siveness, 383
Hindbrain, 350
Hinrichs, M. A., 322
Histogenesis, 185
Histones, 21, 145, 274
Histozones, 296
Holistic viewpoint, 77
Holoblastic cleavage, 329
Holoblastic eggs, 183, 261, 279,
  281, 310, 337
Holometabolous insects, 433,
  454
Holtfreter, H., 509
Holtfreter, J., 316, 320, 346, 349,
  365, 371, 483, 484
Hormonal control:
  of chromosomal puffing, 446
  of metamorphosis, 428, 449
Hormone concentrations, 431
  insect prothoracic, 436
  threshold, 429, 439
    and rate of metamorphosis,
    430
Hormones, 175–176, 204, 423,
  433, 478
  cell differentiation, 478
  and gene activation, 444, 450
  and growth regulation, 175
  and limb development, 431
Hörstadius, S., 255, 258
Humphreys, T., 118
Hyaline layer of egg, 226, 303,
  304, 317
Hyaluronidase, 221

Hybrid embryos, 344
 enzymes, 484, 485
 regeneration in *Acetabularia,* 95
Hybridization, 276
Hydra, 126–139, 159, 206, 448
 cell and tissue differentiation, 132–133
 cellular dedifferentiation, 134
 germ cell differentiation and sexuality, 137–139
 growth and form, 129–130
 individuality, 127
 nematocyst differentiation, 136
 reconstitution of form, 139–140
 tissue maintenance, 132
Hydranth, 159
Hydranth formation, 143
Hydroid, 126, 159
Hydroides, 221–224
Hypertonic seawater, 229
Hypoblast, 326, 369
Hypocotyl, 150
Hypostome, 130, 133, 139, 142
 differentiation, 140
Hypothalamus, 429

Ilyanassa, 204, 288, 290
Imaginal discs, 435, 440
 determination of, 455
 differentiation of, 457
 initiation, 461
Immunological reactions, 475
Implantation of mammalian egg, 328, 329
Implanted imaginal discs, 457
Incomplete metamorphosis, 440
Indeterminate (regulative) egg, 247, 286, 309
Individuality, 127
Individuation fields, 404
Inducer, 140

Inducible enzyme, 18
Induction, 341, 351, 359
 by blastomeres, 341
 and electrolyte balance, 352
 invertebrates, 343
Inductor, 342, 348
Inductor-inhibitor hypothesis, 140, 145
Inequality of division, 216
Infective differentiation, 482
Infraciliature, 79
Initiation of imaginal discs, 461
Inner cell mass of blastocyst, 328, 329, 513
Insect, 120, 189, 209
 chromosomal puffs, 274, 444, 445
 cleavage, 453, 454
 cuticle, 436
 embryonic development, 453
 genetic control of development, 466–469
 germ band, 454
 growth stages, 437
 gynandromorph, 458
 hormonal control of puffing, 446
 imaginal discs, determination, 455, 463
 metamorphosis, 433–443
Integration, 184
 of gastrulation, 315
Intercellular adhesion (cohesion), 9, 120, 365
 development of, 369
Intercellular communication, 119, 372, 450
Intercellular junctions, 102, 369
Intercellular material, 118
 (*See also* Extracellular material)
Intercleavage interval, 261
Interphase nucleus, 230
 in cell cycle, 64
 periods, 238

Interphase nucleus, state, 237
Interspecific grafting, 94
Interstitial cells of hydra, 131, 132, 136
Intestinal villus, 479
Intranuclear electrolyte balance, 450
Invagination, 240, 256, 293, 303–306, 315, 318
 and morphogenesis, 305
 relative adhesiveness hypothesis, 308
 tension hypothesis, 306
Invaginative gastrulation, 308
Inversion:
 of amphibian egg, 286
 of polar axis, 245, 246
Involution, 303, 315, 331
Iodine, 162, 426
Ionizing radiations, 492
Iron accumulation, 480
Isolated blastomeres, 287, 296
Isozymes, 24
Iverson, O. H., 144

Jacob, F., 17, 469
Jacobson, A. G., 411
Jacobson, C.-O., 334, 335
Jacobson, M., 119, 121
Jaffe, L. F., 242
Jelly coat, 221
Jellyfish, 161
Jeon, K. W., 54
Joseffson, L., 258
Junctional communication, 259, 373, 382, 383
Junctional membranes, 450
Juvenile forms, 189
Juvenile hormone, 437, 439, 446, 456

Keratin synthesis, 374

Keyhole limpet, 220

Kidney rudiment, 380

Kinetodesma, 79

Kinetosome, 78

Kinety, 79, 85

King, T. J., 269

Kinin, 152

Kleinsmith, L. G., 506

Kollar, E. J., 377

Kollros, J. J., 430

Konigsberg, I. R., 484

Krebs cycle, 42

Kroeger, H., 444, 445, 458, 459

Labile state, 513

Lallier, R., 258

Lampbrush chromosomes, 213, 275

Lang, A., 156

Larval growth, 442

Larval organisms, 190, 423, 424, 433

Larval skeleton, 297, 307

Lash, J., 484

Lateral mesodermal plate, 332

Leaf blade, 174

Leaf determination, 173

Leaf-initiating growth periods, 171

Leaf primordia, 167, 174
  initiation of, 167

Leakage in cell membrane, 499

Leblond, C. P., 59, 479

Leikola, A., 348

Lenique, H. M., 146

Lens competence, 411

Lens differentiation, 412, 475, 476

Lens epithelium and malignant change, 500

Lens fibercell, 476

Lens induction, 411

Lens inhibitory factors, 412

Lens placode, 410, 475

Lens regeneration, 414

Leukemia, 494

Levels of cell differentiation, 473
  organization, 7, 11, 365

Levi-Montalcini, R., 500

Lewis, E. B., 467, 468

Lillie, F. R., 220, 376

Lillie, R. S., 229

Limb:
  development, 385, 404
  embryonic determination, 385
  gene deficiencies, 395

Limb-bud combinations, 390

Limb disc, 385

Limb induction, 388

Limb morphogenesis, 393

Limb regeneration, 397
  and nerve supply, 403

Limb skeleton, 393

Lindahl, P. E., 255

Lineage cell, 136
  (See also Stem cells)

Lineus, 143

Linum, 151

Lithium, 409

Liverwort, 218

Locke, M., 463

Loeb, J., 229

Loewenstein, W. R., 119, 144, 250, 372, 373

Lorch, I. J., 54

Lumbrinereis, 247

Luminescence, 68

Lund, E. J., 126, 242

Lundblad, M., 146

Lupin, 171

Lymphomas, 494

Lymphosarcomas, 496

Lysine, 258

Lysis, 93

Lysosomes, 35, 37

Lytic enzymes, 221, 223

McCallion, D. J., 348

Macrocrystallinity, 244

Macrogametes, 195, 199

Macromeres, 251, 287

Macronucleus, 80, 89

Macrovilli, 210

Macula adherens, 116

Maggio, R., de, 263

Maintenance of form, 127

Maintenance factor, 391

Male gamete, 155, 195, 198

Male pronucleus, 217, 230, 234, 285

Malignancy, 349, 491

Malignant transformation, 492, 499

Malignant tumors, 491

L-Mallic acid, 218

Mammal:
  cleavage, 329
  developmental timetable, 328
  egg, 200
  fusion of blastomeres, 485, 510
  implantation of blastocyst, 328, 330
  ovum, 210
  sperm, 196
  teratomas, 506–508
  tumors, 493–505
  twinning, 328

Mammary cancer, 505

Mangold, O., 343

Markert, C. L., 296, 297

"Masked-messenger" hypothesis, 262

"Masked" organizer hypothesis, 348

Masked viruses, 504

Matthews, M. B., 121

Maturation divisions, 198, 202
    of the egg, 227
    of sperm cells, 198

Mazia, D., 64

Medusa, 159–161

Meiosis, 152, 153, 163, 198, 202, 203

Melanoblasts, 487

Melanocytes, 486

Membrane, 37–41, 57, 447
    in cleavage, 66
    and cortex expansion during cell division, 66
    of malignant cells, 514
    ruffles (cell locomotion), 115
    structure, 41
    system, 103
    vesicles, 208

Meristem, 162
    growth, 170
    transition to flowering, 177

Meroblastic eggs, 143, 281, 313, 321

Mesenchymal blastema, 400

Mesenchyme, 256, 354

Mesoderm, 313

Mesodermal crescent, 341

Mesodermal maintenance factor, 391

Mesodermal mantle, 358

Mesodermal pouches, 240

Mesodermal somites, 385

Mesodermalizing principle, 350

Mesoglea, role of, 127, 132, 133, 138

Mesomeres, 251

Metabolic gradients, 125
    theory of, 170

Metamorphic development, 425

Metamorphic transformation, 423, 433

Metamorphosis, 271, 297, 298, 337, 423, 433, 455
    biochemical changes in amphibians, 427
    and evolution, 424
    and growth, 442
    hormonal control in amphibians, 428
    muscle reconstruction, 442
    structural changes in amphibians, 427

Methylcholanthrene, 492

Microcephalic embryo, 286

Microciona, 117

Microenvironments:
    and cell differentiation, 146
    and tissue mass effect, 481

Microfibrils, 91, 336, 338

Microgametes, 195

Micromeres, 251, 254, 281, 287, 297, 307, 312

Micronucleus, 78

Micropinocytosis, 60

Micropyle, 226, 234, 285

Microtubules, 91, 337, 338, 379
    in axoneme of Actinosphaerium, 49
    in blood platelets, 51
    in centriole, 53
    in flagellum and cilium, 49
    in mitosis, 62
    in orientation and cell shape, 55, 104, 337
    in reassembly, 49
    in spindle, 63

Microvilli, 210

Midbrain, 350

Midpiece (spermatozoon), 197

Migrating nuclei, 454

Migration of pronuclei, 230

Milaire, J., 389, 393

Mintz, B., 485, 486, 505, 510

Mitchison, J. M., 66

Mitochondria, 37
    in Acetabularia, 96
    in ascidian embryo, 293
    in spermatogenesis, 198
    in spermatozoon, 230
    structure of, 42
    in vitellogenesis, 208, 210

Mitochondrial DNA, 43, 44

Mitosis, 54, 60, 61, 64, 70, 303

Mitotic apparatus, 204
    growth of, 253
    orientation, 251

Mitotic spindle, 64, 204, 231

Mixed cell aggregates, 365–368

Modulation, 75, 133, 399, 474

Mollusk, 189
    centrifugation of egg, 279
    cleavage, 287–289
    fertilization, 220, 221, 233

Molting cycle, 436

Molting hormone, 436

Monod, J., 17, 469

Monroy, A., 228, 263

Moore, A. R., 305

Morel, G. M., 173

Morphogenesis and invaginations, 305

Morphogenetic field, 165, 169, 180, 355, 358–359, 448, 458, 465
    movements, 318, 337
    phase, 238

Morula, 303, 328

Moruloid, 151

Mosaic (determinate) egg, 247, 286, 296, 454

Mosaic mouse, 485

Moscona, A., 120, 367, 369, 377–379

Mosquito, 208

Moss, 218

Motility of spermatozoa, 218

Mouse:
  developmental timetable, 328
  stages of development, 328,
    330
Mucopolysaccharide, 102, 114, 140,
  226, 304, 413
  in cortical granules, 212
Mucoprotein, 35, 104
Multienzyme systems, 43, 77, 485,
  497
Multinucleate myotubes, 484
Multiple embryos, 322
Multipotent cells, 513
Muscle cell, 72
Muscle tube, 485
Mussel, 221
Mutations:
  and morphogenesis, 394
    polyoma virus, 497
    slime mold, 112
Myelin sheath, 40
Myoblast fusion, 485
Myofibrils, 133, 484
Myogenesis, 484
Myogenic mesoderm, 393
Myonemes, 133
Myoplasm, 292
Myosin, 37, 46
Myotome, 313
Mytilus, 221
"Myxomyosin," 51

NaCNS, 256
Nanney, D. L., 13, 38, 74, 77
Nauplius larva, 265
Necrosis, 394
Needham, J., 185, 188, 280
Nematocyst, 130, 136
Nemertean, 143
Neoblasts, 132
Neoplasia, 504

Neotenous larvae, 443
Nephrogenic mesenchyme, 380
Nephrotome, 332
Nereis, 207, 240, 287–288
Nerve axon, 142
  cell growth, 59
  growth factor, 60, 500
Neural competence, 354
Neural crest, 332, 487
  cell migration, 332
  derivatives, 354
Neural differentiation, 343
Neural folds, 331
Neural groove, 309, 331, 337
Neural induction in protochordates,
  343
Neural plate, 294, 309, 331, 332,
  335, 344
Neural retina, 407, 412, 414
Neural trophic factor, 403
Neural tube, 294, 331, 343
Neuralizing principle, 350
Neuroectoderm, 358
Neuronal specificities, 419
Neurons, 479
Neuroplasm, 59
Neurosecretory granules, 140
Neurula, 269, 294, 331
Neurula nuclei, 272
Neurulation, 268, 330
Nicotiana, 151
Nieuwkoop, P. D., 350, 354,
  358
Noncellular slime molds, 103
Nonsexual development (*see* Asex-
  ual development)
Nonviral carcinogenesis, 500
Notochord, 294, 313, 331, 332, 342,
  343, 350
  explantation, 350
  lengthening, 332
Nucellus, 157
Nuclear conjugation, 163

Nuclear membrane, 21, 35, 244
  relation to endoplasmic retic-
    ulum, 63
Nuclear pores, 35, 38
Nucleation centers, 52
Nucleic acid, 213
  synthesis, 272
    during cleavage, 264
Nucleocytoplasmic balance, 74
Nucleocytoplasmic interaction, 58
Nucleocytoplasmic ratio, 239, 259
Nucleolar marker, 270
Nucleoli, as cell markers, 400
Nucleolus, 21, 35, 213, 269
  number of, 214
Nucleotide, 29
  in polyoma virus, 496
  sequence, 15
Nucleus, 21
  effect of cytoplasmic environ-
    ment, 272
  potency, 268
  size, 272
  transplantation, 268
Nurse cells, 208, 214
Nymph, 438

Oikopleura, 294
One gene–one enzyme concept, 24
Ontogenesis, 184, 185
Ontogenetic relics, 424
Oocyte, 200–215
  accessory cells, 206
  amino acid pool, 262
  amphibian, 210, 211
  asymmetry inheritance, 288
  comparison with spermatocyte,
    202
  follicle cells, 206
  fusion, 206
  growth, 205–214

Oocyte:
  mammalian, 188
  maturation, 203
  metabolism, 205
  nucleus, 214
  nucleus transfer, 272, 275
  polar axis, 239
  primary, 193
  relative growth of nucleus, 212
  solitary, 206
  viscosity at maturation, 244
  yolk accumulation, amphibian, 210
  yolk transport, 206
Oogenesis, 202–214
Oogonia, 202, 480
Ooplasmic crescent, 341
Ooplasmic territories, 296, 309
Oparin, A. L., 10
Oparin hypothesis, 9
Operator genes, 17
Operon hypothesis, 16, 18, 45, 469
Ophryotricha, 206
Oppenheimer, J., 194, 339
Optic cup, 409
Optic nerve regeneration, 419
Optic tectum, nerve reestablishment, 418, 419
Optic vesicle, 408, 409, 412
  induction, 412
Optic-vesicle-lens system, 421
Oral primordium (ciliates), 82
Organelles, 37
  basal body, 52
  centriole (and centrosome), 53, 64, 67, 198, 230, 231, 233, 290
  chloroplast, 37, 43, 96, 97
  cilium, 57
  flagellum, 51
  Golgi apparatus, 38, 136, 198–199, 210–211
  lysosome, 35, 37

Organelles:
  mitochondrion, 37, 42, 96, 198, 208, 210, 230
  nucleus, 21
  spindle, 64–67, 234
  yolk platelet, 40, 208, 210, 214, 284
Organization:
  of amphibian blastula, 310
  of the shoot apex, 165
Organization center, 147, 325, 453
"Organizer," 343, 348
Ovaries, 137, 160
Overgrowth-stimulating factor, 499
Ovogenesis, 138
Ovule, 157
Oxidation-reduction gradient, 249
Oxidative metabolism, 257
Oxygen consumption during fertilization, 227

Pancreas, 474
Pancreatic cells, differentiation of, 474
Pancreatic enzymes, 382
Pancreatic rudiment, 382
Papilloma, 502, 504
  virus, 504
Paracrystalline organization, 46, 55
Paramecium, 51
  cilia disassembly and reassembly, 51
  clones, 80
  cortical inheritance, 85
  fission and cell cycle, 68
  giant, multinucleate, 58
  infraciliature, 84–85
Parenchyma cells, 150
Parthenogenesis, 229
Parthenogenetic development, 153
Pasteels, J. J., 284

Patella, 287, 288
Pattern of feather buds, 379
Pattern and fields, 356
Patterns of differentiation, 286
Pectinaria, 64
Pelmatohydra, 136
Pelvetia, 153, 242
Periacrosomal material, 222
Perianth, 178
Periblast, 313
Periclinal division, 173
Perijunctional insulation, 121
Periplasm, 453
Perivitelline space, 226
Perpetual growth of axon, 59
Petals, 179
Petiole, 174
"Phage," 30
Phagocytes, 442
Phagocytosis, 60, 394
Pharynx, 316
Phospholipids, 40, 42
Photoperiodism, 97, 177
Photosynthesis, 68
  rate of, 96
Phragmoplast, 67
Physarum, 68, 103
Physiological polyspermy, 233–234
Picken, L., 104
Pierce, G. B., 506
Pigment cells, 332, 354, 481
Pigmentation pattern, 427
Pigmented epithelium, 408, 410, 415, 416
Pinocytosis, 210
Pinocytotic vesicles, 40
Pituitary hormones and metamorphosis, 428
Plane of mitosis, 374
Plant meristem, 165
Planula, 161
Plasma membrane, 35, 90, 222–242, 284, 290

Plasmagel layer, 279

Plasmalemma, 90

(*See also* Plasma membrane)

Plasmodium, 70

Plastid DNA, 45

Plastids, 37, 44

Plastochrons, 171

Ploidy, 58, 88, 153, 163, 237, 267

Pluteus larva, 253, 257, 263, 298, 426

Pluteus larva reconstitution, 513

Polar axis, 204, 245, 251, 255, 311, 341

   inversion, 245

Polar body, 202, 239, 270

Polar cytoplasm, 313

Polar disc, 328

Polar lobe, 288

Polarity, 237

   of animal eggs, 239

   and centrifugation, 245

   in ciliary fields, 356

   inducer hypothesis, 140

   inversion, 286

   in regeneration, 127

   in tissue, 143

Polio virus, 29

Pollen tube, 158

Polyasters, 233

Polychaete, 189

   centrifugation of eggs, 244

   cleavage and cortical segregation, 287–289

   developmental timetable, 287

   egg size, 223, 247

   fertilization, 221, 222, 226

   oogenesis, 206–208

   polar lobe, 287, 288

   regeneration, 126, 145

   stages of development, 288

Polycyclic hydrocarbons, 492, 495, 501

Polydactyl mutants, 391, 395

Polymerization, 14

   cellulose, 91

   protein, effect of medium, 25

   spontaneous, 9, 14, 34

Polymorphic development, 159

Polyoma virus, 496

Polyp, 161

Polypeptide, 9, 14, 16, 23, 24, 30, 34

   assembly, 9, 16

   chain, 21, 47

   conformations, 47

   sequence, 473

   tertiary protein organization, 23, 24, 34

   virus sheath (TMV), 30

Polyribosomes, 262, 266, 276

   assembly, 45, 64

Polysaccharide, 9

   extracellular, 105

   synthesis in *Acetabularia* cap, 94

   synthesis in sheath of *Dictyostelium*, 112

Polysome, 16

   (*See also* Polyribosome)

Polyspermy, 215, 230, 232, 268, 285

Polyspondylium, 106

Pomataceros, 221

Porphyrin, 10, 14

Postembryonic growth phase, 201

Potency:

   of isolated blastomeres, 249

   of nuclei, 268

Precartilage, 395

Precartilaginous mesenchyme, 414

Precocious determination, 298

Precocious metamorphosis, 426

Premalignant cells, 505

Prepattern in reaggregated tissue, 460

Prepatterns, 457, 465, 470, 510

Preskeletal mesoderm, 393

Presumptive endoderm, 316

Presumptive epidermis, 352

Presumptive mesoderm, 316

Presumptive muscle, 393

Presumptive notochord, 309, 316, 325

Presumptive retina, 408

Presumptive territories, 293

Primary germ layers, 275

Primary induction, 352

Primary-inductor analysis, 351

Primary inductors, 348

Primary mesenchyme, 254, 263, 297, 308

Primary oocyte, 230

Primitive gut, 316

Primitive knot, 324

Primitive shield, 322

Primitive streak, 324, 328, 369, 511

Proamniotic cavity, 328

Procambial tissue, 174

Procaryote cells, 10, 13

Process of self-assembly, 473

Proerythroblast, 480

Progesterone, 204, 330

Prolactin, 429

Prolastids, 44

Proliferation, 130

Pronuclei, migration of, 230, 282

Prospective-lens ectoderm, 411

Protein, 10, 14

   assembly, 24–29, 34

   molecules, 21

   reassembly, 23

   structure, 23, 26

   synthesis, 15, 64, 258, 262

Protein-lipid complex, 41

Proteinoids, 10

Prothalloid embryos, 157

Prothallus, 156

Prothoracic gland, 437, 439

Protists, 149

Protochordates, 189
Protodifferentiated state, 474
Protofilaments, 51
Protopterus, 285
Provascular corpus tissue, 178
Pseudoplasmodium, 109, 114
Pseudopodia, 110, 308
Puff patterns, 445
Puffing, 213, 444
  and genic information, 444
  (*See also* chromosomal puffing)
Pupal diapause, 437
  reorganization, 441
Pupation, 435, 444
Puromycin, 70, 204, 297, 497

Radial symmetry, 178
Rana, 213, 247, 269, 351, 411
Rappaport, R., 67
Rate of cell respiration, 238
  cleavage, 259
  development, 186
  oxygen consumption during
    development, 263
Rathkea, 160
Ratio of cytoplasm and yolk, 239
Rawles, M., 375
Reaggregation, 365, 414
  and assembly, 459
  of tissue cells, 117
Reassembly:
  of amoeba, 54
  of cells, 300
  of collagen fibrils, 24
  of enzymes, 300
  of organisms, 300
  of ribonuclease, 23
  of *Stentor*, 82
  of tissues, 300
  of virus, 300
  (*See also* Self-assembly)

Recapitulation theory, 190
Reconstitution:
  of form, 129, 134
  of limb, 396
  sponge, 300
Red blood cell system, 479, 480
Redifferentiation, 149, 399
Reduction division, 153, 163
  (*See also* Meiosis)
Reductionist viewpoint, 77
Refertilization, 224
Regeneration, 126, 143, 385
  blastema, 404
  eye, 415
  Hydra, 139
  limb, 396
  nerve dependency, 403
  planarian, 144
  polychaete, 126, 145
  optic nerve, 420
  retina, 418
  specific repressors, 144
  Tubularia, 143
Regional differentiation of neural
  tissue, 346
Regional organization of central
  nervous system, 358
Regional specificity of induction,
  350
Regression of tumors, 500
Regulative development, 247, 286
Regulative eggs, 309
Regulator genes, 17
Relative growth, 167
Renal adenocarcinoma, 271
Reniers-Decoen, M., 129
Repressible enzyme, 18
Repressor, 18
Reproductive cells, 149, 155
Respiratory enzyme assemblies, 42
Respiratory metabolism, 238
Reticular membranes, 63

Reticulocyte, 481
Reticulo-endothelial system, 479
Retina, 408, 416
Retinal-cell differentiation, 416
Retinal cytoarchitecture, 416
Retinal determination, 408
Retinal differentiation, 416
Retinal pigment cell, 72
Retinal-rod structure, 40
Revel, J. P., 371
Reverberi, G., 341, 342
Rhizoid, 91, 92, 242
Rhodnius, 437, 463
Rhynchosciara, 441
Ribosomal function, 474
Ribosomal RNA, 16, 213–214, 276,
  474
Ribosomes, 15–17, 208, 210, 213,
  262, 276, 297
  and fertilization, 226
RNA (ribonucleic acid), 15
  amphibian development, 265
  basal body, 52
  chloroplast, 44, 45, 97
  cytodifferentiation, 475
  cytoplasmic, 98, 99
  floral apical meristem, 177
  frog tail resorption, 133
  gastrulation, 276, 298, 344
  histones, 274
  lampbrush chromosome, 275
  masked messenger, 262, 266
  messenger, 15, 18, 98, 226, 262,
    265–266, 344, 474, 476
  mitochondrial, 43
  molting, 437
  nerve cell growth, 60
  oogenesis, 212
  polymerase, 15
  puffing, 445, 447
  rate of synthesis, 70
  regenerating limb mesoderm, 393

RNA (ribonucleic acid):
  ribosomal, 16, 213–214, 276, 474
  stored messenger, 266
  synthesis: arrest of, 481
    in cell cycle, 60, 64, 70
    during development, 276, 297
    suppression in mature oocyte, 226
  transcription, 15
  translation, 15
  transplanted nuclei, 272, 273
  tumor induction, 497, 501
  types of, 15, 16
  viral, 30
Robertson, J. D., 38, 41
Romberger, J. A., 171
Root apex, 165
Root hormones, 176
Rose, S. M., 143–145
Rotation of symmetrization, 285
Rous sarcoma cells, 500
Rous sarcoma virus, 496
Roux, W., 5, 333
Rubin, H., 498, 499
Runnström, J., 255
Runnström-Hörstadius theory, 255
Rutter, W. J., 379, 474

Sabellaria, 288
Saccoglossus, fertilization in, 221–224
Salamander, 153, 210
  stages of development, 310
Salivary glands, 274
Sarcomas, 493, 496, 500
Saunders, J. W., 388, 389, 394
Saunders-Zwilling theory, 391
Saxén, L., 351
Schafer, P., 495
Schultz, J., 275
Schwann cells, 40

Sclera, 413
Sclerotome, 332
Scott, F. M., 300
Scyphistoma, 161
Sea urchin, 189
  animalization of egg, 257
  blastomere separation, 250, 254–256
  cleavage and developmental rate, 252, 262
  cleavage asters, 232
  egg size, 247
  fertilization, 228
  gene products and gastrulation, 276
  gradients in developing egg, 253–258
  larval stages, 248, 249
  sperm, 196
  stages of development, 248–249
  vegetalization, 257
  yolk reduction effect, 256
Seaweeds, 153
Second metamorphosis, 430
Secondary embryo, 343, 345
Secondary inductors, 354
Segmentation (strobilation), 161
Segregation of blastomeres, 253
Selective adhesion, 368, 371
Selective cell adhesion, 118, 368
Self-assembly (reassembly), 11
  ascidian, 300
  blastomeres, 512
  cell (amoeba), 54
  cell (stentor), 83
  cilium, 58
  flagellum, 48
  flatworm, 126
  Hydra, 139–140
  microtubules, 48–50
  protein, 24–29, 34

Self-assembly (reassembly):
  sea urchin embryo, 513
  slime mold, 106–109
  tissue, 375
  Tubularia, 126, 142
  virus, 29–33
  (See also Reassembly)
Self-cataphoresis, 244, 259
Self-differentiation and tissue mass, 509
Self-organization, 375
Self-reorganization, 383
Seminiferous tubules, 479
Sensory bristles, development of, 464
Sepals, 179
Serial nuclear transplantation, 269
Serpula, 247
Sex hormones, 492
Sexual reproduction, 149, 152
Sexual state, 129
Sexuality, 136–138, 160
Sheath material of cellular slime mold, 112
Shell membranes, 215
Shoot apex, 156, 170
Shoot apical meristems, 168
Shope, R. E., 502, 503
Shope rabbit papilloma, 502
Shostak, S., 132
Siredon, 213
Situs inversus, 286
Size:
  of cells, 80, 89
  and form, 146
Skeletal spicules, 307
Skin, 373
Skin wound, 493
Slime mold:
  cellular, 105
  mutants, 112

Slime mold, noncellular, 51

Smith, L. D., 213

Snake venom nerve-growth factor, 500

Snodgrass, R. E., 433

Snow, M., 168

Soma, K., 171

Somatic cells, multipotent, 299

Somatic crossing-over experiments, 461

Somatic plant cells, development of, 155

Somites, 328, 331, 343, 385, 389

Sondhi, K. C., 460

Sorting-out, 117, 365, 369

Sonnenborn, T. M., 77

Spangenberg, D., 162

Spatial pattern, 453

Species-specific factor, 120

Species-specificity, 118

Specific induction, 345

Specific inhibitors, 147

Specificity of neuronal connections, 418

Specificity of sorting-out, 368

Spek, J., 240, 242, 280

Spemann, H., 343, 348, 409, 411

Sperm-activating agent, 218

Sperm aster, 230

Sperm cell, 133, 224

Sperm-cell differentiation, 197

Sperm centriole, 231, 232

Sperm midpiece, 195

Sperm nucleus, 196, 223, 230, 272

Sperm penetration, 223

Sperm structure, 195

Sperm track, 285

Spermatids, 198

Spermatocytes, 198

Spermatogenesis, 138, 198

Spermatogonia, 198, 480

Spermatozoa, 215, 218

storage of, 219

Spermioteliosis, 198

Sperry, R. W., 418, 419

Sphagnum moss, 151

Spiegel, M., 118

Spindle elongation, 67

Spindle formation, 234

Spindle microtubules, 64

Spinocaudal structure, 350

Spiral phyllotaxis, 171

Spiralia, 189, 190

Spirorbis, 208

Spirostomum, 78

Sponges, 117

disaggregation, 117, 119

reaggregation, 117, 119

Spontaneous reassociation, 30

Spore development, 156

Spores, 149, 163

Sporocarp, 106

Sporophyte, 153, 157

Spratt, N. T., 324, 326, 371

Spreading of epithelia, 313, 318, 320, 337

Spruce shoot apex, 172

Stages of feather germ development, 376

Stamen primordia, 179

Stamens, 158

Starfish:

blastula formation, 303–305

electrical coupling of blastomeres, 251

gastrulation, 306–308

larvae, 307

membrane potential, 225

Stebbins, G. L., 174

Steinberg, M. S., 368

Stem cells, 479, 488, 508

and teratomas, 507

Stenoteles (nematocysts), 136

Stentor, 78–88

cortical pattern, 83

grafting, 87

reconstitution, 81

structure, 80

synchronization of graft reconstruction, 88

Stern, C., 425, 457, 465

Stern, H., 13, 38, 74

Stevens, L. C., 507, 508

Stimulator-inhibitor gradients, 469

Stolon, 159

Stomodaeum, 254, 316

Stone, L. S., 316, 416

Stratification of egg contents, 244

Strobilation, 162, 300, 426

Structural genes, 17

Structural proteins, 14, 24

Styela, 280–284, 290, 294

Subcortical cytoplasmic currents, 66

Sublethal cytolysis, 348

Suctorians, 49

Supracellular control, 421

Supracellular matrices, 119

Supramolecular aggregates, 30

Surface coat, 304, 316

Surface volume ratio, 37

and cell division, 61

and cell shape, 103

and membrane systems, 42, 55

oocyte growth, 205

Sussman, M., 114

Swann, M. M., 66

Symbiotic algae, 130

Symmetrization, 281, 286

Synchronization of graft reconstitution in Stentor, 88

Synchronization rhythm in heart cell culture, 119

Synchronous cloning, 72

Synchronous division in *Physarum*, 70

Syncytium, 136

Tadano, M., 246

Tail resorption, 433

Taraxacum, 151

Tartar, V., 81

Taylor, A. C., 366, 375

Teleology, 184

Teleost, 313

Telolecithal egg, 206, 239

Temperature:

and developmental rate, 186

and twinning, 325

Temperature-dependent development, 160

Temperature range for normal development, 238

Teratomas, 491, 494, 506

differentiation of, 507

induction of, 506

Teredo, 289

Termite, 425

Testes, 137, 160

Testicular teratoma, 508

Tetrahymena, 51

biological clock, 68

clones, 80

corticotypes, 87

Thalidomide, 394

Thallus, 242

Theory of metabolic gradients, 125

Theory of recapitulation, 5, 188

Thymidine, 71, 399

Thymine, 71

Thyroid hormone, 429

and basal metabolism 432

Thyroid-stimulating hormone, 429

Thyrotropic-releasing factor, 429

Thyroxine, 162, 428

Thyzanozoon, 247

Tight junctions, 369

Timing of metamorphosis, 440

Tissue assembly, 375

Tissue differentiation, 132

Tissue hierarchy, 365, 368

Tissue hybrid interaction, 365, 368

Tissue interaction, 377

Tissue maintenance, 132, 373

Tissue mass:

and cytodifferentiation, 382

and self-differentiation, 509

Tissue movement, 306, 338

Tissue replacement, 478

Tissue segregation, 333

Tissue-specific differentiation, 73

Tissue-specific hierarchy, 383

Tissue-specific inhibitor, 383

Tissue-specific intercellular adhesion, 371

Tissue specificity, 366

Tissue spreading, 313, 315, 318, 326

Tobacco, plantlet development from pollen grain, 151

Tobacco mosaic virus (TMV), 29, 49

Toivonen, S., 350, 351

Tomopteris, 206

Topographical reorganization during gastrulation, 295

Totipotency of nuclei, 274

Totipotent somatic cells, 150, 156, 163

Townes, P. L., 365, 371

Toxipneustes, 247

Transcription, 15

Transdetermination, 461–463, 470

Transformative development, 161

Translation, 15

Transplantable tumors, 507

Transplantation of nuclei, 268

Trelstad, R. L., 369, 371

Trembley, A., 129

Trinkaus, J. P., 314, 500

Triploid tadpoles, 270

Tritiated thymidine, 54, 272, 416

Triturus, 210, 213, 239, 247, 285, 414

Trochophore larva, 288

Trophoblast, 328, 329

Tropocollagen, 26

Trunk-tail inductor, 345

Trypsin, as vegetalizing agent, 256

Trypsin digestion, 375, 380

Tryptophan, 258

Tubularia, 126, 140, 142, 144, 206, 426

Tumor development, 502

Tumor induction and membrane leakage, 499

Tumor initiation, 502

Tumor myeloma, 487

Tumor progression, 502

Tumor promotion, 502

Tumor-type agglutinans, 500

Tumors, 151, 271, 491, 494

Tung, T. C., 342

Tunica-corpus theory, 173

Tunicates, 155

Twin embryos, 286

Twin larvae, 286

Twinning, 322, 511

low-temperature induction, 325

Twitty, V. C., 487

Tyler, A., 118, 217

Tyler-Weiss hypothesis, 118

Ultraviolet radiation, 322, 492, 493

Ulva, 153

Unequal division of oocyte, 203

Unit membrane, 40, 43

Unity of development, 182
Ursprung, H., 457, 459

Vegetal plate, 305, 306
Vegetalization in sea urchin egg, 256
Vernalization, 177
Vertebral cartilage, induction of, 414
Vertebrate organizer, 491
Viability of spermatazoa, 219
Vinca, 178
Viral DNA, 30, 497
Viral induction of tumors, 495
Viral nucleic acid, 29
Virus assembly, 29
Virus cycle, 30
Virus genic pathways, 31
Virus-induced sarcomas, 495
Virus structure, 30
Vital staining experiments, 240, 249, 254, 257, 313, 408
Vitamin A, 374, 428, 481
Vitelline membrane, 214, 223, 225, 314
Vitellogenesis, 208
  in insect, 208
  in polychaete, 208
  in vertebrate, 209

Vogt, W., 313, 344
von Ubitch, L., 255

Waddington, C. H., 348, 487
Wardlaw, C. W., 165, 168–171, 174
Webster, G., 145
Weismann, A., 266
Weiss, P., 59, 114, 118, 366, 375, 379, 399, 420, 474, 475, 482
Weitland, H. M., 330
Wessells, N. K., 374, 376, 379, 382
Wetmore, R. H., 157
Whitaker, D. M., 242
Whitehead, A. N., 2
Whole-eye regeneration, 415
Wigglesworth, V. B., 463
Wilde, C. E., 481
Williams, C., 435, 441
Wilson, E. B., 290
Wilson, H. V., 117
Wing bud, 389, 390
"Wingless" mutant, 395
Wolff, G., 414
Wolffian regeneration (lens), 414
Wolpert, L., 145, 304
Wound epithelium, 115
Wound healing, 318
Wourms, J. P., 512

Xenopus, 21, 210, 211, 213, 269, 270, 276, 346, 419, 421, 433
Xenoplastic grafting, 459
Xenoplastic transplantation, 356
X-irradiation, 400, 492

Yamada, T., 350
Yellow crescent, 282, 292
Yolk, 182, 201, 206, 313, 348
  distribution gradients, 207
  platelets, 40, 208, 210, 214, 284
Yolk-cytoplasm ratio, 205
Yolk plug, 315, 330
Yolk protein, 208
Yolk sac, 324, 325
Yolk spheres, 210
Yolk syncytium, 313
Yolk transport, 206

Zona adherens, 116
Zona occludens, 116
Zona pellucida, 210, 215, 328, 329
Zona radiata, 210, 214
Zona reaction, 234
Zone of initiation in meristem, 167
Zoothalamnion, 101
Zygote, 149, 153, 157, 163, 223, 224